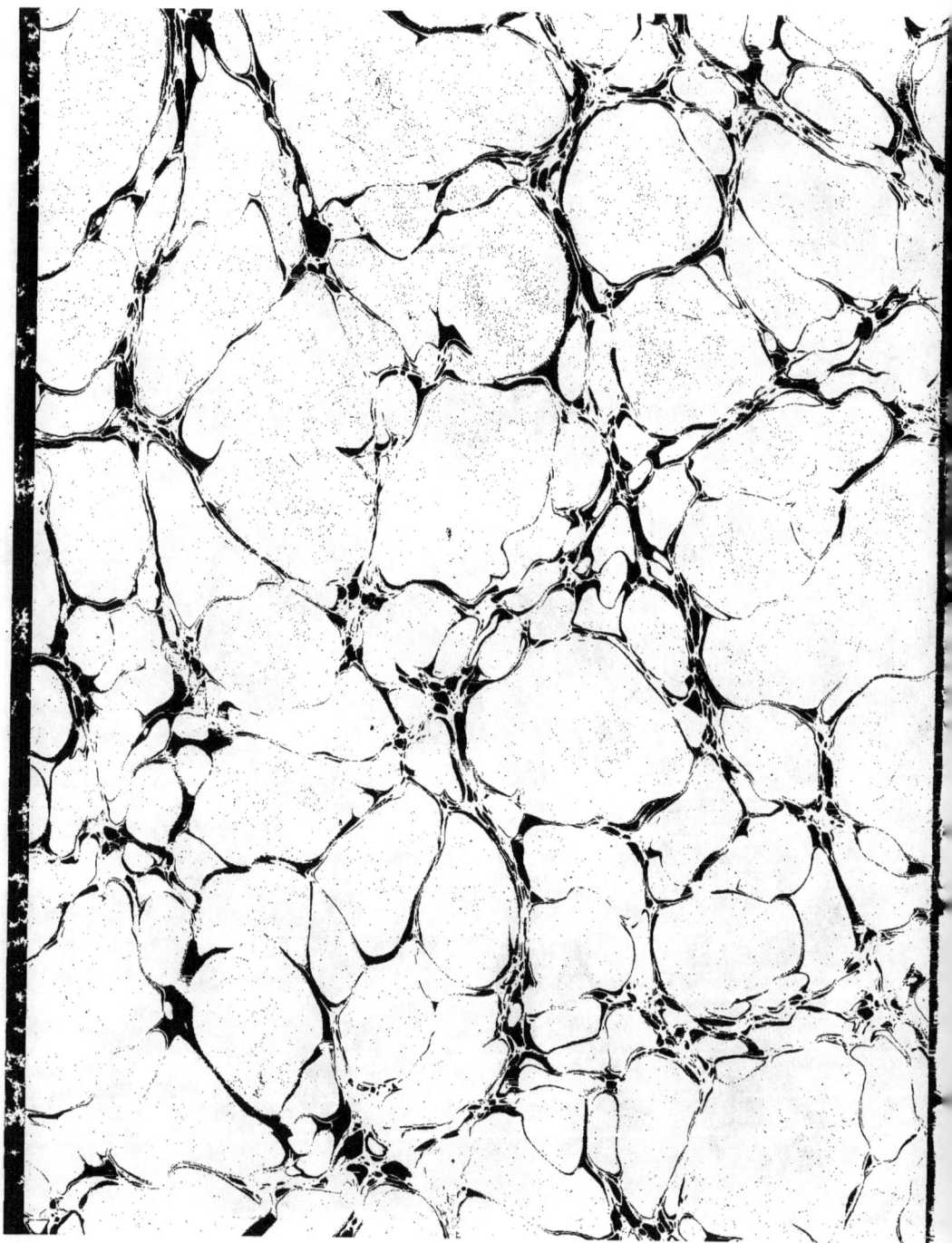

CONGRÈS

GÉOLOGIQUE INTERNATIONAL.

COMPTE RENDU

DE LA

5ME SESSION, WASHINGTON, 1891.

WASHINGTON:
IMPRIMERIE DU GOUVERNEMENT.
1893.

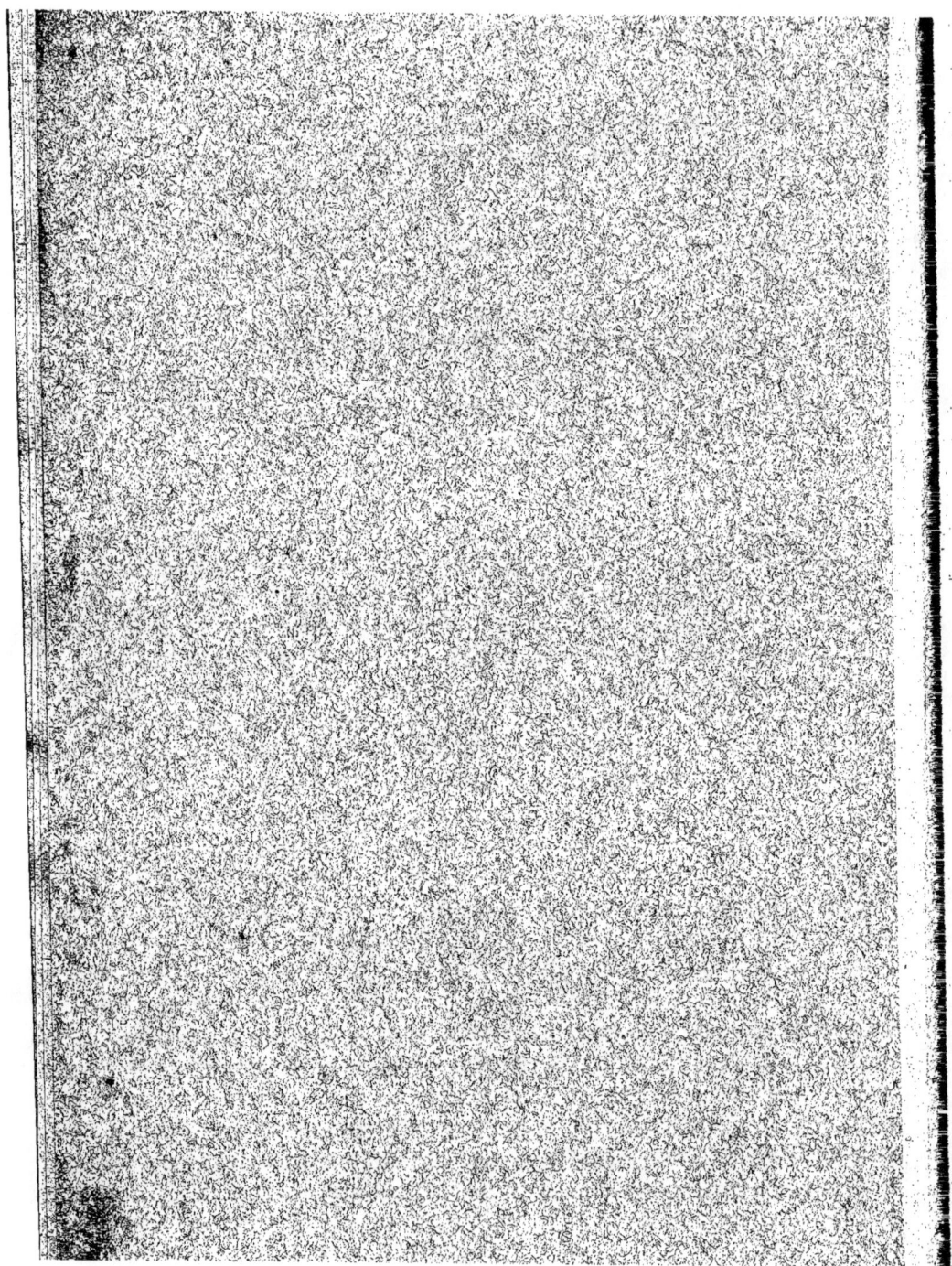

CONGRÈS

GÉOLOGIQUE INTERNATIONAL.

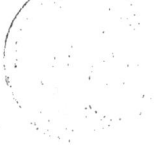

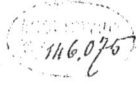

COMPTE RENDU

DE LA

5ME SESSION, WASHINGTON, 1891.

WASHINGTON:
IMPRIMERIE DU GOUVERNEMENT.
1893.

CONGRÈS

GÉOLOGIQUE INTERNATIONAL.

CINQUIÈME SESSION,

1891.

PRÉFACE.

En préparant pour la publication le compte-rendu du cinquième Congrès géologique international j'ai suivi autant que possible l'arrangement adopté dans les comptes-rendus des congrès précédents.

La première partie donne l'histoire du Congrès et des préparatifs qui ont été faits pour la réception de ses membres; elle comprend aussi une liste des membres du Congrès, avec les adresses données par chacun d'entre eux.

La deuxième partie renferme les discours d'ouverture et un aperçu succinct des discussions. Le procès-verbal (en anglais) de chaque séance fut présenté aux membres dès le lendemain matin. Ces procès-verbaux furent rédigés, tour à tour, par deux secrétaires à la fois; d'une part par M. C. Diener et le Professeur G. H. Williams, de l'autre par le Docteur F. Frech et M. J. C. Branner. M. le Professeur H. S. Williams et M. Emm. de Margerie se sont chargés des procès-verbaux des séances du conseil. A la fin se trouvent les procès-verbaux des séances des commissions internationales qui me sont parvenus.

Dans la troisième partie se trouvent deux mémoires sur la corrélation des roches, un troisième sur la corrélation des roches paléozoïques ne m'étant parvenu, malheureusement, que trop tard pour être imprimé. Cette partie renferme aussi les comptes-rendus des discussions faits in extenso et rédigés dans les langues mêmes dans lesquelles celles-ci ont été conduites. Ces comptes-rendus me furent remis par les personnes mêmes qui prirent part à la discussion. Dans les cas où aucun rapport ne m'a été envoyé, j'ai dû me contenter de reproduire les notes brièves des procès-verbaux.

Dans la quatrième partie, enfin, on trouvera des explications géologiques des régions parcourues par les membres du Congrès dans leurs excursions. Celles-ci furent pour la plupart préparées d'avance et distribuées aux membres présents à la session du Congrès; mais on a dû les soumettre à la revision de leurs auteurs respectifs.

Le premier rapport de la commission internationale de bibliographie géologique a été rédigé par le secrétaire de la commission, M. Emm. de Margerie; il sera bientôt imprimé à Paris et distribué séparément aux membres du Congrès par le rédacteur.

v

La rédaction du volume laisse, certainement, beaucoup à désirer, mais je prie les membres de ne pas la juger trop sévèrement. Mon collègue, M Cross, m'a été d'un grand secours dans cette œuvre, et je prends cette occasion de l'en remercier sincèrement. Les autres secrétaires sont domiciliés à trop grande distance de Washington pour m'aider, et aucun géologue qui connaît la langue française n'étant à portée j'ai dû me fier en grande partie à ma propre connaissance de cette langue pour les rédactions françaises.

Mes remercîments gracieux pour corrections et remarques critiques à cet égard sont dûs à M. Emm. de Margerie de Paris, à M. Jules Marcou de Cambridge, Massachusetts, et à MM. A. Harisse du Bureau des Républiques pan-américains et W. W. Rockhill du Ministère des Affaires Étrangères à Washington.

N'oublions pas les remercîments du Congrès dûs au Sénat et à la Chambre des Députés des États-Unis pour leur action unanime en faveur des buts poursuivis par la science géologique. Les deux chambres ont décrété de faire imprimer, aux frais du Trésor, le volume devant vous, et sans ce secours généreux toutes les contributions libérales de quelques membres patriotiques, destinées à défrayer les dépenses du Congrès, n'auraient pas suffi pour faire paraître le beau volume que je vous présente.

S. F. EMMONS,
Secrétaire Général.

TABLE DES MATIÈRES.

TABLE DES PLANCHES

FIGURES.

PREMIÈRE PARTIE.

HISTORIQUE DU CONGRÈS.

HISTORIQUE.

La quatrième session du Congrès Géologique International se tint à Londres du 17 au 22 septembre 1888.

Dans sa cinquième séance, le 21 septembre, le Congrès a accepté l'invitation de la ville de Philadelphie à tenir sa prochaine réunion, en 1891, dans cette ville.

À la séance de clôture du 22 septembre sur l'avis du conseil le choix définitif de la ville a été confié à un comité provisoire, avec pleins pouvoirs, se composant de MM. Dana, Frazer, Gilbert, Hall, Marsh, Newberry, Sterry Hunt et Walcott.

La première réunion de ce comité provisoire américain eut lieu à New Haven, Conn., le 15 novembre 1888, où un comité d'organisation a été élu, avec pouvoir d'augmenter le nombre de ses membres. Ce comité se composait premièrement de MM.:

Ashburner.	* Gilbert.	Newberry.
Branner.	Hall.	* Powell.
Chamberlin.	Heilprin.	Procter.
Cook.	Hitchcock.	Shaler.
Cope.	Hunt.	Stevenson.
Dana.	Leidy.	A. Winchell.
Davis.	Lesley.	H. S. Williams.
* Dutton.	LeConte.	Whitfield.
Frazer.	Marsh.	* Walcott.

Plus tard les personnes suivantes ont été adjointes au comité:

MM. * Baker.	MM. * Emmons.	MM. * Langley.
* Becker.	* Goode.	* McGee.
* Cross.	* Hague.	* Mendenhall
* Dall.	* Hubbard.	* Willis.
* Day.	* Iddings.	* Willetts.

D'autre part le comité a perdu, avant l'ouverture du Congrès, huit de ses membres par le décès de MM. Ashburner, Cook, Winchell et Leidy, et par la démission de MM. Cope, Frazer, Heilprin et Hunt.

*Signifie membres résident à Washington qui formèrent plus tard le Comité local d'arrangements.

3

Dans la troisième séance du comité, le 18 avril, vu des circonstances qui étaient survenues depuis la clôture du Congrès de Londres, il a été résolu de proposer au bureau que la prochaine réunion du Congrès soit tenue à Washington au lieu de Philadelphie. Cette proposition a été soumise au bureau du Congrès, qui l'a adoptée.

Dans la séance du 13 novembre 1890 le comité s'est organisé en choisissant le bureau provisoire suivant:

Président, J. S. Newberry; vice-président, G. K. Gilbert; secrétaires, H. S. Williams, S. F. Emmons; trésorier provisoire, S. F. Emmons.

En outre, il a désigné comme comité local, qui serait chargé des préparatifs pour la réunion du Congrès, les membres résidant à Washington (désignés dans la liste par un *), et a nommé des sous-comités, 1°, pour le programme scientifique, 2°, pour les excursions lointaines, 3°, pour le bureau du Congrès.

Les résolutions suivantes ont été adoptées:

1°. Que la session du Congrès aurait lieu dans la semaine qui commence le 26 août, 1891.

2°. Que les secrétaires seraient autorisés à préparer des circulaires d'information, en français et en anglais, sur l'organisation, le temps et le lieu de réunion du Congrès et sur d'autres points nécessaires, circulaires qui doivent être distribuées aux membres des Congrès précédents et à tous ceux qui pourraient s'intéresser aux travaux du Congrès.

La composition définitive du Comité d'organisation est la suivante:

COMMITTEE OF ORGANIZATION.

HONORARY PRESIDENT.

Prof. S. P. LANGLEY.

HONORARY MEMBERS.

The Secretary of State, Hon. JAMES G. BLAINE.
The Secretary of the Interior, Hon. JOHN W. NOBLE.
The Secretary of Agriculture, Hon. J. M. RUSK.
The Director of the U. S. Geological Survey, Hon. J. W. POWELL.
The Superintendent of the Coast and Geodetic Survey, Hon. T. C. MENDENHALL.
The Director of the Mint, Hon. E. O. LEECH.
The Chief of Engineers, Gen. THOMAS L. CASEY, U. S. A.
The Chief of Ordnance, Gen. D. W. FLAGLER, U. S. A.
The Superintendent of the Naval Observatory, Capt. F. V. McNAIR, U. S. N.
The Superintendent of the Nautical Almanac, Prof. SIMON NEWCOMB.
The Chief of the Hydrographic Bureau, Lieutenant-Commander R. CLOVER, U. S. N.
The Chief of the Weather Bureau, Prof. M. W. HARRINGTON.
The President of the Columbian University, Prof. J. C. WELLING.
The President of the American Association for the Advancement of Science, Prof. A. B. PRESCOTT.

The Ex-President of the American Association for the Advancement of Science, Prof. G. W. GOODALE.
The Ex-Director of the U. S. Geological Survey, CLARENCE KING.
Prof. EDWARD ORTON, State Geologist of Ohio.
Prof. J. M. SAFFORD, ex-State Geologist of Tennessee.
Prof. EUGENE A. SMITH, State Geologist of Alabama.
Prof. JOHN C. SMOCK, State Geologist of New Jersey.
Prof. I. C. WHITE, ex-State Geologist of West Virginia.
Prof. N. H. WINCHELL, State Geologist of Minnesota.
Prof. G. H. WILLIAMS, Johns Hopkins University.
Prof. RAPHAEL PUMPELLY, U. S. Geological Survey.
Prof. W. H. BREWER, Yale University.
Prof. GEORGE J. BRUSH, Yale University.
Prof. A. E. VERRILL, Yale University.
Prof. E. S. DANA, Yale University.
Prof. EUGENE W. HILGARD, University of California.
Prof. ALEXANDER AGASSIZ, Harvard University.
Prof. J. D. WHITNEY, Harvard University.
Prof. JULES MARCOU, Cambridge, Mass.
THOMAS WILSON, Smithsonian Institution.
Prof. LESTER F. WARD, U. S. Geological Survey.
Dr. C. A. WHITE, U. S. Geological Survey.
GEORGE H. ELDRIDGE, U. S. Geological Survey.
Dr. A. C. PEALE, U. S. Geological Survey.

OFFICERS.

Prof. J. S. NEWBERRY, *Chairman.*
G. K. GILBERT, *Vice-Chairman.*
ARNOLD HAGUE, *Treasurer.*
H. S. WILLIAMS, S. F. EMMONS, *General Secretaries.*
WHITMAN CROSS, *Assistant Secretary.*

COMMITTEE.

Marcus Baker.
G. F. Becker.
Dr. J. C. Branner.
President T. C. Chamberlin.
Whitman Cross.
Dr. W. H. Dall.
Prof. James D. Dana.
Prof. W. M. Davis.
Dr. D. T. Day.
Maj. C. E. Dutton.
S. F. Emmons.
G. K. Gilbert.
G. Brown Goode.
Arnold Hague.
Prof. James Hall.
Prof. C. H. Hitchcock.
Gardiner G. Hubbard.

J. P. Iddings.
Prof. S. P. Langley.
Prof. Joseph LeConte.
Prof. J. P. Lesley.
W J McGee.
Prof. O. C. Marsh.
Prof. T. C. Mendenhall.
Prof. J. S. Newberry.
Maj. J. W. Powell.
J. R. Procter.
Prof. N. S. Shaler.
Prof. J. J. Stevenson.
C. D. Walcott.
R. P. Whitfield.
Hon. Edwin Willetts.
Prof. H. S. Williams.
Bailey Willis.

Conformément aux instructions du comité, les Secrétaires préparèrent et distribuèrent les lettres circulaires suivantes (en français et en anglais):

I. COMITÉ D'ORGANISATION.

WASHINGTON, D. C., le 1er janvier, 1891.

MONSIEUR: Le bureau du Congrès géologique international à décidé que la 5e session se tiendra à Washington, D. C. (États-Unis d'Amérique), et la date de la réunion a été fixée au 26 août, 1891.

La session annuelle de l'Association américaine pour l'avancement des sciences et la session d'été de la Société géologique d'Amérique se tiendront la semaine précédente dans la même ville.

La session du Congrès sera suivie de plusieurs excursions organisées en vue de faire visiter aux personnes qui auront participé au Congrès les endroits qui leur sembleront présenter le plus d'intérêt.

Nous venons vous prier, Monsieur, de prendre part aux travaux du Congrès, et, si telle est votre intention, de vouloir bien adresser au Secrétariat du Comité d'organisation votre demande d'inscription comme membre du Congrès.

La cotisation à payer pour chaque membre est fixée à deux dollars et demi ($2.50).

Le reçu du Trésorier donne droit à la carte de membre, ainsi qu'au compte rendu et aux autres publications ordinaires du Congrès.

Le Comité d'organisation fera les démarches nécessaires pour obtenir des compagnies transatlantiques les conditions les plus favorables pour le voyage, aller et retour, aux États-Unis; elle demandera également aux compagnies américaines de chemins de fer des billets à prix réduit pour les excursions géologiques.

Pour que ces négociations puissent aboutir, il est indispensable que le comité connaisse le nombre approximatif des membres qui seront présents, et qu'il puisse dresser par avance, conformément aux désirs exprimés par la majorité des membres, la liste des endroits à visiter. En raison de la variété des points intéressants pour le géologue, et de la longueur des distances, le comité ne serait pas en état, sans ces renseignements, de rédiger une programme d'excursions dans des limites de dépense raisonnables.

Pour ces motifs nous vous prions de vouloir bien remplir l'imprimé ci-joint et de l'adresser, aussitôt qu'il vous sera possible, au Secrétariat du Comité à Washington.

Le programme détaillé des séances, excursions, etc., sera envoyé ultérieurement aux personnes qui auront signifié leur intention de participer aux délibérations du Congrès.

Les cartes de membres seront délivrées à Washington, au Secrétariat du Congrès, contre le reçu du Trésorier, à partir du 19 août.

Le Président,
J. S. NEWBERRY.

Les Secrétaires,
H. S. WILLIAMS,
S. F. EMMONS.

L'imprimé ci-joint accompagnait chaque lettre:

CINQUIÈME CONGRÈS GÉOLOGIQUE INTERNATIONAL.

1. Je désire être inscrit comme membre du cinquième Congrès géologique international, et je remets ci-joint ma cotisation ($2.50).
2. Il est probable que j'assisterai [que je n'assisterai pas] en personne aux sessions du Congrès à Washington.

3. Les phénomènes géologiques qui m'intéressent le plus ou les localités qu'il me serait le plus agréable de visiter au cours des excursions sont: [établir la liste par ordre de préférence]:

Signature [nom et prénom] ———— ————.
[qualité   ] ———— ————.
[demeure   ] ———— ————.

à

M. S. F. EMMONS,
Secrétaire et Trésorier [*provisoire*]
1330 F st., Washington, D. C., États-Unis d'Amérique.

De peur que ces lettres, pour une cause quelconque, n'arrivassent pas à tous les géologues qui devaient les recevoir, la lettre suivante, adjointe aux circulaires, a été envoyée aux rédacteurs de trente-cinq des principaux journaux scientifiques du monde:

WASHINGTON, *le 30 janvier, 1891.*

MONSIEUR LE RÉDACTEUR:

MONSIEUR: J'ai l'honneur d'appeler votre attention sur la circulaire ci-jointe concernant la prochaine réunion du Congrès géologique international, et je vous prie de vouloir bien en communiquer le contenu à vos lecteurs, afin que ceux d'entre eux qui ne l'auraient pas reçue et qui désireraient se faire inscrire comme membres, puissent envoyer leurs adresses au soussigné, qui s'empressera de leur remettre toutes les circulaires qui paraîtront concernant le Congrès.

Agréez, Monsieur, l'assurance de ma haute considération,

S. F. EMMONS,
Secrétaire.

En outre la lettre suivante a été envoyé aux directeurs des grands services géologiques, aux secrétaires des sociétés géologiques et aux musées géologiques dans tous les pays du monde civilisé:

SECRÉTARIAT, WASHINGTON, *30 janvier 1891.*

MONSIEUR: Nous avons l'honneur d'appeler votre attention sur la circulaire ci-jointe, concernant la prochaine réunion du Congrès géologique international, et nous vous prions de vouloir bien en porter le contenu à la connaissance du conseil et de membres de votre service (société ou Musée).

Au nom du Comité d'organisation nous les invitons cordialement à prendre part aux séances du Congrès, ou à y envoyer des délégués, de manière à faire contribuer leurs lumières et le fruit de leurs recherches à la solution des questions mises en discussion.

Le Comité espère que le but dans lequel le Congrès a été organisé leur offrira assez d'intérêt pour les engager, si non à prendre part à la réunion et aux excursions qui la suivront, du moins à se faire inscrire comme membres du Congrès, afin de recevoir le compte-rendu des discussions.

Le Comité fera tout son possible pour pourvoir au confort et au bien-être de ceux qui honoreront de leur présence cette réunion du Congrès.

Veuillez agreér, Monsieur, l'assurance de nos sentiments les plus distingués,

H. S. WILLIAMS,

S. F. EMMONS,

Secrétaires.

Avec cette lettre il a été envoyé la Circulaire I et l'avertissement qui suit:

On soumet, de la part des sous-comités, les avertissements préalables qui suivent. Les routes données pour les excursions ne sont qu' approximatives, et les chiffres de dépense pourront être réduits plus tard.

Excursions lointaines: Ces excursions se feront sur des trains a "Pullman vestibule cars" (wagon-lits), avec "hotel car" (wagon restaurant) attaché à chaque train, de manière que l'on vivra sur le train, indépendamment des hôtels. La dépense sur ces trains, tout compris, n'excédera pas dix dollars ($10) par jour et par personne.

Première excursion, de 20 jours.—De Washington au nord jusqu'à Niagara Falls. De Niagara Falls au nord-ouest, par les grands lacs, les villes de Chicago et St. Paul, les plaines de Dakota, aux geysers du Yellowstone Park. De Yellowstone Park, on ira au sud à travers les plaines de basalte de la vallée de "Snake River," à "Great Salt Lake City." En suite, tournant à l'est, on traversera la chaîne du Wasatch, les plateaux du bassin du Colorado River, et les cañons des Montagnes Rocheuses jusqu'à la ville de Denver, d'où l'on peut retourner directement à Washington ou à New York. Ou bien:

Deuxième excursion, 12 jours de plus.—A Denver on peut organiser une excursion à nombre limité pour visiter les grands cañons du Colorado River. On ira par chemin de fer à Flagstaff dans l'Arizona. De là il faut aller 80 milles (130 kilomètres) en wagon ou à cheval, campant en route, pour arriver aux bords du Grand Cañon (Kaibab division). Les dépenses de cette partie de l'excursion pourront excéder $10 par jour.

Troisième excursion, de 6 à 10 jours.—Une excursion plus courte pourra se faire en allant de Niagara à l'est, descendant la rivière St. Laurent jusqu'à Québec, et retournant à New York par le lac Champlain et la rivière Hudson. De cette manière on traversera un pays classique pour les terrains paléozoïques et archéiques, et pour les phénomènes glaciaires anciens.

Excursions courtes.—Il se fera pendant les sessions du Congrès des excursions à des endroits plus proches de Washington qui ne dureront qu'un ou deux jours.

Logement.—La dépense journalière dans les hôtels et pensions de Washington variera entre $1.50 et $4.00 par jour et par personne.

Vapeurs transatlantiques.—Les prix du passage (aller et retour) sur les vapeurs transatlantiques varient entre $100 et $175 par personne, suivant la location des cabines et la vitesse du vapeur.

Enfin la lettre suivante a été adressée à tous les ministres de pays étrangers qui se trouvaient domiciliés à Washington:

WASHINGTON, *February 16, 1891.*

To ——— ———,

Envoy Extraordinary and Minister Plenipotentiary of ———,

SIR: The International Congress of Geologists, which was first convened at Paris in 1878, and has since met at Bologna (1881), Berlin (1885), and London (1888), will hold its next session in this city during the week commencing August 26, 1891.

The primary object of this Congress is to establish among geologists and geological organizations uniform methods of geological nomenclature, classification, and cartography, which, by facilitating mutual comprehension of the results of geological research by different nations, will tend to increase our knowledge of the structure and resources of the earth.

I have the honor to request, on behalf of the committee of organization of this Congress, that you will graciously transmit to your Government an official notification of the meeting, that it may be made known to such of its departments as may take interest in its labors.

The committee will feel honored if your Government will designate a delegate or delegates, who shall participate in the deliberations of the Congress, and contribute their knowledge and the result of their researches toward the solution of the questions under discussion.

I beg to inclose herewith a form of circular designed for individual geologists, which will explain more fully the details of the meeting of the Congress.

I have the honor to be, sir, your most obedient servant,

S. F. EMMONS,
Secretary.

A sa réunion du 1ᵉʳ avril 1891, le Comité d'organisation a adopté le rapport du sous-comité sur le programme scientifique (voir la Circulaire d'information II, p. 10), et a chargé ce sous-comité d'inviter les géologues qui s'y intéressent spécialement à préparer des communication sur les questions designées dans ce programme.

A tous ceux qui avaient signifié leur intention de devenir membres du Congrès il a été expédiée la lettre suivante:

CONGRÈS GÉOLOGIQUE INTERNATIONAL, CINQUIÈME SESSION, WASHINGTON, 1891.

Circulaire d'information, II.]

WASHINGTON, *le 1ᵉʳ juin, 1891.*

MONSIEUR: Le Comité d'organisation a l'honneur de vous donner les détails suivants sur la cinquième session du Congrès géologique international qui doit avoir lieu dans cette ville au mois d'août prochain:

Lieu d'assemblée.—Les séances du Congrès se tiendront dans les salles de la "Columbian University" au coin des rues 15ᵐᵉ et H (quartier N. O. de la ville), dont l'usage a été gracieusement accordé par la faculté de l'université. On a réservé une grande salle de cours pour les séances du Congrès, et plusieurs salles pour les réunions du Conseil, et pour l'exposition des cartes, des collections de roches et de minéraux. Un service spécial de postes, de télégraphe, et de commissionnaires sera organisé dans l'édifice pendant la semaine des séances, et il y aura un bureau d'information où les membres pourront s'enrôler. Ceux qui arriveront avant le 26 août sont priés d'enregistrer leurs noms au Secrétariat, 1330 F street, pour recevoir leurs cartes de membres.

Les séances de l'American Association for the Advancement of Science, et celles de la Société Géologique d'Amérique se tiendront dans le même bâtiment.

PROGRAMME JOURNALIER DES DIVERSES SÉANCES.

Du 19 au 22 août.—Réunions des diverses sections de l'American Association for the Advancement of Science. Les membres étrangers du Congrès ont été élus membres associés honoraires de l'Association, ce qui leur donne le droit d'assister aux séances, de prendre part aux excursions géologiques et archéologiques dans les environs de Washington, et de profiter, sur les chemins de fer américains, des réductions de prix accordées aux membres de l'Association.

Le 24 et le 25 août.—Séances de la Société Géologique d'Amérique. Les membres étrangers du Congrès sont également invités à assister aux séances de cette société, à présenter des mémoires et à prendre part aux discussions.

Du 26 août au 2 septembre.—Séances du Congrès Géologique International. Les réunions du Conseil se tiendront à 10 heures du matin, et celles du Congrès ordinairement à 11 heures et à 2 heures et demi. Le soir il y aura des réceptions pour les membres du Congrès par les sociétés scientifiques et divers particuliers, et des excursions sur la rivière Potomac. La séance de samedi aura lieu de meilleure heure, afin qu'on puisse faire des excursions géologiques dans l'après-midi, qui pourront être continuées le dimanche par ceux qui le désireront. Des programmes détaillés seront distribués à l'ouverture du Congrès. Les excursions lointaines commenceront le mercredi 2 septembre.

PROGRAMME SCIENTIFIQUE.

Le Comité d'organisation propose comme sujets de discussion pour ce Congrès, outre ceux qui lui ont été laissés par le dernier Congrès, tels que rapports de comités, etc., les thèmes suivants:

    I. La corrélation chronologique des roches clastiques.
        1°. Corrélation d'après données structurales, ou tectoniques, soit:
            a. par données stratigraphiques.
            b. par données lithologiques.
            c. par données physiographiques.
        2°. Corrélation d'après données paléontologiques, soit:

a. de plantes fossiles,		a. de fossiles marins,
	ou	
b. d'animaux fossiles,		b. de fossiles terrestres.

    II. Les gammes de coloriage générales, et autres procédés graphiques.
    III. La classification génétique des roches pléistocènes ou quaternaires.

TRAJET.

Par l'intermédiare de MM. Thomas Cook et fils, le comité s'est arrangé avec certaines compagnies transatlantiques pour des billets à prix réduit. Le prix du passage varie sur chaque paquebot suivant la situation des cabines et le nombre des personnes qui les occupent. Ainsi les prix cités plus bas sont fixés pour l'installation ordinaire, et pour deux personnes dans une seule cabine. Ceux qui désireront une installation supérieure devront payer un supplément de prix.

Les prix fixés à moins de $100, aller et retour, sont accordés à condition que vingt membres ou plus s'embarquent sur un même vapeur. Pour obtenir les prix les plus bas, les membres sont priés de s'adresser le plus tôt possible à MM. Thos. Cook & Son (Ludgate Circus, Londres, Angleterre), et d'indiquer, dans l'ordre de préférence, ceux des bateaux énumérés plus loin qu'ils voudraient prendre, et le genre d'installation qu'ils désirent, en spécifiant s'ils veulent des cabines intérieures ou extérieures, à une, deux, ou trois personnes. Les membres qui se connaissent feront bien de demander d'avance à occuper une cabine ensemble. Les billets de retour sont valables pendant six mois. Les membres se serviront du reçu du Trésorier du Congrès pour se faire reconnaître des agents des paquebots.

Les billets de Londres à Boulogne, pour ceux qui s'embarqueront sur les paquebots Netherlands-American, sont fixés à $3.

Sur les chemins de fer américains le billet de New York à Washington coûte $6.50, de Philadelphie $4, de Baltimore $1.20.

Les membres du Congrès, qui sont en même temps membres de l'American Association for the Advancement of Science, ont droit à une réduction d'un tiers sur les prix réguliers, s'ils retournent au même endroit. Pour obtenir cette réduction, il faut demander au bureau du chemin de fer, en prenant le billet, un reçu de la forme désignée pour les membres de l'Association. En quittant Washington, celui qui possèdera un tel reçu obtiendra, en le présentant avec sa carte de membre de l'Association, un billet de retour au tiers du prix ordinaire.

HÔTELS ET PENSIONS.

On a obtenu pour les membres du Congrès des prix spéciaux dans les hôtels suivants, qui sont tous à cinq minutes ou moins de la "Columbian University":

Arlington Hotel (rendez-vous général).—Dans Vermont avenue, entre les rues H et I (à l'ouest de la 15ᵐᵉ). Système américain. Prix ordinaire $5 par jour, et plus, suivant l'installation. On accordera aux membres du Congrès une réduction d'un tiers sur ces prix.

The Arno.—Seizième rue, entre les rues I et K. Système européen. Pour les membres du Congrès, $1 la chambre, y compris l'usage des bains de l'hôtel. Baignoires privées, $1 extra par jour. Restaurant et café à l'hôtel.

Ebbitt House.—Au coin des rues F et 14ᵐᵉ, près du Bureau Géologique des États-Unis. Système américain. Prix ordinaire $4 par jour. Pour les membres du Congrès, $2.50 par jour, et $1 extra pour une chambre avec baignoire.

The Elsmere.—Rue H, entre 14ᵐᵉ et 15ᵐᵉ. Pension privée. Chambre et pension pour les membres du Congrès à $10.50 par semaine pendant les réunions.

On a fait des arrangements pour obtenir dans des maisons meublées des appartements pour ceux qui le désirent. On peut s'adresser dans ce but à M. Arnold Hague, président du Comité d'entretien, 1330 F street, en indiquant le nombre et la qualité des chambres que l'on désire.

EXCURSIONS.

La grande excursion sera faite au moyen de trains spéciaux, portant 75 personnes chacun, et pourvus pour le confort des voyageurs de tous les perfectionnements, tels que des chambres à coucher et cabinets de toilette, salles à manger, baignoires, salons de coiffure et fumoirs. Le train constituera un hôtel en mouvement, où l'on pourra, à toute heure, circuler librement et sans danger d'une extrémité à l'autre Il ira là où les rails ont été posés dans les régions à visiter, et il s'arrêtera quand on le désirera, quelque soit l'endroit. L'excursion, telle qu'elle est projetée maintenant, durera 25 jours, et elle coûtera $265 par personne, ce qui couvrira toutes les dépenses nécessaires. La route tracée, qui sera donnée en détail dans une autre circulaire, couvre 12 degrés de latitude et 39 degrés de longitude, et elle permet au voyageur de voir les plus beaux paysages et les phénomènes géologiques les plus importants des États de l'Est, de la vallée du Mississipi et de la région des Montagnes Rocheuses, où l'on passera une semaine parmi les merveilles du Yellowstone Park.

Trois excursions plus courtes ont été proposées. Des géologues américains connaissant les régions sont prêts à conduire des groupes de voyageurs. Si un nombre suffisant de membres s'engagent à prendre part à ces excursions, on pourra obtenir de la part des chemins de fer des concessions pour réduire à un minimum les frais de voyage.

(1) A travers la région appalachienne du sud, pour examiner les plis des roches paléozoïques, comprimés d'un façon toute spéciale, et pour visiter les mines récemment ouvertes de houille, de fer, de manganèse, d'étain et d'or.

(2) Aux régions cuprifères et ferrifères du Supérieur, pour examiner les grands développements des roches pré-cambriennes ou algonkiennes.

(3) A travers les régions houillères et pétrolifères de Pensylvanie aux chutes du Niagara, puis par la rivière St.-Laurent jusqu'à Montréal et à Québec, pour retourner en traversant les régions classiques paléozoïques et taconiques de New York et de Vermont.

Les membres qui désireraient examiner des localités ou des horizons géologiques spéciaux sont priés de communiquer leurs désirs aux secrétaires le plus tôt possible. On ne négligera rien pour les contenter. Une courte excursion a déjà été projetée par M. le professeur H. S. Williams pour la semaine avant la réunion du Congrès. Cette excursion permettra aux géologues de voir le développement typique des couches paléozoïques (surtout du Dévonien) de l'État de New York. Un nombre considérable de géologues européens ont déjà signifié leur désir d'y prendre part.

Le Président du Comité d'Organisation,
J. S. NEWBERRY.
Les Secrétaires généraux,
H. S. WILLIAMS,
S. F. EMMONS.

Toutes les correspondances doivent s'adresser à
S. F. EMMONS,
    1330 F street, Washington, D. C.

Au mois de juin, aussitôt qu'on a pu savoir à peu près le nombre de participants à l'excursion aux Montagnes Rocheuses, la circulaire ci-jointe, imprimée sur la carte des États-Unis (voir p. 255), avec désignation de la route qui serait suivi par l'excursion, a été expédiée à tous les membres du Congrès.

EXCURSION AUX MONTAGNES ROCHEUSES DU CONGRÈS GÉOLOGIQUE INTERNATIONAL.

Circulaire d'information III.]

WASHINGTON, D. C., *15 juin 1891.*

Après la clôture du Congrès une excursion partira de Washington pour visiter le "Yellowstone Park" et autres points intéressants au point de vue géologique dans les États de l'Ouest. Elle se fera sur un train spécial dont la marche sera indépendante des trains ordinaires; ce train passera d'une ligne de chemin de fer à une autre sans que les voyageurs aient a se déranger; ils y seront absolument chez eux pendant la principale partie du voyage. Ce train sera composé de wagons-lits (Pullman vestibule cars), d'un wagon-restaurant et d'un wagon composé de salles de lecture, fumoir, baignoire et salon de coiffure, arrangés de manière que les voyageurs puissent passer d'un wagon à l'autre par des passages couverts et sans danger. Un conducteur expérimenté accompagnera le train pour veiller au confort des voyageurs. Des géologues qui ont étudié les régions que l'on visitera expliqueront les phénomènes géologiques.

Le train parcourra près de dix mille kilomètres, et 39 degrés de longitude. Il traversera une vingtaine d'états et territoires des États-Unis, et une des provinces du Canada. La route qui sera suivie par l'excursion principale et celles des excursions supplémentaires sont tracées en rouge sur la carte qui se trouve sur le revers de cette feuille; les itinéraires suivants en donnent des résumés:

ITINÉRAIRE PRINCIPAL.

2 septembre.—Départ de Washington à 9 heures du matin. Pendant la journée on traversera la chaîne des Apalaches, en vue des plis isoclinaux caractéristiques de ces montagnes.

5 septembre.—A l'ouest à travers les États d'Ohio, Indiana et Illinois jusqu'à la ville de Chicago, et de là au nord-ouest dans l'État de Wisconsin. La région des "prairies," des plaines ondulantes composées de terrains dévoniens et siluriens couverts de drift glaciaire de types divers. On verra la "kettle moraine," une moraine de rétrocession de l'ancienne nappe glaciaire.

4 septembre.—Dans l'État de Minnesota, sur les bords du Mississipi. Roches siluriennes et cambriennes, et drift glaciaire. Le lac Pepin, canal d'écoulement pleistocène de l'ancien lac Agassiz. Arrêt de six heures pour visiter les villes jumelles de Minneapolis et St.-Paul. Ceux qui le préfèrent pourront visiter, sous la conduite du Prof. N. H. Winchell, les chutes de St. Antoine, la gorge du Mississipi et la vallée de la rivière Minnesota, qui ensemble servent à mesurer l'âge de l'époque glaciaire.

5 septembre.—A l'ouest dans le Dakota et le Montana, à travers les coteaux du Missouri et les "Great Plains," composées de terrains crétacés et tertiaires horizontaux. On verra une partie des "Mauvaises Terres." La rivière Missouri sera traversée à Bismarck.

6 septembre.—A Cinnabar, sur la rivière Yellowstone, on quittera le train et on prendra des voitures pour le "Mammoth Hot Springs" dans le Yellowstone National Park, à une distance de 11 kilomètres. Dans l'après-midi on visitera à pied les sources thermales, les lacs chauds et les terrasses de tuf calcaire dans le voisinage de l'hôtel.

7 à 12 septembre.—On passera ces six jours dans le Parc, voyageant en voiture d'un endroit à un autre, et passant la nuit a l'hôtel à "Norris Geyser basin," à "Lower Geyser basin," à "Yellowstone Cañon" et à "Yellowstone Lake." MM. Arnold Hague et Jos. P. Iddings serviront de guides géologues pour les phénomènes du Parc. Des hôtels, on visitera à pied des objets d'intérêt dans le voisinage, tels que les Geysers divers, Old Faithful, Giant, Excelsior, etc., les chutes du Yellowstone, les volcans de boue et les falaises d'obsidienne. Il y aura une excursion sur le lac Yellowstone (45 kilom. de long) en bateau à vapeur.

13 septembre.—Reprenant le train on ira à l'ouest aux sources de la rivière Missouri, passant en revue des terrains paléozoïques, algonkiens et archéens. De là on arrivera aux eaux qui descendent à l'océan Pacifique par les rivières Snake et Columbia. Dans l'après-midi on s'arrêtera pendant quelques heures à la ville minière de Butte, où il y a des mines de cuivre et d'argent dans le granit éruptif, dont le produit en 1890 a été évalué à plus de vingt-six millions de dollars. Mr. S. F. Emmons servira de guide pour ceux qui désireront visiter les mines.

14 septembre.—Au sud dans l'Idaho et l'Utah, jusqu'à Salt Lake City. Le matin on côtoiera les marges des grandes plaines de lave de la vallée de la rivière Snake. Des coulées récentes se voient dans les vallées entre des collines de roches mésozoïques et paléozoïques. Ceux qui voudront visiter les grandes chutes de la rivière Snake (Shoshone Falls) quitteront le train spécial à Pocatello le matin, pour le rejoindre deux jours plus tard à Salt Lake City. De la vallée de Snake River on passera dans le bassin intérieur du Grand Lac Salé, sans montée appréciable. On peut suivre les plages de son prédécesseur pleistocène, le lac Bonneville, sur les versants des montagnes qui l'entouraient. On s'arrêtera au canal d'écoulement de ce lac dans la vallée du Snake River. Dans l'après-midi on côtoiera les rives est du Grand Lac Salé, et les côtes ouest de la chaîne Wahsatch.

15 et 16 septembre.—A Salt Lake City et dans les environs. Le Temple, le Tabernacle et autres institutions des Mormons. Mr. G. K. Gilbert guidera des excursions à l'entrée de Little Cottonwood Cañon, dans les montagnes Wahsatch, et à Garfield Landing à l'extrémité sud du lac, pour examiner les moraines et les failles d'âge pleistocène, et les plages du lac Bonneville. A Garfield on pourra prendre des bains de mer dans le Lac Salé, dont les eaux sont si denses que l'on flotte à la surface sans effort.

17 septembre.—Au sud dans les vallées du Lac Salé, du lac Utah à eau douce, par les villages de Mormons, et les canaux d'irrigation au moyen desquels on cultive la terre dans ces régions arides. Au sud-est par une gorge qui traverse la chaîne Wah-satch, où l'on voit des coupes géologiques montant du Paléozoïque jusqu'au Tertiaire. Puis à l'est dans la région des Plateaux dont les eaux s'écoulent dans la rivière Colorado. On traversera la rivière dans l'après-midi, le long d'une vallée monoclinale, bordé au nord par un escarpement imposant de couches crétacées et tertiaires, les "Book Cliffs."

18 septembre.—On passera de la région des Plateaux à couches horizontales dans celle des Montagnes Rocheuses, où les couches sont recourbées, plissées, rejetées par des failles et pénétrées de roches éruptives de types divers. Dans cette région MM. Emmons et Cross expliqueront les phénomènes géologiques. Le matin on suivra la gorge de la rivière "Grand," à travers les terrains tertiaires et les couches à houille du Crétacé jusqu'à Glenwood Springs, où l'on se trouvera dans les roches paléozoïques, et en vue des massifs dioritiques des "Elk Mountains." Arrêt de quelques heures pour prendre des bains thermaux, ou faire des excursions géologiques à pied. De Glenwood on montera les vallées des rivières Grand et Eagle, à travers les plis anticlinaux et synclinaux dans les roches mésozoïques et paléozoïques, jusqu'à Tennessee Pass (3,250 mètres d'altitude), où l'on passera des eaux Pacifiques aux eaux Atlantiques. On arrivera dans l'après-midi à Leadville, la plus grande ville minière du Colorado, dont le produit en argent et en plomb s'élève à plus de \$150,000,000. Arrêt de quelques heures pour visiter les mines.

19 septembre.—Descente au sud le long de la grande vallée de la rivière Arkansas; à droite la chaîne des Sawatch dont les crêtes sont élevées de plus de 4,300 mètres au dessus du niveau de la mer. Puis à l'est par une série de gorges dans l'Archéen, le Paléozoïque et le Mésozoïque, dont la dernière, la "Royal Gorge," a une profondeur de 1,000 mètres, avant que la rivière débouche sur les plaines à Cañon City. Arrêt pour voir les "Hog-Back Ridges," des arêtes monoclinales de couches mésozoïques et paléozoïques, et les gisements de poissons siluriens. A Pueblo, sur les plaines à l'est des montagnes, grandes usines de plomb et de fer. De là au nord, en longeant les flancs des Montagnes Rocheuses. On s'arrêtera la nuit à Manitou Springs, un recoin des montagnes, près de Pikes Peak.

20 septembre.—A Manitou Springs. On pourra faire des excursions au "Jardin des Dieux," à des cavernes en couches calcaires, et au sommet de Pikes Peak (4,312 m.) a pied, à cheval, ou en chemin de fer. Dans la nuit le train ira à Denver.

21 septembre.—A Denver. Ville capitale du Colorado, avec 130,000 habitants, quoique fondée en 1860. En pleine vue des Montagnes Rocheuses, dont une étendue longitudinale de 240 kilomètres est visible avec un ciel serein. Ceux qui voudront s'arrêter plus longtemps dans les montagnes, pourront quitter le train spécial et revenir plus tard par les trains ordinaires.

22 septembre.—A l'est dans les États du Kansas et du Missouri, en traversant une seconde fois les Grandes Plaines.

23 septembre.—A l'est sur les "prairies" de l'Illinois, à travers le canal d'écoulement pléistocène du lac Michigan. Vingt-quatre heures à Chicago, ville de 1,100,000 habitants, fondée en 1837.

24 septembre.—Départ de Chicago à 3 heures de l'après-midi. A travers l'Illinois, l'Indiana et le Michigan. Pays plat, couvert de drift glaciaire.

25 septembre.—A l'est dans la province d'Ontario, Canada. Une plaine ondulante de drift, reposant sur des couches siluriennes. Plages modernes et pléistocènes du lac Ontario. Arrivée aux chutes du Niagara avant midi, départ dans la nuit.

26 septembre.—Au sud dans la vallée de la rivière Hudson le matin, arrivant à New York avant midi. Vues des monts Catskill, de la gorged ans les "Highlands" et des "Palisades" (falaises de diabase).

ROUTE SUPPLÉMENTAIRE À SHOSHONE FALLS.

Pour faire cette excursion spéciale, on quittera le corps principal des voyageurs à Pocatello de bonne heure le 14 septembre et on le regagnera à Salt Lake City dans l'après-midi du 16 septembre.

14 septembre.—De Pocatello à Shoshone par chemin de fer: de Shoshone à Shoshone Falls en voiture, le trajet étant entièrement à travers la plaine de lave de la vallée de la rivière Snake. A Shoshone Falls la rivière fait une chute de 65 mètres dans une gorge profonde, creusée à travers des coulées de laves basaltiques et andésitiques.

15 septembre.—A Shoshone Falls. Dans l'après-midi on ira à Shoshone en voiture.

16 septembre.—De Shoshone à Pocatello et de là à Salt Lake City sur la ligne du corps principal.

ROUTE SUPPLÉMENTAIRE AU GRANDE CAÑON DU COLORADO.

On quittera le corps principal à Pueblo, Colorado, le soir du 19 septembre; on voyagera par chemin de fer (1,180 kilom.) à Flagstaff, Arizona; de là deux jours en voiture (120 kilom.) au Grand Cañon, et on reviendra à Flagstaff et à Pueblo par la même route. Cette excursion prendra dix jours à peu près, mais les arrangements ne sont pas encore assez avancés pour qu'on puisse en annoncer l'itinéraire. Revenus à Pueblo, les excursionistes continueront leur voyage vers l'est sur la route du corps principal, mais à une date plus avancée, et par les trains ordinaires.

De Pueblo la route serpente parmi les plaines crétacées à l'est des montagnes. A Trinidad, des mines de houilles dans le Crétacé, couverts de coulées de laves basaltiques. Au delà des thermaux de Las Végas la route contourne l'extrémité sud des Montagnes Rocheuses, en allant à Albuquerque dans la vallée du Rio Grande. D'Albuquerque à Flagstaff on va à l'ouest à travers une partie de la Région des Plateaux, où les roches varient en âge du Carbonifère au Crétacé. Le plateau est accidenté de falaises, de mesas et de buttes, de deux grands volcans éteints, Mt. Taylor et Mt. San Francisco, et de nombreuses cheminées et cônes mineurs. Les couches présentent plusieurs flexures, typiques du plateau. De Flagstaff au Grand Cañon la route suit la base du Mt. San Francisco, passe parmi des cônes de cendres basaltiques, et ensuite elle traverse une plaine de couches carbonifères jusqu'au bord du cañon. A ce point le cañon a une profondeur de 1,830 mètres, et ses murs sont finement sculptés en alcoves, arcs-boutants, tours et minarets. Le point de vue est vis-à-vis de celui d'où furent esquissées les illustrations du mémoire du capitaine C. E. Dutton sur le cañon.

On s'arrangera pour s'arrêter ou à Santa Fé ou à Albuquerque, villes mexicaines de la vallée du Rio Grande, et à un ou plusieurs des pueblos des indiens. Les voyageurs seront guidés par le major J. W. Powell.

DÉPENSES.

Le prix des billets pour la route principale sera de $265, somme qui couvrira les frais du chemin de fer, des voitures au Yellowstone Park, du logement et de la pension sur le train et aux hôtels.

Les frais de l'excursion supplémentaire à Shoshone Falls seront de $15 environ. L'état des arrangements ne permet pas encore de fixer une somme plus exacte.

Les frais de l'excursion supplémentaire au Grand Cañon n'ont pas encore été déterminés définitivement. Un arrangement provisoire indique qu'ils atteindront la somme de $100, mais on a des raisons de croire qu'il sera possible d'obtenir des termes plus favorables.

Comme les arrangements définitifs dépendront en partie du nombre des voyageurs à transporter, ceux qui pensent y prendre part sont priés de communiquer leur intention à

S. F. EMMONS,
Secrétaire du Comité d'Organisation, Congrès Géologique International,
Washington, D. C.

Enfin, à son arrivée dans la ville de Washington, chaque membre du Congrès a reçu la circulaire suivante:

Circulaire d'information v.]

WASHINGTON, *21 août 1891.*

MONSIEUR: Le Comité d'organisation a l'honneur de soumettre ce qui suit pour le renseignement des membres du cinquième Congrès géologique international qui se trouvent dans cette ville.

Les réunions se tiendront dans le local du "Columbian University," au coin des rues 15me et H. Le président et les directeurs de l'université ont eu la gracieuseté de placer plusieurs salles à la disposition du Congrès pour ses réunions et celles de son conseil. Dans le local on trouvera des salles pour l'exposition des cartes et des minéraux et roches, pour l'écriture et pour l'entretien général. Il y aura un bureau d'information, aussi bien que des services spéciaux de poste, de télégraphe et des commissionnaires, dont la disposition sera indiquée par des placards à l'entrée. Les bureaux du Secrétaire et du Trésorier s'y trouveront également à partir du 22 août, et messieurs les membres sont priés d'y enregistrer leurs noms et adresses. Toutes les lettres pour les membres du Congrès se trouveront dans le bureau de poste spécial.

On peut se procurer de M. Arnold Hague, président du comité d'entretien, les renseignements sur les hôtels, les médailles, les guides, etc.; de M. W. H. Dall, président du comité des salles, ceux sur l'exposition des cartes, échantillons, etc. On devra consulter M. Bailey Willis pour avoir des renseignements sur les excursions courtes. Un agent de MM. Raymond et Whitcomb distribuera les billets pour l'excursion aux Montagnes Rocheuses à partir du 26 août. Un guide géologique pour cette excursion est à l'imprimerie et sera distribué dans quelque jours. Le Guide géologique de Macfarlane, sur les chemins de fer américains, sera en vente à un prix réduit spécial pour les membres du Congrès.

Le Secrétaire,
S. F. EMMONS.

Ceux d'entre les membres du conseil du Congrès précédent (Londres, 1888) qui se trouvaient à Washington se réunirent dans la salle du conseil le mercredi 26 août à 10 heures du matin, et délibérèrent des recommandations faites par le Comité d'organisation au sujet: 1°, de la composition du bureau; 2°, de la langue dont on peut se servir dans les discussions; 3°, de l'ordre du jour.

Le bureau fut choisi tel qu'il fut ensuite soumis au Congrès (p. 52). La question de la langue a soulevé des discussions, quelques membres étant d'opinion que les trois langues, allemande, anglaise, et française, devaient être au même rang dans les discussions. On s'est accordé finalement sur la résolution suivante:

Le Conseil recommande au Congrès que le français continue à être la langue officielle du Congrès. Toutefois, dans la réunion de Washington, l'anglais sera accepté concurremment avec le français. Le volume sera publié en français.

Cette résolution a été interprétée de manière à admettre que les procès-verbaux préliminaires puissent être dressés en anglais d'abord, pour être traduits en français pour la publication définitive; que sur demande les délibérations sur des questions scientifiques, faites dans une langue autre que le français, dussent être traduites aussitôt en français.

En outre il a été décidé d'adopter la succession suivante dans les discussions scientifiques:

1°. Classification des roches pléistocènes.
2°. Corrélation chronologique des roches clastiques.
3°. Coloriage des cartes géologiques.

Enfin on arrêta l'ordre du jour suivant:

PROGRAMME.

MERCREDI, 26 AOÛT.

A 10 heures du matin: Réunion du Conseil.

A 2 heures et demi de l'après-midi: Ouverture du Congrès. Élection des membres du bureau. Discours du Ministre de l'Intérieur, du Président du Congrès et du Président du comité local.

A 9 heures du soir: Réception dans la grande salle de l'hôtel Arlington par la Société Géologique d'Amérique.

JEUDI, 27 AOÛT.

A 10 heures du matin: Réunion du Conseil.

A 11 heures du matin et à 2 heures et demi de l'après-midi: Séances du Congrès.

A 10 heures du soir: Réception par M. et Mme. S. F. Emmons, 1725 H street, et par M. Thos. Wilson, 1218 Connecticut avenue.

451 GE——2

VENDREDI, 28 AOÛT.

A 10 heures du matin : Réunion du Conseil.

A 11 heures du matin et à 2 heures et demi de l'après-midi : Séances du Congrès.

Le soir : Le Musée national sera ouvert pour l'inspection des membres du Congrès.

SAMEDI, 29 AOÛT.

A 10 heures du matin : Séance du Congrès.

A 2 heures et demi de l'après midi : Excursion sur la rivière Potomac à Mount Vernon, le résidence de Washington.   Promenades en voiture autour de la ville.

A 5 heures : reception par le ministre du royaume de Korée et Mme. Ye.

DIMANCHE, 30 AOÛT.

A 8 heures du matin : Excursion géologique sur la rivière Potomac en bateau à vapeur, pour voir les terrains tertiaires de la plaine côtière.

LUNDI, 31 AOÛT.

A 10 heures du matin : Réunion du Conseil.

A 11 heures du matin : Séance du Congrès.

A 9 heres du soir : Réception au local du U. S. Geological Survey, 1330 F street, par le Directeur et les géologues du Survey.

MARDI, 1er SEPTEMBRE.

A 10 heures du matin : Réunion du Conseil.

A 11 heures du matin : Séance de clôture du Congrès.

Le soir : Excursion en bateau à vapeur, fourni par le Comité d'organisation, et dîner à Marshall Hall, sur les bords du Potomac.

MERCREDI, 2 SEPTEMBRE.

Départ des excursions géologiques.

Voici ensuite la composition des bureaux successifs du Congrès :—

A.—COMITÉ FONDATEUR,

formé à la suite de l'Exposition de Philadelphie en 1876 pour l'organisation à Paris, en 1878, d'un Congrès géologique international.

Président.

M. JAMES HALL, à Albany (États-Unis).

Secrétaire.

M. T. STERRY HUNT, à Montréal (Canada).

Membres du comité.

MM. WILLIAM B. ROGERS, J. W. DAWSON, J. S. NEWBERRY, C. H. HITCHCOCK, R. PUMPELLY, J. P. LESLEY (États-Unis et Canada); T. H. HUXLEY, à Londres (Angleterre); OTTO TORELL, à Stockholm (Suède); E. H. DE BAUMHAUER, Haarlem (Hollande).

B.—BUREAU DU PREMIER CONGRÈS, PARIS, 1878.

Président.

M. HÉBERT, membre de l'Institut, professeur à la Sorbonne.

Vice-Présidents.

Angleterre, M. DAVIDSON.
Australie, M. LIVERSIDGE.
Belgique, M. DE KONINCK.
Canada, M. T. STERRY HUNT.
Danemark, M. JOHNSTRUP.
Espagne, M. VILANOVA.
États-Unis, MM. J. HALL et J. P. LESLEY.
France, MM. DAUBRÉE et A. GAUDRY.

Hongrie, M. VON SZABÓ.
Italie, M. CAPELLINI.
Pays-Bas, M. DE BAUMHAUER.
Portugal, M. le colonel RIBEIRO.
Roumanie, M. STEFANESCU.
Russie, M. DE MOELLER.
Suède et Norvège, M. O. TORELL.
Suisse, M. A. FAVRE.

Secrétaire général.

M. JANNETTAZ.

Secrétaires.

MM. BROCCHI, DELAIRE, SAUVAGE, VÉLAIN.

Trésorier.

M. A. BIOCHE.

C.—Bureau du Second Congrès, Bologne, 1881.

Président d'honneur.

M. Q. Sella.

Président.

M. J. Capellini.

Ancien Président.

M. E. Hébert.

Vice-Présidents.

Autriche, M. Mojsisovics.
Bavière, M. von Zittel.
Belgique, M. Dewalque.
Canada, M. T. Sterry Hunt.
Danemark, M. W. Schmidt.
Espagne, M. Vilanova.
États-Unis, M. J. Hall.
France, M. Daubrée.
Grande Bretagne, M. Hughes.

Hongrie, M. von Szabó.
Indes, M. Blanford.
Italie, MM. Meneghini, et de Zigno.
Portugal, M. Delgado.
Prusse, M. Beyrich.
Roumanie, M. Stefanescu.
Russie, M. de Moeller.
Suède, M. O. Torell.
Suisse, M. Renevier.

Secrétaire général.

M. F. Giordano.

Secrétaires.

MM. J. G. Bornemann, Delaire, Fontannes, Pilar, Taramelli, Topley, Uzielli, et Zezi.

Trésorier.

M. Scarabelli Gommi Flamini.

D.—Bureau du Troisième Congrès, Berlin, 1885.

Président d'honneur.

Son Excellence M. le docteur von Dechen.

Ancien Président.

M. Capellini.

Président.

M. Beyrich.

Vice-Présidents.

Allemagne, MM. Credner, Fraas, et
  von Guembel.
Autriche, M. Stur.
Belgique, M. Dewalque.
Danemark, M. Johnstrup.
Espagne, M. Vilanova.
États-Unis, M. James Hall.
France, M. Jacquot.
Grande Bretagne, M. Hughes.
Hongrie, M. von Szabó.

Indes, M. Blanford.
Italie, M. de Zigno.
Norvège, M. Kjerulf.
Pays-Bas, M. van Calker.
Portugal, M. Choffat.
Roumanie, M. Stefanescu
Russie, M. Inostranzeff.
Suède, M. Torell.
Suisse, M. Renevier.

Secrétaire général.

M. Hauchecorne.

Secrétaires,

MM. Bornemann père, Fontannes, Fornasini, et Wahnschaffe.

Trésorier.

M. Berendt.

Membres du conseil.

M. Benecke.
M. Dupont.
M. Ewald.
M. Frazer.
M. Gaudry.
M. Geikie.
M. Giordano.
M. von Hantken.
M. de Lapparent.
M. Lepsius.
M. Mayer-Eymar.

M. von Mojsisovics.
M. Neumayr.
M. Newberry.
M. Pilar.
M. von Richthofen.
M. Strüver.
M. Taramelli.
M. Topley.
M. Williams (H. S.)
M. von Zittel.

E.—Bureau du Quatrième Congrès, Londres, 1888.

Président d'honneur.

M. T. H. Huxley.

Ancien Présidents.

MM. E. Hébert, 1878; G. Capellini, 1881; E. Beyrich, 1885.

Président.

M. J. Prestwich.

Vice-Présidents.

Allemagne, M. K. VON ZITTEL.
Australie, M. F. LIVERSIDGE.
Autriche, M. M. NEUMAYR.
Belgique, M. G. DEWALQUE.
Canada, M. T. STERRY HUNT.
Danemark, M. M. JOHNSTRUP.
Espagne, M. J. VILANOVA-Y-PIERA.
États-Unis, M. P. FRAZER.
France, M. A. DE LAPPARENT.
Grande Bretagne, MM. W. T. BLAN-
　FORD, A. GEIKIE, et T. McK. HUGHES.

Hongrie, M. J. VON SZABÓ.
Indes, M. H. B. MEDLICOTT.
Italie, M. F. GIORDANO.
Norvège, M. H. REUSCH.
Pays-Bas, M. K. MARTIN.
Portugal, M. J. F. N. DELGADO.
Roumanie, M. G. STEFANESCU.
Russie, M. A. INOSTRANZEFF.
Suède, M. O. TORELL.
Suisse, M. E. RENEVIER.

Secrétaires généraux.

MM. J. W. HULKE, et W, TOPLEY.

Secrétaires.

MM. Ch. BARROIS, C. FORNASINI, C. LE NEVE FOSTER, C. GOTTSCHE, A. RENARD,
et G. H. WILLIAMS.

Trésorier.

M. F. W. RUDLER.

Membres du conseil.

M. T. G. BONNEY.
M. A. BRIART.
M. E. COHEN.
M. H. CREDNER.
M. E. DUPONT.
M. J. EVANS.
M. W. H. FLOWER.
M. A. GAUDRY.
M. J. GOSSELET.
M. M. VON HANTKEN.
M. W. HAUCHECORNE.

M. A. HEIM.
M. J. HOOKER.
M. A. ISSEL.
M. J. W. JUDD.
M. R. LEPSIUS.
M. C. LORY.
M. A. MICHEL LÉVY.
M. T. MACFARLANE.
M. O. C. MARSH.
M. E. VON MOJSISOVICS.

M. J. S. NEWBERRY.
M. S. NIKITIN.
M. R. OWEN.
M. A. PILAR.
M. F. VON RICHTHOFEN.
M. T. SCHMIDT.
M. D. STUR.
M. T. TSCHERNICHEFF.
M. E. VAN DEN BROECK.
M. C. D. WALCOTT.

F.—BUREAU DU CINQUIÈME CONGRÈS.

Présidents d'honneur.

MM. JAMES HALL et J. D. DANA.

Anciens Présidents.

MM. E. BEYRICH, 1885; G. CAPELLINI, 1881; J. PRESTWICH, 1888.

BUREAU DU CONGRÈS.

Président.

M. J. S. NEWBERRY.

Vice-Présidents.

Allemagne, MM. K. VON ZITTEL, et H. CREDNER.
Angleterre, M. T. McK. HUGHES.
Autriche, M. E. TIETZE.
Australie, M. F. LIVERSIDGE.
Belgique, M. E. VAN DEN BROECK.
Canada, MM. J. C. K. LAFLAMME et THOMAS MACFARLANE.
Chili, M. F. I. SAN ROMAN.
Danemark, M. F. JOHNSTRUP.
Écosse, M. H. M. CADELL.
Espagne, M. M. F. DE CASTRO.
États-Unis, MM. JOSEPH LECONTE, J. W. POWELL, ET RAPHAEL PUMPELLY.

France, MM. A. GAUDRY et C. BARROIS.
Hongrie, M. J. VON SZABÓ.
Indes, M. F. R. MALLET.
Irlande, M. A. SOLLAS.
Italie, M. G. UZIELLI.
Mexique, M. A. DEL CASTILLO.
Norvège, M. H. REUSCH.
Nouvelle Zélande, M. F. HUTTON.
Portugal, M. J. F. N. DELGADO.
Roumanie, M. G. STEFANESCU.
Russie, MM. T. TSCHERNYSCHEW, F. SCHMIDT, et A. PAVLOW.
Suède, M. GERARD DE GEER.
Suisse, M. H. GOLLIEZ.

Secrétaires généraux.

MM. H. S. WILLIAMS et S. F. EMMONS.

Secrétaires.

MM. J. C. BRANNER, WHITMAN CROSS, C. DIENER, F. FRECH, EMM. DE MARGERIE, et G. H. WILLIAMS.

Trésorier.

M. ARNOLD HAGUE.

LISTE GÉNÉRALE DES MEMBRES DU CONGRÈS.

[Explication : * Indique les membres qui ont été présents à la session ; † les membres décédés.
L'abréviation F. R. S. signifie, Fellow of the Royal Society. F. G. S., Fellow of the Geological
Society of London. F. G. S. A., Fellow of the Geological Society of America. M. S. G. F., Membre
de la Société Géologique de France.]

*ADAMS, FRANK DAWSON, F. G. S. A., lecturer in geology, McGill College, Montreal, Canada.

*AGUILERA, JOSÉ G., sous-directeur du service géologique (Comision geológico) du Mexique, Ville de Mexico.

ALLEN, HENRY A., Esq., F. G. S., geological survey of Great Britain, Museum, 28 Jermyn street, London, S. W.

AMI, HENRY M., A. M., F. G. S., F. G. S. A., paleontologist, geological survey of Canada, Ottawa, Canada.

AMMON, LUDWIG VON, Dr. Phil., Bergamtsassessor, 16 Ludwigstrasse, München, Bayern,

*ANDREAE, ACHILLES, Dr. Phil., Professor an der Universität, Heidelberg, Deutschland.

*APLIN, STEPHEN ARNOLD, Jr., U. S. geological survey, Washington, District of Columbia.

ARGALL, PHILIP, mining engineer, P. O. Box 1095, Denver, Colorado.

ARMACHEWSKY, PIERRE, professeur de minéralogie à l'Université, Kiew, Russie.

ARNAUD, H., M. S. G. F., avocat, Angoulême (Charente), France.

ATKINSON, WHEATLEY JAMES, Esq., F. G. S., 76 Christ-church road, Streatham Hill, London, S. W.

*AYRES, HORACE B., land examiner, land department, St. P. & D. R. R. Co., St. Paul, Minnesota, (now Allamuchy, N. J.)

*BAKER, MARCUS, general assistant, U. S. geological survey, Washington, District of Columbia.

BALDACCI, LUIGI, ingénieur des mines, Ufficio geologico, Roma.

*BARKER, GEORGE F., professor of physics, University of Pennsylvania, Philadelphia, Pennsylvania.

*BARROIS, CHARLES, Docteur-ès-sciences, M. S. G. F., professeur à la Faculté des sciences, 185 rue Solférino, Lille, France.

BARTLETT, JOHN H., mining engineer, P. O. Box 83, Roanoke, Virginia.

BASSANI, FRANCESCO, professeur de géologie et de paléontologie à l'Université royale, Napoli, Italie.

BATHER, FRANCIS ARTHUR, Esq., M. A., F. G. S., British Museum (nat. his.), Cromwell road, London, S. W.

BECKE, FRIEDRICH, Dr. Phil., Professor an der deutschen Universität, Prag, Böhmen.

*BECKER, GEORGE F., Ph. D., F. G. S. A., geologist, U. S. geological survey, Washington, District of Columbia.

BEECHER, CHARLES EMERSON, invertebrate paleontologist, Yale University, New Haven, Connecticut.

BELL, ALEXANDER GRAHAM, 1331 Connecticut ave., Washington, District of Columbia.

BENECKE, E. W., Dr. Phil., Professor der Geologie und der Palaeontologie an der Universität, Strassburg i. E., Deutschland.

BERG, CARLOS, Prof. Dr., Director del Museo Nacional, Buenos Aires, Argentine
Republic.

*BERGEAT, ALFRED, Dr. Phil., Landwehrstrasse, 48, 2, München, 111, Bayern.

BERGERON, JULES, Docteur-ès-sciences, M. S. G. F., 157 boulevard Haussmann, Paris.

BERTKAU, PHILIPP, Sekretär des naturhistorischen Vereins, Maarflach 4, Bonn,
Deutschland.

BERTRAND, M., M. S. G. F., professeur à l'École des mines, 62 boulevard St. Michel,
Paris.

BEYRICH, ERNST, Dr. Phil., F. G. S., Professor der Palaeontologie an der Universität,
Kurfürstendamm 140, Berlin, W.

BIEN, JULIUS, cartographer and publisher, 140 Sixth ave., New York City.

BIRKINBINE, JOHN, mining engineer, 25 N. Juniper st., Philadelphia, Pennsylvania.

BLAKE, WILLIAM P., F. G. S., F. G. S. A., mining engineer, New Haven, Connecticut.

BLANFORD, W. T., LL. D., F. R. S., geological survey of India (retired), 72 Bed-
ford Gardens, London, W.

BLOMSTRAND, CHRISTIAN WILHELM, professeur de chimie et de minéralogie à
l'Université, Lund, Suède.

BLOW, ALBERT A., mining engineer, Leadville, Colorado, (now Denver, Colo.)

*BOARDMAN, Mrs. ALICE L., 38 Kenilworth st., Roxbury, Massachusetts.

BOEHMER, MAX, mining engineer, Leadville, Colorado.

*BOGDANOFF, NICHOLAS, ingénieur des mines, Kamennostrowsky prospect, maison
No. 6, St. Pétersbourg.

BONAPARTE, S. A. le Prince ROLAND NAPOLÉON, M. S. G. F., 10 Avenue d'Jéna,
Paris.

BOORAEM, ROBERT ELMER, mining engineer, 120 W. 58th st., New York City.

*BORNE, GEORG von dem, Dr. Phil., Berneuchen (O. P. D. Frankfurt am O.),
Deutschland.

BORNEMANN, J. G., Dr. Phil.. Eisenach, Deutschland.

BORNEMANN, LOUIS GEORGES, Jr., Dr. Phil., Wartburgchaussee 9, Eisenach,
Deutschland.

BOTTI, Comm. ULDERIGO, membre des Soc. géol. de France et d'Italie, Reggio,
Calabria, Italie.

*BOULE, MARCELLIN, M. S. G. F., agrégé de l'Université, attaché au laboratoire de
paléontologie au Muséum d'histoire naturelle, 17 rue Lacépède, Paris.

BOWMAN, AMOS, F. G. S. A., mining engineer, Anacortes, Washington.

BOYD, CHARLES R., Wytheville, Virginia.

*BRANNER, JOHN C., Ph. D., F. G. S. A., State geologist of Arkansas, Little Rock,
Arkansas, (now Prof. of Geology, Leland Stanford University, Palo Alto,
California).

*BRANNER, Mrs. JOHN C., Little Rock, Arkansas, (now Palo Alto, California).

*BROADHEAD, GARLAND C., F. G. S. A., professor of geology, Missouri State
University, Columbia, Missouri.

BRÖGGER, W. C., Dr. Phil., professeur ordinaire en minéralogie et géologie à
l'Université, Christiania, Norvège.

*BROOKS, Major T. B., Newburg, New York.

BROWN, H. Y. LYELL, government geologist, Adelaide, South Australia.

BULKLEY, FRED G., mining engineer, general manager Aspen Mining & Smelting
Co., Aspen, Colorado.

BUSSE, MAX, Dr. Phil., Bergrath, Coblenz, Deutschland.

*CADELL, HENRY M., Esq., late of the geological survey of Scotland, Grange,
Bo'ness, Scotland.

CALL, RICHARD E., M. Sc., A. M., Des Moines, Iowa.

*CALVIN, SAMUEL, F. G. S. A., professor of geology and structural zoölogy, Iowa State University, Iowa City, Iowa.

*CAMPBELL, HENRY D., A. M., Ph. D., F. G. S. A., professor of geology and biology, Washington and Lee University, Lexington, Virginia.

CANAVARI, MARIO, professeur de géologie au musée d'histoire naturelle de l'Université royale, Pisa, Italie.

*CANNON, GEORGE LYMAN, Jr., instructor in geology and biology, High School, Denver, Colorado.

CAPELLINI, GIOVANNI, F. G. S., professeur de géologie à l'Université, Bologne, Italie.

*CASTILLO, ANTONIO del, directeur du Service géologique (Comision geológico) du Mexique, Coliseo Viejo 21, ville de Mexico.

CASTRO, MANUEL FERNANDEZ, directeur de la Carte géologique d'Espagne, Jorge Juan 23, Madrid, Espagne.

*CHAMBERLIN, THOMAS C., LL. D., F. G. S. A., president of the University of Wisconsin, Madison, Wisconsin (now prof. of geology, University of Chicago, Chicago, Illinois).

*CHAPIN, JAMES HENRY, Ph.D., F. G. S. A., professor of geology and mineralogy, St. Lawrence University at Canton, New York: Meriden, Connecticut.

*CHATARD, THOMAS M., Ph. D., chemist, U. S. geological survey, Washington, District of Columbia.

CHOFFAT, PAUL, géologue attaché à la Commission des travaux géologiques du Portugal, 113 Rua do Arco a Jesus, Lisbonne, Portugal.

*CHRISTIE, JAMES C., Esq., F. G. S., Old Cathcart, near Glasgow, Scotland.

*CHURCHILL, WILLIAM, Esq., London.

*CLAGHORN, CLARENCE R., F. G. S. A., mining engineer, 204 Walnut Place, Philadelphia, Pennsylvania.

*CLARK, WILLIAM B., Ph. D., F. G. S. A., associate prfessor of geology, Johns Hopkins University, Baltimore, Maryland.

*CLARKE, FRANK W., chief chemist, U. S. geological survey, Washington, District of Columbia.

*CLAYPOLE, EDWARD W., D. Sc., F. G. S. A., professor of geology, Buchtel Celloge, Akron, Ohio.

COBB, COLLIER, instructor in geology, Massachusetts Institute of Technology, 94 Brattle st., Cambridge, Massachusetts, (now Chapel Hill, N. C.).

COCCHI, IGNIO, professeur de géologie, Firenze, Italie.

COHEN, EMIL, Dr. Phil., Professor der Mineralogie an der Universität, Greifswald (Pommern), Deutschland.

COLE, GRENVILLE A. J., Esq., F. G. S., professor of geology, Royal College of Science, Dublin, Ireland.

Commission des travaux géologiques du Portugal, 113 Rua do Arco a Jesus, Lisbonne, Portugal.

*COMSTOCK, THEODORE B., F. G. S. A., director Arizona School of Mines, Tucson, Arizona.

CONWENTZ, H., Dr. Phil., Director des naturhistorischen Museums, Danzig, Deutschland.

*COPE, EDWARD D., Ph. D., F. G. S. A., professor of geology and paleontology, University of Pennsylvania, Philadelphia, Pennsylvania.

CORTESI, EMILIO, ingénieur des mines, Ufficio geologico, Roma.

COURTIS, WILLIAM M., mining engineer, 449 Fourth ave., Detroit, Michigan.

COXE, ECKLEY B., mining engineer, Drifton, Pennsylvania.

CRAGIN, FRANCIS W., B.S., F. G. S. A., professor of geology, Colorado College, Colorado Springs, Colorado.

CRAWFORD, J., Government geologist, care U. S. Consul, Managua, Nicaragua.

*CREDNER, HERMANN, Dr. Phil., Professor der Geologie an der Universität, Director der k. sächsischen geologischen Landesuntersuchung, Leipzig, Sachsen.

*CREDNER, RUDOLF, Dr. Phil., Professor der Erdkunde an der Universität, Vorsitzender der geog. Gesellschaft, Greifswald (Pommern), Deutschland.

* CROSS, WHITMAN, Ph. D., F. G. S. A., geologist, U. S. geological survey, Washington, District of Columbia.

* CULVER, GARRY E., professor of geology, University of South Dakota, Vermillion, South Dakota, (now Madison, Wisconsin).

*CUSHING, HENRY P., F. G. S. A., 786 Prospect st., Cleveland, Ohio.

DAGGETT, ELLSWORTH, U. S. surveyor-general, Salt Lake City, Utah.

DAGINCOURT, EM., Docteur-ès-sciences, M. S. G. F., minéralogiste, 16 rue de Tournon, Paris.

*DALE, T. NELSON, F. G. S. A., assistant geologist, U. S. geological survey, Newport, Rhode Island.

* DALL, WILLIAM H., Ph. D., paleontologist, U. S. geological survey, Washington, District of Columbia.

DALTON, WILLIAM H., Esq., F. G. S., late of the geological survey of Great Britain, Museum, 28 Jermyn st., London, S. W.

DANA, EDWARD S., Ph. D., curator of mineralogy, Yale University, New Haven, Connecticut.

DANA, JAMES D., LL. D., F. G. S., F. G. S. A., professor of geology and mineralogy, Yale University, New Haven, Connecticut.

* DARTON, NELSON H., F. G. S. A., assistant geologist, U. S. geological survey, Washington, District of Columbia.

*DARWIN, CHARLES, librarian, U. S. geological survey, Washington, District of Columbia.

*DAVIDSON, Mrs. A. D., Oberlin, Ohio.

DAVIES, ARTHUR MORLEY, Esq., demonstrator in geology, Royal College of Science, South Kensington, London, S. W.

* DAVIS, WILLIAM MORRIS, Ph. D., F. G. S. A., professor of physical geography, Harvard University, Cambridge, Massachusetts.

DAVIS, WILLIAM H., Esq., accountant, Walnut Tree Villa, Somerset road, Tottenham, England.

*DAWSON, JAMES FRANCIS, professor of geology, Georgetown College, Washington, District of Columbia.

*DAY, DAVID T., Ph. D., F. G. S. A., geologist, U. S. geological survey, Washington, District of Columbia.

DAY, JAMES THOMAS, Esq., F. G. S., 12 Albert square, Stepney, London, E.

*DELAIRE, ALEXIS, M. S. G. F., ingénieur civil, 238 Boulevard St. Germain, Paris.

DELGADO, Colonel JOAQUIN F. N., directeur des travaux géologiques du Portugal, 113 rua do Arco a Jesus, Lisbonne, Portugal.

DELVAUX, ÉMILE, vice-président de la Société géologique de Belgique, 216 avenue Brugman, Bruxelles.

DEMARCHI, LAMBERTO, ingénieur des mines, via Napoli 65, Roma.

DEWALQUE, G., Dr. Phil., F. G. S., professeur de géologie à l'Université, Liège, Belgique.

*DIENER, CARL, Dr. Phil., Privatdocent an der k. k. Universität, Marxergasse 24, Wien III, Oesterreich.

DIEST, PETER H. VAN, mining engineer, 1230 Washington ave., Denver, Colorado.

DILLER, JOSEPH SILAS, F. G. S. A., geologist, U. S. geological survey, Washington, District of Columbia.

D'INVILLIERS, EDWARD V., F. G. S. A., geologist and mining engineer, 711 Walnut st., Philadelphia, Pennsylvania.

* DODGE, RICHARD ELWOOD, assistant geologist, U. S. geological survey, 22 Stoughton Hall, Cambridge, Massachusetts.

DOLLFUS, GUSTAVE, M. S. G. F., 45 rue de Chabrol, Paris.

* DOULCET, JEAN, secrétaire d'ambassade, 4 place du Palais Bourbon, Paris.

DRUNEN, JAMES VAN, professeur à l'École polytechnique, 9 rue des Champs Élysées, Bruxelles.

* DUMBLE, EDWIN T., F. G. S. A., State geologist of Texas, Austin, Texas.

* DUNIKOWSKI, EMILE DE, Dr. Phil., Professor an der Universität, Lemberg, Galizien, Oesterreich.

DWIGHT, WILLIAM B., A. M., Ph. B., F. G. S. A., professor of natural history, Vassar College, Poughkeepsie, New York.

* EAKINS, LINCOLN G., assistant chemist, U. S. geological survey, Washington, District of Columbia, (now Denver, Colorado).

EAMES, Dr. R. M., late assistant State geologist of Minnesota, Salisbury, North Carolina.

* EASTMAN, JOHN R., professor of mathematics, U. S. Naval Observatory, Washington, District of Columbia.

* ELDRIDGE, GEORGE H., F. G. S. A., geologist, U. S. geological survey, Washington, District of Columbia.

* ELGUERA, MANUEL, ingénieur, secrétaire de la Commission de Pérou, Washington, District of Columbia.

* EMMONS, SAMUEL FRANKLIN, A. M., F. G. S., F. G. S. A., geologist, U. S. geological survey, Washington, District of Columbia.

* EMMONS, Mrs. S. F. (Sophie Markoe), 1721 H st., Washington, District of Columbia.

ENDLICH, FREDERIC MILLER, Ph. D., mining engineer, Ouray, Colorado.

EUSTIS,. WE. C., Room 32, 55 Kilby st., Boston, Massachusetts.

EVANS, JOHN, LL. D., F. R. S., etc., Nash Mills, Hemel Hempstead, England.

EVERETTE, WILLIS E., Tacoma, Washington.

EYERMAN, JOHN, F. G. S. A., instructor in Lafayette College, Easton, Pennsylvania.

* EYSSÉRIC, JOSEPH, géographe, 14 rue Dupléssis, Carpentras (Vaucluse), France.

FABRE, GEORGES, M. S. G. F., inspecteur des forêts, 26 rue Menard, Nîmes (Gard), France.

FABRI, ANTONIO, ingénieur en chef des mines, R. Corpo delle miniere, Firenze, Italie.

FAIRBANKS, HAROLD W., California State Mining Bureau, San Francisco, California.

* FAIRCHILD, HERMAN LeROY, F. G. S. A., professor of geology, University of Rochester, Rochester, New York.

FALY, JOSEPH, ingénieur principal des mines, Mons, Belgique.

* FARNSWORTH, P. J., A. M., M. D., F. G. S. A., professor of materia medica, State University of Iowa, Clinton, Iowa.

FELIX, JOHANNES, Dr. Phil., Professor an der Universität, Gellertstrasse 3, Leipzig, Sachsen.

FENNELL, CHARLES WILLIAM, Esq., surveyor, Westgate, Wakefield, Yorkshire, England.

* FERNOW, BERNHARD E., chief of forestry division, Department of Agriculture, Washington, District of Columbia.

FIRKET, ADOLPHE, professeur à l'Université, 28 rue Dartois, Liège, Belgique.

* FISCHER, MORITZ, F. G. S. A., curator of the E. M. museum, Princeton College. Princeton, New Jersey.

*FOOTE, ALBERT E., M. D., F. G. S. A., 1224 N. 41st st., Philadelphia, Pennsylvania.

FORRESTER, ROBERT, F. G. S. A., mining engineer, Castlegate, Utah.

*FORSTER, Miss MARY, lecturer on natural science, London (51 West 12th st., New York City).

FOSTER, CLEMENT LE NEVE, Esq., D. Sc., F. G. S., professor of mining engineering, Royal College of Science, South Kensington, London. H. M. inspector of mines, Llandudno, North Wales.

FOX-STRANGWAYS, CHARLES, Esq., F. G. S., geological survey of Great Britain, Museum, 28 Jermyn st., London, S. W.

FRANCHI, SECONDO, ingénieur des mines, via Angelo Brofferio 1, Torino. Ufficio geologico, Roma.

FRAZER, PERSIFOR, D. Sc., F. G. S. A., professor of chemistry in Franklin Institute, Drexel Building, Philadelphia, Pennsylvania.

*FRECH, FRITZ, Dr. Phil., Professor an der Universität, Schuhbrücke 38, Breslau, Deutschland.

*FRICK, JOHN H., A. M., professor of natural science, Central Wesleyan College, Warrenton, Missouri.

FUCHS, THEODOR, Director der geologischen Abtheilung des k. k. naturhistorischen Hofmuseums, Wien, Oesterreich.

*FULLER, HOMER T., Ph. D., F. G. S. A., professor of geology, Worcester Polytechnic Institute, Worcester, Massachusetts.

GALITZINE, PRINCE PROZOROVSKY, maître des cérémonies de la cour de l'Empereur de Russie, musée géologique de l'Université, St. Pétersbourg.

GANNETT, HENRY, geographer, U. S. geological survey, Washington, District of Columbia.

GARWOOD, EDMUND J., Esq., B. A., F. G. S., St. John's College, Cambridge, England. United University Club, Pall Mall, London, W.

*GAUDRY, ALBERT, F. G. S., M. S. G. F., professeur de paléontologie au Muséum d'histoire naturelle, 7bis rue des Saints Pères, Paris.

GEANDEY, FERDINAND, 11 rue de Sèze, Lyon, France.

*GEER, GERARD DE, statsgeolog, Geologiska Byrån, Stockholm, Suède.

GEINITZ, HANS BRUNO, Dr. Phil., F. G. S., Geheimer Bergrath, Professor der Geologie am k. Polytechnicum, Dresden, Sachsen.

GIESECKE, BRUNO, Dr. Phil., Nürnbergerstrasse 12, Leipzig, Sachsen.

*GILBERT, GROVE K., A. M., F. G. S. A., geologist, U. S. geological survey, Washington, District of Columbia.

*GILL, Dr. THEODORE N., paleontologist, Smithsonian Institution, Washington, District of Columbia.

† GIORDANO, FELICE, inspecteur général des mines, R. Corpo delle miniere, directeur du relevé géologique d'Italie, R. Ufficio geologico, via Sta. Susanna 1, Roma.

GLEN, DAVID CORSE, Esq., engineer, 14 Armfield Place, Glasgow, Scotland.

*GOLLIEZ, HENRI, professeur de géologie à l'Université, Lausanne, Suisse.

GOODALE, CHARLES L., professor of botany, Harvard University, Cambridge, Massachusetts.

*GOODE, G. BROWN, assistant secretary of the Smithsonian Institution, Washington, District of Columbia.

GOSSELET, JULES, F. G. S., M. S. G. F., professeur à la Faculté des sciences, 18 rue d'Antin, Lille, France.

*GREGORY, JOHN WALTER, Esq., F. G. S., British Museum (nat. hist.), South Kensington, London, S. W.

*GREGORY, Miss, London.

*GRISWOLD, LEON S., assistant geologist of Arkansas, Little Rock, Arkansas.

*HAGUE, ARNOLD, F. G. S., F. G. S. A., geologist, U. S. geological survey, Washington, District of Columbia.

HAGUE, JAMES D., mining engineer, 18 Wall st., New York city.

*HALE, WILLIAM H., 40 First Place, Brooklyn, New York.

HALFAR, ANTON, Bezirksgeolog der k. preussischen geologischen Landes-Anstalt, Invalidenstrasse 121, III, Berlin.

HALL, CHRISTOPHER W., A. M., F. G. S. A., professor of geology and mineralogy, University of Minnesota, Minneapolis, Minnesota.

*HALL, JAMES, LL. D., F. G. S., F. G. S. A., state geologist and paleontologist, director of State museum of natural history, Albany, New York.

HALL, Capt. MARSHALL, F. G. S., Easterton Lodge, Parkstone, R. S. O., Dorset, England.

*HALLOCK, WILLIAM, Ph. D., physicist, U. S., geological survey, Washington, District of Columbia, (now asst. prof. physics, Columbia College, New York City).

HANKS, HENRY G., F. G. S. A., 1124 Greenwich st., San Francisco, California.

HANSFORD, CHARLES, Esq., 3 Alexandra Terrace, Dorchester, Dorset, England.

†HANTKEN, MAX VON, Professor an der Universität, VI Eötvös-uteza 9, Budapest, Ungarn.

*HARKER, ALFRED, Esq., M. A., F. G. S., demonstrator in geology, University of Cambridge, Cambridge, England.

*HARKNESS, WILLIAM, professor of mathematics, U. S. Naval Observatory, Cosmos Club, Washington, District of Columbia.

*HARRIS, GEORGE F., Esq., F. G. S., 20 Craster Road, Brixton Hill, London, S. W.

*HARRIS, GILBERT D., assistant paleontologist, U. S. geological survey, Washington, District of Columbia.

*HARRIS, THADDEUS WILLIAM, instructor in geology, Harvard University, Cambridge, Massachusetts.

*HATCHER, J. B., U. S. geological survey, Lone Pine, Nebraska. Princeton, New Jersey.

HAUCHECORNE, WILHELM, Dr. Phil., Geh. Bergrath, Director der k. preussischen geologischen Landes-Anstalt und Bergakademie, 44 Invalidenstrasse, Berlin.

*HAY, ROBERT, F. G. S. A., geologist, Junction City, Kansas.

*HAYDEN, Lieut. EVERETT, Hydrographic Office, Navy Department, Washington, District of Columbia.

HAYES, C. WILLARD, Ph. D., F. G. S. A., assistant geologist, U. S. geological survey, Washington, District of Columbia.

*HEAD, M. R., 5467 Jefferson ave., Hyde Park, Chicago, Illinois.

HÉBERT, Madame EDMOND, M. S. G. F., 10, rue Garancière, Paris.

*HENDERSON, JOSEPH J., civil engineer, Kingsbridge, New York.

HENNEQUIN, ÉMILE, colonel d'état-major, directeur de l'Institut cartographique militaire, Bruxelles.

HEUSLER, CONRAD, Geheimer Bergrath, Bonn., Preussen.

HICKS, HENRY, D., M. D., F. R. S., F. G. S., secretary of the geological society of London, Hendon Grove, Hendon, London, N. W.

*HILGARD, EUGENE W., LL. D., Ph. D., F. G. S. A., professor of agriculture and chemistry, University of California, Berkeley, California.

*HILL, ROBERT THOMAS, F. G. S. A., geologist, Washington, District of Columbia.

*HILLEBRAND, WILLIAM F., Ph. D., chemist, U. S. geological survey, Washington, District of Columbia.

HILLS, RICHARD C., F. G. S., geologist, Colorado Fuel Co., Denver, Colorado.

*HITCHCOCK, CHARLES H., Ph. D., F. G. S. A., professor of geology, Dartmouth College, Hanover, New Hampshire.

HOBBS, WILLIAM HERBERT, Ph. D., F. G. S. A., assistant professor of mineralogy, University of Wisconsin, Madison, Wisconsin.

*HOBSON, BERNARD, Esq., F. G. S., lecturer in geology, Owens College, Manchester (residence, Tapton Elms, Sheffield), England.

*HODGES, ALMON D., Jr., mining engineer, geologist of the Commission from Peru, P. O. box 1857, Boston, Massachusetts.

*HOLDEN, LUTHER L., geographer, Jamaica Plain, Boston, Massachusetts.

*HOLMES, JOSEPH AUSTIN, F. G. S. A., State geologist of North Carolina, professor of geology, University of North Carolina, Chapel Hill, North Carolina.

*HOLMES, Mrs. J. A., Chapel Hill, North Carolina.

*HOLMES, WILLIAM H., archæologist, Bureau of Ethnology, Smithsonian Institution, Washington, District of Columbia.

*HOLST, NILS OLAF, statsgeolog, geologiska byrån, Stockholm, Suède.

*HOUGH, WALTER, assistant, U. S. National Museum, Washington, District of Columbia.

HOVELACQUE, MAURICE docteur-ès-sciences, M. S. G. F., 1 rue de Castiglione, Paris.

*HOVEY, Rev. HORACE C., F. G. S. A., 60 High st., Newburyport, Massachusetts.

*HOWELL, EDWIN E., F. G. S. A., 612 Seventeenth st., Washington, District of Columbia.

HOWSE, RICHARD, Esq., F. G. S., curator of museum, Natural History Society of Northumberland, 12 St. Thomas Crescent, Newcastle-on-Tyne, England.

*HUBBARD, GARDINER G., Twin Oaks, Woodley Lane road, 1328 Connecticut avenue, Washington, District of Columbia.

HUBBARD, LUCIUS LEE, Ph. D., petrographer, geological survey of Michigan, Houghton, Michigan.

HUDLESTON, WILFRED H., Esq., M. A., F. R. S., F. G. S., vice-president of the geological society of London, 8 Stanhope Gardens, London, S. W.

*HUGHES, THOMAS MCKENNY, Esq., M. A., F. R. S., F. G. S., Woodwardian professor of geology, Cambridge, England.

*HUGHES, Mrs. T. McK. (Mary Caroline), Cambridge, England.

HULL, EDWARD, LL. D., F. R. S., F. G. S., 20 Arundel Gardens, Notting Hill, London, W.

HUME, WILLIAM F., Esq., F. G. S., 27 Ella Road, Crouch End, London, N.

†HUNT, THOMAS STERRY, LL. D., F. R. S., F. G. S. A., Park Avenue Hotel, New York City.

*HURLBURT, EDWARD, 128 Genessee st., Utica, New York.

HYATT, ALPHEUS, Museum of nat. hist., Boston, Massachusetts.

*IDDINGS, JOSEPH P., F. G. S. A., geologist, U. S. geological survey, Washington, District of Columbia, (now professor of petrology, University of Chicago, Chicago, Illinois.)

INGEN, GILBERT VAN, paleontologist, Poughkeepsie, New York.

INKEY, Baron ADELBERT VON, Bergingenieur, Pressburg, Oesterreich.

INOSTRANZEFF, ALEXANDRE, professeur de géologie à l'Université, St. Pétersbourg.

*JAEKEL, OTTO, Dr. Phil., Privatdocent an der Universität, Invalidenstrasse 43, Berlin.

*JAMES, JOSEPH F., F. G. S. A., assistant paleontologist, U. S. geological survey, Washington, District of Columbia, (now Department of Agriculture.)

JANET, CHARLES, M. S. G. F., ingénieur des arts et manufactures, Beauvais, France.

JENTZSCH, ALFRED, Dr. Phil, Director des Provinzial-Museums, Königsberg, Preussen.

*JOHNSON, H. L. E., M. D., 1400 L street, Washington, District of Columbia.

*JOHNSON, LAWRENCE C., F. G. S. A., assistant geologist, U. S. geological survey, Gainesville, Florida.

JOHNSTON-LAVIS, H. J., M. D., F. G. S., 7 Chiatamone, Napoli, Italie.

JUDD, JOHN W., Esq., F. G. S., professor of geology, Royal College of Science, South Kensington (residence, 31 Ennerdale Road, Kew), London.

KARAKASCH, NICOLAS, conservateur du musée géologique à l'Université, St. Pétersbourg.

KARPINSKI, THEODOR, directeur du Comité géologique de Russie, St. Pétersbourg.

* KAYSER, EMANUEL, Dr. Phil., Professor der Geologie und Palæontologie an der Universität, Marburg, Deutschland.

* KEITH, ARTHUR, A. M., F. G. S. A., assistant geologist, U. S. geological survey, Washington, District of Columbia.

* KEMP, JAMES FURMAN, E. M., F. G. S. A., adjunct professor of geology, Columbia College, New York City.

KENNEDY, GEORGE L., A. M., F. G. S., professor of chemistry, mining and geology, Kings College, Windsor, Nova Scotia, Canada.

* KENNEDY, HARRIS, student at Harvard University, Roxbury, Massachusetts.

* KEYES, CHARLES R., A. M., F. G. S. A., Johns Hopkins University, Baltimore, Maryland.

KEYES, W. S., mining engineer, Pacific Union Club, San Francisco, California.

KEYSERLING, ALEXANDRE, Comte, F. G. S., Memb. de l'Académie impériale des sciences de St. Pétersbourg, Railküll via Reval, Estorie, Russie.

* KIMBALL, JAMES P., Ph. D., F. G. S. A., ex-Director of the U. S. Mint, 1311 New Hampshire ave., Washington, District of Columbia.

* KING, CLARENCE, F. G. S., F. G. S. A., geologist, ex-Director of the U. S. geological survey, 18 Wall st., New York City.

KLOOS, JOHANN HERMANN, Dr. Phil., Professor am Polytechnicum, Braunschweig, Deutschland.

* KNOWLTON, FRANK HALL, M. S., F. G. S. A., assistant paleontologist, U. S. geological survey, Washington, District of Columbia.

* KOENIGS, W., Dr. Phil., Privatdocent an der Universität, München, Bayern.

KORTHALS, W. C., M. S. G. F., Heidelberg, 46 Neuenheimer Landstrasse, Heidelberg, Baden.

* KOST, J., Tiffin, Ohio.

* KRASSNOF, ANDRE NIKOLAEVITSCH, professeur à l'Université de Charkow, Russie Méridionale.

* KROUSTCHOFF, CONSTANTIN DE, Dr. Phil., Musée minéralogique de l'Académie impériale des sciences, St. Pétersbourg.

* KUMMEL, HENRY B., student at Harvard University, Milwaukee, Wisconsin.

KUNZ, GEORGE F., F. G. S. A., mineralogist, 402 Garden st., Hoboken, New Jersey, 11 Union Square, New York.

* LADD, GEORGE EDGAR, A. M., F. G. S. A., geologist, geological survey of Missouri, Jefferson City, Missouri.

* LAFLAMME, J. C. K., A. M., F. G. S. A., professor of geology and mineralogy, Laval University, Quebec, Canada.

LAMBORN, ROBERT H., 32 Nassau st., New York City.

LANDERO, CARLOS F., inspecteur général des mines du Mexique, rue Santa Monica 17, Guadalajara, Mexique.

* LANE, ALFRED C., Ph. D., F. G. S. A., assistant geologist, geological survey of Michigan, Houghton, Michigan.

* LANGDON, DANIEL W., Jr., F. G. S. A., geologist of Chesapeake and Ohio R. R. Co., Cincinnati, Ohio.

LAPPARENT, ALBERT DE, M. S. G. F., professeur à l'Institut catholique, 3 rue de Tilsitt, Paris.

LAPWORTH, CHARLES, LL. D., F. R. S., F. G. S., professor of geology, Mason Science College, Birmingham, England.

*LAWSON, ANDREW C., Ph. D., F. G. S. A., assistant professor of geology, University of California, Berkeley, California.

LECKIE, R. G., mining engineer, care of Londonderry Iron Co., Londonderry, Nova Scotia, Canada.

*LECONTE, JOSEPH, LL. D., F. G. S. A., professor of geology, University of California, Berkeley, California.

LEHAIE, A. HOUZEAU DE, ex-président de la Société belge de géologie, paléontologie et d'hydrologie, Hyon (Mons), Belgique.

LEIGHTON, THOMAS, Esq., F. G. S. Lindisfarm, St. Julian's Road, West Norwood, England.

LEONARD, HUGH, Esq., late chief engineer Indian Public Works Department, 7 Hanover Square, London, W.

LEPSIUS, RICHARD, Dr. Phil., Professor der Geologie am k. Polytechnicum, Darmstadt, Hessen, Deutschland.

LESLEY, J. PETER, LL. D., F. G. S. A., State geologist of Pennsylvania, 1008 Clinton st., Philadelphia, Pennsylvania.

LERCH, OTTO, assistant geologist, geological survey of Texas, Austin, Texas.

*LEVERETT, FRANK, F. G. S. A., assistant geologist, U. S. geological survey, Madison, Wisconsin.

*LEVITZKY, ROMAN, ingénieur des mines, Swietznoj N. 5, St. Pétersbourg.

LICHERDOPOL, JEAN P., professeur de chimie, Strada Domniti 7, Bucharest, Roumanie.

*LINCOLN, DAVID F., M. D., Geneva, New York.

*LINDAHL, JOSUA, Ph. D., F. G. S. A., State geologist of Illinois, Springfield, Illinois.

LIVERSIDGE, ARCHIBALD, Esq., M. A., F. R. S., professor of chemistry, University of Sidney, New South Wales.

LÓCZY, L. DE, Professor an der Universität, Budapest IV, Ungarn.

LOEWINSON-LESSING, FRANÇOIS, Dr. Phil., agrégé à l'Université impériale, musée géologique, St. Pétersbourg.

*LOHEST, MAX, professeur à l'Université, Liège, Belgique.

LOPEZ, MONROY PEDRO, ingénieur des mines, ville de San Luis Potosi, Mexique.

LOUIS, DAVID A., Esq., mining engineer, 77 Shirland Gardens, London, W.

*LUNDBOHM, HJALMAR, statsgeolog, geologiska byrån, Stockholm, Suède.

LUTHE, FREDERIC H., McGregor, Iowa.

LUTHER, GEORGE E., assistant geologist, U. S. geological survey, Madison, Wisconsin.

LYMAN, BENJAMIN SMITH, mining engineer, 708 Locust st., Philadelphia, Pennsylvania.

LYONS, HENRY GEORGE, lieutenant, royal engineers, Kasr el Nil, Cairo, Egypt.

McCALLEY, HENRY, A. M., C. E., F. G. S. A., assistant geologist of Alabama, University P. O., Alabama.

*McGEE, W J, F. G. S. A., geologist, U. S. geological survey, Washington, District of Columbia.

*McGEE, Mrs. W J (Anita Newcomb), Washington, District of Columbia.

McKELLAR, PETER, F. G. S., F. G. S. A., mining explorer, Fort Williams, Ontario, Canada.

*MACFARLANE, THOMAS, chief analyist, Inland Revenue Department, Ottawa, Canada.

*MALLET, FREDERIC RICHARD, Esq., F. G. S., superintendent of the geological survey of India (retired), 18 the Common, Ealing, London.

*MALLET, RICHARD T., Esq., chief engineer Public Works Department of India, Bellerieve, Downshire Square, Reading, England.

MARCOU, JULES, F. G. S., F. G. S. A., 42 Garden st., Cambridge, Massachusetts.

*MARGERIE, EMMANUEL DE, M. S. G. F., attaché au Service de la carte géologique de la France, 132 rue de Grenelle, Paris.

MARIS, CERENATI D., Musée de géologie à l'Université royale, Roma.

MARKBY, JOHN RANDALL, Esq., 7 University st., Tottenham Court Road, London, W. C.

'MARKOE, Miss MARY G., 1721 H st., Washington, District of Columbia.

MARR, JOHN E., Esq., M. E., F. G. S., St. John's College, Cambridge, England.

*MARSH, OTHNIEL C., LL. D., F. G. S., F. G. S. A., professor of paleontology, Yale University, New Haven, Connecticut.

*MARTIN, DANIEL S., professor of geology, Rutgers Female College, 54–56 West 55th st., New York City.

*MASON, T. B. M., Lieutenant U. S. Navy, Navy Department, Washington, District of Columbia.

MATTIROLO, ETTORE, ingénieur des mines, Torino. Ufficio geologico, Roma.

MAURER FREDERICK, Rentner, Heinrichstrasse 109, Darmstadt, Deutschland.

*MEANS, JOHN H., assistant geologist, geological survey of Arkansas, Little Rock, Arkansas.

MEDLICOTT, HENRY B., M. A., F. R. S., F. G. S., director of the geological survey of India, 43 St. John's Road, Clifton, Bristol, England.

*MERRILL, FREDERICK J. H., Ph., D., F. G. S. A., assistant director of the state museum, Albany, New York.

*MERRILL, GEORGE P., M. S., F. G. S. A., curator of geology, U. S. national museum, Washington, District of Columbia.

MICHEL-LÉVY, A., M. S. G. F., directeur du Service de la carte géologique de France, 26 rue Spontini, Paris.

* MINDELEFF, COSMOS, archæologist, Bureau of Ethnology, Smithsonian Institution, Washington, District of Columbia.

MIXER, FREDERICK M., Hotel Broadway, Denver, Colorado.

MOLENGRAAF, G. A. F., Dr. Phil., professeur à l'Université, 1ste Parkstraat 394, Amsterdam, Holland.

*MOSES, THOMAS F., M. D., F. G. S. A., president of Urbana University, Urbana, Ohio.

MOURLON, MICHEL, membre de l'Académie royale des sciences de Belgique, 107 rue Belliard, Bruxelles.

MÜGGE, OTTO, Dr. Phil., Professor an der k. Akademie, Münster, Westfalen, Deutschland.

MUNIER-CHALMAS, M. S. G. F., professeur de géologie à la Sorbonne, Paris.

MUNROE, H. S., professor of mining, School of Mines, Columbia College, 49th street and Madison avenue, New York City.

* MURDOCK, JOHN, librarian, Smithsonian Institution, Washington, District of Columbia.

MUSY, Prof. M., président de la Société fribourgeoise des sciences naturelles, Fribourg, Suisse.

NEFF, PETER, A. M., F. G. S. A., 401 Prospect street, Cleveland, Ohio.

† NEWBERRY, JOHN S., M. D., LL. D., F. G. S., F. G. S. A., professor of geology and paleontology, Columbia College, New York City.

NEWBERRY, WOLCOTT E., mining engineer, Colorado Springs, Colorado.

* NEWELL, FREDERICK H., F. G. S. A., topographer, U. S. geological survey, Washington, District of Columbia.

NEWTON, EDWIN B., Esq., F. G. S., geological survey of Great Britain, Museum, 28 Jermyn street, London, S. W.

NEWTON, RICHARD BULLEN, Esq., F. G. S., British Museum (nat. hist.), South Kensington, London, S. W.

NIEDZWIEDZKI, J., Professor der Mineralogie und Geologie an der Universität, Lemberg, Oesterreich.

NIES, FRIEDRICH, Dr. Phil., Professor der Geologie am k. Polytechnicum, Hohenheim, Würtemberg.

NIKITIN, Prof. Serge, Institut des mines, St. Pétersbourg.

†NORTHROP, JOHN I., School of Mines, Columbia College, New York City.

*NORTON, WILLIAM H., professor of geology, Cornell College, Mt. Vernon, Iowa.

NOVARESE, VITTORIO, ingénieur des mines, Torino. Ufficio geologico, Roma.

NYE, Dr. W., 385 Dearborn avenue, Chicago, Illinois.

OCHSENIUS, CHARLES, Dr. Phil., Privatdocent an der Universität, Marburg, Hessen, Deutschland.

*OGDEN, HERBERT G., U. S. coast and geodetic survey, Washington, District of Columbia.

*ORDOÑEZ, EZEQUIEL, géologue de la Commission géologique du Mexique, ville de Mexico.

ORTON, EDWARD, Ph. D., LD. D., F. G. S. A., State geologist of Ohio, Columbus, Ohio.

*OSANN, ALFRED, Dr. Phil., Professor an der Universität, Heidelberg, Deutschland.

OSBORN, HENRY F., professor of comparative anatomy, Columbia College, New York City.

PALMER, CHARLES S., Ph. D., professor of chemistry, University of Colorado, Boulder, Colorado.

PARONA, CARLO FABRIZIO, professeur de géologie à l'Université, Palazzo Carignano, Torino, Italie.

PARRY, J. M., Aspen, Colorado.

PATTON, HORACE B., Ph. D., F. G. S. A., geologist, geological survey of Michigan, Houghton, Michigan. Golden, Colorado.

*PAVLOW, ALEXIS, professeur de géologie à l'Université, Musée de géologie, Moscou, Russie.

*PAVLOW, Mme. MARIE, Maison Cheremétief, log. 65, Cheremétevsky pereoulok, Moscou, Russie.

PEALE, ALBERT C., M. D., geologist, U. S. geological survey, Washington, District of Columbia, (now 1909 Chesnut st., Philadelphia, Pennsylvania.)

PENFIELD, SAMUEL LEWIS, assistant professor of mineralogy, Sheffield Scientific School, New Haven, Connecticut.

*PENROSE, RICHARD A. F., Jr., Ph. D., F. G. S. A., assistant geologist, geological survey of Arkansas, Little Rock, Arkansas, (now prof. of economic geology, University of Chicago, Chicago, Illinois.)

*PERKINS, GEORGE H., professor of natural history, University of Vermont, Burlington, Vermont.

*PERRY, JOSEPH H., F. G. S. A., teacher of chemistry and geology, 176 Highland st., Worcester, Massachusetts.

*PETTEE, WILLIAM H., A. M., F. G. S. A., professor of mineralogy and economic geology, University of Michigan, Ann Arbor, Michigan.

*PLIENINGER, FELIX, Palaeontologisches Institut, München, Bayern.

PORTER, J. A., 804 Boston Building, Denver, Colorado.

PORTIS, Dr. ALESSANDRO, professeur de géologie à l'Université royale, Roma.

POŠEPNÝ, Prof. F., Bergrath, O. Döbling, Karl Ludwigstr. 62, Wien.

POSTLETHWAITE, JOHN, Esq., F. G. S., Eskin Place, Keswick, Cumberland, England.

POTTER, WILLIAM B., A. M., E. M., F. G. S. A., professor of mining and metallurgy, Washington University, St. Louis, Missouri.

* POWELL, JOHN W., Ph. D., LL. D., F. G. S. A., director of the U. S. geological survey, Washington, District of Columbia.

* PRENTISS, D. WEBSTER, M. D., 1101 Fourteenth st., Washington, District of Columbia.

* PRESCOTT, ALBERT B., President A. A. A. S., director of the chemical laboratory, University of Michigan, Ann Arbor, Michigan.

PRESTWICH, JOSEPH, Esq., M. A., F. R. S., F. G. S., president of the Fourth International Congress of Geologists, Shoreham, Kent, England.

PROCTER, JOHN R., F. G. S. A., State geologist of Kentucky, Frankfort, Kentucky, (now Civil Service Commissioner, Washington, D. C.)

PROESCHOLDT, Prof. H., Dr. Phil., Landesgeolog, Meiningen, Deutschland.

* PROSSER, CHARLES S., M. S., F. G. S. A., assistant paleontologist, U. S. geological survey, Washington, District of Columbia. Topeka, Kansas.

PUGA, GUILLERMO B. Y, président de la Société "Antonio Alzate," professeur de minéralogie et de géologie à l'École préparatoire, ville de Mexico, Mexique.

* PUMPELLY, RAPHAEL, F. G. S. A., geologist, U. S. geological survey, Newport, Rhode Island.

PURGOLD, ALFRED, Blasewitz, Gotha, Deutschland.

* PUTNAM, F. W., professor of archæology and ethnology, Harvard University, Cambridge, Massachusetts.

RAND, THEODORE D., Esq., 17 South Third st., Philadelphia, Pennsylvania.

RADKEWITCH, GRÉGOIRE, agrégé de géologie à l'Université, Kiew, Russie.

RANDOLPH, JOHN C. F., mining engineer, University Club, Madison Square, New York City.

* RATHBUN, RICHARD, curator, U. S. national museum, Washington, District of Columbia.

RAUNHEIM, SALY E., mining engineer, P. O. Box 3, New York City.

RAYMOND, R. W., secty. Am. Inst. Mg. Engrs., 13 Burling Slip, New York City.

* READ, M. A., student, Harvard University, 22 Stoughton Hall, Cambridge, Massachusetts.

RÉCLUS, ÉLISÉE, géographe, Librairie Hachette, Paris.

REID, CLEMENT, Esq., F. G. S., geological survey of Great Britain, Museum, 28 Jermyn st., London, S. W.

REISS, W., Dr. Phil., Kurfürstenstrasse 98, Berlin, W.

* REUSCH, HANS, Dr. Phil., F. G. S., director of the geological survey of Norway, Christiania, Norway.

RICCIARDI, Dr. LEONARDO, professeur de chimie, Reggio, Calabria, Italie.

* RICE, WILLIAM NORTH, A. M., Ph. D., F. G. S. A., professor of geology, Wesleyan University, Middletown, Connecticut.

* RICHARDSON, CLIFFORD, district chemist, Washington, District of Columbia.

RICHE, ATTALE, M. G. S. F., chef des travaux de géologie à la Faculté des sciences, rue de Penthièvre, 11, Lyon, France.

RICKETTS, LOUIS D., Ph. D., mining engineer, 11 John st., New York City. Rawlins, Wyoming.

RIVERO, JOAQUIN L., ingénieur des mines, Zacatecas, Mexique.

* ROCKWELL, Gen. ALFRED P., E. M., Manchester, Massachusetts.

* ROLFE, CHARLES W., M. S., F. G. S. A., professor of geology, University of Illinois, Urbana, Illinois.

ROLKER, CHARLES M., mining engineer, 98 Broadway, New York City.

* ROMBERG, JULIUS, Wilsnackerstrasse 61, Berlin, N. W.

ROSENBUSCH, H., F. G. S., Director der grossh. Bad. geologischen Landesanstalt, Professor an der Universität, Heidelberg, Deutschland.

* ROTHPLETZ, AUGUST, Dr. Phil., Privatdocent an der Universität, Theresienstrasse 96-2, München, Bayern.

RUDLER, FREDERICK W., Esq., F. G. S., geological survey of Great Britain, Museum, 28 Jermyn st., London, S. W.

RUSSELL, ISRAEL C., M. S., F. G. S. A., geologist, U. S. geological survey, Washington, District of Columbia, (now Prof. of Geology, University of Michigan, Ann Harbor, Michigan).

† RUST, WILLIAM P., Trenton Falls, New York.

RUTLEY, FRANK, Esq., F. G. S., lecturer on mineralogy, Royal College of Science, London, S. W.

RUTOT, AIMÉ, conservateur au Musée royale d'histoire naturelle de Belgique, 177 rue de la Loi, Bruxelles.

RYAN, GEORGE, 1143 E. Second South st., Salt Lake City, Utah.

SACCO, FREDERICO, professeur de paléontologie à l'Université, Torino, Italie.

* SAFFORD, JAMES M., M. D., F. G. S. A., professor of geology, Vanderbilt University, Nashville, Tennessee.

ST. JOHN, ORESTES H., F. G. S. A., geologist, Topeka, Kansas.

* SALISBURY, ROLLIN D., A. M., F. G. S. A., professor of geology, University of Chicago, Chicago, Illinois.

* SAN ROMAN, FRANCISCO I., ingénieur en chef de la section des mines et géographie de la direction de travaux publiques de Chili, Egército Libertador 63, Santiago, Chili.

SAUSSURE, Dr. HENRI DE, Tertasse 2, Genève, Suisse.

SCHÄFFER, CHARLES, M. D., F. G. S. A., 1309 Arch st., Philadelphia, Pennsylvania.

SCHEIBE, ROBERT, Dr. Phil., Bezirksgeolog der k. preus. geologischen Landes-Anstalt, Invalidenstrasse 44, Berlin.

* SCHMIDT, CARL, Dr. Phil., professeur de minéralogie et de géologie à l'Université, Münsterplatz, Bâle, Suisse.

* SCHMIDT, FRIEDRICH, professeur à l'Académie impériale des sciences, St. Pétersbourg.

SCHMITZ, le Rév. Père GASPER, S. J., 88 rue St. Gilles, Liège, Belgique.

SCUDDER, SAMUEL H., paleontologist, U. S. geological survey, Cambridge, Massachusetts.

SEELEY, HARRY GOVIER, Esq., F. R. S., F. G. S., professor of geography and geology, Kings College, 25 Palace Gardens Terrace, Kensington, London, W.

SELIGMANN, GUSTAV, Dr. Phil., Schlossrondel 18, Coblenz, Deutschland.

* SHALER, NATHANIEL S., LL. D., F. G. S. A., professor of geology, Harvard University, Cambridge, Massachusetts.

SHERBORN, C. DAVIES, Esq., F. G. S., 540 Kings Road, London, S. W.

SHULAK, Rev. FRANCIS X., professor of mineralogy and natural history, St. Ignatius College, Chicago, Illinois.

* SIHLEANO, Mme. HENRIETTE, licenciée-ès-sciences naturelles, 1 Strada Eldorado, Bucharest, Roumanie.

* SIHLEANO, Prof. STEFAN, 1 Strada Eldorado, Bucharest, Roumanie.

SILVESTER, FRANK W., Esq., Hedges, St. Albans, England.

* SJÖGREN, HJALMAR, professeur de géologie à l'Université, Upsala, Suède.

* SMITH, Miss A. CLEMENTINA, 1727 H st., NW., Washington, District of Columbia.

SMITH, Sir DONALD A., K. C. M. G., Governor Hudson's Bay Company, 22 St. Alexis st., Montreal, Canada.

SMITH, EUGENE A., Ph. D., F. G. S. A., State geologist of Albama, University P. O., Alabama.

SMOCK, JOHN C., Ph. D., F. G. S. A., State geologist of New Jersey, Trenton, New Jersey.

*SÖHLE, ULRICH, Student der Geologie, München, Bayern.

SOLLAS, WILLIAM J., LL. D., F. R. S., F. G. S., professor of geology and mineralogy, Trinity College, Dublin, Ireland.

SOWERBUTTS, ELI, Esq., secretary of the Manchester geographical society, 44 Brown st., Manchester, England.

*SPENCER, J. W., A. M., Ph. D., F. G. S. A., State geologist of Georgia, Atlanta, Georgia.

*STAINIER, XAVIER, docteur-ès-sciences naturelles, Chaussée de Wavre 80, Bruxelles.

*STANTON, TIMOTHY W., F. G. S. A., assistant paleontologist, U. S. geological survey, Washington, District of Columbia.

*STEFANESCU, GRÉGOIRE, professeur de géologie à l'Université de Bucharest, Strada Verde 8, Bucharest, Roumanie.

*STEFANESCU, Mme. MARIA GR., Strada Verde 8, Bucharest, Roumanie.

STEFANESCU, Prof. SABBA, Strada Donnitei 7, Bucharest, Roumanie.

STEFANO, GIOVANNI DI, paléontologue. Roma.

*STEINMANN, GUSTAV, Dr. Phil., Professor der Geologie an der Universität, Freiburg in Baden, Deutschland.

*STEJNEGER, LEONHARD, curator, department of reptiles, U. S. national museum, Washington, District of Columbia.

STELZNER, ALFRED W., Dr. Phil., Professor der Geologie an der k. sächs. Berg-akademie, Freiberg, Sachsen.

*STEVENSON, JOHN J., Ph. D., F. G. S. A., professor of geology, University of the City of New York, Washington Square, New York City.

STIRRUP, MARK, Esq., F. G. S., High Thorn, Bowdon near Manchester, England.

*STOEK, HARRY H., instructor in mining and geology, Lehigh University, South Bethlehem, Pennsylvania, (now Pennsylvania State College, State College, Pennsylvania.)

*STOKES, HENRY N., Ph. D., assistant chemist, U. S. geological survey, Washington, District of Columbia, (now University of Chicago, Chicago, Illinois.)

'STREERUWITZ, W. H. VON, civil and mining engineer, P. O. Box 46, Austin, Texas.

*STRENG, AUGUST, Dr. Phil., Professor der Geologie an der Universität, Giessen, Deutschland.

STRIPPLEMANN, LEO, Bergingenieur, Gen. Director der consolidirten Alkaliwerke in Westeregeln, Provinz Sachsen. Ansbacherstrasse 2, Berlin, W.

STRÜCKMANN, C., Dr. Phil., Amtsrath, Sedanstrasse 3, Hannover, Deutschland.

STÜBEL, ALPHONS, Dr. Phil., 17 Feldgasse, Dresden, Sachsen.

SWAIN, ERNEST, Esq., F. G. S., 21 Ladbroke Road, London, W.

SWAIN, RICHARD T., Esq., 5 Addison Crescent, Kensington, London, W.

SWALLOW, GEORGE C., M. D., LL. D., F. G. S. A., State geologist of Montana, Helena, Montana.

SZABÓ, JOSEPH VON, Dr. Phil., F. G. S., Professor der Mineralogie and Geologie an der Universität, Budapest, Ungarn.

*TABUTEAU, Lieut. Col. A. O. (retired), F. G. S., Brow Hill, Batheaston, Bath, England.

TAFF, JOSEPH A., assistant geologist, geological survey of Texas, Austin, Texas.

TARASSENKO, BASILE, Conservateur du musée de l'Université, Kiew, Russie.

TARR, R. S., assistant professor of geology, Cornell University, Ithaca, New York.

TAYLOR, T. B., Fort Wayne, Indiana.

TELLINI, ACHILLE T., Musée de géologie à l'Université royale, Roma.

TERMIER, PIERRE, M. S. G. F., ingénieur des mines, professeur à l'École des mines, St. Étienne, France.

*TIETZE, EMIL, Dr. Phil., Oberbergrath, Chefgeolog der k. k. geologischen Reichsanstalt, Rasumofskygasse 23, Wien.

*TIFFANY, ASA SCOTT, F. G. S. A., paleontologist and archæologist, Davenport, Iowa.

*TODD, JAMES E., A. M., F. G. S. A., professor of science, Tabor College, Tabor, Iowa, (now Vermillion, South Dakota.)

TOPLEY, WILLIAM, Esq., F. R. S., F. G. S., geological survey of Great Britain, Museum, 28 Jermyn st., London, S. W.

TOUTKOWSKY, PAUL, Conservateur du musée de l'Université, Kiew, Russie.

*TSCHERNYSCHEW, THÉODORE, géologue-en-chef du Comité géologique de la Russie, St. Pétersbourg.

UDALL, JOHN, Esq., F. G. S., 21 Summer Hill Terrace, Birmingham, England.

*UHLER, PHILIP R., president of the academy of sciences, Peabody Institute, Baltimore, Maryland.

*ULRICH, ARNOLD, Dr. Phil., Assistent am geognostisch-palaeontologischen Institut, Strassburg, Elsass, Deutschland.

*ULRICH, E. O., A. M., F. G. S. A., paleontologist, Newport, Kentucky.

ULRICH, GEORGE H. F., Esq., F. G. S., professor of mining and mineralogy, Dunediu, New Zealand.

*UPHAM, WARREN, F. G. S. A., assistant geologist, U. S. geological survey, 36 Newberry st., Somerville, Massachusetts.

UZIELLI, GUSTAVE, professeur de géologie et de minéralogie, via Senese 62, Firenze, Italie.

*VAN DEN BROECK, E., conservateur au Musée royale d'histoire naturelle, 39 Place de l'Industrie, Bruxelles.

*VAN HISE, CHARLES R., M. S., F. G. S. A., geologist, U. S. geological survey, professor of mineralogy and petrology, University of Wisconsin, Madison, Wisconsin.

VENUKOFF, PAUL, professeur à l'Université, Kiew, Russie.

*VERY, FRANK W., professor of astronomy, Western University of Pennsylvania, Allegheny, Pennsylvania.

*VIEDENZ, ADOLF, Bergrath, Eberswalde, Berlin.

VILANOVA Y PIERA, J., professeur de paléontologie à l'Université, calle de San Vincente, Madrid, Espagne.

VINSANI, CONTARDO, professeur de topographie et construction à l'Institut technique royal, Reggio, Calabria, Italie.

VOGDT, CONSTANTIN DE, conservateur au musée géologique de l'Université, St. Pétersbourg.

WADSWORTH, MARSHMAN EDWARD, Ph. D., F. G. S. A., State geologist of Michigan, director of State Mining School, Houghton, Michigan.

*WAHNSCRAFFE, FELIX, Dr. Phil., Geolog der k. preus. geologischen Landes-Anstalt, Privatdocent an der Universität, Chausseestrasse 52ª, III, Berlin, N.

*WALCOTT, CHARLES D., F. G. S., F. G. S. A., paleontologist, U. S. geological survey, Washington, District of Columbia.

WALCOTT, Mrs. C. D. (Helena Burrows), 1746 Q st., Washington, District of Columbia.

WALTER, R. J., 2130 Downing ave., Denver, Colorado.

*WALTHER, JOHANNES, Dr. Phil., Professor der Geologie au der Universität, Jena, Deutschland.

*WARD, HENRY A., naturalist, Rochester, New York.

*WARD, LESTER F., A. M., F. G. S. A., paleontologist, U. S. geological survey, Washington, District of Columbia.

†WATERS, J. H. ERNEST, mining engineer, Telluride, Colorado.

WATSON, GEORGE A., deputy surveyor general (retired), Hendre, Overton Park, Cheltenham, England.

WEED, WALTER H., E. M., F. G. S. A., assistant geologist, U. S. geological survey, Washington, District of Columbia.

*WEEKS, JOSEPH D., mining engineer, Pittsburg, Pennsylvania.

*WEIGAND, BRUNO, Dr. Phil., Oberlehrer, Steinring 1, Strassburg, Elsass, Deutschland.

WEISBACH, ALBIN, Dr. Phil., Professor der Mineralogie an der k. sächs. Bergakademie, Freiberg, Sachsen.

WEISSLEDER, EDMUND, Bergrath, Leopoldshall-bei-Stassfurt, Deutschland.

WELLS, HORACE L., assistant professor of chemistry, Sheffield Scientific School, New Haven, Connecticut.

WELLS, WILLIAM H., Jr., 274 South Ashland ave., Chicago, Illinois.

*WHEELER, C. GILBERT, chemist and geologist, 143 Lake st., Chicago, Illinois.

*WHEELER, H. A., professor of mining, Washington University, St. Louis, Missouri.

WHITAKER, WILLIAM, Esq., F. R. S., F. G. S., geological survey of Great Britain, Museum, 28 Jermyn st., London, S. W.

*WHITE, CHARLES A., M. D., F. G. S. A., paleontologist, U. S. geological survey, Washington, District of Columbia.

*WHITE, CHARLES DAVID, F. G. S. A., assistant paleontologist, U. S. geological survey, Washington, District of Columbia.

*WHITE, ISRAEL C., Ph. D., F. G. S. A., professor of geology, University of West Virginia, Morgantown, West Virginia.

*WHITFIELD, ROBERT P., Ph. D., F. G. S. A., geologist and paleontologist, American Museum of Natural History, 77th st. and 8th ave., New York City.

WIGHT, GEORGE P., Esq., civil servant, War Office, London, S. W.

WIGHT, STANLEY, Brookfield, Massachusetts.

*WILLCOX, JOSEPH, commissioner of the geological survey of Pennsylvania, Philadelphia, Pennsylvania.

*WILLETS, Hon. EDWIN, Assistant Secretary of Agriculture, Washington, District of Columbia.

WILLIAMS, ARTHUR H., Esq., merchant, 385 Holloway Road, London, N.

WILLIAMS, EDWARD H., Jr., E. M., F. G. S. A., professor of mining engineering and geology, Lehigh University, Bethlehem, Pennsylvania.

*WILLIAMS, GEORGE HUNTINGTON, Ph. D., F. G. S. A., professor of geology, Johns Hopkins University, Baltimore, Maryland.

*WILLIAMS, HENRY SLATER, Ph. D., F. G. S., F. G. S. A., professor of geology and paleontology, Yale College, New Haven, Connecticut.

*WILLIAMS, Mrs. HENRY S., New Haven, Connecticut.

*†WILLIAMS, J. FRANCIS, Ph. D., F. G. S. A., assistant professor of geology, Cornell University, Ithaca, New York.

*WILLIS, BAILEY, F. G. S. A., geologist, U. S. geological survey, Washington, District of Columbia.

*WILLIS, Mrs. BAILEY, Washington, District of Columbia.

WILSON, HERBERT M., geographer, U. S. geological survey, Washington, District of Columbia.

*WILSON, THOMAS, curator, U. S. national museum, Washington, District of Columbia.

WINCHELL, HORACE V., F. G. S. A., assistant State geologist, 120 State st., Minneapolis, Minnesota.

* WINCHELL, NEWTON H., A. M., F. G. S. A., State geologist of Minnesota, Minneapolis, Minnesota.
* WINSLOW, ARTHUR, F. G. S. A., State geologist of Missouri, Jefferson City, Missouri.
* WÖHRMANN, Baron SIDNEY VON, Dr. Phil. (de Livonie, Russie), Akademie der Wissenchaften, München, Bayern.
* WOLFF, JOHN E., Ph. D., F. G. S. A., instructor in petrography, Harvard University, Cambridge, Massachusetts.
WOOD, TINGLEY S., mining engineer, Leadville, Colorado.
WOODALL, JOHN W., Banker, St. Nicholas House, Scarborough, England.
* WOODROW, JAMES, professor of geology and mineralogy, University of South Carolina, Columbia, South Carolina.
WOODWARD, HORACE B., Esq., F. G. S., geological survey of Great Britain, Museum, 28 Jermyn st., London, S. W.
WOODWARD, ROBERT S., C. E., F. G. S. A., U. S. coast and geodetic survey, Washington, District of Columbia, (now Columbia College, New York.)
WOODWORTH, JAY BACKUS, 7 Rutland st., Cambridge, Massachusetts.
* WÜLFING, E. A., Dr. Phil., Privatdocent au der Universität, Tübingen, Deutschland.
* YEATES, WILLIAM S., assistant curator, department of mineralogy, U. S. national museum, Washington, District of Columbia, (now State Geologist, Atlanta, Georgia.)
YOUNG, ALFRED C., Esq., 64 Tyrwhitt Road, St. John's, London, S. E.
ZACCAGNA, DOMENICO, ingénieur des mines, Carrara. Ufficio geologico, Roma.
† ZANDT, FERDINAND VAN, mining engineer, Blue Bird Mining Co., Butte City, Montana.
ZEBALLOS, ESTANISLAS S., directeur général des postes et télégraphes, Buenos Aires, République Argentine, (now Argentine Minister, Washington, D. C.)
* ZITTEL, KARL VON, Dr. Phil., Professor der Geologie an der Universität, München, Bayern.
ZLATARSKI, GEORGES N., géologue en chef des mines, Sofia, Bulgarie.

```
Membres présents............................................... 251
Membres absents............................................... 295
                                                              ───
    Total..................................................... 546
```

LISTE CLASSIFIÉE DES MEMBRES ÉTRANGERS.

[*Indique que le membre était présent à Washington.]

Pays.	Membres présents.	Membres absents.	Totaux.
Allemagne	23	35	58
Amérique Centrale	0	1	1
République Argentine	0	2	2
Australie	0	2	2
Autriche-Hongrie	3	8	11
Belgique	3	10	13
Bulgarie	0	1	1
Canada	3	5	8
Chile	1	0	1
Égypte	0	1	1
Espagne	0	2	2
France	7	17	24
Grande-Bretagne	14	54	68
Hollande	0	1	1
Italie	0	24	24
Mexique	3	4	7
Norvège	1	1	2
Nouvelle Zélande	0	1	1
Pérou	1	0	1
Portugal	0	2	2
Roumanie	4	2	6
Russie	9	13	22
Suède	4	1	5
Suisse	2	2	4
Totaux	78	189	267

ALLEMAGNE.

Ammon, L. von ....München.
*Andreæ, A.............Heidelberg.
Benecke, E. W.........Strassburg i. E.
*Bergeat, A............München.
Bertkau, P ....Bonn.
Beyrich, E ....Berlin.
*Borne, G. von dem......Halle a. d. S.
Bornemann, J. G ......Eisenach.
Bornemann, L. G., jr ..Eisenach.
Busse, M ....Coblenz.
Cohen, E ....Greifswald.
Conwentz, H ....Danzig.
*Credner, H ....Leipzig.
*Credner, R ....Greifswald.
Felix, J ....Leipzig.
*Frech, F ....Breslau.
Geinitz, H. B ....Dresden.
Giesecke, B...........Leipzig.
Halfar, A.............Berlin.
Hauchecorne, W.......Berlin.
Heusler, C............Bonn.
*Jaekel, O............Berlin.
Jentzsch, A...........Konigsberg i. Pr.
*Kayser, E ....Marburg.
Kloos, J. H...........Braunschweig.
Korthals, W. C ....Heidelberg.
*Koenigs, W..........München.
Lepsius, R ....Darmstadt.
Maurer, F ....Darmstadt.
Mügge, O............Münster.

Nies, F ....Hohenheim-bei-
                      Stuttgart.
Ochsenius, C ....Marburg.
*Osann, A ....Heidelberg.
*Plieninger, F ....München.
Proescholdt, H ....Meiningen.
Purgold, A ....Blasewitz.
Reiss, W ....Berlin.
*Romberg, J ....Berlin.
Rosenbusch, H........Heidelberg.
*Rothpletz, A..........München.
Scheibe, R...........Berlin.
Seligmann, G.........Coblenz.
*Söhle, U............München.
*Steinmann, G.........Freiburg i. B.
Stelzner, A. W........Freiberg i. S.
*Streng, A............Giessen.
Stripplemann, L ......Berlin.
Strückmann, C........Hannover.
Stübel, A ....Dresden.
*Ulrich, A ....Strassburg i. E.
*Viedenz, A..........Berlin.
*Wahnschaffe, F.......Berlin.
*Walther, J...........Jena.
*Weigand, B...........Strassburg i. E.
Weisbach, A..........Freiberg i. S.
Weissleder, E.........Stassfurt.
*Wülfing, E. A.........Tübingen.
*Zittell, K. von.........München.

AMÉRIQUE CENTRALE.

Crawford, J ....Managua.

RÉPUBLIQUE ARGENTINE.

Berg, C...............Buenos Aires. | Zeballos, Estanislas S..Buenos Aires.

AUSTRALIE.

Brown, H. Y. Lyell ....Adelaide. | Liversidge, A..........Sidney.

AUTRICHE-HONGRIE.

Becke, F ....Prag.
*Diener, C............Wien.
*Dunikowski, E. de .....Lemberg.
Fuchs, T ....Wien.
Hantken, M. von.......Budapest.
Lóczy, L ....Budapest.

Niedzwiedzki, J ......Lemberg.
Pošepný, F ....Wien.
Szabó, J. von ....Budapest.
*Tietze, E ....Wien.
Von Inkey, A ....Pressburg.

BELGIQUE.

Delvaux, E .... Bruxelles.
Dewalque, G .... Liège.
Faly, J .... Mons.
Firket, A .... Liège.
Hennequin, E .... Bruxelles.
Lehaie, A. H. de .... Hyon.
*Lohest, M .... Liège.

Mourlon, M .... Bruxelles.
Rutot, A .... Bruxelles.
Schmitz, G .... Liège.
*Stainier, X .... Bruxelles.
*Van den Broeck, E ..... Bruxelles.
Van Drunen, J .... Bruxelles.

BULGARIE.

Zlatarski, G. N .... Sofia.

CANADA.

*Adams, F. D .... Montreal.
Ami, H. M .... Ottawa.
Kennedy, G. G .... Windsor.
*Laflamme, J. C. K ..... Quebec.

Leckie, R. G .... Londonderry.
*Macfarlane, T .... Ottawa.
McKellar, P .... Fort William.
Smith, D. A .... Montreal.

CHILI.

*San Roman, Francisco I .... Santiago.

ÉGYPTE.

Lyons, H. G .... Cairo.

ESPAGNE.

Castro, F. de .... Madrid.

Vilanova y Piera, J .... Madrid.

FRANCE.

Arnaud, H .... Angoulême.
*Barrois, Ch .... Lille.
Bergeron, J .... Paris.
Bertrand, M .... Paris.
Bonaparte, R .... Paris.
*Boule, M .... Paris.
Dagincourt, M .... Paris.
*Delaire, A .... Paris.
Dollfuss, G .... Paris.
*Doulcet, J .... Paris.
*Eysséric, J .... Carpentras.
Fabre, G .... Nimes.

*Gaudry, A .... Paris.
Geandey, F .... Lyon.
Gosselet, J .... Lille.
Hébert, E .... Paris.
Hovelacque, M .... Paris.
Janet, C .... Beauvais.
Lapparent, A. de ..... Paris.
*Margerie, Emm. de ... Paris.
Michel-Lévy, A .... Paris.
Réclus, E .... Paris.
Riche, A .... Lyon.
Termier, P .... St.-Étienne.

GRANDE BRETAGNE.

Allen, H. A .... London.
Atkinson, W. J .... London.
Bather, F. A .... London.
Blanford, W. T ...... London.
*Cadell, H. M .... Bo'ness.
*Christie, J. C .... Glasgow.
*Churchill, W .... London.
Cole, G. A .... Dublin.
Dalton, W. H .... London.

Evans, J .... Hemel Hempst'd.
Davies, A. M .... London.
Davis, W. H .... Tottenham.
Day, J. T .... London.
Fennell, C. W .... Wakefield.
*Forster, Miss M ...... London.
Foster, C. Le N ...... London.
Fox-Strangways, C ... London.
Garwood, E. J .... Cambridge.

GRANDE BRETAGNE—continued.

Glen, D. C .... Glasgow.
* Gregory, J. W .... London.
* Gregory, Miss .... London.
Hull, M .... Parkstone.
Hansford, C .... Dorchester.
* Harker, A .... Cambridge.
* Harris, G. F .... London.
Hicks, H .... London.
* Hobson, B .... Sheffield.
Howse, R .... Newcastle.
Hudleston, W. H ..... London.
* Hughes, T. McK ..... Cambridge.
* Hughes, Mrs. M. C ... Cambridge.
Hull, E .... London.
Hume, W. F .... London.
Judd, J. W .... Kew.
Lapworth, C .... Birmingham.
Leighton, T .... West Norwood.
Leonard, H .... London.
Louis, D. A .... London.
* Mallett, F. R .... London.
* Mallet, R. T .... Reading.
Markby, J. R .... London.
Marr, J. E .... Cambridge.
Medlicott, H. B .... Bristol.

Newton, E. T .... London.
Newton, R. B .... London.
Postlethwaite, J ...... Keswick.
Prestwich, J .... Shoreham.
Reid, C .... London.
Rudler, F. W .... London.
Rutley, F .... London.
Seeley, H. G .... London.
Sherborn, C. D .... London.
Silvester, F. W .... St. Albans.
Sollas, W. J .... Dublin.
Sowerbutts, E .... Manchester.
Stirrup, M .... Bowdon.
Swain, E .... London.
Swain, R .... London.
* Tabuteau, A. O .... Bath.
Topley, W .... London.
Watson, G. A .... Cheltenham.
Whitaker, W .... London.
Wight, G. P .... London.
Williams, A. H .... London.
Woodward, H. B ..... London.
Woodall, J. W .... Scarborough.
Udall, J .... Birmingham.
Young, A. C .... London.

HOLLANDE.

Molengraff, G. A. F .... Amsterdam.

ITALIE.

Baldacci, L .... Roma.
Bassani, F .... Napoli.
Botti, U .... Reggio,Calabria.
Canavari, M .... Pisa.
Capellini, G .... Bologna.
Cocchi, I .... Firenze.
Cortese, E .... Roma.
Demarchi, L .... Roma.
Fabri, A .... Firenze.
Franchi S .... Torino.
Giordano, F .... Roma.
Johnston-Lavis, H. J .. Napoli.

Maris, C. D .... Roma.
Mattirolo, E .... Torino.
Novarese, V .... Torino.
Parona, C. F .... Torino.
Portis, A .... Roma.
Ricciardi, L .... Reggio,Calabria.
Sacco, F .... Torino.
Stefano, G. di .... Roma.
Tellini, A .... Roma.
Uzielli, G .... Firenze.
Vinsani, C .... Reggio,Calabria
Zaccagna, D .... Carrara.

MEXIQUE.

* Aguilera, J. D .... Ville de Mexico.
* Castillo, A. del ...... Ville de Mexico.
Landero, C. F. de ..... Guadalajara.
Lopez, M. P .... Ville de San Luis

* Ordoñez, E .... Ville de Mexico.
Puga, G. B. y .... Ville de Mexico.
Rivero, J. L .... Zacatecas.

NORVÈGE.

Brögger, W. C .... Christiania. | *Reusch, H .... Christiania.

NOUVELLE ZÉLANDE.

Ulrich, G. H. F ....Dunedin

PÉROU.

Elguera, M............ Washington (Peruvian Legation).

PORTUGAL.

Choffat, P ....Lisbonne.   Delgado, J. F. N ....Lisbonne.

ROUMANIE.

Licherdopol, J. P ......Bucharest.   * Stefanescu, G ....Bucharest.
* Sihleano, S............Bucharest.   * Stefanescu, M. G.......Bucharest.
* Sihleano, H ....Bucharest.   Stefanescu, S..........Bucharest.

RUSSIE.

Armachewsky, P ......Kiew.   Nikitin, S ....St. Pétersbourg.
* Bogdanoff, N ....St. Pétersbourg.   * Pavlow, A.............Moscou.
Galitzine, P...........St. Pétersbourg.   * Pavlow, M ....Moscou.
Inostranzeff, A ....St. Pétersbourg.   Radkewitch, G ....Kiew.
Karakasch, N ....St. Pétersbourg.   * Schmidt, F............St. Pétersbourg.
Karpinski, T ....St. Pétersbourg.   Tarassenko, B ....Kiew.
Keyserling, A. C.......Raiküll.   Toutkowsky, P........Kiew.
* Krassnof, A. N.........Charkow.   * Tschernychew, Th ....St. Pétersbourg.
* Kroustchoff, C. de .....St. Pétersbourg.   Venukoff, P ....Kiew.
* Levitzky, R...........St. Pétersbourg.   Vogdt, C. de...........St. Petersbourg.
Loewinson-Lessing, F. St. Pétersbourg.   * Wöhrmann, B. S. von ..Livonia.

SUÈDE.

Blomstrand, C. R ......Lund.   * Lundbohm, H ....Stockholm.
* Geer, G. de ....Stockholm.   * Sjögren, H ....Upsala.
* Holst, N. O ....Stockholm.

SUISSE.

* Golliez, H.............Lausanne.   * Schmidt, C............Bâle.
Musy, M ....Fribourg.   Saussure, H. de ....Genève.

LISTE DES DÉLÉGUÉS ET DES INSTITUTIONS REPRÉSENTÉES.

[L'étoile (*) indique les Institutions qui out souscrit pour le Compte Rendu.]

ALLEMAGNE:
   München.—Le gouvernement de Bavière, *K. Von Zittel.*
AUTRICHE-HONGRIE:
   Wien.—K. K. geologische Reichsanstalt, *E. Tietze.*
          K. K. geographische Gesellschaft in Wien, *C. Diener.*
BELGIQUE:
   Le gouvernement de, *M. Lohest, E. Van den Broeck.*
   Bruxelles.—La Commission géologique de Belgique, *E. Van den Broeck.*
          Le Musée royal d'histoire naturelle de Belgique, *E. Van den Broeck.*
          *La Société belge de géologie, paléontologie, et d'hydrologie, *E. Van den Broeck.*
          La Société royale malacologique de Belgique, *X. Stainier.*
   Liège.—La Société géologique de Belgique, *M. Lohest, X. Stainier.*
CANADA:
   St. John.—Natural History Society of New Brunswick.
CHILI:
   Le gouvernement de, *F. I. San Roman.*
ÉTATS-UNIS:
   The American Association for the Advancement of Science, *A. B. Prescott, F. W. Putnam, J. J. Stevenson, W J McGee, G. K. Gilbert.*
   The American Institute of Mining Engineers, *S. F. Emmons.*
   New York City.—The American Geographical Society.
   Chicago, Ill.—The Geological Society of Chicago, *W. R. Head.*
   Minneapolis, Minn.—Minnesota Academy of Natural Sciences, *C. W. Hall.*
   Poughkeepsie, N. Y.—Vassar College, *W. B. Dwight.*
   Washington, D. C.—The Smithsonian Institution, *G. P. Merrill.*
          The National Geographical Society, *G. G. Hubbard.*
          * The Anthropological Society of Washington, D. C., *W J McGee, T. Wilson, W. H. Holmes.*
FRANCE:
   Paris.—Le Ministère de l'Instruction publique, *M. Boule.*
          Service de la carte géologique detaillée de la France, *Ch. Barrois.*
          La Société géologique de France, *A. Gaudry.*
          La Société de géographie de Paris, *Emm. de Margerie.*
          La Société des ingénieurs civils, *T. B. M. Mason.*
ITALIE:
   Roma.—Ministero di Agricoltura, Industria et Commercio.
   Napoli.—L'Academie royale des Sciences.
MEXIQUE:
   * *Ville de Mexico.*—La Société "Antonio Alzate."
PÉROU:
   Le gouvernement de, *A. D. Hodges, jr., C. Ochsenius.*
PORTUGAL:
   Lisbonne—La Commission des travaux géologiques de Portugal.

ROUMANIE:

Bucharest.—Le Ministère de l'Agriculture, Industrie, Commerce et Domaines, G. Stefanescu.

RUSSIE:

Le gouvernement de, Th. Tschernyschew.

St. Pétersbourg.—Le Comité géologique de la Russie, F. Schmidt.

*Ekatérinebourg.—La Société des Sciences Naturelles.

*St. Pétersbourg.—Le Musée géologique de l'Université de St. Pétersbourg.

Moscou.—La Société Impériale des naturalistes, A. Pavlow.

La Société Impériale Russe de Géographie, K. de Kroustchoff.

SUISSE:

Le gouvernement de, H. Gollies, C. Schmidt.

DEUXIÈME PARTIE.

PROCÈS-VERBAUX DES SÉANCES DU CONGRÈS.

PROCÈS-VERBAUX DES SÉANCES.

La séance est ouverte à 2h. et demi de l'après-midi.

M. T. McK. HUGHES, Vice-Président anglais du Congrès, dit qu'il a été invité à ouvrir la séance, en l'absence de son ami, le distingué professeur Prestwich, Président du Congrès de Londres, qui, vu l'état de sa santé, n'a pu le faire lui-même. M. Hughes espère que son ami a pourtant devant lui bien des années à consacrer au travail, et que, toutes les fois que la distance et les fatigues d'un long voyage ne s'y opposeront pas, Professeur Prestwich pourra présider de nouveau les séances et faire profiter encore ses collègues des lumières que lui ont acquise sa longue expérience.

M. Hughes déclare la séance ouverte, et annonce que l'on va procéder à l'élection du Bureau.

La liste ci-après, que le conseil, conformément aux règlements, avait préparée d'avance, est votée par acclamation.

51

BUREAU DU CONGRÈS.

Présidents d'honneur.

JAMES HALL, [J. D. DANA.]*

Anciens Présidents.

[G. CAPELLINI, 1881], [E. BEYRICH, 1885],
[J. PRESTWICH, 1888.]

Président.

[J. S. NEWBERRY.]

Vice-Présidents.

Allemagne, K. VON ZITTEL, H. CREDNER.
Angleterre, T. McK. HUGHES.
Autriche, E. TIETZE.
Australie, [A. LIVERSIDGE].
Belgique, E. VAN DEN BROECK.
Canada, J. C. K. LAFLAMME, THOS. MAC-FARLANE.
Chili, F. I. SAN ROMAN.
Danemark, [F. JOHNSTRUP].
Écosse, H. M. CADELL.
Espagne, [M. F. DE CASTRO].
États-Unis, JOSEPH LECONTE, J. W. POWELL, RAPHAEL PUMPELLY.
France, A. GAUDRY, CH. BARROIS.

Hongrie, [JOS. VON SZABO].
Indes, F. R. MALLET.
Irlande, [W. J. SOLLAS].
Italie, [G. UZIELLI].
Mexique, A. DEL CASTILLO.
Norvège, H. REUSCH.
Nouvelle Zélande, [F. W. HUTTON].
Portugal, [J. F. N. DELGADO].
Roumanie, G. STEFANESCU.
Russie, Th. TSCHERNYSCHEW, F. SCHMIDT, A. PAVLOW.
Suède, GERARD DE GEER.
Suisse, H. GOLLIEZ.

Secrétaires généraux.

H. S. WILLIAMS, S. F. EMMONS.

Secrétaires.

J. C. BRANNER,
W. CROSS,
C. DIENER,

F. FRECH,
EMM. DE MARGERIE,
GEO. H. WILLIAMS.

Trésorier.

ARNOLD HAGUE.

* Les noms des membres absents sont indiqués par une accolade. [ ]

Sur la proposition du Prof. HUGHES, M. LECONTE, Vice-Président,
préside, et prononce le discours suivant en anglais:

MESDAMES ET MESSIEURS: En l'absence du Président du Congrès,
M. le professeur Newberry, qui, je regrette de le dire, est retenu chez
lui par l'état de sa santé, je suis appelé, d'une manière inattendue et
à court délai, à vous adresser quelques paroles de bienvenue. Je me
trouve fort embarrassé de cette mission soudaine, mais surtout du
grand bonheur qui m'est ainsi conféré. Cependant, j'espère que, vu
les circonstances, vous ne vous attendez pas à écouter un long discours.

On sait que l'idée d'un congrès international de géologues est américaine; et, si je ne me trompe, c'est en 1876 qu'elle est venue au Dr.
Sterry Hunt. Il était donc juste que toutes les premières réunions,
jusqu'à celle-ci qui est la cinquième, eussent lieu en Europe, où la
géologie prit naissance, et où elle trouve encore ses plus grands
développements. Après avoir été tenu d'abord à Paris en 1878, à
Bologne en 1881, à Berlin en 1885 et à Londres en 1888, ce cinquième
Congrès revient au pays qui l'a vu naître. Nous, géologues américains,
remercions bien sincèrement nos collègues étrangers d'avoir répondu
si cordialement à notre invitation. C'est peut-être la première fois
qu'il y a presque autant de membres de différents pays que de nation-
aux. Cela est dû en partie, non seulement au nombre comparativement
restreint de géologues aux États-Unis, mais aussi, je puis dire principalement, à l'empressement montré par les premiers à se rendre auprès
de nous.

Il est bon de rappeler ici brièvement les buts principaux dans lesquels les congrès se réunissent. Ce sont 1°, de fixer une nomenclature
des grandes divisions et sous-divisions des roches, et des époques qui y
correspondent, afin que les géologues de tous les pays puissent se comprendre; 2°, d'être d'accord sur les signes conventionnels servant à
représenter les faits géologiques, et particulièrement d'adopter une
gamme de coloriage générale pour l'indication des terrains d'époques
différentes, de façon qu'une carte géologique puisse être lue à première
vue, comme une page imprimée ou une feuille de musique; 3°, de discuter librement les questions complexes de géologie sur lesquelles il
existe encore des divergences d'opinion; 4°, de procurer à ses membres
l'occasion de se connaître personnellement, de s'apprécier, et de se ten-
dre une main fraternelle, ce qui est de dernier mais non le moindre but
que le Congrès désire atteindre.

De ces propositions, les deux premières sont, sans contredit, de la
plus grande importance, quoiqu'elles soient cependant celles que l'expérience a démontré comme étant les plus difficiles à adopter. Nous

avons été forcés d'être moins ambitieux dans nos vues. C'est comme qui dirait vouloir être d'accord sur un langage universel. Ces choses là ne se produisent pas du jour au lendemain ; il leur faut du temps pour grandir, car elles ne viennent que graduellement et par évolution. Mais souvenons-nous que le facteur principal de cette évolution se trouve dans les discussions de ce Congrès. Espèrons donc qu'elles en seront le point culminant.

Je ne désire pas vous fatiguer par un rapport détaillé des sujets qui ont été soumis à vos délibèrations. Ils sont déjà suffisamment indiqués sur le programme qui est entre vos mains ; mais il me paraît opportun, à l'occasion de cette première réunion en Amérique, d'ajouter quelques mots touchant les traits caractéristiques les plus saillants de la géologie américaine tels que vous pourrez peut-être les observer, même dans le parcours rapide d'une excursion.

Si nous comparons une carte d'Europe avec une carte d'Amérique, nous sommes frappés de la complexité des contours de l'une et de la simplicité relative de l'autre. L'Europe est profondément découpée par des mers intérieures qui détruisent son intégrité continentale, tandis que l'Amérique est une unité continentale dont les bords seulement sont dentelés. Il existe une différence semblable dans la structure orographique. L'Europe est traversée par des chaînes de montagnes qui, s'étendant dans plusieurs directions, la divisent en plusieurs bassins séparés. En Amérique, au contraire, nous avons dans l'Est une série de chaînes parallèles dont les eaux s'écoulent dans l'Atlantique ; une autre dans l'Ouest tributaire du Pacifique, et un grand bassin intérieur dont les eaux s'écoulent par le Mississipi dans le Golfe du Mexique. De plus, le continent est divisé transversalement vers le milieu par une chaîne basse de montagnes dont les eaux s'écoulent vers le nord dans l'océan Arctique, et vers le sud dans le Golfe du Mexique.

Comme l'histoire donne la clef des conditions sociales et politiques, de même la géologie donne celle des formes continentales et des structures orogéniques. En Europe, les oscillations de la croûte terrestre dans l'histoire géologique ont été si grandes et si variées, qu'elles équivalent à des créations et à des destructions fréquentes du continent. En Amérique, au contraire, le développement, du moins depuis le commencement de l'époque cambrienne, s'est presque constamment porté vers son achèvement. Toutefois, des hésitations, quelques incertitudes et quelques oscillations se sont produites, particulièrement à la fin des grandes époques, mais, pris dans son ensemble, ce développement a continué sa marche régulière comme si, doué de la pensée, il savait ce qu'il voulait et était déterminé à atteindre son but. Maintenant, en conséquence de cette croissance uniforme, les faits géologiques sont plus clairement démontrés par la nature, et plus aisément interprétés par le géologue. Les problèmes géologiques sont réduits à leur plus simple expression, et leur solution est comparativement facile. Ceci est particulièrement vrai en ce qui concerne les grandes étendues de l'Ouest, non seulement à cause de la simplicité de la structure, mais aussi par suite de la date récente des changements. Les signes ici sont si évidents que celui qui court, même avec la vitesse d'un train express, peut les déchiffrer sans peine. Il est dit qu'*à brebis tondue Dieu mesure le vent*. Nous pouvons faire ici l'application de ce proverbe; car la complexité des problèmes géologiques a été mesurée à raison du petit nombre et de l'inexpérience relative de nos géologues. En vérité le champ est vaste et les travailleurs sont peu nombreux; mais la moisson est abondante et mûre pour la récolte. Autrement, il eut été impossible d'arriver aux résultats déjà obtenus par les géologues américains. Ils ont été la conséquence, non d'une application ou d'une faculté exceptionnelle, mais des occasions remarquablement avantageuses que nous possédons.

Parmi les effets dus au développement régulier de notre continent, je vais citer brièvement quelques-uns de ceux qui pourront être observés pendant le trajet rapide des excursions:

1. *La continuité des assises.*—Les discordances de stratification dans les couches et les lacunes, si fréquentes en Europe, sont en grande partie, pratiquement ou complètement, comblées dans l'Ouest. Ceci est particulièrement vrai pour les vastes lacunes entre le Paléozoïque et le Mésozoïque et entre le Mésozoïque et le Cénozoïque. Une série continue peut être observée dans les couches des plaines du Sud-Ouest depuis le Carbonifère par l'intermédiaire du Permien jusque dans le Triasique. De même le Laramie des Plaines et de la Région des Plateaux, avec le groupe de Téjon de la Californie, comble entièrement la lacune entre le Crétacé et le Tertiaire.

2. *Massifs surélevés.*—L'élévation régulière sans plissements dans la Région des Plateaux a donné lieu à un système de fissures nord-sud, ayant dans certains cas une centaine de milles de longueur, par lesquelles la croûte terrestre est brisée en massifs ou blocs séparés, dont quelques-uns se sont soulevés en un seul morceau, et d'autres se sont affaissés, produisant ainsi des dislocations verticales de plusieurs milliers de pieds. De telles dislocations sont connues des géologues de tous les pays, mais dans cette région elles sont sur une si grande échelle et d'une date si récente, qu'elles forment des escarpements de faille de 1,000 à 2,000 pieds de hauteur.

3. *Soulèvement des montagnes.*—Dans la Région des Plateaux et du Grand Bassin, M. Gilbert a découvert et décrit un nouveau type de chaîne de montagnes produites, non par une poussée latérale et des plissements, mais par une inclinaison des grands massifs dont nous venons de parler Le côté supérieur de ces massifs forme une chaîne

.de montagnes et le côté inférieur une vallée. L'exemple le plus frappant de ce fait est la Sierra Nevada. Il se peut que des montagnes de ce genre se trouveront dans d'autres endroits, mais probablement très modifiées par des érosions subséquentes.

4. *Coulées de laves.*—Dans la Région des Plateaux et du Bassin, il existe, particulièrement dans le Nord-ouest des États-Unis, d'énormes étendues, de 10,000 et même de plus de 100,000 milles carrés, couvertes de lave d'une épaisseur de 2,000 à 4,000 pieds, et qui sont formées de coulées successives superposées. On a suggéré que cette lave était sortie, non de cratères, mais des grandes fissures déjà citées. Vous en verrez probablement quelques-unes; et vos yeux exercés pourront déterminer si cette suggestion est bien fondée. De pareilles coulées de lave se trouvent aux Indes sur une grande échelle, mais en Europe il n'y en a que des débris.

5. *Mouvement continental.*—On n'ignore pas que le Quaternaire a été l'époque d'un mouvement très étendu de la croûte terrestre dans plusieurs régions, mais nulle part il n'a été, ni si grandiose, ni si imposant qu'en Amérique. Les preuves de ce soulèvement se rencontrent dans les lits de rivières se continuant sous la mer au delà des rivages actuels. Le soulèvement commença à une époque Tertiaire récente, et atteignit son point culminant vers le commencement du Quaternaire, et son amplitude n'était pas moins de 2,500 à 3,000 pieds. De pareils phénomènes se rencontrant sur les deux côtes, il est évident que le soulèvement a affecté tout le continent. Les preuves d'un affaissement sont très distinctes, mais pas plus qu'en Europe.

6. *Calotte glaciaire.*—L'existence d'une calotte glaciaire est liée étroitement à ce dernier par son époque et possiblement par son effet. Ceci a été complètement démontré pour la première fois en Amérique. Les contours de sa limite la plus éloignée et les temps d'arrêt de sa rétrogression ont été tracés pas à pas avec la plus grande exactitude.

Tels sont quelques-uns des traits caractéristiques de la géologie américaine. Je n'en ai mentionné qu'un petit nombre et seulement ceux qui se rapportent à la géologie structurale et dynamique, parce qu'ils seront les plus en vue dans le cours d'un trajet rapide. J'ai essayé de faire voir que tous ces traits sont liés au développement régulier du continent américain. Vous en remarquerez probablement quelques-uns. Nous espérons, ou, pour mieux dire, nous sommes certains de profiter largement de vos observations éclairées et de votre grande expérience.

Et maintenant, Messieurs les membres du Congrès et particulièrement nos collègues étrangers, au nom d'une grande nation, au nom de sa capitale, au nom des savants et surtout des géologues de l'Amérique, nous vous saluons et vous souhaitons la bienvenue.

La parole est à M. GARDINER G. HUBBARD, Président de la Société nationale de Géographie:

MESSIEURS: En ma qualité de président du Comité local d'arrangements je souhaite la bienvenue aux membres du Congrès International des Géologues à l'occasion de son cinquième anniversaire triennal.

Quatre réunions ont déjà eu lieu, à Paris, à Bologne, à Berlin, et à Londres. Celle-ci, la cinquième, rassemble ses membres des nations du Vieux et du Nouveau Monde dans la capitale de l'Amérique. A l'artiste, à l'antiquaire, à l'archéologue, à l'historien de la civilisation ou des époques réculées, l'Amérique a bien peu à offrir. Mais au géologue, quel champ vaste et rempli d'intérêt lui ouvre-t-elle! C'est ici qu'il peut étudier plus facilement et plus sûrement la base même de notre planète. Que sont pour lui les plus grands monuments de l'antiquité, les vestiges préhistoriques des habitants de la terre et les imposantes cathédrales construites, il y a des siècles, par la main des hommes? Pour lui mille ans sont comme un jour. Il cherche les premiers commencements du monde, avant que les bases des montagnes ne furent posées, lorsque "la terre était sans forme et vide, et que les ténèbres étaient sur la face des eaux." Ici, au Nouveau Monde, jeune dans sa civilisation mais vieux par sa formation géologique, il peut tracer, plus clairement et avec plus de certitude qu'ailleurs, la structure géologique de notre globe, ses longues périodes géologiques variées, et étudier plus soigneusement l'étendue, ainsi que les effets, de l'action glaciaire.

Le Comité d'excursions a réglé l'itinéraire d'un voyage dont, nous l'espérons, vous ferez tous partie, car nous croyons qu'il surpassera en intérêt général et scientifique ceux qui ont été faits jusqu'à présent. Vous visiterez le Niagara, merveilleux par son histoire et sa formation géologique, et incomparable par ses beautés et sa grandeur. Vous verrez la plus longue chaîne de montagnes qui s'étend, presque sans interruption, depuis l'océan Arctique jusqu'à l'océan Antarctique; un plateau plus riche en fossiles que toute autre partie du globe. Là se trouve le cheval fossile, prototype de son espèce, quoiqu'à l'époque de la découverte de l'Amérique il eut déjà disparu du continent. Là est située la vallée la plus riche et la plus grande du monde, entre les Montagnes Appalaches et les Montagnes Rocheuses, allant du nord au sud et entièrement dans la zone tempérée. Vous irez au Parc du Yellowstone, où M. Hague, qui a passé bien des années à l'explorer, vous servira de guide. Près de ce parc, à Butte City, sont les plus grandes mines de cuivre du monde, aussi bien que des mines d'argent fameuses. Au delà des Montagnes Rocheuses sont les filons du Comstock, la Vallée de Yosemite et le crâne merveilleux de Calavéras, décrit par cet éminent géologue, Bret Harte. M. Gilbert vous conduira à travers le Grand Bassin des Montagnes Rocheuses dans la vallée de l'Utah, par le lac Salé et la rivière Jordan, et vous montrera les plaines, autrefois déserts arides, qui aujourd'hui fleurissent comme un jardin au milieu desquelles la ville de Zion est si admirablement placée. M. Emmons sera votre guide pendant le voyage sur la ligne du chemin de fer Rio Grande Western, ce qui vous donnera l'occasion de passer devant la Montagne de la Sainte Croix jusqu'à Leadville et au "Divide," situé à une hauteur de 11,000 pieds. En traversant le grand plateau vous franchirez le Colorado, la rivière de cañons les plus considérables du monde. A 300 milles plus loin, à l'extrémité sud du plateau, vous reverrez le Colorado, qui ne sera pas alors au niveau de ce plateau mais à un mille audessous, car dans sa course il s'est frayé un chemin à travers les roches à une profondeur de plus de 5,000 pieds. C'est là où le Major Powell viendra à votre rencontre. Demandez-lui de vous raconter son admirable voyage de la source à l'embouchure de cette rivière, passant par tous ces cañons et au milieu de tant de dangers, le premier et seul parcours qui ait jamais été fait de toute la longueur de ce fleuve.

Le Nouveau Continent vous souhaite la bienvenue au nom des formations géologiques les plus anciennes, les plus facilement étudiées, ainsi que les plus intéressantes du globe; et vous prie, à votre retour dans le Vieux Monde, de lui présenter ses hommages respectueux. Nous vous recevons bien cordialement dans la Capitale de notre pays, et notre seul regret est que cette réunion a lieu pendant nos vacances, époque où le Président et la plupart des membres de son Cabinet sont absents. Le Geological Survey fait partie du Département de l'Intérieur. Le haut fonctionnaire qui dirige cette branche du Gouvernement, l'Honorable John W. Noble, que nous honorons tous, se trouve aujourd'hui parmi nous. J'ai le plaisir, maintenant, de vous le présenter.

Le Secrétaire de l'Intérieur se lève et prononce le discours suivant:

Messieurs les Membres du Congrès International des Géologues:

Mesdames et Messieurs: C'est de grand cœur que vous êtes accueillis aux États-Unis et à Washington par le peuple et par le Gouvernement de ce pays.

Le projet de former votre organisation a d'abord reçu son application pratique le jour de l'anniversaire de notre Indépendance par la Société Américaine pour l'Avancement des Sciences à Buffalo, de sorte qu'à cette réunion on peut dire que de nouveau on vous souhaite la bienvenue chez vous. Mais quel est le pays qui seul peut prétendre au droit d'être la patrie de la science ou de ceux qui s'y sont devoués? Il suffit de contempler cette vaste assemblée, composée de membres distingués et illustres du Vieux et du Nouveau Monde, pour voir que vous ne pouvez être considérés comme des étrangers en aucun lieu où les connaissances utiles sont désirées et où l'humanité se développe. Dans le domaine de

la pensée les frontières des nations cessent d'exister, et même l'amour
de la patrie se transforme et s'élève jusqu'à l'amour du prochain. Vos
luttes sont amicales et vos victoires sans taches. Il n'y a aucun pays
où vous ne soyez reçus avec empressement, de même qu'il n'y a pas
de peuple qui ne bénisse vos travaux et vos découvertes. Mais plus
encore êtes-vous les bienvenus dans cette grande République où nos
institutions, nos progrès et nos espérances sont fondées sur la diffusion
des connaissances utiles. Votre science, bien que la plus jeune, occupe
déjà un des premiers rangs parmi ses aînées par l'exercice de la pensée
librement exprimée et par ses nobles aspirations. Vous avez gagné
votre place "*mente et malleo*." Nous vous accueillons donc chaleureuse-
ment sur le sol et dans la capitale de la plus jeune des nations, qui
reçut en héritage le travail, et dont la force est le savoir.

Ce Gouvernement a été généreux, on peut même dire, munificent,
dans ses libéralités et ses encouragements pour l'avancement de la
géologie. Le Département de l'Intérieur a sous sa vaste juridiction le
"Geological Survey," présidé par un de vos membres, son savant et
habile Directeur. C'est à ce Survey qu'est remis le soin d'examiner
les roches, les minéraux, les métaux et les sols de notre pays, ainsi que
la préparation des cartes topographiques nécessaires à la représentation de la distribution de ses ressources naturelles. Il comprend des
sous-divisions géographiques, géologiques, paléontologiques, et autres
qui s'y rapportent. Les allocations pour son maintien comptent parmi
les plus fortes accordées aux bureaux du Gouvernement, car l'on
reconnaît que c'est sur les renseignements précis qu'il fournit, au sujet
du développement économique et efficace de nos richesses naturelles,
que dépend en grande partie la prospérité de notre peuple.

C'est éprouver une grande satisfaction que de savoir que votre
science a été généreusement reconnue et puissamment aidée par d'autres nations, qui sont représentées ici par un grand nombre d'entre
vous. Mais c'est surtout pour vous, Messieurs, un motif de félicitation
d'avoir si bien réussi à former cette association internationale, qui vous
met à même d'échanger les connaissances et les résultats que vos gou-
vernements respectifs vous aident à acquérir, en contribuant à vos
études spéciales par tous les moyens qu'offrent la civilisation de notre
époque.

Notre ère est remplie de l'idée de la dépendance réciproque des
nations. Vous formez l'avant-garde de ce mouvement, et votre Congrès en est la plus noble représentation. Cette libéralité constitue
l'existence même de la science; elle protège les arts, le commerce et
l'industrie. Puisse-t-elle, Messieurs, grandir avec nous et augmenter
ses forces en même temps que les nôtres, jusqu'à ce que, comme les
couches diverses des différentes régions du globe produisent les sols
essentiels à l'existence humaine en chacune d'elles et contribuent au

bien-être et au progrès de tous, dans l'avenir, tous les peuples de la
terre ne forment qu'une seule famille et que la fraternité et l'intelligence, non la force ni l'inimitié, soient les sources de notre prospérité

Puisse votre séjour au milieu de nous vous être agréable; vos
voyages aux merveilles de notre continent vous fournir des données
aussi nouvelles qu'intéressantes, et votre retour dans vos patries respectives avoir lieu sans encombres fâcheux. En terminant permettez-
moi, Messieurs, de vous le dire une seconde fois: Soyez les bienvenus!

A la demande du Congrès, M. J. W. POWELL, Directeur du Geological
Survey des États-Unis, fit les remarques suivantes:

MESSIEURS LES DÉLÉGUÉS: Je n'avais pas l'intention de prendre
la parole en cette occasion. Le Président du Congrès, et M. Gardiner
Hubbard, le Président du Comité local, vous ont parlé; M. le Secrétaire de l'Intérieur vous a souhaité la bienvenue au nom de la nation.
Que puis-je ajouter de plus aux sentiments qu'ils ont si éloquemment
exprimés? Rien; à moins peut-être qu'en ma qualité de Directeur du
Geological Survey des États-Unis et au nom de mes collègues, il me
soit permis d'ajouter que c'est avec un vif plaisir et une profonde
satisfaction que nous vous accueillons ici.

M. T. McK. HUGHES prend la parole:

MESSIEURS: Je ne puis résister à l'invitation de répondre aux gra-
cieuses paroles de bienvenue que M. le Secrétaire de l'Intérieur, et ceux
qui ont parlé avec lui, viennent de prononcer. Je suis convaincu
qu'aucun de nous n'est étranger à la science géologique de ce grand
pays. La générosité avec laquelle votre gouvernement a doté ses
splendides institutions scientifiques est des plus surprenantes; et les
ouvrages qu'elles publient comme résultats de leurs travaux sont dans
des proportions aussi étonnantes. Et en outre, en conséquence de votre
grande libéralité à distribuer ces œuvres aux bibliothèques publiques
de l'Europe, nous sommes tous plus ou moins au courant du mouvement géologique de ce pays. Tout récemment encore nous fûmes
surpris de recevoir un grand volume in-octavo, qui contenait la description détaillée des travaux qui se faisaient ici. Il nous est impossible,
Messieurs, de lire aussi rapidement que vous écrivez.

Nous éprouvons une douce émotion en contemplant la noble capitale
de cette grande République, qui a été le port de salut d'un si grand
nombre de personnes appartenant à toutes les nationalités représentées ici.

Je n'ignore pas que dans ces dernières années vous avez eu quelques
difficultés au sujet des émigrants, craignant qu'il n'en débarque trop
sur vos rives hospitalières; mais je suis convaincu que ces émigrants
géologiques, bien que j'aie entendu dire que des arrangements aient

été faits pour les envoyer dans l'Ouest, finiront par retourner à Washington.

Au nom de mes compatriotes et de beaucoup d'amis de toutes les parties de l'Europe rassemblés ici, permettez-moi de vous offrir nos remercîments les plus sincères pour votre accueil bienveillant, et pour les aimables paroles qui l'accompagnaient.

M. ALBERT GAUDRY s'exprime dans les termes suivants :

Les géologues qui parlent la langue française me chargent de remercier le Gouvernement des États-Unis, la ville de Washington et tous les savants américains pour leur bonne hospitalité. Parmi tant de merveilles qu'a vu notre dix-neuvième siècle, une des plus grandes, assurément, est le développement actuel des États-Unis. A côté des hommes qui se vouent à l'agriculture, au commerce, à l'industrie, enrichissant ce pays et s'enrichissant eux mêmes, il y a ici des savants désintéressés. Ils s'exposent à toutes les fatigues, compromettant parfois leur santé, leur fortune; ils vont partout, jusque dans le cœur des Montagnes Rocheuses, là où il y a quelque nouveauté à découvrir : nous sommes heureux de serrer la main de tels hommes. Du reste, Messieurs, vous n'êtes pas des inconnus pour nous. Grâce à la générosité avec laquelle vous distribuez vos magnifiques volumes, la géologie américaine est devenue internationale. Depuis longtemps, nous vivons avec vos livres, nous sommes confidents de vos pensées scientifiques; en arrivant ici nous croyons trouver de vieux amis. Merci pour votre générosité, qui nous a initiés à vos belles découvertes avant que nous soyons venus les voir sur place.

M. LeConte annonce que le Congrès se réunira le lendemain à 11 heures du matin, et déclare la séance levée.

DEUXIÈME SÉANCE—JEUDI 27 AOÛT 1891.

[Présidence du Prof. JOSEPH LECONTE.]

La séance est ouverte à 11 heures 40 minutes du matin.

Aucun rapport n'ayant été présenté par les comités, la question mise à l'ordre du jour est la *classification génétique des dépôts pléistocènes.*

M. CHAMBERLIN ouvre la discussion en démontrant qu'il est possible de classifier ces dépôts d'après trois bases: 1° Structurale; 2° Chronologique; et 3° Génétique. Ensuite il explique en détail une brochure imprimée, exposant son système de classification génétique, qui avait été préalablement distribuée aux membres du Congrès.

M. GAUDRY prend la parole, et dit que dans le bassin de Paris il y a deux horizons distincts qui se distinguent par des faunes différentes; l'un indique un climat froid, et l'autre un climat chaud. Cependant, nos géologues n'ont pas pu déterminer laquelle de ces deux faunes a précédé l'autre. En Angleterre il y a les mêmes divergences de vues. En Allemagne il n'existe qu'une faune quaternaire qui indique un climat froid, tandis qu'en Italie la faune du climat froid est absente.

M. CREDNER dit que la plaine du nord de l'Allemagne contient des dépôts étroitement liés à ceux du Pléistocène d'Amérique, et que la classification du Prof. Chamberlin est admirable et parfaitement applicable à l'Allemagne.

M. DE GEER approuve la classification proposée par le Prof. Chamberlin. Pendant bien des années, il a appuyé l'adoption d'une classification de ce genre pour la Scandinavie. On pourrait suggérer quelques légères modifications afin de l'adapter aux conditions scandinaves; telles que, par exemple, faire une classe séparée pour les dépôts marins; et réduire, peut-être, les classes IV et V du Prof. Chamberlin en subdivisions de la classe III, puisque les formations semblent être fréquemment accidentelles ou locales. Il approuve la distinction suggérée entre les *Œsars* et les *Kames:* les premiers sont généralement radiaux et les seconds périphériques, eu égard à la distribution de la calotte glaciaire.

M. HUGHES explique que les classifications chronologiques et génétiques sont plus ou moins théoriques. Leur valeur dépend de l'exactitude avec laquelle on interprète les caractères structuraux et autres, y comprenant la paléontologie, et tout ce qui se peut observer directement. Les conclusions de M. Gaudry amènent à une classification

chronologique qui n'est pas incompatible avec celle de M. Chamberlin; c'est-à-dire que dans d'autres régions des dépôts aussi divergents
en structure peuvent être synchroniques et leurs différences peuvent
provenir de différences d'origine.

Il exprime l'opinion que ce que l'on appelle la période glaciaire, en
Angleterre au moins, à été unique et continue, interrompue seulement
par des oscillations. Ces oscillations étaient semblables en nature à
celles que l'on peut observer aujourd'hui au bout du glacier du Rhône
ou d'une autre masse de glace pareille, mais différaient en degré à raison
des quantités beaucoup plus grandes et des espaces de temps beaucoup
plus longs dont il s'agit, lorsqu'on étudie les traces des conditions glaciaires précédentes dans les climats tempérés. Sans citer les cas particuliers dont il a parlé ailleurs, il croit que le placement erroné de certains dépôts (par exemple, de dépôts paléolithiques) aux époques prégla-
ciaires et interglaciaires provient d'erreurs dans la classification génétique. Des couches remaniées contenant souvent des débris de dépôts
glaciaires plus anciens ont été ainsi considérées à tort comme d'origine
glaciaire directe.

Se reportant à l'origine du "boulder clay," il explique un phénomène
que l'on déjà cité—c'est-à-dire que dans une partie du drift glaciaire
on trouve des blocs striés en abondance, tandis que dans une autre partie
du même dépôt il n'y en a point. Si la glace dépasse la source des
matières solides (c'est-à-dire, les roches qui s'élèvent au-dessus de sa
surface) alors tous les débris qui sont tombés sur sa surface s'enfoncent
graduellement à travers les crevasses, etc., et deviennent striés au fond
de la glace. Mais si le drift est formé principalement de matières qui
ont été transportées sur la glace, comme dans le cas de glaciers très
courts, alors il n'y aura que très peu de galets striés.

Il pense aussi que les noms de *Ösars*, *Eskers* et *Kames* étaient appliqués au moins à deux types distincts d'accumulation de matières, souvent primitivement d'origine glaciaire; mais comme ces noms n'étaient
que des termes équivalents et généraux, dans des langues différentes,
pour les mêmes classes de phénomènes, il préférait employer un terme
qualificatif, tel que "radial-Kame," etc., au lieu de donner à un mot
vieux et suranné un sens nouveau et restreint. Il explique les creux
dans les "pitted plains" comme dûs à des interruptions entre les collines et chaînons du drift Esker.

M. WAHNSCHAFFE appuie la classification chronologique, et la considère comme applicable aux dépôts quaternaires du nord de l'Allemagne.
Ces dépôts commencent par des sables et des graviers pré-glaciaires
contenant la *Paludina diluviana*, qui est une forme encore vivante,
et des *Lithoglyphus naticoides*. Au-dessus d'eux vient une moraine
de fond typique, couverte de sables et de graviers stratifiés, renfermant la faune diluvienne bien connue; et à celle-ci succède ensuite le

"till" supérieur, considéré aujourd'hui comme moraine de fond de la seconde époque glaciaire.

M. CREDNER fait observer que la présence du sable entre deux moraines de fond indique un mouvement rétrograde, et une seconde avance de la glace. En Allemagne les sables ainsi intercalés sont toujours d'une étendue limitée, et ne prouvent pas l'existence d'une véritable époque interglaciaire. Les couches de sable entre les moraines ne sont pas continues, mais locales. On ne peut donc leur donner la signification qui leur est attribuée par M. Wahnschaffe.

M. PAVLOW demande la permission de faires quelques remarques générales sur les principes de la classification en question. Il faut d'abord s'entendre sur ce que nous comprenons sous le nom de Pléistocène. Est-ce seulement l'époque glaciaire, ou faut-il y comprendre tous les dépôts formés depuis le Tertiaire jusqu'à présent? Dans le dernier cas, il serait utile d'établir quelques subdivisions fondamentales, et de renoncer pour quelque temps aux détails. Ensuite il faut déterminer ces subdivisions principales. Si l'on veut les établir sur une base chronologique, on soulèvera les questions d'une ou de plusieurs périodes glaciaires, etc. Mais nous sommes ici pour discuter la classification génétique des dépôts pléistocènes; et ses subdivisions fondamentales doivent être basées sur les agents principaux qui ont pris part à la formation de ces dépôts. Non seulement la glace, qui est sans doute le plus important, mais aussi les courants d'eau, les eaux stagnantes et la végétation des tourbières, les neiges, les pluies et les averses, etc. Il voudrait exposer les résultats de ses études sur les dépôts quaternaires de son pays, et la classification qui y est en usage; mais d'abord, il préfère écouter les opinions de ses collègues sur les questions générales qu'il a exposées.

M. DE GEER est d'accord avec M. Wahnschaffe, que la classification chronologique est, du moins localement, possible. Il reconnaît aussi deux époques glaciaires, qui sont dues à deux grandes oscillations. Celles-ci ne peuvent pas toujours être séparées, comme en Russie par exemple. C'est pourquoi il vaudrait mieux commencer par une classification génétique, puisqu'elle est moins embarrassante à appliquer sur le terrain.

M. WAHNSCHAFFE, répondant à l'assertion de M. Credner, qu'il n'y a pas de preuve d'une période inter-glaciaire dans le nord de l'Allemagne, pense qu'une faune diluvienne entre les deux "tills" est une preuve suffisante.

M. CREDNER fait observer qu'on n'a pas trouvé de squelette complet, mais seulement des ossements épars qui ont pu être charriés et déposés avec les graviers.

M. WAHNSCHAFFE répond que les ossements trouvés dans ces graviers sont relativement volumineux comparés aux graviers eux mêmes;

et que, pour cette raison, ils n'ont pas pu être transportés d'une grande distance.

M. SHALER dit qu'il est possible que des dépôts organiques se rencontrent très près de la nappe de glace, ce qui permet un entrelacement de dépôts organiques et glaciaires.

M. GILBERT remarque que J. C. Russell a fait l'observation qu'en Alaska, où le mouvement de la glace est très lent, elle peut être recouverte de terre végétale, et même de forêts dans lesquelles des animaux, tels que les ours, existent de nos jours.

M. DIENER dit qu'il ne voit pas une preuve positive de périodes interglaciaires dans les couches de sable intercalées. Dans les Alpes autrichiennes, des moraines, qui ne datent pas de plus de vingt ans, sont couvertes de pâturages, et dans le Caucase, les rhododendrons poussent au bord même de la glace.

M. HOLST signale deux moraines qui sont séparées par du sable intercalé; et il pense qu'elles ont pu être formées, toutes les deux, par la même nappe de glace. La fonte de la glace laisse une moraine de fond non-oxydée (bleue), couverte d'une moraine supérieure oxydée (jaune). Ceci se rencontre aussi dans le nord de la Suède, où il n'y a pas d'indication de mouvement rétrograde de la glace.

M. DE GEER remarque qu'il ne comprend pas comment trente ou quarante pieds de sables stratifiés peuvent se trouver entre deux moraines d'un même glacier. Les couleurs sont quelquefois le contraire de celles qui ont été citées par M. Holst, et les blocs dans les deux moraines sont de provenances différentes.

M. CHRISTIE décrit la coupe de tourbe et de boue glaciaire (silt), comprise entre deux couches de " till," qui se trouve sur les bords de la rivière Clyde, au-dessus de Glasgow.

M. CADELL fait une description de cinq couches distinctes de " till" qui se trouvent dans le fond préglaciaire d'une rivière dans l'est de l'Écosse; il mentionne un autre lit de rivière rempli de gravier grossier, provenant des roches qui se rencontrent plus au nord dans l'Écosse, gravier qui a été couvert d'une couche plus récente de " boulder clay."

M. MCGEE insiste sur l'importance de la configuration du sol pour l'interprétation des procédés géologiques. Toute classification primaire doit être génétique. Il discute ensuite, d'une manière détaillée, le système de classification des dépôts pléistocènes qui suit:

Classification des formations pléistocènes.

A. Aqueuses:
    1. Au-dessous du niveau de base (base-level).
        a. Marines.
        b. D'estuaires.
        c. Lacustres.

451 GE——5

2. Au niveau de base.
    a. Littorales.
    b. Marécageuses.
    c. Alluviennes (certaines terraces, etc.).
3. Au-dessus du niveau de base.
    a. Torrentielles.
    b. Éboulis (y compris les *playas*).
B. Glaciaires :
    1. Directes. (Chamberlin, classe I.)
    2. Indirectes. (Chamberlin, classes II à V en partie.)
C. Aqueo-glaciaires. (Chamberlin, classes II à V en partie.)
D. Éoliennes. (Chamberlin, classe VI (?).)
E. Volcaniques :
    1. Directes.
        a. Coulées de lave.
        b. Cônes de cendres.
        c. Tufs, couches de lapilli, etc.
    2. Indirectes.
        a. Couches de cendres.
        b. Couches de lapilli.

M. CHAMBERLIN, en terminant la discussion, dit que l'application d'une classification chronologique est très difficile, et qu'une telle classification pourrait même être un obstacle aux observations et à la recherche du vrai. Cette classification est le dernier but des études glaciaires, mais nous ne sommes pas encore en mesure de l'atteindre. Les sous-sols rouges et oxydés ne sont pas développés dans les latitudes du nord. Les dépôts organiques entre les couches glaciaires sont abondants dans l'ouest, mais ne sont pas restreints à un seul niveau. Beaucoup de faits d'érosion et de géologie physique indiquent que la période glaciaire en Amérique a été largement différenciée et d'une durée prolongée. Combien d'époques distinctes elle a embrassé, c'est ce que nous ne savons pas encore.

M. COPE dit qu'une abondante faune tropicale se trouve dans les couches à "Equus" qui, si elles appartiennent à une époque interglaciaire, indiquent pour cette époque un climat très chaud. A cette faune en succède une autre vraiment boréale. Ici l'on trouve tous les matériaux nécessaires pour faire une subdivision chronologique des dépôts pléistocènes.

Les Secrétaires,
C. DIENER.

Le Président,
J. LeCONTE.
                              G. H. WILLIAMS.

TROISIÈME SÉANCE—VENDREDI 28 AOÛT 1891.

[Présidence de M. LeCONTE.]

La séance est ouverte à 11 heures 40 du matin.

M. EMMONS annonce que le procès-verbal de la dernière séance est imprimé et prêt à être distribué à chacun des membres du Congrès, et propose, afin d'éviter de perdre du temps inutilement, que l'on se dispense de la lecture par le Secrétaire; mais que les membres soient priés de lui transmettre par écrit les corrections qu'ils désirent faire à leurs remarques, pour être insérées dans le compte-rendu final, et que, s'il n'y a aucune objection, le procès-verbal imprimé, sujet toutefois à ces corrections, soit considéré comme étant adopté.

Le procès-verbal est adopté, après corrections du Prof. Pavlow et du Dr. Holst.

La question mise à l'ordre du jour est la *corrélation chronologique des roches clastiques.*

M. POWELL annonce qu'une série de mémoires sur la corrélation des roches a été préparée, comme bulletins du Geological Survey, et qu'ils seront distribués aux membres du Congrès aussi rapidement que les exemplaires viendront de chez l'imprimeur.

M. HUGHES remarque que la question de classification et de nomenclature des roches cambriennes et siluriennes avait été discutée, mais non amenée à une conclusion, par le Congrès précédent. Cette question implique aussi la discussion de la priorité d'une détermination correcte de la succession et du groupement, qui doivent être sujets aux mêmes règles applicables à d'autres nomenclatures systématiques. Il ne croit pas qu'il y ait aucun fait nouveau qui l'affecte sérieusement, et comme il a donné ses opinions tout récemment dans la "Vie de Sedgwick," pour sa part, il ne veut pas forcer le Congrès à considérer la question, surtout en l'absence de ceux qui s'opposent à ses vues.

M. GILBERT ouvre la discussion en présentant une classification générale des méthodes de corrélation.

Les couches sont classées localement par la superposition en séquence chronologique. La corrélation géologique est la chronologie des couches qui ne sont pas en séquence visible. Pour faciliter la discussion, les méthodes de corrélation sont arrangées en dix groupes dont six physiques et quatre biotiques.

Méthodes physiques de corrélation.

1° *Par continuité visible.*—L'affleurement d'une couche est tracé d'un point à l'autre, et les différentes parties sont mises en corrélation les unes avec les autres.

2° *Par similitude lithologique.* Cette méthode, autrefois si répandue, est employée quand les distances sont courtes.

3° La corrélation par la *similitude de séquences lithologiques* est d'un usage important quand les localités comparées se trouvent dans les limites de la même province géologique; mais elle n'est pas sûre, en passant d'une province à une autre.

4° La corrélation par les *dislocations ou discordances de stratification* est d'un usage limité, surtout si on l'emploie avec d'autres méthodes. L'habitude de s'en servir, dans le cas où les localités sont éloignées les unes des autres, donne lieu à des doutes.

5° Les dépôts peuvent aussi être mis en corrélation par leurs *rapports simultanés à quelque évènement physique*, comme, par exemple, une plage soulevée avec les dépôts lacustres qu'elle entoure; une plaine de niveau de base (base-level) avec une couche subaqueuse adjacente; et les dépôts alluviens, littoraux et subaqueux qui se trouvent dans des relations topographiques convenables. La corrélation des dépôts glaciaires du Pléistocène s'appuie largement sur un épisode climatique qui est supposé avoir été produit par une cause générale quelconque.

6° La corrélation des couches peut se faire par la comparison des changements qu'elles ont subies à la suite de certains effets géologiques que l'on croit continus. Les anciens et les nouveaux dépôts glaciaires dans des régions différentes peuvent être raccordés selon l'étendue relative de l'érosion et de la désagrégation atmosphérique qu'ils ont subies. L'induration et le métamorphisme semblent fournir des preuves d'ancienneté, mais elles doivent céder aux évidences d'un autre genre. Le métamorphisme occupe une place importante dans la corrélation des roches précambriennes auxquelles la plupart des autres méthodes ne sont pas applicables.

Ces méthodes physiques se modifient selon la distribution géographique des climats et des procédés de changement géologiques.

Méthodes biotiques de corrélation.

7° Une faune ou une flore nouvellement découverte est comparée à une série typique de faunes ou de flores au moyen des espèces qu'elle possède en commun avec chacune d'elles.

8° Elle est aussi comparée au moyen des formes représentatives, ou par genres et familles.

7° et 8°. Ces comparaisons s'affermissent, si deux ou plusieurs séries de faunes en séquence se trouvent systématiquement liées à celles d'une série typique.

9° La corrélation de deux faunes ou de deux flores autrement liées se fait d'après leurs relations aux faunes et aux flores actuelles de leurs localités. Cette méthode a été appliquée par Lyell aux roches tertiaires.

10° La corrélation des faunes peut se faire d'après leurs rapports aux épisodes climatiques, eu égard à leurs localités. Par exemple, des coquilles boréales, trouvées dans des latitudes au-dessous de leur emplacement actuel, sont censées appartenir à l'époque glaciaire.

En général, la limitation d'une corrélation exacte au moyen des méthodes biotiques provient des faits qui résultent d'une distribution géographique. Les corrélations à courte distance sont préférables à celles qui sont éloignées.

La corrélation biotique au moyen de fossiles d'espèces diverses peut avoir des valeurs diverses. Généralement la valeur d'une espèce pour la corrélation est en raison inverse de sa durée, et directe quant à son étendue dans l'espace. La valeur d'un groupe biotique dépend, 1° de l'étendue occupée par son espèce dans le temps et dans l'espace; 2° du degré de conservation des individus de son espèce.

M. von Zittel, s'occupant des méthodes biotiques, exprime son opinion sur la valeur relative des plantes et des animaux pour la corrélation. Les plantes fossiles sont relativement peu importantes. Parmi les animaux on peut distinguer ceux qui sont marins, lacustres ou terrestres. De ces classes les invertébrés marins sont de la plus grande valeur pour la corrélation. Les vertébrés fossiles changent rapidement, caractère utile pour la corrélation; mais beaucoup de terrains en sont entièrement dépourvus. Par exemple, il ne s'en trouve presque pas dans les couches alpines dont l'âge correspond à celui des couches qui contiennent la faune de mammifères du bassin de Paris. Dans certains dépôts lacustres les invertébrés peuvent être absents; et, dans ce cas, la faune vertébrée est le guide le plus sûr.

M. de Geer insiste sur l'importance d'une comparaison numérique entre les différentes espèces. Le chiffre précis d'individus qui se trouvent dans une couche donnée est d'une grande valeur.

M. Marsh dit qu'il est d'accord, en général, avec les conclusions énoncées par M. von Zittel; toutefois, il attache une importance particulière aux fossiles vertébrés. Dans les couches mésozoïques et tertiaires des Montagnes Rocheuses il a trouvé que les vertébrés étaient les meilleurs guides pour la corrélation. Ceci est dû, en partie, à ce que les invertébrés manquent totalement, ou qu'ils sont lacustres. En 1877, il avait nommé une suite d'horizons d'après le genre vertébré le plus caractéristique qui se trouvait exclusivement dans chacun d'eux.

Il présente un tableau chronologique des terrains de l'Amérique basé sur leurs vertébrés fossiles et où il y a été tenu compte de toutes les découvertes récentes.

M. HUGHES parle de la tendance actuelle et croissante vers une classification naturelle, prenant la place des systèmes artificiels basés sur un seul caractère ou un seul groupe d'animaux ou de plantes. Les preuves sont complexes et embrassent une variété considérable de rapports divers. Il fait voir les exceptions aux conclusions normales déduites de la superposition, des caractères lithologiques et de la similitude des séries. Il nous faut un système de critéria si variés que, si l'un d'eux ou plusieurs manquaient, nous puissions en employer d'autres. Tous les genres de preuves sont utiles; positives, négatives et circonstancielles.

Il n'est pas d'accord avec ceux qui veulent déprécier la valeur des plantes fossiles comme moyens de corrélation. Il est vrai que dans l'état actuel de la science il est difficile de s'en servir, parce que nous ne trouvons presque jamais que des fragments de plantes. D'autre part lorsqu'on ne trouve que des débris de coquille cassés, ou d'ossements, le même doute n'existe-t-il pas? La difficulté n'est pas inhérente à la nature des fossiles, mais à leur état de conservation, qui est ordinairement très imparfait.

M. POWELL insiste sur la nécessité de la spécialisation pour les géologues qui s'occupent de la corrélation. Les preuves dérivées des faits biotiques et physiques pourraient apparemment ne pas être d'accord, mais, pour que l'on puisse obtenir un résultat satisfaisant, il faut que ces deux genres de preuves soient en harmonie. Il démontre, par son expérience personnelle, comment l'identification de formations synchroniques pourrait être faite sur une grande étendue en employant concurremment les méthodes physiques et biotiques.

M. McGEE remarque que dans la plaine côtière (coastal plain) des États-Unis on n'emploie qu'une corrélation physique. Les bases s'accordent avec celles qui ont été esquissées par M. Gilbert, sauf quelques légères modifications et une addition importante, ainsi qu'il suit:

Pour la corrélation locale	*Continuité visible.* *Similitude lithologique.* *Similitude de suites.*
Pour la corrélation dans une province	*Dislocations orogéniques considérées comme indices de géographie et de topographie.*
Pour la corrélation avec les provinces voisines	*Rapport avec les incidents physiques, y compris les mouvements continentaux, le transport des matériaux, la configuration des terres, etc.*
Pour la corrélation générale	*Homogénie ou identité d'origine.*

Par la corrélation établie sur ces bases, l'histoire physique d'une portion considérable du continent peut être déterminée, de manière qu'une carte géographique, et même topographique, de chaque épisode qui se produit dans l'accroissement d'un continent puisse être assez exactement dessinée. Une fois que ces épisodes sont clairement définis, et que les fossiles trouvés dans les couches auront été étudiés, il sera possible de déterminer définitivement la distribution géographique des organismes pendant chaque épisode; alors la paléontologie pourra être établie sur une base nouvelle et plus élevée.

M. DAVIS fait voir qu'il est facile de déchiffrer l'histoire géologique, non seulement par l'examen des dépôts successifs qui se sont formés, mais aussi par l'étude des phénomènes d'érosion et de dégradation. Comme exemple de cette méthode, il explique une coupe topographique à l'ouest de la ville de New York. Dans celle-ci, nous avons la preuve de l'existence d'une ancienne "*peneplain*," ou surface de base (*base level*) de la période crétacée. Cette surface a été subséquemment soulevée (plus vers l'ouest que vers l'est) à la fin de la période crétacée ou au commencement de l'ère tertiaire. Elle a été disséquée depuis par l'excavation des vallées plus récentes. La région basse de la vallée de l'Hudson est citée comme exemple de cette dissection récente.

M. CLAYPOLE considère que les différentes méthodes de corrélation géologique varient beaucoup dans leur valeur. Il est improbable que les vestiges de plantes ou de mammifères atteindront jamais la perfection qui se trouve dans ceux de la faune marine invertébrée. Cette dernière est pour le géologue ce qu'une triangulation de première ordre est pour le géodésien. Elle indique les divisions principales, qui sont ensuite subdivisées au moyen d'autres fossiles, tels que les plantes et les vertébrés.

M. VAN HISE parle des méthodes de corrélation employées pour les roches précambriennes, qui se trouvent dans des localités très éloignées les unes des autres aux États-Unis, et qui, autant que cela est connu jusqu'à présent, sont presque entièrement dépourvues de fossiles. Les données physiques seules peuvent servir à faire la corrélation de ces roches. L'expérience a démontré que, parmi tous les criteria physiques, la discordance de stratification est de beaucoup le plus important. D'autres caractères physiques, tels que le degré d'induration, le métamorphisme et les relations avec les roches éruptives, n'ont de valeur que pour la subdivision des roches dans une région déterminée et limitée, mais ne peuvent être employées avec aucune certitude pour identifier les formations synchroniques dans les régions très séparées. L'idée que les caractères lithologiques peuvent établir l'époque géologique d'une série de couches a retardé la subdivision des roches précambriennes. Les recherches de M. Pumpelly et d'autres, faites dans l'est des États-Unis, ont démontré que les roches siluriennes, dévoniennes, et même carbonifères, sont devenues aussi complètement cristallines, par suite de certaines conditions physiques, que les plus anciennes roches de l'Ouest. Puisque les preuves biologiques font défaut pour établir la corrélation des roches précambriennes, et puisque la similitude des caractères physiques ne justifie pas la corrélation des séries de roches dans les localités très éloignées, on a jugé nécessaire d'abandonner l'usage des noms tels que "*Huronian*" et "*Keweenawan*," comme termes universels. Les vestiges des êtres organiques ne manquent pas dans les roches précambriennes, et il faut espérer que les paléontologistes réussiront à différencier plusieurs périodes au-dessous de la cambrienne, comme celle-ci l'a été de la silurienne.

Les Secrétaires,
O. DIENER,
G. H. WILLIAMS.

Le Président,
J. LeCONTE.

QUATRIÈME SÉANCE—SAMEDI 29 AOÛT 1891.

[Présidence de M. GAUDRY, puis de M. VON ZITTEL.]

La séance est ouverte à 10 heures du matin.

M. EMMONS présente le procès-verbal imprimé de la dernière séance, qui est adopté, après corrections verbales du professeur von Zittel et de M. Van Hise.

M. DELAIRE, au nom du prince Roland Bonaparte, présente deux communications, au sujet des phénomènes du glacier d'Aletsch, et d'une excursion géologique en Corse.

M. CH. BARROIS, au nom de M. Michel-Lévy, présente une communication relative à l'histoire géologique des volcans de l'Auvergne, contenant une classification des roches éruptives représentées par des symboles.

La discussion sur la corrélation est ensuite reprise.

M. HILGARD insiste sur l'importance de l'abondance ou de la rareté des espèces pour la corrélation des couches. Il pense qu'une évaluation du nombre des espèces devrait être faite. Il est aussi d'avis que, comparées à la faune marine, les plantes ont peu de valeur pour la corrélation, en raison de leur distribution locale, de leur proximité accidentelle à l'eau, de leur transport et de leur conservation. Les plantes ne peuvent être employées que quand de grandes étendues auront été étudiées.

M. VON ZITTEL est appelé à prendre la présidence.

M. WARD continue la discussion dans laquelle il développe deux des principes les plus généraux de corrélation au moyen de plantes fossiles, comme suit:

I. Que les grands types de la végétation sont caractéristiques des grandes époques géologiques.

Ce principe est applicable en comparant les dépôts d'âges différents, considérablement séparés, quand la stratigraphie n'est pas décisive. Par exemple, un petit fragment d'une plante carbonifère prouve conclusivement que les roches dans lesquelles elle se trouve sont paléozoïques, ou bien, encore, une simple feuille dicotylédone établit qu'elles doivent être aussi récentes que la période crétacée.

II. Que pour des dépôts qui ne sont pas d'un âge si différent, comme, par exemple, lorsqu'ils sont dans le même système, ou dans la même série géologique, beaucoup de matériaux sont nécessaires pour fixer leur position au moyen des plantes fossiles.

L'oubli de ce principe a été la cause de beaucoup d'erreurs faites par les paléobotanistes, et a contribué le plus à discréditer la paléobotanique. Les géologues ont fait des demandes extravagantes aux paléobotanistes, qui à leur tour ont fait violence à la vérité en essayant d'y satisfaire. D'un autre côté, lorsque les matériaux sont abondants, les plantes fossiles ont souvent corrigé les erreurs stratigraphiques, et ont résolu des problèmes relatifs à l'âge géologique, dont la solution paraissait impossible par d'autres genres de preuves.

M. WALCOTT explique la corrélation des roches cambriennes de l'Amérique du Nord en se servant de deux grandes cartes. Les principes suivis aujourd'hui, dit-il, sont les mêmes qui furent employés par le Geological Survey de New York avant 1847: à part que ces principes ont été tant soit peu modifiés par la théorie de l'évolution. Les données physiques et biotiques sont toutes les deux utiles dans la corrélation des roches cambriennes des provinces de l'Atlantique, des Appalaches, des Montagnes Rocheuses, et du Bassin intérieur. Dans toute la province des Appalaches, les données physiques suffisent à établir la corrélation du Cambrien inférieur depuis le Vermont jusqu'à l'Alabama, mais elles sont insuffisantes pour un raccordement avec celui qui se trouve dans la vallée du Saint Laurent. La corrélation des dépôts du synclinorium des Appalaches, avec ceux de celui des Montagnes Rocheuses, a été faite exclusivement par des données biotiques; et, pour la grande étendue du Cambrien supérieur sur tout le continent, les données biotiques ont servi pour la corrélation de celui des Montagnes Rocheuses avec ceux de l'intérieur et des Appalaches.

Les corrélations qui ont été faites indiquent que, pendant la période du Cambrien inférieur et moyen, un grand continent existait dans l'Intérieur; et que les sédiments cambriens s'accumulèrent dans des dépressions à l'ouest des Appalaches et des Montagnes Rocheuses. Pendant le Cambrien supérieur, l'intérieur du continent s'affaissa sous l'océan et les grès du Cambrien supérieur y furent déposés, en même temps que les calcaires cambriens des provinces des Montagnes Rocheuses et des Appalaches. Les résultats de ces corrélations ajoutent un chapitre nouveau à l'histoire de l'évolution du continent nord-américain.

M. HALL fait mention des difficultés que l'on rencontra, même dans l'État de New York, à l'époque des premiers essais de corrélation des roches. Il insiste sur l'importance qu'il y a à considérer, et les caractères physiques, et les faunes des couches. Dans certains cas cependant, les caractères physiques des roches changent considérablement en passant d'une région à une autre: les grès sont remplacés par des calcaires, et les calcaires se transforment en schistes; l'épaisseur des couches peut aussi varier beaucoup. Les fossiles sont d'une valeur inégale dans les corrélations de ce genre; les Lamellibranches sont des formes qui se trouvent près des côtes, mais n'existent pas dans les eaux profondes;

c'est pourquoi elles ne sont pas aussi utiles pour la corrélation que les Brachiopodes, dont la distribution est plus répandue.

M. H. S. WILLIAMS insiste sur les rapports des espèces avec les conditions de dépôt. L'abondance d'une espèce varie avec le milieu, et l'étude de la corrélation devrait comprendre celle de ces conditions. Les grès déposés près des rivages peuvent avoir une faune différente de celle d'une calcaire qui a été déposée en même temps mais loin des côtes; et un changement de faune peut être occasioné par les conditions du dépôt même. L'âge des couches devrait être déterminé par la comparaison des espèces du même genre, plutôt que par celle des genres différents. Il y a des centres d'abondance qui montrent une grande variabilité dans leurs caractères; en dehors de ces centres les espèces présentent des variétés qui pourraient être appelées extra-limitrophes, et qui ne sont pas typiques, quoiqu'elles aient été souvent publiées comme types.

M. FRECH dit que dans la comparaison de la faune du Paléozoïque moyen de l'Europe avec celle de l'Amérique du Nord, il se trouve deux points d'un intérêt spécial:

A. L'identité de quelques horizons comparativement petits.

B. Les différences bien plus grandes qui existent dans ces mêmes couches.

Les faunes semblables sont:

1. Celles du Niagara et des schistes de Wenlock.

2. La Rhynchonella des calcaires de Tully, et les Goniatites des couches soi-disant de Naples, dans le Dévonien supérieur.

3. Les Goniatites à la base du Carbonifère dans l'État de l'Iowa, en Espagne, et au centre de l'Allemagne. Les fossiles d'Hamilton sont d'un intérêt particulier, parce que nous avons sur le Rhin, dans ce qu'on appelle les *Lenneschiefer*, une faune avec le même faciès. Mais, quoique ces roches fussent déposées dans des conditions semblables, le nombre d'espèces identiques dans les deux pays est très petit; et il y a beaucoup de genres dans chacun d'eux qui ne se trouvent pas dans l'autre. Tout le Dévonien inférieur manque dans la Russie d'Europe et une partie aussi dans le centre de l'Allemagne; mais le grand changement physique, qui vint après, explique suffisamment les différences qui caractérisent la jonction du Dévonien et du Silurien.

M. CH. BARROIS ne croit pas qu'il soit possible de comparer en détail les formations paléozoïques de l'Europe et de l'Amérique. Quelques zones individuelles de la série américaine peuvent être raccordées avec les niveaux européens, mais il est presque impossible d'établir particulièrement l'identité d'autres niveaux qui leur sont adjacents.

M. VAN HISE parle de la distribution, du caractère et de la succession des sédiments précambriens de l'Amérique du Nord. Toutes les roches d'une date antérieure à la faune Olenellus sont considérées comme étant précambriennes. Elles renferment d'épaisses couches de schistes charbonneux, des fossiles variés et distincts, et beaucoup d'autres indices d'une vie exubérante. Si l'on parvient jamais à y trouver une faune qui sera aussi différente peut-être de la faune cambrienne que celle-ci l'est de la faune silurienne, il y aura lieu de lui donner un autre nom. Il existe dans beaucoup de régions de l'Amérique des roches sédimentaires précambriennes de grande épaisseur et peu altérées. Dans un nombre considérable de localités, ces roches ont été séparées en séries par des discordances de stratification très générales, et ces séries ont été ensuite divisées en étages. Les régions précambriennes les plus importantes sont le lac Supérieur et le lac Huron, l'Arizona centrale, le New Brunswick, la Terre-Neuve, et la partie sud-ouest du Montana. Comme exemple démonstratif de ces successions, on peut citer la première. Dans cette région l'ordre descendant est:

1° Le grès du lac Supérieur (Potsdam), discordance de stratification;
2° Keweenawan, discordance;
3° Huronien supérieur, discordance;
4° Huronien inférieur, discordance;
5° Base de roches cristallines.

Chacune de ces séries est divisée en plusieurs étages.

Dans une même région il est possible de faire la corrélation des séries et des étages sur une base physique. Dans des régions différentes, les séries présentent des caractères lithologiques variables ainsi que des successions dissemblables. En raison de l'absence d'une faune précambrienne bien connue il est impraticable, à présent, de faire des corrélations dans les régions très éloignées. De là le terme Algonkien, qui a été proposé par le U. S. geological survey comme devant embrasser toutes les roches clastiques précambriennes. Aucun géologue praticien en Amérique ne soutient aujourd'hui l'indivisibilité du précambrien dans toutes les régions. Si les conclusions qui précèdent sont correctes, la théorie de la succession invariable, supportée par Hunt et développée presque entièrement dans le laboratoire, n'a aucune valeur. Qu'elle est inexacte sur un ou plusieurs points fondamentaux, c'est là un fait démontré par l'ordre des roches observé dans chaque région où il existe des successions assez complètes.

M. PUMPELLY confirme les conclusions de M. Van Hise en ce qui concerne ses propres observations sur les terrains cités. Il réfère particulièrement à celles qu'il a faites dans les Montagnes Vertes, où, dans une localité, le métamorphisme a complètement masqué le caractère originel des roches et, par suite, rendu impossible une corrélation par les caractères lithologiques. Comme exemple, il cite une formation qui est un quartzite sur un point, un gneiss blanc contenant des fels-paths nouveaux sur un autre, un conglomérat sans aucune structure schisteuse ailleurs, et un micaschiste dans une quatrième localité.

M. BARROIS, à propos des observations de M. Van Hise, dit qu'il n'existe pas de base générale, soit biologique ou lithologique, pour la corrélation des roches précambriennes de l'Europe avec celles de l'Amérique du Nord, et que même les termes appliqués à ces roches sont sujets à des malentendus. Il est certain que les divisions employées en France ne peuvent pas être raccordées avec celles qui sont aujourd'hui en usage aux États-Unis. La corrélation générale ne peut, jusqu'à présent, être basée sur les discordances de stratification; l'autopsie est la seule base sur laquelle il soit possible d'établir des comparaisons.

Il indique certains parallélismes entre l'histoire des schistes cristallins d'Amérique, telle que l'a fait connaître M. Pumpelly, et les roches gneissiques de Brest où les schistes argileux cambriens sont transformés en des gneiss d'un aspect archéen, tandis que les quartzites fossilifères, qui alternent avec ces gneiss, sont devenus des quartzites cristallins. Pour arriver à un accord commun au sujet des roches cristallines, les géologues doivent examiner le terrain collectivement.

M. WHITE est appelé à donner son opinion, mais, la discussion s'étant trop écartée de la question telle qu'elle avait commencé, il s'excuse de ne pas y prendre part.

M. COPE discute la question au point de vue général, en s'attachant spécialement à la valeur des Vertébrés pour la corrélation, surtout pour la corrélation intercontinentale. Il démontre qu'il y a une diversité prononcée dans les faunes vertébrées actuelles des continents, et que c'est plutôt dans l'étendue verticale qu'horizontale que l'on doit en chercher les variations. Cette étude fait voir que certaines régions ont été envahies par des faunes etrangères; comme, par exemple, une fois une faune de l'Amérique du Sud a envahi l'Amérique du Nord et s'en est ensuite retirée, tandis qu'une autre fois une faune nord américaine s'est répandue dans l'Amérique du Sud où ses traces se trouvent encore. Il incline à croire que certaines formes de vertébrés, au lieu d'avoir une origine unique, ont pu prendre naissance séparément sur divers points de la surface du globe et évoluer ensuite parallèlement. Nous avons le parallélisme dans des régions séparées, mais il est défectueux dans le Laramie.

M. GILBERT est d'opinion que l'on doit se servir de plusieurs méthodes de corrélation. Il doute de la certitude qu'offre la corrélation des roches non-fossilifères en se basant sur le degré de modification des caractères originels, même lorsqu'il s'agit de localités voisines. Il pense aussi que l'abondance ou la rareté de formes fossiles est comparable aux différences lithologiques, et considère la présence seule d'une espèce comme aussi importante que son abondance pour établir la corrélation.

M. VAN HISE explique qu'en Europe on n'a pas souvent fait de distinction entre l'Algonkien et l'Archéen parce que là, comme dans les Appalaches, il y a eu, à plusieurs reprises, de forts mouvements dynamiques plus récents.

M. COPE ajoute que la vie organique dans ses progrès sur la terre diffère des minéraux et des roches en ce qu'elle possède des lois, ce qui lui donne un élément d'indépendence.

Des annonces sont faites par le Secrétaire, M. EMMONS, et la séance, levée à 1 heure, est ajournée au lundi 31 août, à 11 heures du matin.

Les Secrétaires,
F. FRECH,
J. C. BRANNER.

Le Président pro tem.,
K. VON ZITTEL.

CINQUIÈME SÉANCE—LUNDI 31 AOÛT 1891.

[Présidence de M. JAMES HALL.]

La séance est ouverte à 11 heures 25 minutes du matin.

M. EMMONS présente le procès-verbal de la dernière séance, qui est adopté; il fait aussi des annonces relatives aux excursions, etc.

La question mise à l'ordre du jour est celle des *gammes de coloriage générales* et d'autres procédés graphiques.

M. POWELL montre un tableau sur lequel est représentée la gamme de coloriage dont se sert le Geological Survey des États-Unis; il en explique les méthodes, et donne les raisons de leur application. Les couleurs adoptées pour les roches d'époques différentes sont:

Période.	Couleur.	Marque.
1. Néocène	Orange	N.
2. Éocène	Jaune	E.
3. Crétacée	Jaune vert	K.
4. Jura-Trias	Bleu vert	J.
5. Carbonifère	Bleu	C.
6. Dévonienne	Violet	D.
7. Silurienne	Pourpre	S.
8. Cambrienne	Rose	C.
9. Algonkienne	Rouge	A.

Les couleurs sont employées pour désigner les périodes géologiques, et les hachures ou signes coloriés indiquent les assises (*formations*); quant aux divisions d'ordre inférieur, la distinction en est ordinairement reléguée dans le texte. Le nombre de signes pour désigner les assises peut être augmenté indéfiniment, tout en suivant un système déterminé.

M. VAN HISE fait observer que les roches archéennes se distinguent par une impression brune en dessous, et que les roches métamorphiques d'un âge connu reçoivent la couleur des roches non altérées qui y correspondent.

M. WILLCOX montre que dans le plan présenté par le Major Powell les couleurs ne sont pas distribuées également d'après la gamme chromatique.

M. POWELL explique que l'on n'avait pas essayé de choisir des couleurs également distribuées conformément à la gamme chromatique, mais d'employer celles qui pouvaient être le plus facilement reconnues.

M. CADELL demande pourquoi on ne s'est pas servi du noir et du gris.

M. POWELL répond qu'on a employé le bleu au lieu des noirs pour le Carbonifère; que les noirs causent des malentendus en ce qui concerne la houille, qui se trouve dans le Crétacé et dans le Tertiaire, aussi bien que dans le Carbonifère.

M. CHRISTIE pense que le noir est souvent très incommode parce qu'il rend illisibles les détails de la carte.

M. CADELL dit que les cartes dressées par le Geological Survey de la Grande-Bretagne sont coloriées à la main; et que le système suivi par le Geological Survey des États-Unis ne pourrait, pour cette raison, être employé économiquement.

M. POWELL explique que le système du Geological Survey des États-Unis est très économique, une fois que les signes sont transférés sur les pierres lithographiques.

M. HUGHES pense qu'il est difficile de trouver une gamme de nature à satisfaire tout le monde. La permanence des couleurs, la facilité de les appliquer, et la distinction avec laquelle elles montrent ce qui est désiré, doivent être prises en considération. Celle qui remplit le mieux les conditions voulues survivra. Il remarque ensuite que parmi les membres du Congrès beaucoup, surtout ceux qui vont prendre part à la grande excursion désirent vivement que tout le temps, qu'il sera possible d'économiser sur l'ordre du jour, soit consacré à l'explication des phénomènes géologiques américains récemment rapportés des frontières du Geological Survey par l'avant-garde de nos travailleurs scientifiques.

Sur la proposition de M. POWELL, le programme pour l'après-midi est modifié, afin que la géologie du pays, qui sera traversé par ceux qui participeront à la grande excursion, puisse être sommairement décrite par ceux qui la connaissent le mieux.

La séance est ajournée à 2 heures et demi de l'après-midi.

En se réunissant à 2 heures et demi, sous la présidence de M. LeConte, M. CHAMBERLIN, M. GILBERT, le Major POWELL et M. EMMONS font de courtes conférences sur la géologie du pays qui doit être parcouru pendant l'excursion aux Montagnes Rocheuses.

La séance est levée à 4 heures 40 minutes de l'après-midi.

Les Secrétaires,
J. C. BRANNER,
F. FRECH.

Le Président,
JAMES HALL.

SÉANCE DE CLÔTURE—MARDI 1ᵉʳ SEPTEMBRE.

[Présidence de M. LeConte.]

La séance est ouverte à 11 heures et quart du matin.

M. Emmons présente le procès-verbal de la dernière séance, dont la rédaction est adopté. Il fait ensuite quelques annonces relatives aux différentes excursions qui ont été projetées.

Sur la proposition de MM. H. S. Williams et de Margerie, il est procédé à la nomination d'une *Commission internationale de bibliographie géologique*. Sont nommés membres MM. :

Frech	Allemagne.
Gilbert	Amérique du Nord.
Golliez	Suisse.
Gregory	Grande Bretagne.
De Margerie	France.
Reusch	Scandinavie.
Steinmann	Amérique du Sud.
Tschernyschew	Russie.
Tietze	Autriche-Hongrie.
Van den Broeck	Belgique.

Cette commission, qui sera permanente, est autorisée à s'adjoindre de nouveaux membres en nombre illimité. Ses fonctions principales consisteront: 1°, à dresser et à publier le plus tôt possible une liste des *bibliographies géologiques partielles* qui existent déjà; 2°, à provoquer, de la part des sociétés géologiques et des services géologiques, fonctionnant dans les pays qui n'ont pas encore été l'objet d'un travail de cette nature, la préparation de catalogues détaillés des ouvrages concernant la géologie de leurs territoires respectifs; et 3°, à étudier les moyens qui permettraient d'arriver, désormais, à la centralisation méthodique de la *Bibliographie courante*. La commission devra présenter, à la prochaine session du Congrès, un rapport faisant connaître l'état de ses travaux.

M. Emmons, au nom du conseil, annonce que les géologues suisses, par l'entremise de leurs délégués, MM. Golliez et Schmidt, invitent le Congrès à se réunir en Suisse en 1894. [*Applaudissements.*]

M. Golliez ajoute qu'en conséquence du peu de temps qui s'est écoulé depuis la décision prise de ne pas rassembler le prochain Congrès à Vienne, il n'a pas été en mesure de se concerter avec tous les géologues de son pays, parce que beaucoup d'entre eux se trouvent actuellement en campagne; mais, qu'il prenait sur lui de donner l'assurance que la réception la plus chaleureuse serait faite au Congrès, non seulement par tous les géologues de la Suisse, mais aussi par le Gouvernement Fédéral qui, malgré qu'il n'avait pas pris l'initiative dans la question, ainsi que le Gouvernement des États-Unis, s'était déclaré prêt à coopérer avec ses géologues, si l'on jugeait à propos de choisir la Suisse pour la réunion du Congrès en 1894. Au point de vue géologique, la Suisse a toujours été internationale par son caractère, puisque les géologues de tous les pays s'y sont rencontrés pour étudier ses montagnes; elle est donc, sous ce rapport, singulièrement adaptée à être le lieu de réunion d'un congrès de géologues.

Par un vote unanime, la Suisse est choisie pour la réunion du prochain Congrès, et les géologues dont les noms suivent sont nommés membres du Comité d'organisation, avec l'autorité de s'adjoindre d'autres membres, et de fixer la date, ainsi que la ville, où le Congrès se réunira en 1894:

Dr. A. HEIM,   Prof. FR. LANG,
Prof. E. RENEVIER,   Prof. H. GOLLIEZ,
Dr. A. BALTZER,   Dr. G. SCHMIDT.

M. PUMPELLY propose que les remercîments du Congrès soient transmis aux géologues suisses, par l'entremise de MM. Golliez et Schmidt, pour leur invitation.

M. POWELL appuie la proposition de M. Pumpelly et donne l'assurance aux géologues suisses que les Américains éprouveront le plus grand plaisir à visiter la Suisse, qu'ils considèrent comme la Mecque de tous les voyageurs et de tous les géologues.

La proposition est adoptée à l'unanimité.

M. EMMONS lit ensuite la lettre suivante:

COMITÉ GÉOLOGIQUE DE LA RUSSIE, ST. PÉTERSBOURG, le 1er Août 1891.

Le Comité géologique de Russie a l'honneur d'informer (le Congrès) que, par ordre de Sa Majesté l'Empereur, il est autorisé à déclarer le consentement du Gouvernement russe à l'invitation de la VII me session du Congrès géologique international à St. Pétersbourg.

En invitant les géologues de toutes les nations à visiter la Russie en 1897, le Comité géologique fera tout son possible pour rendre leur séjour dans notre pays plein d'intérêt scientifique et agréable.

A. KARPINSKY,
Directeur du Comité géologique.

M. TSCHERNYSCHEW, au nom des géologues de la Russie, invite cordialement les membres du Congrès à se réunir à St. Pétersbourg en 1897, et déclare que la libéralité de l'Empereur leur assure une réception des plus hospitalières, et mettra les géologues russes à même de leur procurer de longues et intéressantes excursions dans toutes les parties de ce vaste empire.

M. PUMPELLY propose un vote de remercîments à S. M. l'Empereur, à M. Karpinsky et aux géologues russes pour leur invitation, transmise par M. Tschernyschew, à réunir le Congrès de 1897 à St. Pétersbourg.

M. CHAMBERLIN appuie cette proposition, et parle des grandes attraits qu'offrent la Russie et sa géologie aux géologues de tous les pays; et il ajoute qu'elles seront particulièrement intéressantes pour les Américains, qui y trouveront de nombreux points de ressemblance avec la géologie des États-Unis.

La proposition est adoptée à l'unanimité.

M. EMMONS suggère que les remercîments du Congrès à l'Empereur et à M. Karpinsky leur soient transmis par le télégraphe.

Cette suggestion est adoptée par acclamation.

Aucune autre question importante ne demandant l'action du Congrès, le Président, M. LECONTE, prononce les paroles suivantes:

MESDAMES ET MESSIEURS: Je n'ai que quelques remarques à faire avant d'annoncer la clôture de ce Congrès. Le moment de notre séparation est arrivé et nous pouvons, à bon droit, demander quel a été le résultat de notre réunion.

Les deux premiers jours ont été consacrés à discuter la classification des roches pléistocènes, une question trop complexe et trop difficile pour être complètement discutée, et sur laquelle il reste encore beaucoup à dire. Une partie seulement a été discutée, c'est-à-dire la classification basée sur la mode de formation, quoique l'on ait touché à la classification chronologique, avec laquelle la première est étroitement liée. Puis, nous avons considéré le sujet général de la corrélation chronologique des roches des différents pays, question qui présente beaucoup de complexité et de difficulté. Le sujet qui vint ensuite a été le coloriage des cartes et l'unification des figurés. Le Major Powell nous a décrit le plan qu'il a trouvé le plus utile pour les États-Unis, et les objections faites contre son adoption ont été impartialement discutées. Enfin, nous avons eu une description sommaire de la géologie des régions que vous visiterez dans vos excursions. Ceci n'exige aucune remarque spéciale.

Maintenant on peut demander qu'avons-nous accompli, qu'avons-nous résolu? Je pense qu'aucun savant ne fera une telle demande. Les questions scientifiques ne sont pas décidées par le vote de la majorité. [*Applaudissements.*] Nous ne montons pas à l'assaut de la citadelle de la vérité; nous y faisons des approches graduelles. Or, dans les discussions que nous avons écoutées, il n'y a aucun doute que tous les sujets ont été traités sous toutes leurs faces, aussi bien que possible. Des différentes opinions émises, les meilleures seulement survivront, et encore non sans subir d'abord de grandes modifications, ainsi que cela arrive dans chaque évolution. Mais, comme j'ai déjà remarqué dans mon discours d'ouverture, le facteur le plus important de cette évolution est la discussion.

Il convient, maintenant, d'exprimer les sentiments de reconnaissance que nous éprouvons, aux personnes qui ont tant contribué à rendre cette réunion si agréable, et je cède la parole à M. le professeur Hughes.

M. HUGHES s'exprime en ces termes:

M. LE PRÉSIDENT: Vous venez de remarquer avec justesse, qu'un vote de remercîments était dû aux personnes qui ont aidé à rendre ce Congrès ce qu'il a été. Je consens volontiers à prendre ma part à ce devoir, car c'est une tâche bien agréable d'offrir nos remercîments et d'exprimer les sentiments que nous ressentons. Toutefois, il se mêle un certain regret à ce plaisir en ce qu'il nous montre que le moment de notre séparation approche. Nous regrettons aussi que les paroles nous paraissent trop faibles pour faire voir la profondeur de nos sentiments.

A qui donc adresserons-nous ce vote de remercîments? Ce n'est pas à un seul individu qu'il doit être offert, mais à plusieurs. Cependant il y en a un que nous devons premièrement signaler et qui est peut-être le plus digne représentant de l'hospitalité et de l'accueil cordial que nous avons reçus, ainsi que de la liberté de discussion qui a présidé à nos séances. Je parle de son excellence le Ministre de l'Intérieur. C'est un homme si admirablement doué que, lorsque dans la crise suprême de ce pays vous avez eu besoin de quelqu'un ayant l'expérience des hommes, jointe au courage et aux grandes capacités, vous vous êtes adressés à lui. Dans cette crise, l'honorable M. Noble prit les armes, et donna à sa patrie sa force et tout son dévouement. Le sentiment qu'il y avait des hommes aussi braves que patriotes à cette heure sombre, nous fait ressentir un certain plaisir à nous rappeler cette triste et terrible époque. J'espère, et c'est bien sincèrement, qu'il ne repassera jamais par de pareilles épreuves. La question est heureusement résolue, et aujourd'hui il dirige le grand Département de l'Intérieur, ce vaste intérieur que nous allons bientôt visiter. Il a, en outre, sous son contrôle, ce qui nous intéresse le plus, le Geological Survey des États-Unis.

A l'ouverture de ce Congrès, M. Noble est venu généreusement au devant de nous pour nous souhaiter la bienvenue au nom de son Gouvernement. C'est pourquoi j'éprouve une vive satisfaction, en demandant au Congrès de s'unir dans un vote de remercîments à l'honorable M. Noble, et au grand pays qu'il représente si dignement.

M. VON ZITTEL appuie la proposition de M. Hughes, qui est adoptée à l'unanimité.

M. POWELL se lève pour répondre au nom de M. Noble, et prononce les paroles suivantes:

L'Honorable Secrétaire de l'Intérieur m'a chargé de présenter ses hommages aux personnes qui sont ici, et d'exprimer ses vifs regrets de ne pouvoir assister à cette séance; car, au moment où il se préparait à s'y rendre, une importante affaire d'intérêt publique l'a retenu.

Pendant ces trois dernières semaines nous avons eu l'occasion de jouir de plusieurs réunions scientifiques, et nous sommes heureux que le Secrétaire de l'Intérieur a pu prendre part à celle-ci. Pendant ce temps huit sociétés scientifiques ont tenu leurs séances à Washington, et celle du Congrès Géologique International les termine dignement. Des mémoires ont été lus par des savants venus de toutes les parties du monde, qui ont donné lieu à un échange de vues très étendu. Outre ces séances journalières, nous avons eu d'autres réunions dont quelques-unes étaient sociales; et nous sommes fort réjouis d'une occasion aussi favorable de nous rencontrer et de discuter avec des personnes si distinguées dans les sciences, et si éminemment douées de talents aussi grands que variés.

M. TIETZE fait part des regrets qu'éprouvent les membres du Congrès de quitter Washington, où ils ont été accueillis si cordialement et avec tant de bienveillance; et il propose que les remercîments du Congrès soient offerts au président et aux directeurs de la "Columbian University," dont le local a servi aux séances.

M. GREGORY fait quelques remarques à l'appui de la proposition de M. Tietze, qui est adoptée à l'unanimité.

M. CREDNER s'exprime en ces termes:

M. LE PRÉSIDENT: C'est avec un vif plaisir que je me lève pour proposer un vote de remercîments à ceux qui nous ont permis de visiter leurs musées et leurs institutions, ainsi qu'à toutes les personnes qui ont tant contribué à nos plaisirs et à rendre notre séjour à Washington si agréable. C'est d'abord au Directeur du Geological Survey des États-Unis qu'il est dû, pour les nombreuses marques de bienveillance et l'aide qu'il a donné au Congrès; ensuite, aux professeurs Langley et Brown Goode de l'Institut Smithsonien et du Musée National; au Cosmos Club et à l'University Club, qui ont tant fait pour le bien-être matériel, et pour procurer des relations sociales aux membres du Congrès; à la Société Géologique d'Amérique, si jeune et déjà si importante, ainsi qu'à ses directeurs; à M. et Mme. Emmons et à M. Wilson, qui ont eu l'amabilité de nous admettre dans leurs intérieurs et nous permettre ainsi de jouir de la vie intime américaine; au Chargé d'Affaires de Corée et a Mme. Yé. pour leur gracieuse réception; au Comité qui avait arrangé une excursion sur les bords historiques du Potomac, qui restera un souvenir ineffaçable de notre séjour dans cette ville; à M. Frank Thompson, premier vice-président de la Compagnie des chemins de fer de la Pensylvanie, pour le voyage gratuit aux Appalaches qu'il a offert; enfin à tous ceux qui ont contribué aux plaisirs des membres du Congrès, je propose un vote de remercîments.

M. de MARGERIE appuie la proposition de M. Credner, qui est adoptée à l'unanimité.

M. PAVLOW exprime la grande appréciation des membres du Congrès, et l'assurance des souvenirs agréables qu'ils emporteront, des collections qu'ils ont admirées dans les musées, ainsi que des nombreuses marques de bienveillance dont ils ont tous été l'objet pendant leur séjour à Washington. Il termine en proposant un vote de remercîments au Comité d'organisation.

La proposition de M. Pavlow est votée à l'unanimité.

M. GILBERT, répondant au nom du Comité d'organisation, dit qu'en organisant cette réunion le Comité a été guidé par l'expérience acquise dans des Congrès antérieurs où une tendance vers l'évolution s'était déjà manifestée. Dans les réunions précédentes on se flattait de l'espoir d'arriver à établir des règlements uniformes pour la terminologie et la cartographie géologique, ainsi qu'à décider les questions de nomenclature. Un grand nombre de questions ont été soumises au vote et des décisions importantes furent rendues. Mais plus tard, on découvrit que les règles qui avaient été approuvées étaient impuissantes à contrôler la pratique employée par les géologues de tous les pays; et dans bien des questions on trouva impraticable d'arriver à un accord général. Un sentiment conservateur se développa à Londres, qui porta le Congrès à s'abstenir de tout vote sur les questions scientifiques. A cette occasion, ceux qui assistaient aux sessions réalisèrent les grands avantages provenant des rapports personnels et des conférences avec leurs collègues des autres pays, de sorte que l'intérêt de ces réunions ne disparut pas avec leurs fonctions législatives.

C'est pour cette raison que notre Comité prépara un programme qui attirerait la discussion seulement sur des sujets de grande importance, s'occuperait des relations sociales et des conversations scientifiques entre les membres, et, surtout, qui vous procurerait l'occasion de voir une géologie familière aux Américains, mais qu'il était néanmoins désirable de faire connaître au monde entier.

M. de GEER rappelle au Congrès que, quoique M. Pavlow ait déjà exprimé nos remercîments au Comité d'organisation, quelques paroles spéciales étaient dues au Secrétaire et au Trésorier, Messieurs Emmons et Hague, qui ont accompli la plus grande partie de la tâche qui incombait au Comité. Ils ont eu à envoyer des centaines de circulaires aux quatre coins du globe; ils nous ont tenus au courant de tout ce qui allait avoir lieu, et nous ont renseignés sur les moyens de profiter des avantages qui nous étaient offerts. Ils ont travaillé pour nous, et nous n'avons eu simplement qu'à voir, qu'à entendre et qu'à jouir. Enfin, ils ont compris notre mauvais anglais; j'espère qu'ils le comprendront encore, surtout maintenant.

M. STEFANESCU appuie la proposition de M. de Geer, qui est adoptée à l'unanimité.

M. EMMONS, en son nom et pour M. HAGUE, remercie Messieurs de Geer et Stefanescu de leurs paroles bienveillantes. Ils se sont efforcés, en vérité, d'assurer le bien-être des membres du Congrès, et d'anticiper leurs désirs; et si quelque chose leur à manqué, cela est dû plutôt à l'inexpérience qu'à l'absence de bonne volonté. Ils n'ont eu, ni l'un ni l'autre, l'occasion d'assister aux séances des Congrès précédents et il se peut que des membres européens se soient aperçus du manque de formalité et de cérémonial auxquels ils sont habitués. Qu'ils aient la bonté de se rappeler que ce sont nos coutumes américaines, et que si nous attachons moins d'importance au faste extérieur, nous avons le cœur chaud; dans le cas où quelque chose leur aurait manqué à cet égard, nous espérons bien sincèrement que notre accueil cordial fera compensation.

La parole est à M. GAUDRY.

MESSIEURS: Nous regrettons que le professeur Joseph Prestwich, Président du quatrième Congrès international de géologie, n'ait pu venir ici pour installer le bureau du Cinquième Congrès; mais nous sommes heureux qu'il ait délégué à sa place le professeur Hughes, si apprécié par tous les géologues.

Nous regrettons aussi que le professeur Newberry ne soit pas dans cette enceinte pour nous présider, ainsi que nous l'avions espéré. En étudiant les beaux travaux qu'il a publiés récemment sur les poissons fossiles, nous avons pu nous rendre compte de la vigueur de son esprit; malheureusement ses forces physiques ne lui ont pas permis de se rendre à notre appel. Pour nous consoler, nous avons choisi un autre éminent géologue, le professeur Joseph LeConte. Au nom de tous nos confrères, je le remercie du talent et de la bienveillance avec laquelle il a dirigé nos réunions.

Il y a treize ans, nous organisions à Paris le premier Congrès international de géologie; mes amis, Messieurs Delaire et Barrois, qui sont ici et qui étaient secrétaires de ce premier Congrès, pourront vous dire que nous n'étions pas sans inquiétude pour son succès. Grâce à Dieu, grâce à vous, il a complètement réussi. Nous ne devons pas oublier que, si c'est à Paris que le premier Congrès a été organisé, c'est en Amérique qu'on en a conçu tout d'abord le projet. C'est d'ici qu'est partie l'idée généreuse et féconde de réunir ensemble les différents membres de la grande famille des géologues. Je crois être l'interprète des sentiments de nos confrères de tous les congrès internationaux de géologie, soit présents, soit absents, en adressant le plus cordial remerciment aux savants américains, premiers inspirateurs de nos congrès.

Les réunions de Bologne, de Berlin, de Londres ont bien réussi, et celle de Washington n'a pas été moins satisfaisante. A la vérité, nous n'avons pas fait de règlements nombreux. Mais les congrès antérieurs en avaient établi beaucoup. Il en est des règlements comme de la plupart des choses : *un peu faut, pas trop n'en faut.* Prenons garde de trop gêner nos allures personelles. Nous devons respecter la liberté de la science.

Ce qui importe surtout, c'est de rendre la science le plus élevée possible ; notre domaine est immense, puisque nous faisons l'histoire de toute la terre ; nous devons tâcher de donner à nos esprits une ampleur égale à celle de ce vaste domaine que nous sommes chargés d'explorer. Or, par la force des choses, chacun de nous est entraîné à étudier des points spéciaux ; pour faire des œuvres originales, on doit concentrer ses forces sur une partie seulement de la science. Quelques-uns parmi nous se plaisent dans la recherche des vertébrés, parfois étranges ou gigantesques, qui peuplèrent nos continents dans les temps passés ; d'autres s'attachent aux invertébrés, créatures plus humbles, mais qui rendent de plus grands services aux géologues pour la détermination des âges de la terre. D'autres encore considèrent les développements successifs des flores. Ceux-ci préfèrent les terrains primaires, qui doivent nous révéler les origines de la vie ; ceux-là aiment mieux les terrains secondaires ou tertiaires, qui leur montrent le monde dans un état plus avancé, ou même les couches quaternaires, où ils scrutent les mystères des origines de l'humanité. Beaucoup de nos confrères font de la géologie physique ou chimique. Nous avons raison d'opérer cette division du travail, seulement il faut qu'à certains moments nous rassemblions tous les produits de notre activité ; nous devons mettre en commun nos pensées. Chacun de nous est peu de chose ; mais l'ensemble de nos connaissances forme un merveilleux faisceau ; telle est l'œuvre de nos congrès internationaux de géologie.

Et puis, Messieurs, nous avons du bonheur à travailler ; mais nous en avons aussi à nous aimer les uns les autres. Je crois être le plus vieux des géologues qui ont traversé l'Atlantique pour venir à ce Congrès ; j'ai rencontré bien des travailleurs pendant ma vie, et, je vous le déclare dans toute la sincérité de mon âme, plus on voit, plus on connaît les hommes de science, plus aussi on les aime. Nous avions depuis longtemps appris à admirer les géologues américains ; nous venons d'apprendre à les aimer. Soyez sûrs qu'en rentrant dans notre vieille Europe nous emporterons un cher souvenir des membres du Congrès international de géologie de Washington.

M. LeConte répond :

Comme je ne comprends pas bien le français, je n'ai pu apprécier toutes les aimables et flatteuses remarques du professeur Gaudry ; mais, si j'en ai saisi quelques-unes seulement, l'expression de son visage m'a aidé à deviner les autres. L'approbation qui vient d'une pareille source doit être précieusement conservée; une approbation accordée par des hommes que j'ai appris à honorer depuis longtemps, et que je suis fier aujourd'hui de compter parmi mes amis. Si, comme président, j'ai quelque peu réussi à présider d'une manière satisfaisante à mes collègues, ce n'a pas été en raison de mes aptitudes pour cette haute fonction, n'ayant jamais étudié ce que l'on appelle les règlements parlementaires, mais parce que j'ai eu la bonne fortune de présider une assemblée exceptionelle. Nous ne sommes pas une réunion d'hommes politiques, mais un corps de savants. Nous ne cherchons pas nos intérêts personnels, mais la vérité. Nous ne luttons pas afin de remporter une victoire sur nos semblables, mais pour vaincre la Nature et la dévoiler. Comme vous le voyez, Messieurs, il est inutile d'employer un marteau pour accomplir la tâche qui est devant nous, quoique cependant le géologue se serve de cet instrument, mais pour un tout autre usage. Vous appartenez à ceux dont la mission est d'étudier les lois de la nature, qui devient ainsi une loi pour vous. Il m'a donc été facile de présider cette assemblée. Je vous remercie sincèrement de votre cordiale coopération, qui a tant contribué à aider une personne si peu qualifiée à remplir les fonctions que vous avez bien voulu lui confier.

S'il n'y a absolument aucune autre question à soumettre au Congrès, il ne me reste plus que le devoir d'en prononcer la clôture. Nous ne pouvons pas retarder ce qui est inévitable. En terminant, je ne puis m'empêcher de comparer cette organisation à celle du corps humain. Ses éléments se sont rassemblées de toutes les parties du monde pour constituer une unité organisée, exubérante de vie, remplie de joies et d'avantages mutuels, pendant un court espace de temps, pour finir par se dissoudre et rejoindre ensuite les matières dont elle est composée; mais son esprit, du moins nous l'espérons, sera immortel. Je vous annonce donc, Messieurs, la fin du Cinquième Congrès International, et nous formons tous le vœu, qu'il puisse se réunir avec un nouveau corps, dans la Suisse, en 1894.

Les Secrétaires.
F. FRENCH,
J. C. BRANNER.

Le Président,
J. LECONTE.

COMMISSION INTERNATIONALE DE LA CARTE GÉOLOGIQUE D'EUROPE.

PROCÈS-VERBAL DE LA SÉANCE TENUE À SALZBOURG,

LE 3 AOÛT 1891 A 10 HEURES.

[Présidence de M. BEYRICH.]

Présents: MM. BEYRICH et HAUCHECORNE, de Berlin; MOJSISO-
WICS, de Vienne; GIORDANO, de Rome; RENEVIER, de Lausanne.

Et en outre avec voix consultative M. CAPELLINI, de Bologne, président de la Commission internationale de nomenclature.

M. HAUCHÉCORNE expose les nombreux essais qui ont été faits pour
représenter, sur les feuilles d'Angleterre, Scandinavie et Russie, en
même temps, le sous-sol stratigraphique et les terrains superficiels.
Il présente une feuille prussienne sur laquelle il a pu effectuer cette
double représentation.

Sur sa proposition la Commission adopte en principe, pour les vastes
régions où tout est recouvert de terrains superficiels de transport, de
représenter le sous-sol par des hachures espacées, de la couleur du
terrain voulu, sur le fond jaunâtre du Pléistocène, et par lesdites hachures,
encore plus espacées, sur le fond blanc de l'alluvion. Autant que
possible ces hachures seront dirigées horizontalement ou verticale-
ment, pour les distinguer des hachures obliques beaucoup plus fines,
appliquées aux subdivisions des terrains.

La feuille C.IV (N. Allemagne, etc.) est complète à l'exception de la
Scanie, qui doit être exécutée d'après ce procédé. Aussitôt qu'elle
aura été ainsi complétée, on en commencera le tirage définitif. M.
HAUCHECORNE espère pouvoir la faire paraître dans le courant de
l'hiver. Trois autres feuilles C.V (Suisse, etc.), B.III et B.IV (Iles bri-
tanniques et N. France) sont presque prêtes pour la gravure, et pourront
paraître l'année suivante.

La Commission décide de se réunir dorénavant chaque été dans
quelque localité un peu centrale. Elle fixe d'ores et déjà *Lausanne*
pour son point de réunion en août ou septembre 1892. M. HAUCHE-
CORNE s'engage à y présenter les épreuves des trois feuilles C.V, B.III
et B. IV.

M. DAUBRÉE s'étant retiré de la Commission, celle-ci se complète,
sur la proposition de M. RENEVIER, en appelant M. MICHEL-LEVY à y
représenter la France.

Le Secrétaire,
E. RENEVIER, *professeur.*

COMMISSION INTERNATIONALE DE BIBLIOGRAPHIE GÉOLOGIQUE.

[PROCÈS-VERBAL DE LA PREMIÈRE SÉANCE.]

Une première réunion de la Commission internationale de bibliographie géologique a eu lieu au cours de l'excursion dans les Montagnes Rocheuses, le 20 septembre, 1891, dans un des wagons du train spécial, entre Manitou et Denver (Colorado).

Étaient présents: MM. FRECH, GILBERT, GOLLIEZ, DE MARGERIE, REUSCH, STEINMANN, TSCHERNYSCHEW, TIETZE et VAN DEN BROECK. M. H. S. WILLIAMS assistait également à la séance.

M. GILBERT est nommé par acclamation président de la Commission, et M. DE MARGERIE, secrétaire. M. DE MARGERIE centralisera la correspondance pour l'Europe, et se chargera de faire parvenir à M. GILBERT tous les documents destinés à l'impression.

Le secrétaire rappelle que le but à atteindre est triple: Il consiste 1°, à dresser la liste *des bibliographies géologiques qui existent déjà;* 2°, à faire l'inventaire des parties de la littérature spéciale qui n'ont pas encore été l'objet d'un semblable dépouillement méthodique, de manière à arriver à la mise au clair, une fois pour toutes, de la *bibliographie rétrospective*, et 3°, à procéder à l'enregistrement périodique (chaque année, par exemple) de la *bibliographie courante.*

Au sujet de la bibliographie rétrospective, M. GOLLIEZ annonce à la Commission que la commission géologique de son pays est en train de préparer une *bibliographie géologique de la Suisse*, dont l'achèvement demandera probablement quelques années.

M. TSCHERNYSCHEW signale l'existence d'une catalogue de même nature sur le nord de la Russie, travail encore inédit dont il est l'auteur.

Enfin M. VAN DEN BROECK appelle l'attention sur la *Bibliographie générale de la Belgique*, qui comprendra la liste de tous les documents concernant la géologie de cette contrée, publiés dans le cours du XIXe siècle.

Après une courte discussion la Commission décide de limiter provisoirement ses efforts à la préparation de la liste des bibliographies géologiques partielles déjà existantes. Chacun des membres de la Commission devra exécuter la partie du travail relative au pays qu'il représente. Pour l'Espagne, l'Italie et le Portugal, pays n'ayant pas envoyé de représentants à Washington, le Comité s'adressera aux directeurs des services géologiques fonctionnant dans ces trois États.

M. TIETZE veut bien se charger de la bibliographie des Balkans, et
M. DE MARGERIE s'efforcera de combler les lacunes qui pourraient
exister dans la collection de documents réunie par les membres du
Comité, pour l'Asie, l'Afrique et l'Océanie. Les manuscrits devront
être adressés au Secrétaire avant Pâques 1892, pour être imprimés à
la suite du Compte-rendu de la réunion de Washington.

La liste projetée comprendra le titre détaillé des catégories d'ouvrages
suivantes:

1° *Bibliographies régionales ou locales* (exemples: Bibliographie géologique de l'Italie — Bibliographies géologiques des comtés de l'Angleterre, par Whitaker — Catalogue des publications des Surveys américains, par Prime).

2° *Bibliographies systématiques* ou relatives à une groupe de faits
determinés (exemples: Bibliographie des diverses classes de roches,
inserée dans la Pétrographie de Rosenbusch — Bibliographie du Juras-
sique supérieur, par Neumayr — Glaciers, volcans, etc.).

3° *Bibliographies personnelles* (catalogues des publications d'un même
auteur), comme il en est imprimé souvent dans les notices nécrologiques
(exemples: Autobiographie d'Ami Boué — Royal Society's Catalogue of
Scientific Papers).

4° *Catalogues de cartes géologiques* (exemple: Mapoteca geologica
Americana, de Marcou).

5° *Bibliographies géologiques annuelles*, soit générales (exemples: Geological Record — Annuaire géologique, de Dagincourt — Revue de géologie, par Delesse et de Lapparent), soit spéciales (exemples: Revue géologique suisse, par E. Favre et Schardt — Bibliothèque géologique de la
Russie, par Nikitin — Record of American Geology, de Darton).

6° *Tables générales des périodiques spéciaux ou séries* (exemples:
Repertorium des Neuen Jahrbuch für Mineralogie — Index des publications de la Société géologique de Londres, par Ormerod — Table des
Paleontographica — Liste des cartes publiées dans le Quarterly Journal,
de R. Bliss).

7° Enfin, *Catalogues des bibliothèques spéciales* imprimés (exemples:
Catalogue of the Library of the Geological Society of London — Catalogue des ouvrages géologiques se trouvant dans les bibliothèques de
la Belgique, par Dewalque).

Au point de vue de l'extension à donner au travail, la Commission croit
devoir exclure pour le moment les documents purement minéralogiques
ou paléontologiques; les indications ayant trait à la pétrographie, à la
géographie physique, à la géologie appliquée, aux eaux minérales et à
l'archéologie préhistorique seront au contraire acceptées. Toute latitude est d'ailleurs laissée à cet égard aux collaborateurs, la rédaction
se reservant le droit d'ajouter ou de retrancher aux manuscrits en vue
d'assurer à la publication l'uniformité désirable.

Les bibliographies manuscrites importantes, dont l'existence pourrait être connue des membres de la Commission, seront signalées à la place convenable, avec l'indication du nom et de l'adresse de l'auteur.

La publication aura lieu en français; toutefois les manuscrits pourront être rédigés dans la langue du pays d'origine, et traduits ensuite par les soins de la rédaction.

Les titres devront toujours être dans la langue de la publication originale; ils ne seront suivi d'une traduction française que dans le cas où ils appartiendraient à une langue autre que l'anglais, l'allemand, l'italien ou l'espagnol. Les indications de nom d'auteur, de lieu de publication (avec le nom de l'éditeur, s'il s'agit d'un ouvrage séparé), de date, de format et de nombre de pages devront être aussi exactes et aussi détaillées que possible; on indiquera en outre le nombre approximatif d'entrées contenues dans chaque bibliographie, en y joignant quelques renseignements sommaires sur sa nature (exemples: Catalogue alphabétique par nom d'auteurs — Catalogue classé par ordre de dates — Simple liste de titres — Chaque article est suivi d'un résumé — Le nombre des planches n'est pas indiqué — etc.). On notera également les dates extrêmes des publications cataloguées dans les bibliographies (exemple: 1810 à 1885).

Pour faciliter le classement définitif des matières, en vue de l'impression, les collaborateurs sont engagés à préparer leur travail sur des fiches séparées.

Le Secrétaire,
EMM. DE MARGERIE.

TROISIÈME PARTIE.

COMPTE-RENDU DES SÉANCES DU CONGRÈS.

A.—COMMUNICATIONS PRÉSENTÉS PAR LEURS AUTEURS SUR L'INVITATION DU COMITÉ D'ORGANISATION.

    A. La corrélation par plantes fossiles. Par L. F. WARD.

    B. The Pre-Cambrian rocks of North America. By C. R. VAN HISE.

B.—COMPTE-RENDU DE LA DISCUSSION SUR LA CORRÉLATION DES ROCHES.

C.—COMPTE-RENDU DE LA DISCUSSION SUR LA CLASSIFICATION DES DÉPÔTS PLÉISTOCÈNES.

D.—COMPTE-RENDU DE LA DISCUSSION SUR LA COLORIAGE DES CARTES.

E.—COMPTE-RENDU DES EXCURSIONS GÉOLOGIQUES.

A.—COMMUNICATIONS SUR LA CORRÉLATION DES ROCHES.

A. PRINCIPES ET MÉTHODES D'ÉTUDE DE CORRÉLATION AU MOYEN DES PLANTES FOSSILES.*

Par LESTER F. WARD.

Dans tout ouvrage sur la corrélation géologique, soit au moyen des fossiles, soit au moyen de la stratigraphie, la doctrine moderne d'homotaxis doit être, ce me semble, attentivement considérée, puisqu'il est maintenant bien reconnu que les formes identiques n'indiquent pas nécessairement l'identité d'époque.

Dans le huitième chapitre de son grand ouvrage sur la paléontologie M. Pictet pose le principe général suivant: "Les terrains contemporains ou formés à la même époque renferment des fossiles identiques. Réciproquement; les terrains qui contiennent des fossiles identiques sont contemporains." †

M. Schimper, dans son Traité de Paléontologie Végétale, accepte ce principe et l'applique aux plantes fossiles de la manière suivante: "Les terrains contemporains ou formés à la même époque renferment des flores, sinon complètement identiques, du moins homologues, et par conséquent: Les terrains qui renferment des flores identiques ou homologues sont contemporains." ‡

Neuf ans après l'apparition de la deuxième édition de l'ouvrage de Pictet, cité ci-dessus, et sept ans avant l'apparition du premier tome de Schimper, savoir, le 21 février 1862, le professeur Huxley, président de la Société géologique de Londres, exprima dans son discours annuel des idées en grand désaccord avec celles qui précèdent, idées dont on a cependant de plus en plus reconnu la justesse avec les développements successifs de la science géologique. Quoique, dans ce discours, le professeur Huxley ne citât pas les propositions de Pictet, et qu'il se bornât à faire un exposé beaucoup moins absolu de la position des paléontologues, il essaya de résoudre le problème de la prétendue

---

* En anglais par l'auteur. American Geologist, Vol. IX, January, 1892, p. 34.
† Traité de paléontologie, etc., par F. J. Pictet, 2me éd., tome I, Paris, 1853, p. 100.
‡ Traité de Paléontologie Végétale, etc., par W. P. Schimper, tome I, Paris, 1869, p. 100.

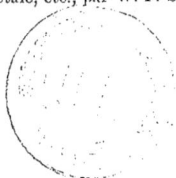

contemporanéité des dépôts qui contiennent des fossiles identiques, et s'exprima dans les termes suivants:

"La succession implique du temps; les parties inférieures d'une série de roches sédimentaires sont certainement plus anciennes que les parties supérieures; et quand on proposa l'idée d'âge comme équivalente de l'idée de succession, il n'était pas surprenant qu'on considérât la correspondance en succession comme correspondance en âge, ou contemporanéité; et, en effet, tant qu'on ne parle que de l'âge relatif, la correspondance en succession est la correspondance en âge; c'est la contemporanéité relative.

Mais il eût valu beaucoup mieux pour la géologie qu'un mot aussi peu exact et aussi ambigu que "contemporain" eût été exclu de son vocabulaire, et qu'on se fût servi à sa place de quelque terme, qui exprimât une ressemblance de relation des séries et qui exclût tout à fait la notion du temps, pour exprimer la correspondance dans la position de deux ou de plusieurs séries de stratifications.

Dans l'anatomie, où il faut constamment parler de correspondance de position, on la dénote par le mot "homologie" et par ses dérivés; et pour la géologie (qui n'est après tout que l'anatomie et la physiologie de la terre) il serait peut-être utile d'inventer quelque mot unique, tel que l'homotaxis (similitude d'ordre), pour exprimer une idée essentiellement semblable."

Le terme "homotaxis," ainsi introduit dans le vocabulaire géologique, a été généralement accepté, et il est employé constamment, même par ceux qui ne se sont pas donné la peine de s'informer de son origine. Les géologues considèrent les rapports stratigraphiques et lithologiques, et les paléontologues les formes organiques qui y ont rapport. Quant à ces derniers, ils sont d'accord, je crois, pour considérer comme *homotactique* * deux flores ou faunes qui possèdent un assez grand nombre d'espèces, ou identiques ou étroitement alliées, ou qui contiennent à un degré considérable les mêmes types de vie.

Je partage complètement l'opinion du docteur Newberry, qu'on pourra donner aux plantes fossiles une grande valeur dans la corrélation des couches géologiques, et aussi qu'il ne restera aucun désaccord entre les fossiles animaux et les fossiles végétaux, quand ces derniers seront parfaitement connus. La difficulté a toujours été que la science de la paléobotanique est dans un état incertain et changeant, et qu'on a manqué de principes précis pour l'application des données paléobotaniques. Je ne prétends pas que cette science soit arrivée a son plus haut degré d'utilité; elle est encore, pour ainsi dire, dans l'enfance. Cependant il existe à présent un assez grand nombre de faits qui en font une aide précieuse de la géologie.

Je me propose, dans ce mémoire, de signaler d'abord les *principes* qui, selon moi, devraient régler l'étude de la paléobotanique comme aide à la corrélation géologique, et ensuite d'expliquer les *méthodes* que j'ai adoptées pour leur application.

---

*La forme "homotactique" paraît être plus régulière qu'"homotaxial," comme plusieurs auteurs l'écrivent.

I. Principes.

Quelque vraie que soit la loi de l'homotaxis, et que des flores semblables puissent avoir fleuri dans différentes parties de la terre à des époques différentes, il est aussi vrai néanmoins que l'existence de pareilles flores dans les différentes parties du monde a une grande importance pour la question de l'âge des stratifications dans lesquelles ces flores se rencontrent. C'est-à-dire que, quoique ces flores puissent avoir fleuri à des époques différentes, la différence entre les époques auxquelles elles ont crû ne peut pas avoir été très grande, car quoiqu'on ne puisse pas affirmer une exacte identité d'âge, on peut cependant dire avec certitude que les stratifications qui contiennent des flores semblables ont nécessairement été déposées à des époques peu distantes l'une de l'autre au point de vue chronologique.

Les grands types de végétation sont caractéristiques des grandes époques géologiques, et il est impossible que les types d'une époque aient existé à une autre époque. Il arrive souvent, par exemple, dans une région très tourmentée, que le géologue stratigraphique se trouve fort embarrassé pour déterminer la position relative de certaines roches. Les géologues ne se fient plus absolument à l'apparence de la roche pour en déterminer l'âge, et les roches carbonifères peuvent avoir une si grande ressemblance avec celles des terrains tertiaires qu'il est impossible de les distinguer lithologiquement. Dans de telles circonstances un seul fossile caractéristique trouvé sur place suffit pour trancher la question. Il se peut que le fossile ne soit qu'un fragment, non déterminable spécifiquement, mais, si sa parenté à un grand type de végétation est certaine, c'est aussi concluant que si on savait à quelle espèce il appartenait.

Supposé, par exemple, qu'on trouve une feuille dicotylédone dans un terrain qu'on supposait auparavant appartenir à la période carbonifère; la seule découverte de cette feuille met le paléobotaniste à même de décider d'une manière certaine que la classification première du terrain était inexacte. En revanche, une seule cicatrice de Lepidodendron ou de Sigillaria trouvée dans une couche, qu'on supposait appartenir à la période tertiaire, ou à la période mésozoïque, serait également concluante contre cette supposition.

Le fameux cas des stratifications de Chardonet, département des Hautes-Alpes, étudiées par Élie de Beaumont en 1828, et attribuées positivement à la période mésozoïque, mais dans lesquelles M. Brongniart trouva des plantes fossiles des genres Calamites, Sigillaria et Lepidodendron, est un des meilleurs exemples de ce principe. Et quoique la science de la paléobotanique fût alors si jeune que Brongniart lui-même était disposé à admettre que ces genres pourraient se trouver dans les terrains mésozoïques, néanmoins, longtemps avant sa mort, on savait que cela serait impossible, et aucun paléobotaniste n'hésitérait à présent à dire au géologue qu'il s'était certainement trompé dans sa stratigraphie.

Mais dans la détermination des couches moins séparées chronologiquement ce procédé n'est pas suffisant, et on ne peut se fier ni à de maigres documents ni à des spécimens uniques et fragmentaires. Dans de tels cas pour être certain il faut avoir une quantité de faits; c'est-à-dire, un nombre considérable de bons échantillons de plantes fossiles pour que le paléobotaniste puisse exprimer son opinion sur l'âge exact du dépôt dans lequel ils se trouvent. La plupart des erreurs sérieuses qu'on a faites, erreurs qui ont fait tomber en discrédit la science de la phyto-paléontologie, ont été causées par l'oubli de ce principe. Les géologues purement stratigraphiques n'ont aucune idée de ces lois, et le paléobotaniste est obligé de traiter leurs idées presque de la manière avec laquelle il traiterait les idées de personnes peu scientifiques. Ils lui apportent sans cesse de simples fragments et des échantillons isolés qui ne sont peut-être pas spécifiquement déterminables, et ils veulent qu'avec de tels matériaux le paléobotaniste leur dise la période précise à laquelle ces fragments appartiennent. Cela est impossible, et le paléobotaniste qui veut tirer des conséquences absolues de tels matériaux est certain de se tromper.

Au contraire, là où il existe une quantité suffisante de faits, la paléobotanique devient aussi concluante pour les formations rapprochées que pour les plus éloignées. Prenons à l'appui de notre thèse les argiles de Gay Head, Massachusetts. Il n'y a dans ce pays aucun endroit plus attrayant pour un géologue stratigraphique que Gay Head. Nous y trouvons des argiles bigarrées, admirablement stratifiées, descendant en pente vers l'océan et sans cesse rafraîchies par l'action de la mer et du temps. Elles forment des falaises presque à pic, visibles à l'œil comme un objet brillant qu'on voit à une grande distance en mer. Un des premiers ouvrages des géologues américains fut d'essayer de résoudre le problème de l'âge de ces falaises. Les annales de ces recherches remontent à cent ans et les noms des géologues les plus célèbres des États-Unis y figurent. Comme résultat de tout ce travail stratigraphique dans un champ si engageant, le professeur Shaler a enfin annoncé à plusieurs reprises dans les deux dernières années, que ces argiles sont de l'âge tertiaire (miocène ou pliocène). Presque au moment où le savant professeur rendit sa décision publique, un jeune paléobotaniste, Mr. David White, visita cet endroit et passa un été à y faire une collection de plantes fossiles. On y avait préalablement trouvé quelques fragments, dont un ou deux avaient été figurés dans les ouvrages du docteur Hitchcock. Mais ces échantillons peu importants n'avaient pas de valeur. Les spécimens n'étaient pas nets, et on n'en pouvait rien conclure. Ce jeune homme fit une grande collection de beaux spécimens de plantes fossiles. Il les envoya à Washington, les classa et les détermina. Ils se trouvent être ni plus ni moins que les types des argiles d'Amboy dans le New Jersey, et ils représentent par conséquent le milieu de la période crétacée.

Cette découverte a fait plus que démontrer que les stratifications (celles à plantes du moins) à Gay Head sont de l'âge crétacé; car elle établit aussi que les argiles d'Amboy dans le New Jersey, en s'étendant à l'est et en reparaissant à Long Island, continuent encore plus loin dans cette direction et se trouvent probablement au-dessous de la plupart des dépôts glacials de Block Island, des Elizabeth Islands, de Martha's Vineyard et de Nantucket. C'est ainsi que la paléobotanique a résolu une question à laquelle la géologie stratigraphique n'aurait probablement jamais répondu. Il y a beaucoup d'autres exemples de ce principe, mais celui-ci doit suffire.

Il y a aussi à considérer un autre principe, qu'on ne peut négliger et dont l'oubli a souvent induit en erreur non seulement la géologie mais aussi la paléontologie. Il est impossible, il faut l'avouer, d'évaluer trop haut l'importance de la classification exacte et systématique des plantes fossiles. Les doutes qui existent sur la vraie nature de beaucoup d'objets végétaux trouvés dans les couches de la terre ont produit un grand scepticisme dans l'esprit de bien des gens relativement à la valeur de la paléobotanique. Les botanistes en particulier, qui se sont occupés des plantes fossiles, sont généralement bien désappointés. Accoutumés à voir devant eux la structure entière de la plante, toutes ses parties et tous ses organes, non seulement ceux de la végétation mais aussi ceux de la reproduction, ils n'ont que peu de patience pour étudier les matériaux qui constituent la plupart des collections de plantes fossiles. Les géologues sont ordinairement disposés à approuver les opinions des botanistes et à partager avec eux leur disposition à condamner la paléobotanique.

Il y a deux réponses à toutes ces objections. Il y a une réponse au botaniste, et il y en a une tout autre au géologue. La réponse au botaniste est que quand on se rend compte des conditions dans lesquelles on trouve ces échantillons, ils contiennent en réalité une grande quantité de renseignements, assez exacts et assez certains relatifs à l'histoire passée de la végétation. Outre le fait avéré qu'on trouve à certains endroits de la terre des flores fossiles presque égales en nombre aux espèces vivant actuellement dans les mêmes localités, il faut encore ajouter la raison suivante, que la paléobotanique nous apprend à étudier plus soigneusement les restes fragmentaires que. nous trouvons; elle stimule nos facultés d'observation des faits que nous possédons, et elle a contribué beaucoup à notre connaissance de la botanique proprement dite. Il est dans l'habitude des botanistes, par exemple, de reproduire les feuilles d'une manière si négligente que le paléobotaniste n'est pas en état de décider à quel genre elles appartiennent.

Cela tient principalement à ce qu'ils ignorent généralement la nervation exacte des feuilles, et qu'ils se bornent à les reproduire au point de vue artistique, tout simplement pour l'effet. La paléobotanique a enseigné aux botanistes que la nervation des feuilles est importante et qu'ils devraient la représenter dans tous les cas possibles. Nous sommes redevables aux plantes fossiles de la découverte que la nervation a un rang générique, tandis que la forme, à laquelle les botanistes se fient surtout, n'a généralement qu'un rang spécifique. Par conséquent les feuilles qui ont une nervation visible ne sont pas inutiles dans la détermination des espèces; elles ont au contraire beaucoup de valeur, et dans beaucoup de cas on peut par elles préciser les genres avec certitude.

En continuant à répondre au botaniste, on peut aussi insister avec justesse sur ce point, que, dans le cas de presque toutes les formes douteuses aussi anciennes que les terrains crétacés, il n'est pas à supposer qu'on puisse déterminer les genres par une comparaison avec les genres des plantes vivantes. On doit compter que les genres que nous avons à étudier dans ces stratifications anciennes sont éteints, et tout ce qu'il faut rechercher avec soin c'est l'évidence de leur parenté avec les plantes modernes.

La réponse au géologue est encore plus concluante; à vrai dire, il n'a le droit de faire aucune objection quelle qu'elle soit. Il lui est en effet indifférent que le paléobotaniste ait donné le nom exact à une forme ou non; cette question est importante au point de vue biologique seulement, mais pas du tout au point de vue géologique. Mais ce qui est important, géologiquement parlant, c'est que la forme dont il s'agit soit distinctement reconnue, qu'elle soit soigneusement représentée et que ce qu'on a trouvé soit bien exprimé en une langue exacte et expressive. Il faut qu'il n'y ait aucun doute possible sur son identité, quand la même forme reparaît dans une localité différente. C'est là un point essentiel pour le géologue. Si la forme, quelle que soit sa vraie nature, a quelque chose de défini et de distinct, quelque chose qu'on puisse reconnaître partout et n'importe où, et si elle est caractéristique d'un certain horizon ou d'une certaine localité elle devient proportionnellement utile pour fixer l'âge de quelque autre dépôt où elle se trouve. Si elle ne se trouve que dans deux localités ou à deux points à la surface de la terre, on en peut conclure légitimement, quoique pas d'une manière absolue, que les deux terrains sont d'un âge assez semblable, la preuve du contraire faisant défaut. Mais si l'objet est abondant et caractéristique de quelque groupe ou de quelque horizon bien connu, c'est alors qu'il devient d'une grande importance comme fossile caractéristique, indépendamment du degré de notre connaissance de sa vraie nature botanique.

II. MÉTHODES.

Je me propose en deuxième ligne d'indiquer la méthode générale que j'ai adoptée pour l'application de ces principes à la corrélation géologique à l'aide des plantes fossiles.

Généralement parlant, cela consiste, sans doute, à comparer des flores semblables et à en déduire une similitude d'âge. Mais il faut se rendre compte de beaucoup de circonstances limitantes. Si les localités où se trouvent des flores semblables ne sont pas trop éloignées géographiquement, on peut en déduire avec plus ou moins de certitude que leur âge n'est pas très différent. Par exemple, quand nous trouvons que la flore du terrain houiller de Richmond est très semblable à celle du terrain houiller de la Caroline septentrionale, la conclusion que ces deux terrains houillers sont d'un âge semblable est tout à fait légitime. Et même quand nous trouvons, en grand nombre, dans les Trias du New Jersey, du Connecticut et du Massachusetts les mêmes espèces que nous rencontrons en Virginie et dans la Caroline du Nord, la conclusion que les couches qui contiennent ces plantes ont été déposées à peu près à la même époque depuis le Massachusetts jusqu'à la Caroline du Nord, ne peut pas être très inexacte.

Au fur et à mesure que ces flores semblables s'éloignent géographiquement l'une de l'autre, la probabilité d'une similitude d'âge et de dépôt s'affaiblit, mais elle reste forte tant que les deux localités sont sur le même continent, ou tant qu'il y a des indications qu'une mer continue s'étendait autrefois depuis l'une jusqu'à l'autre, ou qu'il existait des lacs ou des estuaires semblables dans les deux parties du continent à la même époque.

Le même cas se présente quand nous comparons le Trias des États de l'Est avec celui du Nouveau Mexique, de l'Arizona, et de l'Amérique centrale. Quoique les flores de ces deux parties éloignées du continent américain soient assez différentes, des géologues éminents ont argué que la présence d'un grand nombre d'espèces identiques et d'un faciès semblable dans le type des plantes indiquent l'existence précédente d'une grande mer triasique d'un âge presque uniforme. Cette mer s'étendait depuis le Nouveau Mexique jusqu'au Honduras, et était d'un âge géologique peu différent de celle qui couvrait la partie orientale du continent, depuis le Massachusetts jusqu'à la Caroline du Nord. On pourrait faire application de ce même principe à beaucoup d'autres époques dans l'histoire géologique.

Quelle est donc la méthode spécifique adoptée lorsqu'on compare les flores? On peut la définir brièvement: Préparation et discussion des tableaux de distribution proportionnelle. Ainsi que je l'ai déjà fait remarquer, plus la flore d'un groupe est complète, plus la comparaison sera exacte. Par conséquent la première chose à faire est de dresser une liste complète de toutes les plantes fossiles trouvées dans le groupe dont il s'agit. On peut en premier lieu regarder cette liste d'espèces comme tout à fait inconnue géologiquement.

S'il y a plusieurs localités, terrains, ou bassins distincts qu'on suppose être d'un âge semblable, il faut d'abord énumérer séparément les espèces ou les formes qui se trouvent dans chacun d'eux et faire des comparaisons pour déterminer à quel degré elles sont identiques ou semblables pour les florules différentes. C'est ce que j'ai appelé, dans mon ouvrage sur le Trias, *distribution américaine.* Le nombre de formes communes à deux de ces terrains indiquera la ressemblance botanique de ces deux florules. Ainsi, dans la flore triasique des États-Unis à présent connue, le tableau suivant indique le nombre des espèces communes à deux ou plusieurs bassins:

	Terrains.			
Terrains.	New Jersey et Pennsylvanie.	Virginie.	Caroline du Nord.	Nouveau Mexique et Arizona.
Vallée du Connecticut...	5	5	6	1
New Jersey et Pennsylvanie..	......	7	10	2
Virginie..		......	20	2
Caroline du Nord...			......	2

Mais, puisque le nombre des espèces qui se trouvent dans les différents bassins est très différent, ces chiffres pourrraient laisser une impression fausse. Ce qu'on désire apprendre c'est la prépondérance relative dans une florule quelconque des espèces communes aux autres florules. La seule manière d'exposer ces rapports est de se servir d'un tableau proportionnel ayant cent pour base. Pour les bassins triasiques des États-Unis je présente ces renseignements dans la forme suivante:

Bassins ou terrains.	Trouvées dans—	Limitées à—	Communes à—et à quelque autre bassin.	Pour cent dans les autres bassins.
Vallée du Connecticut.	23	13	9	39
New Jersey et Pennsylvanie	18	5	13	72
Virginie	56	34	22	39
Caroline du Nord..	52	25	27	52
Nouveau Mexique et Arizona	13	11	2	15

Étant donné le fait bien connu qu'un terrain nouvellement exploré fournit des plantes fossiles dont la majorité des formes est inconnue à la science, il faut admettre que la proportion d'espèces communes ainsi mise en lumière est assez grande pour justifier la conclusion qu'elles établissent la contemporanéité complète ou approximative du dépôt, en faisant exception pour le bassin occidental.

La seconde chose à faire est de déterminer combien d'espèces ont été trouvées dans d'autres localités et sous d'autres horizons. C'est ce que j'ai nommé leur *distribution étrangère*. Pour faire ressortir ces détails on rédige un tableau divisé en colonnes pour les différentes localités étrangères. Ces colonnes sont arrangées dans l'ordre géologique ascendant, et contiennent les noms des espèces trouvées dans la localité à comparer. Alors on indique par quelque marque caractéristique dans des colonnes convenables les espèces des autres localités et des autres horizons qu'on a déjà décrits. On enregistre ainsi l'histoire géologique de chaque espèce sur la ligne même où l'espèce se trouve indiquée.

Un tel tableau de distribution des espèces d'un groupe qu'on étudie est excessivement simple et élémentaire, et jusqu'ici n'exige aucune explication. Mais il y a d'autres circonstances à considérer. Dans toutes les formes inférieures, comprenant principalement les cryptogames, les cycadées et les conifères, il faut se rappeler en première ligne que la connaissance de la nature de ces formes anciennes ne suffit pas pour affirmer avec certitude leurs rapports génériques. Elles sont pour la plupart éteintes et portent en conséquence les noms de genres éteints. En deuxième lieu, comme tous les paléontologues le savent aujourd'hui, les formes anciennes de vie sur le globe étaient moins définies, ou selon le langage scientifique moderne, moins complètement différenciées, que dans les temps actuels. Par conséquent on les transpose sans cesse d'un genre à un autre, et d'une famille à une autre, au fur et à mesure du développement de l'évidence scientifique.

Or, il ne reste aucun doute que les formes de vie plus récentes, soit animales, soit végétales, se sont développées des plus anciennes, et ces phases de transition dans les annales paléontologiques corroborent cette vérité beaucoup plus fortement qu'aucun fait dans l'existence des faunes et des flores vivantes du globe. En admettant alors que les flores postérieures dérivent des antérieures, il est de la plus grande importance, non seulement pour le botaniste, mais aussi pour le géologue, de remonter aux relations de parenté des plantes, ce qu'on peut faire avec beaucoup de succès. C'est pourquoi, dans la préparation d'un tableau de distribution proportionnelle, il ne faut pas se borner exclusivement aux espèces qui se trouvent dans le groupe à comparer. En effet ces formes anciennes sont si variables qu'il serait impossible de le faire avec certitude. On s'égarerait beaucoup si on se laissait guider exclusivement par les noms des espèces. Les formes diffèrent dans les diverses localités, mais les divergences sont si légères que le jugement personnel du descripteur invaliderait un tel calcul. Les uns rapprocheraient les formes semblables des différentes localités, les autres les sépareraient comme des espèces distinctes, et l'histoire de la nomenclature de la paléobotanique n'est qu'une série de ces déterminations, apparemment contradictoires, mais qui en réalité ne font que montrer que les formes sont plus ou moins étroitement alliées, quoique elles ne puissent jamais s'accorder à tous les égards.

C'est donc ici la partie difficile dans la préparation d'un tableau de distribution, de savoir choisir non seulement les espèces que tout le monde admet être identiques sous deux ou plusieurs horizons, mais aussi celles qu'on croit alliées à celles du groupe dont il s'agit. Il y a danger, d'un côté, d'omettre des espèces alliées importantes, et d'autre part, d'introduire comme alliées des espèces qui n'ont en réalité aucune relation entre elles. Sans insister sur les détails de cette partie difficile du travail, il faut supposer que le paléobotaniste, s'il est habile dans son art, possède le jugement nécessaire à distinguer les espèces vraiment parentes de celles qui ne le sont qu'apparemment. Aux espèces dont tout le monde admet l'existence sous plus d'un horizon ou dans plus d'un endroit et que nous appellerons brièvement, espèces *identiques*, il faut donc maintenant ajouter les espèces qui ont des rapports avec celles du groupe, et que nous appellerons espèces *alliées*. Mais c'est introduire ici un nouvel élément dans le tableau, puisqu'il est évidemment impossible d'indiquer ces rapports par la méthode employée pour indiquer l'identité. Il faut se servir d'une ligne séparée pour chacune de ces espèces alliées; il faut aussi employer une colonne spéciale à gauche pour indiquer leurs noms. Il est alors possible d'employer pour les espèces alliées la même marque que pour les espèces identiques, et de faire la distribution dans les colonnes de la manière ci-dessus décrite.

Un tableau comme celui-là est utile quand on veut suivre une espèce particulière et déterminer sa présence dans les autres localités et sous les autres horizons. Mais il n'offre pas une vue d'ensemble des rapports de la flore; il faut condenser les détails qu'il contient sous une forme plus commode.

La première mesure à prendre pour arriver à une telle condensation serait probablement d'arranger toutes les espèces selon l'ordre géologique ascendant des formations dans lesquelles elles se trouvent aussi. Il est instructif dans un tel tableau de montrer les espèces identiques et alliées séparément pour chaque formation; c'est-à-dire, de classer ensemble toutes les espèces qui se trouvent dans la première ou plus basse formation, dans la deuxième, dans la troisième, etc., et ainsi de suite pour l'étendue entière du groupe. De cette manière on met en pleine lumière les horizons sous lesquels le plus grand nombre d'espèces se trouvent, et cela non seulement par le nombre même des espèces, mais aussi par le nombre proportionnel de celles qui leur sont identiques avec celles qui ne leur sont qu'alliées.

On peut généraliser cette condensation encore davantage en réduisant la liste des espèces à la forme numérique; c'est-à-dire, en donnant le nombre qui se trouve sous chaque horizon sans écrire les noms des espèces. De même que dans le tableau précédent les mêmes espèces se répètent plusieurs fois, ainsi dans le tableau actuel les chiffres des colonnes comprendront ces duplications. Jusqu'ici nous n'avons considéré le sujet qu'au point de vue géologique, mais il est aussi important pour la géologie que pour la botanique de faire quelque classification des types principaux compris dans une flore quelconque. Le dernier tableau étant bien court, il est possible d'y introduire une classification de cette nature. En discutant la flore triasique, que j'ai prise comme base de mes remarques sur la méthode poursuivie, j'ai exposé respectivement dans ce tableau le nombre de fougères, d'équisétacées, de rhizocarpées, de cycadées et de conifères, trouvées sous chaque horizon, ceux-ci étant les seules types représentés dans la distribution étrangère de cette flore.

Le mode de raisonnement sur l'âge d'une formation par cettte espèce d'évidence est important à considérer. La manière ordinaire est de ne préparer qu'un tableau semblable en étendue à celui qui a été décrit d'abord, souvent sans y tenir compte des espèces alliées, et alors de continuer par la discussion de chaque espèce dans ses rapports avec l'âge du groupe. Les données ainsi envisagées ne sont pas absolues mais relatives, et les conclusions qu'on en tire peuvent quelquefois induire en erreur. J'ai souvent fait remarquer que les erreurs importantes commises par Heer et par Lesquereux en plaçant les couches à plantes américaines trop haut dans l'échelle géologique étaient dues à cette manière de raisonner. Ces géologues comparaient la flore du groupe du Dakota, celle du groupe Laramie et toutes les flores plus récentes des États-Unis, avec celles du Miocène d'Europe et du Tertiatre des régions polaires. Ils insistaient fortement sur le fait qu'il se trouvait dans ces formations des espèces qu'on ne pouvait que difficilement, ou pas du tout, distinguer des espèces américaines. Un tel argument a peu de valeur à cause de l'abondance immense des flores tertiaires qui étaient en question. La flore tertiaire d'Europe comprend des éléments des flores antérieures. Cette flore est si bien conservée que le nombre de ses formes pré-tertiaires, qui persistent dans le Tertiaire, est beaucoup plus grand que le nombre de formes pré-tertiaires qu'on a trouvées jusqu'ici dans les formations plus basses, lesquelles ont fourni relativement peu de plantes. En conséquence une comparaison du groupe américain Laramie, par exemple, avec le Tertiaire européen seulement, sans considérer la flore crétacée européenne, et sans noter la continuation des types crétacés dans l'époque tertiaire, était bien décevante, et finit par créer l'impression générale, encore répandue en Europe, que notre groupe Laramie est de l'âge tertiaire. Ayant remarqué cette fausse argumentation, j'ai été le premier, et à ma con naissance je suis encore le seul, qui aie essayé de donner une énumération des espèces crétacées des plantes fossiles dans le but d'en faire une comparaison avec celles du groupe Laramie des États-Unis.*

Le système que je viens de décrire empêche cette erreur. Il compare en effet des formes déterminées avec celles de toutes les formations parmi lesquelles se trouvent quelques unes de ces espèces quelles qu'elles soient. Par conséquent, quand on veut en tirer des déductions on se trouve constamment arrêté dans la considération d'un horizon particulier par les faits relatifs aux horizons supérieurs et inférieurs. Par exemple, en traitant de la flore triasique dans le tableau numérique ci-dessus décrit, si on limitait son attention ou à l'Oolite ou au Lias on pourrait en conclure que l'une ou l'autre de ces formations était voisine de celle á déterminer. Mais en jetant un regard tout le long de la colonne jusqu'au Rhétique on y trouve une proportion beaucoup plus grande des mêmes espèces, soit identiques, soit alliées. Cette ressemblance était si frappante que le professeur Fontaine en conclut que le Rhétique devait être considéré comme le plus proche voisin du terrain houiller de Richmond, quant à l'âge géologique. Mais les recherches plus récentes de Stur, recherches dont le professeur Fontaine ne pouvait avoir connaissance, et qui ont été faites dans le Keupérien de Lunz en Autriche et dans les autres flores keupériennes d'Europe qui sont presque du même âge, surtout celles de Raibl en Carinthie et de Neue Welt en Suisse, les susdites recherches ont démontré que la flore keupérienne d'Europe, quoique beaucoup moins abondante, contient un plus grand nombre de formes triasiques américaines que la flore rhétique de Franconie, de Scanie, de Brunswick, etc. Conséquemment, quoique très peu de formes américaines se trouvent sous aucun horizon plus bas que celui-ci, nous sommes cependant obligés de conclure que ce keupérien supérieur de Lunz en Autriche ressemble plus à celui des dépôts triasiques à plantes américaines qu'aucun autre sur la terre.

On pourrait maintenant se demander, si en réalité tout cela prouve quelque chose. Quand deux flores aussi anciennes que le Trias et aussi éloignées l'une de l'autre que le sont l'Autriche et la Virginie s'accordent d'une manière si remarquable dans les formes qu'elles contiennent, est-il légitime d'en conclure que la date de déposition de l'une a été la même que celle de l'autre? Certainement non. Et cependant si de pareils faits ne prouvent pas qu'il y a eu sur les deux rives de l'océan

---

*Synopsis of the Flora of the Laramie Group. Sixth Ann. Rept. of the U. S. Geol. Surv., pp. 445–514.

Atlantique une époque qu'on peut regarder à tous les égards comme simultanée, il faut avouer que toute évidence paléontologique est sans valeur. La vérité qu'il faut toujours se rappeler, c'est que la corrélation établie par des données de cette sorte est homotactique, mais qu'elle n'est pas nécessairement chronologique. Il y a peut-être des raisons qui ont fait que les mêmes types se sont produits à une époque plus récente dans une de ces localités que dans l'autre, mais la science jusqu'à présent ne nous donne aucune explication de la cause de ces avances ou de ces retards. Une chose que nous savons d'une manière certaine, c'est que des flores semblables ont existés à un moment donné dans deux parties de la terre très-éloignées l'une de l'autre, et, en attendant que quelques autres faits capables de détruire cette conclusion soient découverts, il est non seulement sans danger mais utile au géologue de considérer les deux dépôts comme appartenant au même âge géologique. Il y a certaines limites entre lesquelles cela doit être vrai, et quand il admet ces limites le paléontologue peut tirer ses conclusions avec autant de confiance qu'il le pouvait avant que la loi d'homotaxis ait été formulée.

B. THE PRE-CAMBRIAN ROCKS OF NORTH AMERICA.[*]

By C. R. VAN HISE.

Occupying vast expanses of country in North America is a basement crystalline complex. In many areas, between this basal complex and the Cambrian are great thicknesses of sedimentary rocks. In numerous regions these rocks have been divided into series, and in certain of them the series have been separated into formations. The clastics of one region have often been equated with those of other regions. It is the purpose of this paper to very briefly summarize the facts as to the occurrences of the pre-Cambrian rocks, to state the methods which have been used in the past in their subdivisions and correlation, and to indicate the nature and limitations of the criteria which experience has shown to be most valuable in this work.

Names applied to pre-Cambrian rocks.—In the early days of American geology Lyell's name Primary or Primitive was more widely applied to the ancient rocks than any other. Among the older geologists this name, including under it in a general way the pre-fossiliferous or metamorphic rocks, was used by Akerly, Alexander, Booth, Dewey, Ducatel, Eaton, Emmons, Hitchcock (Edward), Jackson, Mather, Mitchell, Percival, Rogers (H. D.), Rogers (W. B.), Silliman, Tuomey, Vanuxem, and others. It was nearly universal in 1820, and was applied as late as in the forties.

The term Primitive in the United States was gradually superseded by Azoic. Used by Adams as early as in 1846, in the literature of the fifties and sixties it very widely occurs, and has not yet disappeared. Among more prominent geologists in whose writings it may be found are Adams, Cook, Crosby, Emmons (E.), Frazer, Hitchcock (C. H.), Hitchcock (E.), Kerr, Rogers (H. D.), Safford, Whitney, Wadsworth, and others. In its earlier use Azoic was made to cover all rocks which were apparently destitute of life, without reference to whether they were older than the fossiliferous rocks or not. It was thus applied by by Adams and the elder Hitchcock. With Rogers the Azoic included non-fossiliferous rocks which are younger than the Hypozoic or gneissic

---

[*] This paper is condensed from the final chapter of a correlation paper upon the Archean and Algonkian, published as a Bulletin of the United States Geological Survey. If it is desired to verify any of the statements made, this may be done by referring to the summary of literature contained in the Bulletin.

110

series proper. Ordinarily, however, the term was used to cover all pre-Silurian rocks, the Silurian being then regarded as the base of the fossiliferous systems. It was thus applied by Foster and Whitney to the Lake Superior region, and the Azoic was held by them to be structurally indivisible. While the rocks of the Primitive and Azoic were early subdivided into lithological divisions, there was little or no attempt to apply stratigraphical methods to them. Later the Azoic was subdivided by certain geologists into Laurentian, Huronian, etc.

The work of Logan and Murray marks in America the beginning of a truly structural study of the ancient rocks. They found in different places in Canada pre-Cambrian rocks which they mapped in detail. The two areas in which this work was begun were the north shore of Lake Huron and the Laurentide mountains. With scientific spirit they applied to the rocks of these areas no terms which involved any theory of origin or equivalence, but gave the rocks the name of the localities, in this following one of the fundamental principles of good structural work. Having no fossils for guides, they built up a succession on the north shore of Lake Huron by following formations in continuous exposure, by lithological likenesses of exposures separated by short intervals, by a like order of formations in different localities, and by the use of an unconformity which was held to occur between the so-called Huronian sediments and the underlying crystalline rocks.

In Logan's work upon the Laurentian the same methods were used as far as practicable, but on account of the complicated structure of the region his success was here much less conspicuous. The difficulty of the district drove Logan to take the one characteristic formation, the limestones, as horizons to follow and to serve as planes of reference in working out the structure. But even this guide was not a certain one, as Logan never became quite sure as to the number of limestones present. As the study of the Laurentides continued the rocks were divided into two divisions, a lower Laurentian free from limestone, and an upper Laurentian containing the limestones. The two were held by Vennor, and by Selwyn for a time, to be unconformable. As the area studied in the Laurentide mountains widened, a new formation was found, a laminated gabbro. It was recognized as being largely composed of labradorite or anorthite and so was first called anorthosite or Labradorian, and afterwards Norian. The contacts of this formation with the other formations of the Laurentian were recognized as not those of conformity. In these early days it was naturally supposed that all laminated rocks, whatever their character, were sedimentary, and as in certain places the Labradorian appeared to cut across or overlap the old Laurentian it was designated as upper Laurentian, and what had before been called upper Laurentian was designated middle Laurentian. When the eruptive character of the Labradorian was shown, the Canadian survey returned to the first uses of the terms Upper Laurentian and Lower Laurentian.

In comparing the Huronian and Laurentian, it appears that the principle used in reaching the conclusion that the Original Upper Laurentian, separated by a great distance from the Original Huronian and nowhere in contact with it, is the older, was the metamorphic character of the former as compared with the latter, which in the early work of Logan and Murray was called a non-metamorphic series. The lithological likeness of the gneisses and granites of the Lower Original Laurentian to the granites and gneisses, called Lower Laurentian, unconformably underlying the Huronian doubtless was the reason for placing these as equivalents.

In the later work of Logan and Murray the names Huronian and Laurentian were applied to regions far distant from the original areas, the guiding principles for so doing being wholly lithological likeness and degree of metamorphism. Working under these principles, as granites and granite-gneisses were so abundant in the original Laurentian and were totally absent in the original Huronian, it became customary with these authors to refer to granitoid areas as Laurentian, while sedimentary series containing quartzites, limestone, or dark fine-grained schists were referred to the Huronian, and this reference was frequently made when the series as a whole was very much more crystalline than the Original Huronian. The only exception to a reference of all pre-Cambrian rocks to the Huronian and Laurentian were the series now known as Keweenawan and Animikie. These were recognized as resting unconformably upon the so called Huronian of Lake Superior, while the Keweenawan was seen to be of a wholly different lithological character from the Lake Huron rocks. These series were called the upper copper-bearing series, the Original Huronian often being called the lower copper-bearing series.

We find these two geologists, Logan and Murray, starting with scientific principles, laboriously studying year after year the detailed occurrences of the rocks in the midst of a forest-covered wilderness, until their inductions built up the Original Huronian and Laurentian series. In their later work of a very much less detailed character over vast areas the terms were applied somewhat indiscriminately, and in such a way as to imply that below the upper copper-bearing rocks there are only two systems, one of which is equivalent to the Original Huronian and the other of which is equivalent to the Original Laurentian.

The terms Huronian and Laurentian were gradually adopted by geologists working on the United States side of the boundary, so that in recent years, with the exception of Archean, they have been the most widely used of all terms for designating the ancient rocks. If the masters, Logan and Murray, departed in their later work from strict scientific methods in their use, this departure was as nothing compared with the extremes to which later geologists of America have gone. By many geologists coarse-grained granites and granite-gneisses were designated as Laurentian, without much reference to evidence as to whether they were intrusive rocks of far later age. In applying the term Huronian the methods followed were even worse. Sometimes authors took a green color to be a characteristic feature of the Huronian, and here referred all the green schists; others took a laminated structure to be characteristic of the series, and here referred all the laminated rocks, including even coarse-grained laminated gneisses; another took the volcanics associated with the Huronian to be its characteristic feature and so called various pre-Cambrian volcanic series Huronian; others regarded metalliferous rocks as the important feature of the Huronian.

Lying at the root of all this work is the assumption that rocks of certain kinds are characteristic of a definite period of the world's history, and that if rocks could be found, which were really like the Huronian and Laurentian in lithological character, they should be referred, respectively, to these series.

As to the relations between the Laurentian and Huronian, it was plainly believed by Logan, Murray, and the other early geologists that where the two came in contact they were unconformable, although oftentimes the structural relations which obtained were admittedly obscure. That was the position of the present director of the Canadian survey as late as 1879, who used the term Huronian not only to include rocks that had been before here placed, but to cover all of the Upper Laurentian and the Upper Copper-Bearing series, thus greatly expanding the system. In recent years this survey has held that the Huronian and Laurentian are always conformable, and that often the former grades downward into the latter, and this is the position which has been taken by many geologists of the United States, both in the East and in the West.

The Lake Superior region furnishes a rather marked exception, as do certain others, to the indiscriminate and unwarranted use of the term Huronian. This region is so near and is connected in such a way with the original Huronian of Lake Huron, that it was possible to make a strong case of probability in favor of the equivalence of the clastic rocks of the two regions. The Lake Superior Huronian was divided into formations upon the same principles used in mapping the original Huronian. While the term Laurentian was applied to the pre-Huronian rocks on the north shore of Lake Huron and about Lake Superior, it was recognized by a number of geologists that this was a variation from its application in the original Laurentian area.

As geological knowledge increased, and as the theories involved in

451 GE——8

the terms Primitive and Azoic were more and more attacked, in order to avoid a theory of origin, the term Archean was proposed for the ancient rocks by Dana in 1872. This term rapidly grew in favor. By its use not only the advantage of a theory of origin was avoided, but in common with Primitive and Azoic it was not necessary to subdivide the ancient rocks into Laurentian and Huronian, and thus imply a correlation with the rocks of other regions. In the early rapid work of the far West, detailed observations usually stopped at the base of the fossiliferous series, and it was convenient to regard all the remaining rocks as a unit, and to cover this unit the term Archean was adopted. After a more detailed study of certain regions, the terms Laurentian and Huronian were applied to subdivisions of the Archean. This term Archean also found early favor by the Canadian survey to include these two divisions of pre-Cambrian rocks.

Eozoic was another term suggested to replace Azoic, when it was thought by many that the rocks once supposed to be destitute of life were not really so. This was used to a considerable degree in the sixties and seventies, and retains a place in literature to the present time. This term implies a theory just the opposite of Azoic.

As already said, the theory involved in referring all pre-Cambrian rocks to the Laurentian and Huronian is that there was in pre-Cambrian times this invariable succession. The theory was carried to the extreme by Hunt and his school, who held that before Cambrian time there are six rock systems which are universal and are separated by unconformities. These are, from the base upward, Laurentian, Norian, Arvonian, Huronian, Montalban, Taconian. Of these terms Norian was devised to include the laminated gabbros, the so-called Upper Laurentian of Logan. Arvonian was imported from Wales, where it was applied by Hicks to a series of acid volcanics. Montalban came from the White Mountain region in New Hampshire, where a series of gneisses were thought to be of different lithological character from the Laurentian and Huronian and to overlie them. Taconian was introduced by Ebenezer Emmons to cover a series of fossiliferous rocks which was supposed to be earlier than the base of the Silurian.

Besides the terms given, others have been used to some extent, but they are of little importance. Among these may be mentioned Hypozoic, Prozoic, and Pyrocrystalline.

As the metamorphic theory gained force it became the habit of many geologists to refer to old crystalline or semicrystalline rocks as metamorphic, assuming that they are all produced by the alteration of sediments of some kind. This went so far as to include perfectly massive rocks, such as diabases, gabbros, granites, etc., among the metamorphics. Recently the term has also been applied to rocks recognized as laminated eruptives, but this is not the use referred to. This term metamorphic had the advantage of saying nothing as to age or correlation, but in escaping this difficulty another theory was accepted which, so far as its assumption is concerned, was quite as bad.

In many cases local names have been applied to formations or series in order to avoid any theory of age or correlation. The most conspicuous example of this kind is that of the Keweenaw series of Lake Superior. More recently Lawson has proposed the terms Keewatin and Coutchiching for certain series northwest of Lake Superior, and to include these two he proposes the Ontarian system. In the Grand Canyon the local names Chuar, Grand Canyon, and Vishnu have been applied to pre-Cambrian series which there occur. Comstock has proposed the terms Burnetian, Fernandian, and Texian for series which are found in Texas.

This tendency to return to the use of local names in recent years is plainly a reversion to scientific methods which were never departed from by certain geologists. This class have declined to use any term for the ancient rocks which involves a theory of origin or succession, but have divided the rocks which they found in their respective districts into lithological divisions or into local formations. Conspicuous among early geologists of this class are Jukes, Percival, and Lieber.

Recently Irving has proposed that there be placed below the Paleozoic group another group of coordinate value, for which the term Agnotozoic or Eparchean is suggested. This term cuts out of the Archean a large class of rocks which have before been here included. Finally, the name Algonkian has been brought forward by the U. S. geological survey for a systematic place opposite Agnotozoic or Eparchean.

In the following discussion Cambrian is defined as extending downward to the base of the Olenellus fauna. The pre-Olenellus clastics and their equivalent crystallines are called Algonkian, and the completely crystalline rocks below the Algonkian are denominated Archean. The reasons for these usages will appear in the following pages. The stratigraphical terms, group, system, and series correspond with the usage proposed by the International Geological Congress. The same is true of the chronological divisions, era, and period. Formation is a lithological subdivision of a series.

The character of the Archean.—There is an essential unity in the character of the complex of rocks which is the oldest known in America. This statement covers all the areas in which the rocks are demonstrated to be exceedingly ancient. It includes the basal complex of Arizona, between which and the Tonto sandstone is a clastic system 15,000 feet thick, separated into three series by unconformities, and these again separated from the Tonto above and the basal complex below by great unconformities; it includes the basal complex of the Wasatch and certain of the ranges of Nevada, between which and the Olenellus Cambrian is a great unconformity and a thick series of quartzites; it includes the basal complex of southwestern Montana, between which and the Olenellus Cambrian is 12,000 feet of unaltered slates and a thick series of crystalline rocks of clastic origin, the two being probably separated by a great unconformity; it includes the basal complex of Texas, between which and the Cambrian is an unconformity, at least one, and perhaps two, thick series of clastic rocks; it includes the basal complex of the Lake Superior region, between which and the Potsdam sandstone is an enormous system of clastics many thousands of feet thick, separated by unconformities into three series, and the whole bounded above and below by unconformities; it includes the basal complex of the north shore of Lake Huron, between which and the Cambrian is a clastic series 18,000 feet thick, bounded above and below by unconformities; it includes the basal complex of the Original Laurentian area, between which and the Cambrian is a clastic series estimated to be many thousands of feet thick; it includes the basal complex of Hudson's Bay, between which and the Cambrian are almost certainly two, and perhaps three, series of clastics separated by unconformities; it probably includes the basal complex of Newfoundland, between which and the Olenellus Cambrian is a series of clastics 12,000 feet thick, and above this a great unconformity; it includes much of the great area of northern Canada known as Laurentian, between which and the Cambrian in various districts are clastic series.

In all of these regions, in which the basal complex is vastly older than the Cambrian, it consists of a most intricate mixture of nearly massive rocks, among which granite and granite-gneiss are predominant; of gneissic and schistose rocks, all of which are completely crystalline, and so folded and contorted that nowhere has any certain structure ever been made out over considerable areas. The granites and basic eruptives may occupy considerable areas; the gneisses may be regularly laminated and grade into the granites; the crystalline schists may occupy the outer zones of an area; they may all be confusedly intermingled, schists, gneisses, and granites alternately predominating; sometimes the schistose rocks appear in dike-like forms in the granites, at other times the massives are in dike-like forms in the schists; at still other times the alternations of granite, gneiss, and schists are quite uniform and persistent for considerable areas. The granites usually show a rough lamination, which may not appear in hand specimens but which is evident in large masses.

The minerals in the granitic rocks generally show evidence of dynamic action; they do not have the clear cut, definite relations characteristic of the later plutonic rocks. In the chief mineral constituents of the rocks there is essential uniformity in all of the areas, although certain less common minerals may be found in one area which have not been discovered in another. Orthoclase and acid plagioclase feldspar, quartz, hornblende, muscovite, and biotite are the standard minerals. To describe accurately the appearance of the rocks of the basal complex is exceedingly difficult, but any one who examines a series of specimens from the various areas will perceive the truth of the statement made as to the essential likeness of the rocks from different regions. A suite from any one of the regions which has been personally examined, if unlabelled, could by no possibility be asserted not to come from any other.

The unparalleled intricacy of the structure of this complex; the generally laminated arrangement of its parts; and the broken and distorted forms of the constituent minerals are evidence of repeated dynamic movements of the most powerful character. Further, the basal complex is not only recognized by its positive but its negative characters. Nowhere in it is a persistent thick formation of quartz-schist (although vein-quartz is abundant), of limestone or marble, of a graphitic schist, of a conglomerate either volcanic or sedimentary. If sandstones and limestones, or other surface materials have been a part of this system, the profound and varied mutations through vast lapses of time have wholly obliterated all evidence of their presence.

Besides the areas mentioned in which these most ancient rocks occur, there are many other areas in which there are, between the Cambrian and the basal complex, great series of clastic rocks, although the evidence at hand in favor of vast age for the basement complex is less than in the cases cited. In a third class of areas no definite evidence in the nature of intervening series shows that, between the Cambrian and the basal complex, has intervened an era or even a period.

Because of the unique lithological character of this fundamental complex in all these regions, and because of the essential likeness in structure prevailing, we have ground for grouping these rocks together, whether exactly of the same age or not. Lithological arguments for correlation may well be distrusted; but the exceedingly strange, varied and complex lithological and structural characters of this system, the like of which we have no evidence has been duplicated anywhere in later times, is an argument of great weight. In the complexity of its parts and the implications of its structure, it gives evidence of vast antiquity.

In Algonkian, Cambrian, Silurian, Devonian, and even later times, completely crystalline schists have been produced over large areas; but, while often in these series no evidence now remains of clastic characters, they rarely, if ever, closely resemble this fundamental complex. A clastic series was in the beginning of its history of necessity a shale, a sandstone, a limestone, a chert, or some other form of sediment, often containing carbonaceous material. Cementation, metasomatism, dynamic action may have profoundly changed any of these deposits. A limestone may have been transformed into a crystalline marble, or, if impure, into a hornblende schist containing scarcely a remnant of original carbonate. A cherty carbonate of iron may have become an actinolite-magnetite-schist. Carbonaceous shaly material may have become a graphite-schist, but if such a rock is represented in the fundamental complex, what has become of the carbon? A sandstone may have become a granular quartzite or a foliated, micaceous quartz-schist. But that a great quartzite formation like those of the Huronian of Lake Superior, or the pre-Olenellus of the Wasatch, can have become wholly obliterated by any process short of fusion is almost inconceivable. It has been said that in this fundamental complex, throughout its whole vast area, there are found none of these rocks. In its positive as well as its negative qualities it is a unit. While it can not be considered demonstrated that all of its areas are of the same age, it may then be accepted that in North America is a system of granites, gneisses, and crystalline schists which are the oldest rocks of North America, and which have representatives in many areas throughout the United States, although most widespread and abundant in Canada. That such a basal system exists is no new idea; but it has not generally been recognized that between it and the Cambrian there elapsed an era in which were alternating cycles of the deposition of systems of rocks and of vast erosion intervals.

As here used the term Archean is restricted to this fundamental complex. It is no longer possible to regard as a unit or treat together all the pre-Cambrian rocks. The rocks included in the fundamental complex are everywhere called Azoic or Archean. The crystallines and semi-crystallines above this complex, often called Archean, must be distributed from the Devonian or earlier to the pre-Cambrian. It is clear that if Archean is to remain a serviceable term it must be restricted to some unit. Such a unit is the fundamental complex, and to it this term is most appropriate.

Origin of the Archean.—As to the origin of the Archean rocks, three different views are prominent: (1) The Archean may be considered as metamorphosed detrital rocks; (2) It may be considered as igneous, but later in origin than certain of the pre-Cambrian clastics with which it is in contact; (3) it may be considered as igneous and representing a part of the original crust of the earth, and therefore earlier than any sedimentaries. A modification of this theory is suggested under the topic "Delimitations of the Archean" (p. 123).

1. Those who believe in the detrital origin of the Archean, as above

defined, will not question the conclusion reached as to the age of the fundamental complex; for to produce results so different from any known metamorphic clastic series, must not only imply great age, but probably sediments which were originally deposited under different conditions than those of later times. This school, while believing in the detrital origin of the Archean as a whole, are conscious that it has been cut again and again by eruptives of all kinds; that the supposed clastics have thereby been profoundly metamorphosed by contact and dynamic action, and oftentimes have been so changed that the place can not be pointed out where the intrusives end and the clastics begin.

If this sedimentary view of the origin of the Archean is correct, as no universal break in geological continuity can be accepted, it should be found that between the Archean and the clastics there are somewhere gradations. The manner in which the finely laminated schists and gneisses vary into the coarsely granitoid phases has been admirably described by Jukes in the rocks of Newfoundland; by Lieber and E. Emmons in the rocks of the southern Appalachians; by Hitchcock in the rocks of Massachusetts and Vermont; by Marvine and Stevenson in the rocks of Colorado; by King, Hague, and S. F. Emmons in the rocks of the Fortieth parallel, and by Lawson in the rocks about Lake Superior. Most of these writers and many others, including Selwyn, approaching the problem from the side of the clastic rocks, have regarded the coarsely crystallized rocks as produced by metamorphism, although in the more granular rocks the process has gone so far as to produce aqueo-igneous fusion. Those who have maintained this origin for these rocks have recognized the fact that they have locally acted as eruptives, although in general the material is thought not to have moved far. Marvine so clearly saw that the facts could be explained in two ways that he says that while he regards the whole as metamorphosed sedimentary rocks, another observer, approaching the field from a different direction, where the evidences of intrusive nature are most manifest, would reach the conclusion that the whole is eruptive. Hitchcock and Stevenson and most of the others are in practically the same position.

2. All or a part of the Archean is considered as of igneous origin, but later in age than the pre-Cambrian clastics. The facts of those who have described downward gradations from unmistakable clastics into a crystalline complex have by this school been differently interpreted. It has declined to apply the term metamorphism to a product which has become fluent, and has insisted upon the essentially igneous character of it. Lawson is conspicuous as having recently strongly put this side of the question; but it is noteworthy that Winchell, belonging to the first school, and Lawson to the second, have had essentially the same facts before them, both having done their work in the

same region. The difference is one of definition and emphasis rather than ideas. Both schools regard the granite-gneiss as material which has resulted from a change in the condition of original sediments and has not moved far.

This theory that the Archean, or a part of it, is the liquefied floor of the clastic rocks has the objection that it is an unverified hypothesis. When once a sedimentary rock has become fluent and is wholly free to newly crystallize, how shall the material be identified? To state that such material has not moved far is pure assumption. If the fusion theory is true the average composition of the unfused part of the clastic series and the subjacent material should agree. To obtain average analyses of rocks which vary widely in mineral character within short distances is not easy, but is a thing which must apparently be attempted if this theory is to be maintained, for the writer sees no other way in which an attempt can be made to verify the hypothesis.

Another class of geologists, noting the contact relations between the granitic rocks and the clastics, hold that the former, called by others Lower Laurentian or Archean, are eruptives of later age than the clastics with which they are in contact, without attempting to give any theory as to the source of the material. Here are included Hawes, Hall, Mather, Foster, Whitney, Wadsworth, Rominger, Herrick, and others. Rominger distinctly recognized the granites and granite-gneisses of this kind on the south shore of Lake Superior as the subjacent rocks upon which the Huronian rests. Herrick saw the same relations with reference to his granitic and schistose groups on the north shore. In the last two cases the facts before the writers are precisely the same as those of the geologists of the second school; but by giving no explanation of the source of the material for the granite-gneisses, they have escaped the difficulty of the unverified assumption that these eruptives represent fused sediments. They fail to explain what has become of the floor upon which the clastics are deposited. Some floor they must have had. Where these eruptive contacts are found the floors have disappeared, and if so, the eruptives, if extraneous, must be considered to have eaten up or absorbed it.

3. That the Archean is an igneous rock earlier than any of the sedimentaries is apparently the position of E. Emmons, Lieber, and others. These careful observers not only maintained the igneous origin of the granite-gneiss of the southern Appalachians, but traced the gradations between basic schistose rocks and massive eruptives, including hornblende-schist in unmistakable dykes, and drew the correct conclusion, lately regarded as a new discovery, that such rocks are sometimes metamorphosed eruptives. While the major portion of the granite-gneiss and associated rocks were considered older than the oldest clastics, later intrusives of a similar character were recognized. This theory

that the fundamental complex is igneous is apparently that of Geikie, as to the Archean of Great Britain, and of many German geologists, as to the basal complex of Germany, among whom Lehmann and Roth are conspicuous. Indeed, the latter maintain the igneous origin of all the pre-Cambrian rocks, and Geikie* says of the true Archæan of Great Britain that it not only contains no material which gives any evidence of ever having been sedimentary material of any kind, but it farther contains no material which can be considered a surface volcanic, while in it there are many rocks which are certainly plutonic eruptives.

The geologists of this third school, with the second school, recognize the igneous character of the granite-gneisses having irruptive contacts with the clastic series, but they decline to recognize these rocks as Archean. Such rocks are eruptives. Their age is to be designated precisely as are eruptive rocks which cut Cambrian, Silurian, or Devonian strata.

As bearing in favor of the really igneous character of the Archean is the fact that no case has been demonstrated, except possibly that of the marbles, of the production of a perfectly massive crystalline rock from a clastic without intervening fluidity. Metamorphism, whether the original rock is a massive eruptive or a stratified sedimentary, produces a laminated or schistose rock. If a granitic structure can be taken as evidence of eruptive origin, and we know many eruptive rocks do have such a texture, a very strong case can be made for the eruptive origin of the larger part of the fundamental complex. The line of argument is precisely analogous to that by which the whole has been held to be sedimentary. There are complete gradations from the most completely schistose and laminated phases to the most massive phase. Also bearing in favor of a truly eruptive character for the basal complex is the fact that the rocks referred, in the first part of this section, to the Archean are more nearly simulated by igneous rocks which have eruptive contacts with ancient clastics than by any recognizable metamorphosed sedimentaries.

It may be said that the actual gradations between the Algonkian and Archean in certain places are evidence that the latter are not igneous rocks earlier than the former; that gradations can be explained between subsequent intrusives and clastics, but not between igneous rocks and sedimentaries of later age. It has, however, been shown that as a consequence of powerful dynamic action two unconformable series, one of which is composed of materials from the other and therefore resembles it in composition, may have developed conformable secondary structures and gradations consequent upon the induced crystalline character of the clastic series, the original structures being simultan-

---

* Recently Geikie has announced the presence of possible clastics in one area closely associated with the Hebridean gneiss.

eously obliterated. Also, recently, Pumpelly has ascertained that sub-aerial disintegration of the earlier series is a powerful assistance in the production of such gradations.

Whatever the origin of the fundamental complex, it is plain that the parts of any given area of it are not all of the same age. The dikes which everywhere cut it are the pipes through which have passed the later eruptives. At the time of the intrusion of these eruptives, large lakes of liquid material may have formed which crystallized as bosses, causing the Archean to contain considerable masses of rocks of really later age. Where these rocks are predominant, the material must be classified as a later eruptive; where they are subordinate to the Archean material, they are often difficult to separate from it, although really later in age. Between the areas which rank as eruptives of later age and the genuine Archean, there are doubtless gradations. Along the zone of contact, if the mass of later eruptive be great, there might be an area which could equally well be placed with the fundamental complex or with the later eruptive. Between the Archean and later eruptives, as between the Archean and undoubted sedimentaries, there are gradations.

The problem of the relations of the Archean, as a whole, to the overlying clastics is the same as that within the Archean itself. The finely laminated crystalline schists and gneisses, and the granite-gneisses and granites with which they are associated, have contacts in every respect analogous to those occasionally found between the Archean complex and the clastic series. For example, the rocks heretofore called Archean on the north shore of Lake Huron comprise two parts. One part is older than and lies unconformably below, yielding fragments to, the Original Huronian. The other part has relations with the clastics of the character just considered, with transition phenomena. If this material is an extraneous intrusive it is a post-Archean eruptive. If *in situ* it represents a portion of the pre-Huronian floor completely metamorphosed by selective metamorphism, or by aqueo-igneous fusion, it can fairly, according to the first and second schools, be called a part of the Archean.

It is plain from the great diversity of opinion as to the origin of the Archean rocks, and from the fact that many of the opinions are beliefs rather than verified conclusions, that we have no definite knowledge upon parts of the subject. That there are comparatively few or no wholly massive rocks in this complex is precisely what would be expected under any theory. Its history is too long. Whether originally igneous or aqueous, it could not be hoped that there would be found the characteristic lithological forms of igneous or aqueous agencies. Many or all of these rocks have not only been subject to the movements which have taken place since Paleozoic time, but to the movements which have occurred in the far greater length of previous time—if not too deeply buried to be beyond the influence of the outer foldings, in which case they were buried beyond the crushing strength of rocks, were latently plastic, and were probably at a high temperature. If originally massive and igneous in the ordinary sense, dynamic action has obliterated the regularity of the arrangement of the constituent particles and has given them a more or less laminated or schistose structure. If sedimentary, all trace of that original sedimentary structure has been obliterated by the repeated foldings, contortions, and perhaps high degree of heat, to which they have at various times been subjected. Of a necessity, through this complex have passed all subsequent eruptives. Doubtless, at various places and times in its history, parts of it have become practically liquid, and from this condition it has again crystallized in the forms characteristic of eruptives.

Delimitations of the Archean.—It is generally accepted that the Archean has no limit downward. It is the oldest system and surely includes, if such rocks exist, all of the original crust of the earth. But as denudation progresses material far within the earth approaches its surface, not by intrusion, but by gradually rising as a whole. Before reaching the surface the material has become crystalline. This original crystallization may have taken place in or even later than Algonkian time; hence, if these rocks are to be considered as belonging to the age in which they crystallized, the Archean grades below into the Algonkian, even as it is believed in places to grade above into the Algonkian. The truth of this position is not lessened by the fact that the earth, as a whole, subject to sudden strain, acts as a rigid body. Even if true, it is equally certain that the crust of the earth, under continued strain, adapts itself to it, thus showing real plasticity. But in any case it can not be assumed that the rock material deep within the earth, under pressure far beyond the crushing strength of any known material and at a high temperature, exists as crystallized minerals. We only know that it has these forms when the material rising by erosion nears the surface.

The upper limit of the Archean is not easy to define, and the task is rendered more difficult because geologists are not agreed upon the origin of the Archean. If either the sedimentary or the subcrustal fusion theory of its origin be accepted, there will be found gradations from rocks constituting the ancient complex described, to rocks having like relations with clastics of Algonkian and post-Algonkian time. Upon either of these theories, if sedimentary rocks are only buried deep enough they will pass into crystallines by progressive metamorphism or by subcrustal fusion, just as do rocks of Cambrian and post-Cambrian age. This the elder Hitchcock so clearly saw that he distinctly said that the Laurentian granites and gneiss of Vermont are probably, in part at least, not older than the fossiliferous series. If the Archean be made to include all the thoroughly crystalline rocks below pre-Cambrian clastics, it includes rocks the age of which varies from Algonkian to pre-Algonkian. The anomaly is perhaps best met by making a more or less arbitrary division between Archean and post-Archean crystallines. The natural theoretical plane to choose is the beginning of life; that is, to include in the Archean all truly azoic rocks. While this suggestion has a plausible sound, we must believe that the dawn of life was very gradual and that its traces in its early stages are exceedingly sparse, so that there would be great, if not insuperable, difficulties in its practical application.

If the third theory, that the Archean includes only pre-sedimentary rocks, be correct, its upward limit is easy to define; the Algonkian begins at the time of the deposition of the first sedimentaries. But there are those who deny the existence at the present time of any such ancient rocks, although such concede the existence at one time of such rocks, and that the Archean as thus defined represents a vast lapse of time in the history of the earth. This denial of the present existence of any rocks of greater age than the oldest sedimentaries is of course a pure unverified assumption, defended on the ground of probability. If the original crust of the earth be defined as including more than the first outer skin, it is a question whether the converse proposition is not equally probable. Even if the position be true, the school that believe in the igneous origin of the Archean would still have a large mass of rocks for the Archean by shifting their grounds so as to include in it all the material which, in the slow process of inward crystallization, has now reached the surface of the earth, not by intrusion in the rocks above, but by erosion. This position would, however, be controverted by those who regard such rocks as plutonic and belonging to the age of their equivalent sedimentaries. But in the nature of the case it is not possible to designate the particular age to which these rocks belong. That there exists upon the surface of the earth a part of the original crust of the earth, or its downward continuation by later cooling, can hardly be doubted; and since these can never be assigned to any definite period of sedimentation, they might well be considered as Archean. At any rate, they are a class by themselves which, if not here placed, can not be referred to any of the geological periods. Further, this class of rocks, when in contact with detritals of whatever age, by the very hypothesis of their origin must rest unconformably below them. The coincidence that so frequently, if not always, between the ancient sedimentaries and the basal complex there is a great hiatus, might be urged as evidence of the truthfulness of this hypothesis.

The banded and contorted granite-gneiss, which serves as a background for the Archean, may not improbably be the part which has the origin above suggested, while the other parts of the complex may be due to subsequent intrusives, the whole being kneaded into their present extraordinary complex relations by repeated dynamic movements and other metamorphic influences. This igneous theory of the origin of the Archean, modified so as to include the pre-sedimentary original crust, if any remains, and the deeper crust which has reached the surface by denudation, perhaps more nearly covers the facts than any other as to the relations of the Archean to subsequent rocks, its complex lithological character, the relations of the rock phases to each other, and the long history written in the strained, altered, and broken mineral constituents. It accords with the idea held by Irving, Bonney, and others, that this earliest crystalline complex was produced under conditions differing from those of the rocks of any subsequent period.

While it is impossible to make a wholly satisfactory theoretical definition of the Archean, it is frequently easy in the field to say with a great degree of probability what rocks are Archean and what post-Archean. For instance, in the Arizona region, above the typical Archean complex, there is the most profound unconformity, upon the upper side of which the rocks are readily recognized clastics. From the writer's point of view the same thing is true for a large part of the Lake Superior region. In the Uinta and Wasatch mountains, again, below the quartzites of probably Algonkian age is a great unconformity, and then appears the implicated Archean. In certain other regions the separation of the Algonkian and the Archean is a matter of exceeding difficulty. As representative of this class of cases may be taken the Front range of Colorado, along the east side of which are unmistakable clastics, with an apparent gradation between them and the crystalline complex. In the Appalachians, again, where for the most part the oldest clastic rocks recognizable are Cambrian, it can not be said whether the crystalline complex below is Algonkian or Archean. Here the separation of the Cambrian from the pre-Cambrian has only been accomplished by a minute and laborious study. The two appeared to be in conformity and to grade into each other. It is only recently that this gradation has been shown by Pumpelly to be consistent with a great unconformity between the two. The causes producing these gradations between the Cambrian and pre-Cambrian in Massachusetts (post-Cambrian dynamic action and pre-Cambrian disintegration) may also be found to explain the conformities and gradations between the Algonkian and Archean.

While it is then not easy to define the Archean, it is plain that the discrimination in the field between Archean and Algonkian is a real one and should continue to be applied, even if its exact theoretical meaning can not be said to be certainly known.

Stratigraphy of the Archean.—In characterizing the Archean the methods applicable to its subdivision are clearly pointed out. If no part of it is demonstrably sedimentary, structural methods can not apply. Its subdivisions must be made upon purely lithological grounds. If a part of it is demonstrably eruptive, the relative ages of these parts may often be ascertained, and all are necessarily newer than the part not recognized as eruptive, else these could not be shown to have this origin. Many attempts to apply stratigraphical methods to parts of the Archean have been made, but they have not thus far been successful. Such attempts have been based upon the belief that foliation represents sedimentation; but, even working upon this erroneous basis, it has been stated that the structures are so complicated that little progress has been made. So far as attempts to apply stratigraphical methods to this fundamental complex are concerned, the conclusions of Whitney and Wadsworth are very largely true. If their review of the Azoic rocks had been confined to this basement system, which is perhaps truly Azoic, the conclusion as to its indivisibility on a structural basis would have plausibility.

Necessity for a group between Cambrian and Archean.—The Olenellus fauna is taken as the base of the Cambrian. The reasons for thus delimiting the Cambrian below are fully considered by Mr. Walcott[*] and will be summarized on a subsequent page. His results are accepted. The Cambrian fauna in development is far, some biologists say nine-tenths of the way, up the life column. This statement, if accepted, implies a prior life of vast duration. Just as another period of life has succeeded the Cambrian, another has preceded it. The progress of paleontologic knowledge has of late been downward. Before there was a recognized Cambrian there was a well known Silurian, and it is probable that, when all parts of the world become geologically known, other faunæ will be discovered below the Cambrian, as distinctive in character as the Cambrian is from the Silurian. If this be done, definite information will be available to correlate rock series of different parts of the world in the time place between the Cambrian and Archean. If the condition of the globe were such that life existed in pre-Cambrian time, it also was such that stratified rocks could be deposited not unlike those of later times, so that the only question which arises is whether any of these stratified rocks now remain in such a condition as to be recognizable. Such intervening clastic series do exist below the Olenellus fauna in many regions in North America, and in some cases the volumes of rock and great intervening erosions represent a lapse of time which not inaptly may be compared with all subsequent time. If geological history were to be divided into three approximately

---

*Correlation Papers—Cambrian, by C. D. Walcott, 1891. Bulletin 81 U. S. Geological Survey, 447 pp. 3 pls.

equal divisions, these divisions would not improbably be the time of the Archean, the time of the clastic series between the Archean and the Cambrian, and the time of the Cambrian and post-Cambrian.

It is imperative that some term shall be available to cover the great mass of rocks between the Cambrian and Archean. Irving was the first to realize and urge the necessity for such a term, and proposed for it Agnotozoic. This term implies the existence of life in this system, and the evidence upon this point is conclusive. Life is indicated by the presence of thick beds of graphitic limestones, beds of iron carbonate, and by great thicknesses of carbonaceous shales which are represented by graphitic schists in the more metamorphosed phases of the rocks. It has been urged by Whitney, Wadsworth, and others that the limestone and graphitic schists may have an origin other than organic. Whitney and Wadsworth have gone so far as to say that there is no valid evidence of life in any pre-Potsdam rocks. This was, however, before it was definitely recognized that the Potsdam is Upper Cambrian and that an abundant Cambrian life extends far below. If it were true that these limestones and ore beds are no evidence of life, and it may be admitted that another origin is possible without implying that it is probable, it will hardly be maintained that the hydrocarbons, which occur so abundantly in the little metamorphosed shales of the Huronian about Lake Superior, are other than of organic origin; and if so, the graphitic schists, which stand in the same great system in the geological column, are in all probability but these hydrocarbonaceous slates in more altered condition. However, we are not obliged to depend upon the presence of these varieties of rocks as the only evidence of life. Whether the *Eozoon Canadense* found in the Original Laurentian of Canada is of organic origin will not be here discussed. Its literature is voluminous, and it is a question which concerns the paleontologists. It is doubtless true that many of the specimens which have been called Eozoon are results of the forces of crystallization; but admitting this, it does not follow that all of the material called Eozoon is of this character. Passing by this question, the pre-Cambrian fossils described by Walcott in the Grand Canyon of the Colorado include—

A minute Discinoid or Patelloid shell, a small Lingula-like shell, a species of Hyolithes, and a fragment of what appears to have been a pleural lobe of the segment of a trilobite belonging to a genus allied to the genera Olenellus, Olenoides, or Paradoxides. There is also an obscure Stromatopora-like form that may or may not be organic.

A lingula-like shell has been found by Winchell in the pipestones of Minnesota. Selwyn has described tracks of organic origin in the Animikie (Upper Huronian) series of Lake Superior. Murray, Howley, and Walcott found several low types of fossils in the pre-Olenellus clastics of Newfoundland.

That these fossils are of organic origin can not be doubted. But while many will admit the clastic character of the great groups of rocks considered, and the organic origin of the forms mentioned, as well as the carbon of the carbonaceous shales and schists, they will say that these are merely evidences that the rocks in which they lie are Cambrian. The reply to this is that it is a question of nomenclature. If it is premised that all clastic and fossiliferous rocks more ancient than the Olenellus horizon are Cambrian, it is useless to try to prove that there are pre-Cambrian clastic rocks which bear life. It is, however, necessary to recognize that the carrying downward of the term Cambrian to cover not only the great thicknesses of rocks which are now included but all pre-Olenellus clastics, will probably make the Cambrian as great as, or greater than, all the subsequent periods put together. That this is inadvisable is plain, and the clastic rock masses below the Olenellus fauna are so enormous that the proposal to introduce a general term like Agnotozoic as the equivalent of Paleozoic, Mesozoic, Cenozoic, to cover this great group, is a conservative one. Irving foresaw that the term would be objected to, because sooner or later the life will become to a greater or less degree known, and he suggested as an alternative for Agnotozoic, Eparchean, in contradistinction to Archean, which was reserved by him to cover the fundamental complex. As the character of the life of this group is already beginning to be known, it seems to me that the term Proterozoic, considered for the place by Irving but rejected, is preferable to either Agnotozoic or Eparchean.

In a conference of the members of the United States Geological Survey called by the Director at Washington, these terms were discussed with reference to atlas-sheet mapping, although there was no question on the part of any one as to the necessity for some such term. Recognizing the impracticability of correlation of the different pre-Cambrian clastic series of different provinces, and recognizing the fact that for use in mapping a uniform plan must be adopted, it was suggested that a term of the same class as Cambrian, Silurian, and Devonian should be selected for rocks here included, and to occupy this place the term Algonkian was proposed and accepted. The proposed general scheme of classification for the lower part of the geological column is then as follows:

Paleozoic.....................	Carboniferous, Devonian, Silurian, Cambrian.
Agnotozoic, or Proterozoic......	Algonkian.
Archean	Archean.

The introduction of the term Algonkian has been objected to by some on the ground that it will supersede the older term Huronian. In answer to this it may be said that Huronian has not been generally used as Algonkian is defined, and it therefore does not supersede this term. Huronian will be retained for certain of the clastic series of Lake Superior and Canada, as well as for rocks in an equivalent position in other parts of North America and Europe, if such equivalence can be determined, just as before Algonkian was introduced. The Huronian will stand as one of the great series of rocks which together make up the Algonkian.

Delimitations of the Algonkian.—The farther back we go in the history of the world for any given region, the more frequent have been the changes through which a rock stratum has passed, and therefore there is increasing difficulty in determining bounding planes with sharpness, although in different regions rocks of the same degree of metamorphism may differ vastly in age. The truth of this is well illustrated by comparing the eastern and western regions of the United States. In the former, powerful dynamic movements have occurred until late in Paleozoic time, as a result of which the Cambrian, Silurian, and Devonian rocks over large areas have not been separated from the pre-Cambrian. How much more difficult would one expect it to be to separate the pre-Cambrian clastic series from the Archean? In parts of the west, where no close folding has occurred since Cambrian time, it is easy to separate the Cambrian from the pre-Cambrian; and in regions in which metamorphosing influences have not been at work for a still longer time, it is easy to separate the pre-Cambrian clastics from the Archean. But in other regions this separation is made with the greatest difficulty, and doubtless over large areas this will never be satisfactorily done. Just as in the Appalachians, in parts of which it may be impracticable to separate the Cambrian rocks from the pre-Cambrian clastic series, if such exists, so it will be for a long time impossible to decide in some regions upon sharp boundary lines between the pre-Cambrian clastics and the Archean. Giving full force to this position, it is no reason why the discrimination should not be made where it can be.

Recent work in petrography has demonstrated that dynamic movements and environment, not time, are the important element in the obliteration of clastic characteristics. Dynamic movements also destroy the evidences of discordances between series where there have been real unconformities. This destruction of the evidence of structural breaks comes about largely as the result of a real parallelism of bedding, caused by the close folding, but far more important than this is the production of a common cleavage and foliation with the simultaneous development of crystalline schists from the newer series. As a consequence basal conglomerates are oftentimes almost the only means

451 GE ——9

of discriminating between the newer and the older series, and if the metamorphosing influences are powerful enough to destroy the pebbled character of such beds, changing them into schists or gneisses, as has occurred in many places in the Cambrian of the Appalachians, this means of detecting a break between series is also lost. The problem is rendered still more difficult because of the fact that oftentimes when there is a real unconformity there has been originally no basal conglomerate. At many localities in the Far West the basement fossiliferous series are built up of the constituent minerals of the underlying rocks rather than of large fragments of them, and, even when not folded, have sometimes so closely simulated the original rocks that geologists have been at a loss to determine at what plane the clastics end and the crystallines begin. If it has proved difficult to separate the unfolded clastics from underlying crystallines, how much more difficult must it of necessity be to separate series that have together been subjected to intense and perhaps repeated dynamic actions.

The Algonkian has been defined as including all recognizable pre-Cambrian clastics and their equivalent crystallines. In the consideration of the character, origin, and delimitation of the Archean the lower limit of the Algonkian has been given. Its basal plane is the lowest of the recognizable clastic rocks. It has been seen that there are great differences in the case of recognition of the basal Algonkian plane in different regions. In the Uinta Mountains, in the Grand Canyon region of Arizona, in portions of the Lake Superior region, in the Original Huronian region, and elsewhere, between the Algonkian and the Archean there are great unconformities, above which are the readily recognizable clastic rocks, and below which are the thoroughly crystalline basal complexes. Even in many regions in which there have been repeated foldings since Archean time, and in regions obscured by eruptive activity, it is perfectly clear that a large part of the rocks are clastic and belong with the Algonkian, while other parts have all the characteristics of the fundamental complex. Occupying an intermediate position are occasionally found areas of rocks which can not certainly be placed with the Algonkian or Archean, but this difficulty is not peculiar to this separation any more than to other generally recognized planes, such as that separating the Cambrian and Silurian in folded districts. Many of the members of the Canadian Geological Survey have described the Huronian and Laurentian as conformable, with gradations between the two. This apparent accordance and gradation is, in many cases, due to the fact that placed with the Huronian are many rocks which would, under the usage of the terms here proposed, be regarded as Archean. In other cases there are apparent conformities and gradations between undoubted clastics and the underlying rocks having all the characteristics of the fundamental complex. The significance of these gradations is discussed in another place (p. 125), where it was seen that they are not inconsistent with genuine structural breaks.

It has been stated that the reasons for placing the base of the Cambrian at the Olenellus fauna are considered by Walcott, and that his results are here accepted. It is, however, to the point to consider whether this horizon answers equally well for the upper limit of the Algonkian. Evidently all the arguments brought forward by Walcott for placing this fauna as the base of the Cambrian apply as well for considering the horizon below this uppermost Algonkian; for the widespread character, both European and American, of the Olenellus fauna makes it a particularly easy one to identify, and therefore valuable for the purposes of discrimination. In the Lake Superior region, and in many other localities, above the upper Algonkian are unconformities; the first of the Cambrian being middle or upper. In other regions, as well as in Newfoundland, the upper Algonkian is marked by an unconformity, and the formation immediately above bears the Olenellus fauna. This is the most favorable and clear case. In the Wasatch and several other ranges of Utah and Nevada, in British Columbia, and probably in the Southern Appalachians, below the Olenellus fossiliferous Cambrian are conformable series of quartzites and slates of great thicknesses. Are these lowest Cambrian or uppermost Algonkian? May not the Olenellus fauna in the future be found to extend downward through a greater or less thickness of these apparently barren rocks? If in any region the fauna be found to extend downward for a long way it is probable that species and genera characteristic of the Olenellus horizon, as now known, will drop out and others appear which are different. The Olenellus would thus grade into a pre-Olenellus fauna. Such a gradation will doubtless somewhere be found, while in other regions the change from an Olenellus fauna to one of a pre-Olenellus type may occur abruptly. In either case there will finally appear a fauna which is not the present known Olenellus fauna, but which is as different from it as is the Cambrian from the Silurian (Ordovician). As the term is here used such a fauna is pre-Cambrian, and the rocks containing it are Algonkian. In the following paragraphs great barren inferior series conformably below the known Cambrian are placed with the Algonkian on the ground of probability. The presence of an abundant lower Cambrian life at a certain horizon within the conformable succession, with apparent complete absence of life in immense thicknesses of rocks conformably below, which, so far as lithological character is concerned, are equally likely to bear fossils, throws the weight of evidence in favor of the Algonkian age of these rocks. It is, however, more than probable that some part of the conformable downward extensions of the Cambrian, which are here referred to the Algonkian, will in the future be found to belong with the post-Algonkian. The newest Proterozoic or Algonkian rocks of different regions may stand in different positions, just as the superior rocks of the Paleozoic may in any given region be Cambrian, Silurian, Devonian, or Carboniferous.

Difficulties in Algonkian stratigraphy.—Since, among the pre-Cambrian clastics, paleontology is not yet available in correlation, it is exceedingly difficult to make widespread subdivisions of the Algonkian, such as are made in later time. The difficulty is further increased by the unequal metamorphism, in different regions, of series of the same age. The Algonkian is in just such a position, as regards wide correlation of its constituent series, as would be the Paleozoic and Mesozoic, if their known fossil contents were so small as to be useless for the purposes of correlation. The structure of individual districts and regions could be worked out and the formations correlated, but the attempt to equate the Cambrian, Silurian, Devonian, or Carboniferous of one region with rocks of the same age in a far distant one would be an almost hopeless undertaking. In the Carboniferous the beds of coal would serve as an important guide, but, if implicitly followed and no fossils were available, the Triassic of Virginia, the Carboniferous of the central United States, and the Cretaceous of the west would be placed together. If the iron carbonate formation of the Algonkian were followed as a guide (which appears to be the most characteristic of any one bed in the Lake Superior region), the results would probably be as far from the truth.

We may perhaps go so far, in some cases, as to correlate series which occur in different districts of the same region, when a set of characteristic formations forming the series occur in like order, and the series, as a whole, is in the same relative position to overlying and subjacent series, one or both of which are known to be identical in both districts. It is probable, when several pre-Cambrian series occur of the same general character, with like relations to each other and to the Archean and Cambrian, and not so far apart as to be outside of the same geological basin, that a provisional correlation is warranted. While then it is not practicable to subdivide the Algonkian into general systems which shall cover the whole of North America, it is often possible so to do in a single geological basin, or in adjacent basins in which the relations of the separate formations and series can be worked out.

Before considering the principles applicable to the subdivision and correlation of pre-Cambrian clastics, the successions found in several of the regions will be recited. No attempt can be made in my limited space to give detailed evidence for the conclusions stated. For this it will be necessary to refer to the correlation paper on the Archean and Algonkian, published by the United States Geological Survey.

The Original Laurentian and associated areas.—In this region are Algonkian rocks at the following localities: Hastings District, Lake Nipissing, Ottawa River, Upper St. Lawrence River, and the Adirondacks. The Grenville area of the Ottawa is the Original Laurentian type district, and the one mapped in most detail. While the maps do not connect these areas, the similarity of their clastic rocks is such as to indicate a probable present or former continuity, with the exception, perhaps, of those of Lake Nipissing. The clastics consist of interstratified limestones, quartzites, conglomerates, green slates and schists, mica-schists, hornblende-schists, and regularly bedded gneisses, together estimated to be several thousands of feet thick. Associated with these are diabasic and chloritic rocks, both massive and schistose. In the Hastings series there are also found considerable areas of peculiar volcanic clastics. Below the clastic series is a great complex, in every respect like the Archean as above defined. This latter system, usually called Lower Laurentian, occupies the main area of the region, and the clastics are in a series of troughs within it. What the relations are between the clastics and the fundamental complex has not been definitely made out, although Vennor believes that between the two is an unconformity, and that the clastics are infolded patches; the evidence given for this is, however, rather meagre, and doubtless almost as good a case could be made out with present facts for the theory of an irruptive contact between the granite-gneisses and the clastics. The labradoritic rocks (gabbros) found in this region need not be considered in the stratigraphical succession, as they are eruptives of later age than the clastic series. Besides this eruptive, other acid and basic eruptives cut the bedded succession.

To the clastics Logan, Murray, Vennor, and all who have worked in this region recognized that ordinary stratigraphical methods could be applied. The persistence of the bands of limestones is such as to enable them to be traced for long distances. Although the problem was a difficult one, a detailed mapping, with sections, has been submitted for a small part of the area. The structural relations and correlations which Vennor first gave differ greatly from his final ones, and it may be that, even in the areas in which detailed mapping was attempted, serious mistakes have been made; but if this be true, the region is in no respect different from any other in which the structure is difficult. All of the pre-Cambrian rocks here found were supposed by the Canadian geologists to be lower in the geological column than the Huronian of Lake Huron. Upon the last point no positive evidence is at hand. The two rock series do not come together. In the western Hastings district of the Laurentian the clastics, in lithological character, degree of crystallization, and amount of folding, are intermediate between the Laurentian and Huronian of the type areas. At first the

Hastings clastics were correlated by Vennor with the Huronian, and with this correlation certain of the official Canadian geologists now agree, but afterwards they were traced with breaks of not very great distances to the Original Laurentian and have always been thus mapped.

The Original Huronian.—The Original Huronian of the North Channel of Lake Huron consists of comparatively little altered quartzites, slates, slate-conglomerates, greywackes, cherts, and limestones, having a total thickness of 18,000 feet, counting considerable masses of interstratified greenstone, which are recognized as eruptives. Recent observations render it probable that these rocks are to be divided into two unconformable series, the lower of which is 5,000 feet thick and the upper 13,000 feet thick. The first would thus be properly designated as Lower Huronian and the second as Upper Huronian. Although cut locally by later granites, the lowest member of the inferior series is separated from the basement complex by a great unconformity. The upper members rest unconformably below the Potsdam sandstone. The upper series is so gently folded that the careful work of Logan and Murray enabled them to map these rocks in detail and to work out their structure. This has been done for a considerable district with as much certainty and accuracy as in many areas among the fossiliferous rocks. The rocks were divided into a number of formations which were found to be persistent throughout the district. They were traced as a broad belt for several hundred miles in a general direction northeast. Along the Canadian Pacific Railway as far as Sudbury, and in the vicinity of Sudbury, more than a general study of this great area has been made. The clastic rocks of this part of the region have the same general character as the type district, but in the Sudbury district there are peculiar contemporaneous volcanic clastics. In the little-studied remainder of the region, as mapped, numerous granitic and gneissic areas are included, some of which may be subsequent intrusives, but many of which probably represent the underlying Archean.

Lake Superior region.—In the Lake Superior region, between the Archean and the Potsdam sandstone, the great Algonkian system is subdivided into three series, which are separated by very considerable unconformities. The lowest series is closely folded, semicrystalline, and consists of limestones, quartzites, mica-slates, mica-schists, schist-conglomerates, and ferruginous beds, intersected by basic dykes, and in certain areas also by acid eruptives. It includes volcanic clastics, often agglomeratic, and perhaps its most characteristic rock is a green, chloritic, finely laminated schist or slate. The thickness of this series has not been worked out with accuracy, but it is probably more than 5,000 feet. As the term Huronian has been for many years applied not only to the Upper Huronian series, but to this inferior series about Lake Superior, it is called Lower Huronian.

Above this series is a more gently folded one of conglomerates, quartzites, shales, slates, ferruginous beds, interbedded with and cut by greenstones, the whole having a maximum thickness of at least 12,000 feet. In the Animikie district has been found a fossil track, and in the Minnesota quartzites lingula-like forms, as well as an obscure trilobitic looking impression. Carbonaceous shales are abundant. In its volume, degree of folding, and little altered character the Upper Huronian is in all respects like the upper series of the Original Huronian, and can be correlated with it with a considerable degree of certainty. Above the Upper Huronian is the great Keweenawan series, estimated at its maximum to be 50,000 feet thick, although its average thickness is much less. Its lower division consists largely of basic and acid volcanic flows, but contains thick beds of interstratified sandstones and conglomerates, especially in its upper parts. The upper division, 15,000 feet thick, is wholly of detrital material. The fragmental material is largely derived from the volcanics of the same series.

The unconformities which separate the Lower Huronian from the Upper Huronian, and that which separates the latter from the Keweenawan, each represent an interval of time sufficiently long to raise the land above the sea, to fold the rocks, to carry away thousands of feet of sediments, and to depress the land again below the sea. That is, each represents an amount of time which, perhaps, is as long as any of the periods of deposition themselves. In parts of the region the lowest clastic series rest unconformably upon the fundamental complex, but in certain areas the relations have not been ascertained. The upper of the three clastic series, the Keweenawan, rests unconformably below the Potsdam sandstone.

In the Lake Superior region it has been possible, with a considerable degree of certainty, to refer the detached areas of pre-Cambrian clastic rocks to one or another of the three series mentioned, although there have been sharp differences of opinion with reference to certain of the areas. It has been possible farther to subdivide the series into formations, some of which have a widespread extent within the region. The best results in correlating the subdivisions within the series have been reached in the Penokee and Animikie districts.

Correlations of series have been in this region based upon unconformity, upon lithological similarity, upon the belief that the greater dynamic movements which have affected the region have been widespread and upon degree of crystallization of the rocks. The correlation of the formations within a given series have been based upon lithological characters and upon a similar succession of like beds.

Newfoundland.—In Newfoundland is a clear case of a great series of rocks, of perhaps 10,000 feet thickness, referred to the Huronian by Murray, which is a part of the Algonkian. Here is found the Olenellus fauna in the basal Cambrian rocks, and these are separated by an unconformity from the underlying clastic series, in which, however, have been discovered two or three fossils of a low type. What the relations of this lower slate series are to the crystalline granite-gneiss, which has been referred to the Laurentian, is uncertain. No evidence is available showing that lower than this slate series are any clastics. Certain of the granites of the island of Newfoundland are intrusive, of later age than the slates, some of them being as recent as Carboniferous; so that it is not impossible that many of the granites, syenites, and porphyries referred to the Laurentian may be of far later age.

Grand Canyon of the Colorado.—The Algonkian in the Grand Canyon region is represented by three series, the Chuar, the Grand Canyon, and the Vishnu. The Chuar series consists of shales and limestones over 5,000 feet thick. It contains a fauna of a pre-Cambrian type, including at least five distinct forms. The Grand Canyon series is of sandstones, with basic lava flows in its upper parts, and is nearly 7,000 feet thick. The Vishnu series consists of bedded quartzites and schists, cut by intrusive granite, and is known to be at least 1,000 feet thick, but how much thicker has not been determined as it has not been measured to its base. The Chuar rests upon the Grand Canyon, and between the two is a minor unconformity. The Grand Canyon rests upon the Vishnu, and between the two is another unconformity. The Chuar and Grand Canyon sediments are wholly unmetamorphosed, while the Vishnu sediments are indurated quartzites and semi-crystalline schists. Between these series, as a whole, and the underlying Archean complex is a very great unconformity. Between the Chuar and the Tonto sandstone (Upper Cambrian), there is another unconformity sufficient to have caused the cutting across of at least 10,000 feet of the flexed beds of the Grand Canyon and Chuar series. In this region is the fullest known succession of Algonkian rocks in the United States, with the exception of the Lake Superior region. The statement would not be warranted that the series here found stand as the equivalent of like series in the latter region, but there is a remarkable lithological likeness both in the detrital and eruptive material of the Chuar and Grand Canyon to the Keweenawan. Also these series occupy a position of unconformity below the Upper Cambrian as does the the Keweenawan, and are separated by an unconformity from a series of quartzites and quartz-schists which are analogous to the Huronian. This latter, the Vishnu series, is not well known, so that it is unsafe to assert whether it is nearer like the Lower or the Upper Huronian of the Lake Superior region.

Other Algonkian areas.—Besides the foregoing regions in which one or more series of Algonkian rocks occur, there are many regions in North America of which the same is true, but space does not permit that an attempt be made to give the succession for each.

About Hudson bay there are at least two series, the upper of which is very like the Keweenawan of Lake Superior. In the Eastern townships, in New Brunswick, in Nova Scotia and Cape Breton are found pre-Cambrian clastic rocks which have been referred to the Huronian and Upper Laurentian. In the Black Hills is a great series of pre-Cambrian clastics which are in many respects like the Lower Huronian of Lake Superior. In Missouri are various isles, composed in part of clastic rocks, overlain unconformably by the horizontal Cambrian. In Texas is a great succession of pre-Cambrian clastics which are divisible, according to Comstock, into three series. In the Medicine Bow mountains of Wyoming; in southwestern Montana; in the Uinta mountains; in the Wasatch Mountains; at Promontory ridge, Antelope and Fremont islands of Salt Lake; in several of the desert ranges; in the Front range and in the Quartzite mountains of Colorado; and in other ranges of the Cordilleras are found one or more series of pre-Cambrian clastics. These are sometimes a downward continuation of the Olenellus Cambrian; at other times they are separated from the Cambrian by an unconformity. In some cases the clastics appear to grade downward into the Archean, and in other cases are clearly unconformable upon the Archean. In the great region of northern Canada, pre-Cambrian clastics are known to exist at various districts, but as yet they have been little studied. In the Appalachian region of the United States, it also appears that it will ultimately be shown that there are representatives of the Algonkian system.

Subdivisions of Algonkian.—The foregoing makes apparent the propriety of introducing the term Agnotozoic or Proterozoic to cover the series between the Paleozoic and Archean. The desirability of dividing the group into several systems is also apparent, but it is equally apparent that our limitations of knowledge at the present time make it impossible to do this for the whole of North America; hence, as with Archean, it is unavoidable that a single system term shall be used for the Agnotozoic group, and, as already explained, Algonkian is given this place. However, the major subdivisions of this Algonkian system, in volume of rocks and time duration, are equivalent to the systems of the Paleozoic.

Principles applicable to Algonkian stratigraphy.—In the stratigraphical work of the past, methods have oftentimes been defective. Instead of giving close lithological descriptions of a series of rocks, noting carefully the relations of the different strata (in case they are found to have strata) to one another, and giving a detailed account of the relations which actually obtain between the series considered and surrounding series, writers have too often called the rocks of regions far distant from the original localities to which the terms have been applied, Laurentian, Huronian, etc. Sometimes this is done on the ground that a series as a whole has a certain color, which has been thought to be characteristic of the period. At other times the abundance of volcanic material has been the reason for the reference. Again, quartzites from the Appalachians to the Black Hills have been correlated on no other ground than lithological likeness, as though thick sandstone formations, which subsequently have been cemented to quartzites, have been produced but once in the history of the world. To some geologists the degree of crystallization has been the controlling fact. In other cases the occurrence of some mineral or association of minerals has been the ground upon which the reference of the containing formation is made to some specific period.

As a natural result of work of this kind, rocks of a certain lithological character or degree of crystallization or containing certain constituents have been called Arvonian, Huronian, Norian, Laurentian, as the case may be, which have afterwards proved to be high in the Paleozoic. Other series which are now known to be pre-Cambrian have been called Triassic because of a prevailing red hue.

Any of the above characteristics may be a valuable guide in a given district or, with qualifying and guiding facts of a different character, in a region; but it is their use in an indiscriminate manner, on the assumption that rocks of a given time are everywhere alike, that is protested against. When it is everywhere recognized that, considering the continent as a whole, age is no guide to the chemical or mineral composition, texture, color, degree of crystallization, or any other property of a formation, or vice versa, we shall be on the way to use the properties of rocks in districts and regions as guides to age. For the most part this principle has been recognized, if not practically, at least theoretically; but at one point this is less true than with the others. Degree of crystallization, because often so useful in a district, has been used by many as a general guide in correlation, although the elder Hitchcock, Rogers, Adams, and others gave early warning against such practices. It can not be too strongly insisted upon that contact action of great masses of eruptive rocks, and dynamic action accompanying this, or dynamic action without accompanying volcanic activity, are prime and perhaps the chief causes in the majority of cases for the production of crystallization, not age and depth of burying, although these may be contributory causes and at times the predominant ones.

No one would think of maintaining that in post-Cambrian times a rock of a certain composition is of a definite age; neither would any one think of referring a rock to the Devonian, Silurian, or Cambrian upon the degree of its crystallization. To suppose that the plane of the basal Cambrian is a magic one, below which new conditions of sedimentation prevailed and an entirely different set of principles apply in stratigraphy, is to assume a revolution in the conditions of the world at this time for which there is not one particle of warrant. Those who believe in evolution must believe that for eras of time before the Cambrian there were cycles of deposition of the various classes of sedimentary rocks and the slow evolution of life to the high degree of perfection and the great variety of types, including all important branches except the vertebrates, found in the Olenellus fauna.

It may be said that the foregoing applies equally well to the separation of the Algonkian and Archean. Revolutionary methods can not be applied here more than elsewhere. To this it can only be said that this plane is the most remote and difficult to define of any. It may be that it is wrongly defined. Without question it will in the future be much more accurately defined. Rocks now placed in the Archean will be found to be Algonkian, just as series are being found to be Cambrian, Silurian, or Devonian in the Appalachians, which have commonly been regarded as Huronian or Laurentian. While the distinctions made may not be complete, they are based upon the knowledge available, are not dogmatic, and do not, it appears to the writer, contradict the law of uniformity. The law of uniformity does not imply that the causes now at work have always had the same relative value. Igneous rocks are now a far less abundant geological product than are sedimentary rocks. That these relations are true for any past time can no more be assumed than it can that organic and mechanical sediments have the same relative volume for the pre-Cambrian that they have for the Cretaceous or Tertiary. If the generally accepted hypothesis as to the origin of the globe represents the facts, in all probability rocks of igneous origin must become relatively more important in very ancient times. If we but go back far enough they may become predominant; and in still earlier time, for continental areas, they may be the only rocks.

The problem then of the stratigraphy of the pre-Cambrian clastics is a problem to be treated precisely as that of the Paleozoic. It is, upon the whole, a more difficult problem; for, while in any particular locality it can not be premised from the degree of crystallization that the rocks are pre-Cambrian or post-Cambrian, taking the world as a whole, the rocks become more crystalline in passing to lower series. So that it is to be expected and is the fact, that a greater proportion of pre-Cambrian rocks than of the Paleozoic are highly crystalline. This, however, is not the most serious difficulty with which pre-Cambrian stratigraphy has to contend; it is the sparseness of the remains of life in definitely recognizable forms. It has been seen that a beginning of a pre-Cambrian fauna has already been found, and when it is remembered how rapidly definite paleontological knowledge has extended downward in the past decade, it may be reasonably hoped that before long assistance will be derived from paleontology in the classification of the pre-Cambrian rocks, but it can not be expected that fossils will ever be so important and controlling a guide as in the post-Cambrian; for probably the farther we go back from the Cambrian, the sparser and sparser will the recognizable life-remains become.

As a result of the average greater crystalline character of the pre-Cambrian rocks and the frequency in them of secondary structures, the principles of working out stratigraphy, in regions in which cleavage-foliation occurs with partial or total obliteration of stratification foliation, are of the utmost importance. The failure to clearly recognize and apply these principles has left the crystalline Cambrian and post-Cambrian series of the Appalachians in a state of confusion for many years. Within the last decade, by a recognition and a close application of them, a new start has been made in the study of this difficult region.

That it will not do to regard slaty cleavage or foliation as bedding has long been recognized, although it has often occurred that in regions in which this has been distinctly stated, it has also been said that bedding and cleavage do correspond without any evidence of such correspondence being given. When tangential thrust is the cause of cleavage or foliation, the general principle appears to be that these structures accord with bedding when they are produced by the accommodations of the beds over one another, with consequent shearing parallel to the beds, but they do not accord with bedding when the structures are produced as a result of the mashing or flowage of the whole mass of strata. Slipping parallel to the bedding is likely to occur when the beds differ in strength, i. e., are composed of more or less indurated alternating layers of shale, grit, sandstone, and limestone. Mashing is likely to occur when a thick shale formation is subjected to tangential thrust. In certain series both mashing and shearing parallel to the beds may occur in different degrees. Either may be predominant, and thus there may be all gradations between cleavage or foliation parallel to and transverse to the bedding, or both may occur together or closely associated.

It is not impossible that deeply buried beds may become thoroughly crystalline, with foliation and bedding parallel, when superincumbent pressure and metasomatic changes or laccolitic intrusions are the predominant forces. In such cases the structures of the recrystallized rock will naturally conform to the bedding, and it is probable that differences in the original characters of the layers will be preserved in the metamorphosed rock. A mica-schist or gneiss thus derived from a shale or an arkose, now showing in its interior structure no evidence of clastic character, might be underlain by a quartzite which was produced by the cementation of a quartzose sandstone. This quartzite at the present time would reveal its detrital origin, while the mica-schist or gneiss might not. Some such causes, combined, perhaps, with the shearing parallel to the bedding, seem to explain the thick beds of mica-schist (the structures of which unmistakably correspond with bedding), which overlie the quartzites of the Penokee and Marquette series of Michigan and Wisconsin. A crystalline series of the origin suggested, which subsequently reached the surface by denudation, might be folded, but not sufficiently to produce a new secondary structure, when the different bands of different character would truly represent sedimentary beds. Some such explanation seems to fit the gently folded, thoroughly crystalline mica-schists and gneisses of the Blue Ridge west of Old Fort, North Carolina. In the Adirondacks, where the schistose structure and bedding of the Algonkian rocks appear to correspond, the shearing of the beds over one another and the great laccolites or batholites of gabbro are probably the chief cause of the metamorphism.

In crystalline series, when but one structure can be found, it is safe only to assume that it is foliation. Not only will it not do to use cleavage for working out structure, but an actual regular alternation of mineral constituents in schistose and gneissoid rocks can not be regarded as any evidence of sedimentation. The great series of regularly banded gneisses, in which alternate zones of nearly pure feldspar and other zones in which the bisilicates are concentrated, if taken as due to sedimentation would result in the conclusion that the thickness of the beds in which these structures occur is incredibly great. In thoroughly schistose rocks it is manifest that the best and most reliable means, upon which to base a conclusion as to strikes and dips, is to find contacts between thick beds of rock of a fundamentally different character, as a layer of quartzite, quartz-schist or mica-schist, with limestone, or either of these with gneiss.

The clearest, briefest, and most comprehensive enunciation of the principles applicable to a formation, in which are cleavage-foliation and stratification-foliation, known to the writer, is that of Dale and Pumpelly, which is here quoted in substance, slight modifications being made to fit the change of setting.

I. Lamination may be either stratification-foliation, or cleavage-foliation, or both; or sometimes "false bedding." To establish conformability the conformability of the stratification-foliations must be shown.

II. Stratification-foliation is indicated by; (a) the course of minute but visible plications; (b) the course of the microscopic plications; (c) the general course of the quartz laminæ, whenever they can be clearly distinguished from those which lie in the cleavage planes.

This statement was made with reference to a particular district. It is of course wholly possible that some other substance should play the same role as quartz. In the application of these criteria it must be premised that the parting is not a second or third cleavage. If an earlier cleavage existed, the criteria might give the direction of this first one rather than the bedding, which might have become obliterated at the time of the development of the first cleavage.

III. Cleavage-foliation may consist of; (*a*) planes produced by or coincident with the faulted limbs of the minute plications; (*b*) planes of fracture, resembling joints on a very minute scale, with or without faulting of the plications; (*c*) a cleavage approaching slaty-cleavage, in which the axes of all the particles have assumed either the direction of the cleavage or one forming a very acute angle to it, and where stratification-foliation is no longer visible. (*d*) A secondary cleavage, resembling a minute jointing, may occur.

IV. The degree and direction of the pitch of a fold are indicated by those of the axes of the minor plications on its sides.

V. The strike of the stratification-foliation and cleavage-foliation often differ in the same rock, and are then regarded as indicating a pitching fold.

VI. Such a correspondence exists between the stratification and cleavage-foliations of the great folds and those of the minute plications, that a very small specimen properly oriented gives, in many cases, the key to the structure over a large portion of the side of a fold.

The first four of these six principles are taken from the standard authorities, but the last three, I believe, are new and are due to Pumpelly. It is to be noted that the statement of these principles has been inspired by a study of a region which has proved to be wholly Cambrian or post-Cambrian.

The principles of lithological correlation enunciated by Irving are as follows, except that Series is here substituted for Group, and Algonkian for Huronian, so as to make the terminology correspond with that used in this paper:

Lithological characters are properly used in classification—

(1) To place adjacent formations in different series, on account of their lithological dissimilarities, when such dissimilarities are plainly the result of great alteration in the lower one of the two formations and are not contradicted by structural evidence, or, if used as confirmatory evidence only, when such dissimilarities are the result of original depositional conditions.

(2) To collect together in a single series adjacent formations because of lithological similarities, when such similarities are used as confirmatory evidence only.

(3) To correlate series and formations of different parts of a single geological basin, when such correlations are checked by stratigraphy and particularly by observations made at numerous points between the successions correlated.

They are improperly used—

(1) To place adjacent formations in different series, on account of lithological dissimilarities, when such dissimilarities are merely the result of differences in original depositional conditions, and when such evidence of distinction is not confirmed by or contradicted by structural and paleontological evidence.

(2) To collect in a single series adjacent formations because of lithological similarities, when such similarities are not confirmed by or are contradicted by other evidence.

(3) To establish general correlations between the clastic series of different geological basins, except possibly when the gneissic and true crystalline-schist basement formation of one region is compared with the similar basement formation of another.

(4) To establish and determine any world-wide subdivisions of the non-eruptive basement crystallines (i. e., those which underlie the clastic series, here called Algonkian) at least until very much more definite evidence of the existence of such subdivisions be gathered than has hitherto been done.

In applying these principles it must not be forgotten that a bed of one character may thin out and disappear; may gradually change from a limestone to a shale, from a shale to a sandstone or conglomerate; and that sometimes the change may be abrupt, as, perhaps, upon the opposite sides of an axial ridge, one side of which faces toward the ocean and the other toward an interior sea. All formations, however widespread, terminate somewhere. A single formation of a certain lithological character can only be assumed to be the same bed in a district, when it has been demonstrated to be persistent over a wide area. When several characteristic formations occur in a definite order in different parts of the same district, the probability that they are of identical age is greater than with a single formation, found to be alike at separate points.

The principles applicable to correlation by unconformities are given by Irving as follows:

The structural breaks called unconformities are properly used in classification—
(1) To mark the boundaries of the rock series of a given region.
(2) To aid in establishing correlations between the formations of different parts of a single geological basin.
(3) To aid in the establishment of correlations between the series of regions distantly removed from one another; but caution is needed in attempting such correlations in proportion as the distances between the regions compared grow greater.
They are improperly ignored—
(1) When the evidence they offer as to separateness is allowed to be overborne by anything but the most complete and weighty of paleontological evidence.

Irving's discussion leading to these principles shows that oftentimes unconformities are the most widespread and important of any of the means available to obtain starting planes for comparisons, and that they have the place of first importance in making the major subdivisions for a region of the pre-Cambrian clastic rocks. An erosion interval can only occur as a result of the raising of a district above the sea, a time of degradation, and then a depression below the sea; and if there is a true uncomformity there must also have been an orographic movement and erosion long enough continued to truncate the folds. The erosion interval, if extended over a large area, implies a considerable time break; while the unconformity, if it is marked, can hardly be less than regional in extent. When the newer series is undisturbed, an unconformity is one of the easiest of phenomena to detect, but, more frequently than not among the pre-Cambrian rocks, the older and newer series have again been folded, and this folding has oftentimes gone so far as to produce a cleavage or foliation, which cuts across both older and newer series and makes their most prominent structure in absolute conformity. Even if this degree of folding has not occurred, and the process has not gone far enough to produce prominent secondary structures, the discordance in angle of inclination is more likely to be overlooked than when the series are in an undisturbed condition.

Since unconformities are so valuable in structural work, it is important that the principles be clearly recognized upon which they may be established in disturbed regions. This subject has been discussed at length by Irving, and from his paper the substance of much which follows is taken. An unconformity between series implies a difference in number of orographic movements with intervening erosion. This difference in number may be one or more than one. Even when the difference of orographic movements to which the series have been subjected is but one, the time-gap between the two must have been very considerable, and it may have been of vast duration. Consequently discordant series may differ in degree of consolidation, in the development of cleavage and foliation, and in their relations to eruptives. At the beginning of the deposition of the newer series, basal conglomerates are often formed.

Hence as guiding phenomena in the discovery of unconformities we have (1) ordinary discordance of bedding; (2) difference in the number of dynamic movements to which the series have been subjected; (3) discordance of bedding of upper series and foliation of lower; (4) relations with eruptives; (5) difference in degree of crystallization; (6) basal conglomerates; (7) general field relations.

(1) In cases of ordinary discordance of bedding nothing need be said, except to state that unconformities should not be inferred from a single small contact, where the apparent discordance may be due to false-bedding or to local currents or very local minor disturbances. Also, the discordances caused by faulting may be mistaken for an unconformity if care is not taken. The amount of evidence for the unconformity should be sufficient to show a real discrepancy of bedding for a considerable area.

(2) Difference in the number of dynamic movements to which the series have been subjected is often an important means of determining unconformities. In order that an unconformity shall occur the older series must have been subjected to at least one more orographic movement than the newer. In the most favorable case the older series has undergone two or more orographic movements while the newer series has undergone but a single one. When the lines of these movements are in the same direction and result in folding, the only difference between the two series consists in steepness of inclinations; but in case the earlier movements were in a different direction from the last, the older series will show a compound series of folds due to the resultant effect of the two or more movements, while the newer series will be simply folded. As a matter of course, in this discrimination, bedding must be used rather than foliation. Oftentimes it will happen that the

latest movement has produced a prominent cleavage or foliation, which is common to both older and newer series; and under these circumstances the real discordance which may exist between the two series is particularly apt to be overlooked, and a district will be described as having a simple monoclinal structure, or one in which the series is reproduced by faulting, when evidence is at hand for two or more discordant series.

The orographic movements, instead of producing folding, may cause jointing or faulting—these results, as suggested by Willis, being perhaps due to insufficient load. These phenomena are, however, serviceable in discovering an uniformity, for the sets of faults or joints produced in the older series, before the newer series was formed, are not found in the latter. When there have been later orographic movements which have produced faults or joints in both series, the unconformity will be shown by the presence in the older series of two sets of joints or faults in different directions, provided the directions of thrust were different, while the newer series will be affected by joints or faults only in a single direction. If the jointing or faulting is in the same direction in both the newer and older movements, they will not be of much service in detecting an unconformity, the only difference being their greater frequency in the older series.

When one of the orographic movements has resulted in folding and the other in faulting or jointing, the combination of phenomena are as easily used to detect an unconformity as when effects of the same kind are produced by both movements.

(3) Discordance of the bedding of an unfoliated series with the cleavage or foliation of an adjacent series may be taken as evidence of unconformity, if the former is such that it would take on cleavage or foliation as readily as the latter; for, whatever the origin of the altered series, the development of cleavage or foliation, which must have developed before the new series was deposited, required much time. An unconformity could not be inferred from the fact that a heavy formation of quartzite or of limestone cuts across the cleavage or foliation of an argillite or mica-schist, for clayey rocks very much more readily take on secondary structures. In the same series it often happens that more massive beds escape foliation, which may be prominently developed in other members. But if a formation with slaty or schistose structure is overlain by another formation without secondary structure, which from its composition is as likely to take on foliation as the underlying formation, a discordance, while not demonstrated, is a probability for which other evidence should be sought.

(4) Eruptive rocks are often an important guide in determining structural discordances. These are valuable when the older series has passed through an epoch of eruptive activity before the newer series

451 GE——10

was deposited. In such cases, bosses, contemporaneous or intrusive beds, volcanic fragmental material, or dikes may occur in the older series which nowhere are associated with the newer. It is possible of course that eruptives may penetrate the inferior members of a series and never reach the higher formations, but if it is found that the supposed inferior series is associated with abundant material of igneous origin, which never passes beyond a certain plane, it is almost demonstrative evidence of the later age of the newer series. A notable instance of this is found in the Doe river section of eastern Tennessee, where the granitic rocks, supposed to be older than the associated clastics, are cut by very numerous schistose dikes which never intrude the latter. It might be reasonably inferred, if it were not for these dikes, that the granitic rock is an eruptive later than the clastics (although the absence of contact phenomena would be against this), but as the basic dikes are unquestionably intrusives of later age than the granites, and yet never cut the slates, this explanation can not possibly apply. Evidence of this kind is particularly decisive if the dikes are traced up to the plane of contact and have been found to be eroded or disintegrated, as is the case in the Stamford dike at Clarksburg mountain, Massachusetts, described by Pumpelly, which enabled this author to positively determine, what had been believed before, that the granitoid gneiss is unconformably under the Cambrian quartzite.

(5) Closely connected with (3) and (4) is degree of crystallization as a guide to unconformities. It has been seen that crystalline character is often taken on in proportion as dynamic action occurs. When the folding which has affected only the older series has been severe, it as a whole will be more crystalline in character than the newer. Also the presence of igneous material is often a potent factor in the production of crystalline character. As, however, recrystallization is also produced by metasomatic change, this criterion must be used with caution and as a cause to search for other evidences of an unconformity, rather than alone as a basis upon which to infer an unconformity. But even difference of amount of metasomatic change, if the rocks are equally likely to be affected by these processes, may be evidence of difference in age. In determining degree of crystallization the modern petrographical methods serve one of their most useful purposes, since many rocks, which in exposure or in hand-specimen appear to be about equally crystalline, are shown in thin section to be of a fundamentally different character. A completely crystalline rock sometimes can not be discriminated macroscopically from one which is merely indurated by cementation. For instance, a thoroughly crystalline granite and a recomposed rock built up of the débris of this granite, especially when the particles are in the form of individual minerals rather than pebbles, present much the same appearance in mass; but a glance at sections

of the two under the microscope shows the thoroughly crystalline, inter-locking character of the one and the clastic character of the other. Another case quite as marked is the discrimination between much foliated eruptive rocks, which have passed over into fissile schists, and ordinary argillaceous slates and graywackes. In the latter class the particles of quartz and feldspar may be seen with their oval forms as regular as the day in which they were deposited, while in the other case an entirely different appearance is presented.

(6) Basal conglomerates are one of the most important means of de-termining a plane of unconformity, but it must be clearly shown that the conglomerate is really a basal one. Conglomerates may occur in other positions than at basal horizons, and it will not do to assume that an unusually conglomeratic layer is basal. A conglomerate is likely to be basal when the major portion of the débris is derived from the immediately subjacent member; but even here the exception must be made that, in case this subjacent member is a surface igneous rock, the presence of the conglomerate is no evidence of a time break. If, however, the igneous formation is of such a character as does not originate except as a deep-seated rock, the fact that it is at surface and yields fragments to the overlying formation is evidence of a time gap. Also evidence of a break is just as decisive when the underlying rock has a foliation which has been produced prior to the deposition of the conglomerate. This may be determined from the fact that fragments broken from a foliated rock are apt to be longer in the direction of lam-ination, and, when deposited in the overlying series, they naturally lie with their foliation at an angle to that of the underlying series. It mat-ters not whether the foliated rock of the inferior series be of sedimentary or of igneous origin. If sedimentary, a long time has been required to obliterate evidence of its fragmental character; if igneous, its foliation shows the effect of long-acting forces. While basal conglomerates are often found, they are also often absent where other evidence shows that there are discordant relations between two series. This absence is explained, at least in some cases, by Pumpelly's disintegration theory, the encroaching shore-line finding a set of disintegrated rocks in which the mass is ready to yield particles of the constituent min-erals rather than pebbles.

(7) General field relations are often sufficient to establish discordant relations between series, when all other lines of evidence are lacking. When, in a region, immense stretches of rocks of one series are always found in an undisturbed condition, while an adjacent series is always disturbed, discordant relations may be inferred. This is particularly evident when the horizontal series fills bays in the older rocks, or is found as inliers surrounded by the other rocks. Again, the general field relations may establish an unconformity even if both series are

disturbed. One case of this is the occurrence of a uniform belt of stratified rocks which, perhaps with a monoclinal structure and a somewhat uniform strike and dip, runs for great distances, the rocks of the adjacent unconformable series being here of one kind and there of another kind. The evidence for the unconformity in this case is still farther emphasized if the lower series, instead of having a simple structure, is folded in a complex manner. General field relations may betray unconformity even when the newer series has been folded in a more complex manner, as, for instance, having been subjected to two orographic movements, the first of which placed it in a monoclinal attitude, and the second of which, at right angles to this first force, gave it a fluted structure. The lower series, instead of having this regular structure, being subjected to still earlier orographic movements, would be more irregular in its foldings and faultings, and the difference in simplicity of structure of the two series would increase in proportion to the number and intensity of the earlier movements. However, as the movements which have affected the newer series increase, the difficulty of discovering discordances by general field relations is increased. These and other cases of general field relations which show unconformity may not appear to the observer while doing detailed work, since no contact or other ordinary indication of unconformity is found, but strongly appear when the work is platted. To the mind of the writer, general field relations of the kinds above cited are more decisive evidence of unconformity than almost any kind of local relations. When the local proofs above considered, combined with general field relations, unite to show an unconformity, as is generally the case if the work takes advantage of all the facts available, the accumulated evidence for discordant relations, even in difficult and folded regions, is oftentimes decisive.

Unconformities have been frequently inferred on insufficient ground. This has sometimes resulted from regarding surface igneous rocks as sedimentary, and the basal conglomerate overlying such a formation as evidence of unconformity. More frequently the misinterpreted evidence of unconformity is a discordance in foliation; sometimes cited as occurring in actual contact, but at other times being a discordance only in the strike and dip of foliation at some distance. A contact discordance of foliation or bedding may occur as a result of faulting. The strikes and dips of banded and contorted schists and gneisses often vary so greatly within short intervals that a difference of this kind can not be taken as indication of discordances. This error has occurred because it has been assumed that cleavage-foliation accords with sedimentation. When it is practically, not theoretically, recognized that this structure is secondary, may be produced in either sedimentary or igneous rocks, and that it generally does not correspond over large areas with bedding, such evidence will cease to be used as indication of structural discordance.

The application of the foregoing principles demands that in working out the structure of the crystalline formations the ground must be gone over in detail. No single section will be adequate to give a proper idea of the structure, nor will it do to consider that, as a result of several or a dozen sections, the structure of a large district may be worked out and the formations mapped, as has been too frequently done. Formations which outcrop in one section may not be exposed in another, or a formation between one section and the next may entirely change its character or disappear altogether. If the district is a difficult one, the only safe way is to take advantage in the field of all available overground and underground facts, and to collect abundant material for supplementary office work. When only one structure is present and the character of that is in the least doubtful, it must not be assumed to be bedding. In regions in which the exposures are infrequent, it may be impossible to work out the structure of crystalline rock series which have a true detrital succession, but it is better for the present and the future that no structure be presented than a false one.

After working out the structure of one district in a geological basin, adjacent districts may be mapped much more rapidly. Under proper checks the lithological character of individual beds may be assumed to remain the same. Sets of like formation, occurring in the same order, may be assumed to be the same group of formations. And perhaps most useful of all are discordant relations between series. In correlations from region to region, without the assistance of paleontology, it will probably not be possible to carry the analogy farther than series; and in far-distant regions even the general lithological likeness and similarity of position of series is not sufficient warrant for placing them opposite each other in the time column.

If the foregoing principles are true, it is plain that in working out the structure of a new region local names should be applied to the formations and series. When the time comes that fuller knowledge enables them to be safely correlated with the series to which classical names have been applied, this may be done, and the local names will not be less serviceable to designate particular parts of these general series.

The area in North America in which detailed mapping has been done, with a resultant proper understanding of the structural relations of the pre-Cambrian, is surprisingly small. Scarcely a crystalline area on the continent has escaped the rapid geologist, who has passed over a region and, upon a few facts of uncertain value, published structural conclusions which are not to be verified by future work. The districts carefully studied include the Original Huronian of Lake Huron, several small areas about Ottawa and between the Ottawa River and Lake Ontario, a few small areas about the Lake Superior region, a small part of western Massachusetts, and a part of Maryland. Even in these districts, the work at many points is rather old and to a certain extent unsatisfactory. Before reliable maps can be obtained, this old work must be thoroughly revised in the light of the recent advances in the methods of study of the crystalline rocks. By this it is not implied that the more general work is not of superlative value and of necessity must precede the more accurate accounts. A beginning has been made in American pre-Cambrian stratigraphy, but the great mass of work remains for the future.

B.—DISCUSSION SUR LA CORRÉLATION DES ROCHES CLASTIQUES.

Mr. GILBERT opened the discussion by presenting a general classification of methods of correlation.

Strata are locally classified by superposition in chronologic sequences. Geologic correlation is the chronology of beds not in visible sequence. For convenience in discussion, methods of correlation are classed in ten groups, of which six are physical and four biotic.

PHYSICAL METHODS OF CORRELATION.

1. *Through visible continuity.*—The outcrop of a stratigraphic unit is sometimes traced for a considerable distance. The structural identity of its parts is thus established and their chronologic identity is inferred. Strictly speaking, the chronologic indentity thus determined is closely approximate rather than absolute, but it is so far superior to that demonstrated by any other method that it is usually regarded as absolute. Unfortunately the field of application of this method is narrowly limited. No stratigraphic unit is of universal extent and none remains both entire and distinct for long distances. In one direction it may grow thin and disappear, in another divide, in another gradually approximate in character to the units above and below until it can no longer be distinguished. Moreover, through faulting and through erosion many beds are rendered discontinuous, and through subsequent deposition of various kinds outcrops are buried and continuity is concealed. In these various ways the use of the method is obstructed and it becomes necessary to resort to others affording less definite results.

2. *Through lithologic similarity.*—In this method rocks of the same character occurring at different places are inferred to have the same age. Before close attention had been given to the processes of deposition the method was used widely and without discrimination, but this use was greatly restricted when it came to be well understood that at each geologic date the conditions of deposition have varied from place to place and the characters of deposits have varied with them. The method is now obsolescent except for cases in which distances are short. Where

151

the first method fails because outcrops are partly covered the second is employed to correlate the rocks of neighboring exposures.

3. *Through similarity of lithologic sequence.*—Where two or more beds at one locality correspond in lithologic character and thickness with the beds observed in another locality and succeed one another in the same order, the correlation of the strata of the two localities is more trustworthy than when the correspondence is confined to a single bed. This method has great use in the tracing of terranes from point to point within the same geologic province. It is sometimes used in comparing the strata of two provinces, but the propriety of such use may be questioned.

4. *Through unconformities.*—The time intervals corresponding to unconformities render them of great value in classification of local stratigraphic columns, and almost of necessity the parts into which they divide local columns are compared one with another in the discussion of correlations. If the same time intervals were everywhere recorded the problem of correlation would be greatly simplified, and there is great temptation to postulate the universality of unconformities. Unfortunately there is no foundation for such a postulate. Emergence and submergence appear to be local elements of geologic history and unconformity has only local value as a means of correlation. This method, like the preceding, is often valuable within a single geologic province and usually misleading when employed in the comparison of two provinces.

5. *Through the simultaneous relations of diverse deposits to some physical event.*—A beach, for example, may be correlated with lacustrine deposits encircled by it and with alluvial deposits of a neighboring stream. A base level plane may be correlated with a continuous subaqueous deposit. A marginal moraine may be correlated with overwash gravels on one side and a ground moraine on the other. In the Pleistocene, glacial deposits of various kinds are widely correlated with reference to a climatic episode assumed to arise from some general cause. The use of this method implies large knowledge of the conditions under which the deposits, or other features correlated, were produced, and for this reason it has thus far been applied only in cases where the phenomena compared are of so recent origin as not to be concealed by subsequent deposits.

6. *Through comparison of changes deposits have experienced from the action of geologic processes supposed to be continuous.*—For example, the older and newer drift deposits of different regions are correlated according to the relative extent of their weathering and erosion. As recent deposits are not usually indurated, as older deposits are usually indurated, and as the oldest deposits are metamorphosed, induration and metamorphism give presumptive evidence of age, but this pre-

sumption is not strong and yields to other evidence. In the case of pre-Cambrian rocks, to which biotic methods of correlation do not apply, relative metamorphism is sometimes appealed to as evidence of relative age, and tentative correlations are made by this method.

It may be said of physical methods in general that their application is limited by considerations growing out of the stratigraphic distribution of geologic processes and of other physical conditions. The distribution and rate of sedimentation are determined by diastrophic movements, and the character of local sedimentation is largely influenced by the same movements. So long as the causes and laws of diastrophic change are unknown the legitimate use of physical methods of correlation will necessarily be local only.

BIOTIC METHODS OF CORRELATION.

7. *Through relative abundance of identical species.*—In the application of this method a series of faunas belonging to one geologic province and bearing a definite relation to a local stratigraphic system is assumed as a standard, and with this standard a fossil fauna of some other province is compared. If a large number of the species of the new fauna are found to be common to one of the standard faunas, and comparatively few species of the new fauna are common with other faunas of the standard series, correlation is made with the fauna exhibiting the greater number of common species. In case the common species constitutes a large percentage of either of two faunas correlated great confidence is felt in the correlation. If the percentage is small less confidence is felt, because it not infrequently happens that a new fauna holds a few species in common with each of several standard faunas[*] differing in age.

8. *Through relative abundance of allied or representative species.*—When a new fauna to be compared possesses few or no species identical with species of the standard faunas consideration is given to allied forms, as, for example, to those belonging to the same genus, and correlation is otherwise made in the same way. Results by this method do not command the same confidence as results by the preceding.

7a and 8a. It sometimes happens that two or more faunas occurring in successive deposits within the same province are compared with a standard series of faunas. In case their independent consideration leads to their several correlations with faunas of the standard series having the same order of sequence, then the individual correlations strengthen one another and the concurrent result is accepted with cor-

[*] The word "fauna" in this paragraph and those which follow should be understood to include species of plants as well as species of animals. A term is needed for this function.

responding confidence. The case is similar to that of correlation through similarity of lithologic sequence (3) as compared with correlation through lithologic similarity (2).

9. *Through comparison of faunas with present life.*—This method was proposed by Sir Charles Lyell for the classification of Tertiary deposits. He put into one category every fauna the number of whose living species bore a certain ratio to the total number of its species living and extinct. The standard percentages he indicated for the principal divisions of the Tertiary no longer obtain because the opinions of students as to what constitutes specific identity have undergone change, but the general principle to which he appealed—that the later Tertiary faunas are more closely related to living faunas than are the early Tertiary faunas—is still sparingly used in correlation. The underlying postulate that faunal change proceeds in all districts and under all conditions at appproximately the same rate is open to serious doubt, and the method can not claim high precision, but it nevertheless has a certain utility, as it is available in some cases to which no other methods apply.

10. *Through the relations of faunas to climatic episodes.* —This method is used only in the discrimination of Pleistocene deposits. It has been determined in numerous instances that during the Pleistocene boreal forms migrated to lower latitudes, and a Pleistocene date is therefore inferred for faunas indicative of a colder climate than now exists in the localities of their discovery.

As in the physical, so in the biotic methods of correlation, the principal limitations arise from the phenomena of geographic distribution. The only methods applicable to the faunas of the earlier periods are the seventh and eighth, and the difficulties inherent in their use for the correlation of faunas widely separated are illustrated if attempt is made to apply them to living faunas. Assuming that the faunas of opposed hemispheres were as diverse in Mesozoic and Paleozoic time as they are now the determination of world-wide synchrony is manifestly fraught with difficulty.

The strength and the weakness of biotic correlation are illustrated when the correlation of a new fauna is independently attempted by means of subfaunas of which it is constituted. By reason of the specialization of paleontologic work it frequently happens that the vertebrates, the invertebrates, and the plants of a new fauna are separately compared with standards. In some cases the results are accordant, in other cases discordant, and the fact that discordance often results has tended to diminish the confidence with which unqualified determinations by paleontologists are received. It has sometimes been found, however, that as data multiply and biotic correlations are revised there is a tendency toward harmony in the results obtained by the students of different biotic groups.

Some species are long lived, surviving not only from epoch to epoch but from period to period; others are short lived, terminating their existence within a single epoch, or, so far as the record tells us, within a single age. A short lived species is diagnostic of the epoch or age to which it belongs and is valuable for purposes of correlation. A species continuing through many ages without change is of little value. Some species are widely dispersed, occurring in different continents or in different oceans; others are confined to a small geographic province. Widely distributed species serve the purposes of correlation; highly localized species are rarely of use. In general the value of a species for purposes of correlation is inversely as its range in time, and directly as its range in space. Unfortunately the species with the widest range in space are apt to range widely in time also.

The value of a biotic group for purposes of correlation depends (1) on the range of its species in time and space, (2) on the extent to which its representatives are preserved.

Mr. VON ZITTEL spoke in reference to the biotic methods and gave his opinion of the relative value of plants and animals for purposes of correlation. He regarded plants as relatively unimportant. Among animals those which are marine, lacustrine, and terrestrial may be distinguished. Of these classes marine invertebrates are most valuable for purposes of correlation. The vertebrates change rapidly, but are frequently altogether wanting. For instance, almost no vertebrates occur in the Alpine beds corresponding in age to those which contain the mammalian fauna of the Paris basin. In certain lacustrine deposits invertebrates may be absent, and in such cases the vertebrate fauna is the surest guide.

Mr. DE GEER points out that statistical analyses of the fossil faunas, with reference not only to the number of species but also with regard to that of individuals, afford valuable data for a more objective and far-reaching study of the different localities, of their varying bathymetrical and climatical conditions, and of the succession in which the species that have been preserved appear and disappear in each locality. It is evident that two faunas which, from a bare enumeration of the species, may seem to be almost identical, may very well be quite distinct as regards the really characteristic and predominant species. On the other hand, faunas with but few species in common may be more allied than would appear from mere lists of species, if those which are common to the two are the most predominant ones.

In rocks of slight coherence which can be thoroughly searched the fossils in a given cubic mass should be counted, in harder rocks those in a square surface; and, even where neither method is possible, it is always better to give the number of individuals found than to give only the names of the different species, if only to demonstrate the imperfection of the paleontological record.

He has for several years employed such statistical analyses of Pleistocene and Cretaceous faunas in Sweden, and is convinced that this method of investigation is important, wherever one attempts to pass from the first rough approximations in biotic correlation to a more exact tracing of the different horizons.

MR. MARSH expressed his agreement in general with the conclusions communicated by Prof. von Zittel, but would give special weight to vertebrate fossils.

In 1877, he had endeavored to bring together some results of his researches, in the Rocky mountain region and in other parts of the country, relating to the succession of vertebrate life.* This led to a comparison of the relative value of the three different groups of fossils—plants, invertebrates, and vertebrates—in marking geological time. In examining the subject with some care, he found that, for this purpose, plants are not satisfactory witnesses; that invertebrate animals are much better; but that vertebrates afford the most reliable evidence of climatic and other geological changes. The subdivisions of the latter group, and, in fact, all forms of animal life, are of value in this respect, mainly according to the perfection of their organization, or zoological rank. Fishes, for example, are but slightly affected by changes that would destroy reptiles or birds, and the higher mammals succumb under influences that the lower forms pass through in safety. The special applications of this general law, and its value in geology, readily suggest themselves.

In accordance with this principle, he next attempted to define the principal geological horizons in the West which he had personally investigated, and then taking in each the largest and most dominant vertebrate form which characterized it, used the name for the horizon. In the same way, some of the principal horizons of the East were named, and the whole brought together in a section to illustrate vertebrate life in America.†

The names thus given to various horizons were not intended to replace those already applied, but merely to supplement them, and by new evidence, to clear up those in doubt. The same principle had long before been found to work admirably in Europe, where certain characteristic invertebrate fossils, especially ammonites, had served to mark definitely various subdivisions of a single formation. The wider application of the principle to vertebrate fossils, from their earliest known appearance to the present time, has already helped to complete the record of vertebrate life in America, and rendered an equal service to systematic geology.

---

* Introduction and Succession of Vertebrate Life in America. Address before the American Association for the Advancement of Science, Nashville, Tenn., August 30, 1877.

† The same address, frontispiece.

Cenozoic.		Recent. Quaternary.	Tapir, Peccary, Bison, Llama. *Bos, Equus, Megatherium, Mylodon.*
	Tertiary.	Pliocene.	Equus Beds. — *Equus, Tapirus, Elephas.* Pliohippus Beds. — { *Pliohippus, Tapiravus, Mastodon, Procamelus, Aceratherium, Bos, Merotherium.*
		Miocene.	Miohippus Beds. — *Miohippus, Diceratherium, Thinohyus, Protoceras.* Oreodon Beds. — *Oreodon, Eporeodon, Hyænodon, Hyracodon, Moropus.* Brontotherium Beds. — { *Brontotherium, Brontops, Allops, Titanops, Titano- therium, Moschippus, Elotherium.*
		Eocene.	Diplacodon Beds. — *Diplacodon, Epihippus, Amynodon.* Dinoceras Beds. — { *Dinoceras, Tinoceras, Uintatherium, Palæosyops, Oro- hippus, Hyrachyus, Colonoceras.* Heliobatis Beds. — *Heliobatis, Amia, Lepidosteus.* Coryphodon Beds. — { *Coryphodon, Eohippus, Hyracops, Lemurs, Ungulates, Tillodonts, Rodents, Serpents.*
Mesozoic.		Cretaceous.	Laramie Series, or Ceratops Beds. — *Ceratops, Triceratops, Claosaurus, Ornithomimus, Mammals, Cimolomys, Dipriodon, Sclenacodon, Nan- omys, Stagodon, Birds, Cimolopteryx, Coniornis.* Fox Hill Group. Colorado Series, or Pteranodon Beds. — { Birds with Teeth, *Hesperornis, Ichthyornis.* *Mosasaurs, Edestosaurus, Lestosaurus, Tylosaurus.* *Pterodactyls (Pteranodon).* *Plesiosaurs.* Dakota Group.
		Jurassic.	Atlantosaurus Beds. Baptanodon Beds. Hallopus Beds. — { *Dinosaurs, Brontosaurus, Morosaurus, Diplodocus, Stegosaurus, Camptosaurus, Ceratosaurus.* Mam- mals, *Dryolestes, Stylacodon, Tinodon, Ctenacodon.*
		Triassic.	Otozoum, or Connec- ticut River, Beds. — { First Mammals (*Dromatherium*). Dinosaur Footprints. *Anchisaurus, Ammosaurus.* Crocodiles (*Belodon*).
Palæozoic.		Permian.	Nothodon Beds. — Reptiles (*Nothodon, Sphenacodon*).
		Carboniferous.	Coal Measures, or Eosaurus Beds. — First Reptiles (?) *Eosaurus.* Subcarboniferous, or Sauropus Beds. — First known Amphibians (Labyrinthodonts), *Sauropus.*
		Devonian.	Dinichthys Beds. — *Dinichthys.* Lower Devonian.
		Silurian.	Upper Silurian. Lower Silurian. — First known Fishes.
		Cambrian.	Primordial.
Archæan.		Huronian.	No Vertebrates known.
		Laurentian.	

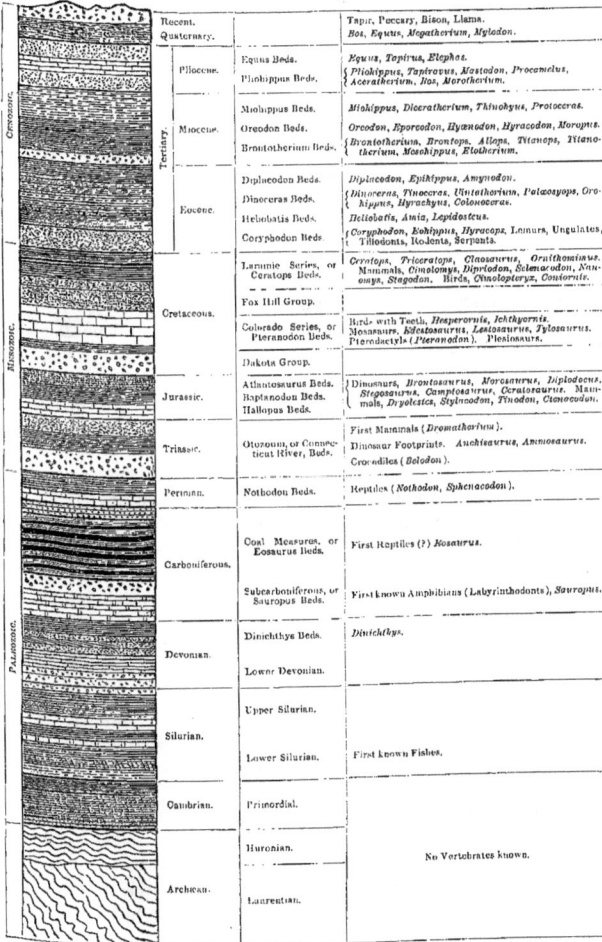

SECTION TO ILLUSTRATE VERTEBRATE LIFE IN AMERICA.

Since this method of defining geological horizons by vertebrate fossils was first used by him in 1877, many important discoveries have been made, especially in the West, and much information bearing on the subject has been obtained from various quarters. In 1884, he revised and extended the first section for his monograph on the *Dinocerata*, and it seems fitting on the present occasion to bring together once more some of the later evidence, and place on record the more important horizons now known to him by personal exploration, or by other investigations which he has verified.

The accompanying section is designed to represent in outline, in their geological order, the successive horizons at present known with certainty from characteristic vertebrate fossils. The correlation of these horizons with those determined on other evidence is important, and considerable progress in this direction has already been made, but the results can not be presented here.

In comparing the present section with the one first published, it will be noticed that no vertebrates are yet known in the Archean or Cambrian, but a single fortunate discovery in Colorado has recently carried back the first-known appearance of fishes from the Lower Devonian to the Lower Silurian, or more specifically, from the Schoharie Grit to the Trenton.

The next point of importance is in the Triassic, in the horizon of the Connecticut river sandstone, where so many footprints have been found and attributed to birds. Recent discoveries in these beds have shown that at least three distinct forms of carnivorous Dinosaurian reptiles, all of moderate size, lived at that period and doubtless did their share in leaving footprints behind them. In two of the skeletons secured, the bones of the hind feet are still in position, and in life could have made some of the footprints previously discovered.

Near the base of the Jurassic, a new horizon may now be defined as the Hallopus beds, as here alone remains of the remarkable reptile, *Hallopus victor*, have been found. Another diminutive Dinosaur, *Nanosaurus*, occurs in the same strata. This horizon is believed to be lower than the Baptanodon beds, although the two have not been found together. The Hallopus beds now known are in Colorado, below the Atlantosaurus beds, but quite distinct from them.

The Baptanodon beds have been found at many localities in Dakota, Wyoming, and northern Utah, everywhere beneath the Atlantosaurus beds, and having below them, at various localities, a series of red beds, which may, perhaps, contain the Hallopus horizon, but are generally regarded as Triassic.

Besides the two species of *Baptanodon* described by him, the next vertebrate in importance, in the same horizon, is a small Plesiosaur, which may be called *Pantosaurus striatus.** One specimen only has been found in northern Wyoming.

The Atlantosaurus beds of the upper Jurassic are now known to be one of the best marked horizons yet discovered. They have been traced for more than 400 miles along the eastern flank of the Rocky Mountains, and nearly everywhere contain great numbers of fossil vertebrates, especially gigantic Dinosaurs and other reptiles, as well as many diminutive mammals of primitive types. The same deposits have been found on the western slope, with the Baptanodon beds beneath them.

The most remarkable of the new horizons recently determined are the Ceratops beds, in the Laramie series, at the top of the Cretaceous. This horizon is as strongly marked as that of the Atlantosaurus beds, and has now been traced for nearly 800 miles along the eastern base of the Rocky Mountains. Toward the north, it is underlaid by marine Cretaceous strata containing Fox Hill fossils, but further south, various older formations are found immediately beneath it. The overlying strata, when present, are usually of Tertiary age. The Fort Union Eocene beds on the upper Missouri, the Brontotherium beds of the Miocene in Wyoming, and further south in Colorado the Pliohippus beds of the Pliocene, may be seen immediately above. The vertebrate fauna of the Ceratops beds is remarkably rich and varied. The gigantic horned Dinosaurs, named by the author the *Ceratopsidæ*, especially abound, and determine the horizon with accuracy. Other Dinosaurs are numerous; and a few birds, and various mammals, some of them of Mesozoic types, have also been secured.

In the various horizons of the Tertiary, as repeated in the present section, p. 157, no changes of importance are required, as more recent discoveries fully confirm their value and accurate determination.

Mr. HUGHES spoke of the present and growing tendency toward a natural classification. The evidence is complex and includes a considerable variety of diverse relations. He pointed out exceptions to the normal conclusions deduced from superposition, lithological character, and similarity of sequence. We must have a system of criteria so varied that if one or more fails others can be employed. All classes of evidence are useful, both positive, negative, and circumstantial. He did not join in the general depreciation of the value of vegetable remains for purposes of correlation. It was true that in the present state of our knowledge there were difficulties, because we rarely found anything but fragments; but, where only bits of broken shell or chips of bone were found, the doubt as to determination was equally great. The difficulty was not inherent in the nature of the remains but in their generally imperfect state of preservation.

Mr. POWELL spoke of the necessity of specialization on the part

* The generic name here used may be substituted for the one first proposed (*Parasaurus*), which proves to be preoccupied.

of geologists engaged in the work of correlation. The evidence derived from physical and biotic facts might apparently disagree. But that a satisfactory result may be reached, these two classes of evidence must be brought into harmony. He cited an example from his own experience of how an identification of synchronous formations might be made over a wide area through a union of physical and biotic methods.

Mr. McGee set forth in some detail the principles of correlation developed in the coastal plain of the Atlantic and Gulf slopes of the United States. The province is large, extending in a broad, meandering zone over 15 degrees of latitude and 25 degrees of longitude. Some of the formations of the province are chiefly littoral and devoid of fossils, while others are fossiliferous; and it became necessary to devise methods of physical correlation in the investigation of the unfossiliferous formations. Subsequently it was found that the extension of the same methods to the widespread fossiliferous formations of the province is not only practicable but yields a physical history so definite as to indicate the geographic distribution of organisms and entire faunas during past ages, and thus to relieve paleontology, in this favorably conditioned province, from the necessity of postulating uniformity in the distribution of past life. Accordingly, physical correlation alone is employed in the surveys of the Coastal plain. The bases of correlation accord with those outlined at the opening of the discussion with certain minor modifications and an essential addition, i. e., correlation by homogeny.

LOCAL DISCRIMINATION AND CORRELATION.

Visible continuity.—The formations of the Coastal plain range in age from Pleistocene to early Cretaceous or late Jurassic, and are little disturbed in attitude, little altered in texture; all, indeed, are commonly unconsolidated, and lithifaction is local and exceptional. Thus, the littoral deposits of different ages are so closely similar as sometimes to be discriminable with difficulty. In such cases the deposits are discriminated wherever they display distinctive physical characteristics, and traced by visible continuity through different parts of the same exposure or through adjacent exposures in which the distinction fades. Thus, visible continuity is useful for local and sublocal discrimination, and at the same time for the correlation of the varying phases of variable deposits; and, moreover, in so far as it facilitates correlation, the method is useful as an index to the relations of the variable deposits and the agencies by which and conditions under which they were laid down.

Lithologic similarity.—Many of the formations are inconstant in composition, and contain elements common to associated deposits; yet by concentrating attention on specific features they may be traced, despite constantly diminishing prominence, into sections which on first inspection baffle discrimination. Moreover, certain formations are characterized by peculiar lithologic characters by which they may commonly be discriminated wherever found (e. g., the Columbia formation alone carries a bed of local and sublocal boulders; the Lafayette formation contains an essentially peculiar matrix of red loam made up chiefly of residuary clays from contiguous provinces; the Potomac formation alone (generally) includes an arkose, etc.), and although discriminations made in this way require checking they are locally useful.

Similarity of sequence.—The formations of the Coastal plain are commonly thin and variable in composition, so that the divisions into which they may be separated are seldom traceable for distances; but local bedding is frequently repeated in neighboring sections and thus facilitates correlations of formations within limited areas; and, moreover, the repetition of the formations themselves (which are often so thin that they may be treated as beds) in neighboring sections frequently facilitates the correlation of sequences over considerable distances, so that by comparison of sequence the several formations in a given drainage basin may be correlated in contiguous drainage basins and so from place to place throughout large portions of the province. But the sequences of beds, like the physical relations, are used chiefly as local bases for a more widely applicable system of correlation.

CORRELATION THROUGHOUT THE PROVINCE.

The formations of the province are commonly separated by unconformities or related physical breaks, and these unconformities, viewed singly or in sequence, may be used, like the formations, simply as physical features in correlating the sequences (formations and unconformities) exposed in neighboring or more distant sections; but each unconformity in the Coastal plain series is found to stand for a definite attitude of the continent at a period roughly fixed by the formations above and below. Accordingly each unconformity is made a subject of comparative study; it is traced from place to place and eventually to remote parts of the province in order to ascertain (1) whether the ancient surface which it represents is rugose, thereby indicating that the land stood high above baselevel and that degradation proceeded rapidly; (2) whether the ancient surface is smooth, thereby indicating that the land stood near baselevel and was slowly planed; (3) whether the ancient surface is in some places rugose and in others smooth, thereby indicating that the land was warped during the period of degradation which the surface represents; or whether (4) the relations of the ancient surface to tide level and to the present surface is variable, thereby indicating that it is deformed with respect to associated formations or the present surface, etc. Thus the physical breaks are treated as indices

451 GE——11

of former topography, which is gradually reconstructed ideally or graphically, and which in turn elucidates the geography of the past; and in this way definite pictures are gained of different stages in the physical development of the continent. Then, since interpretation and correlation are reciprocal, the successive stages determined for contiguous areas in the province may be compared, and their consonance or dissonance serves to facilitate or limit correlation of the breaks themselves and of the formations which they separate. In this way unconformities are employed for correlation not simply as physical features but as records of important events in geologic history; and the correlation represents juxtaposition of events rather than of the simple phenomena in which they are recorded.

CORRELATION WITH CONTIGUOUS PROVINCES.

The Coastal plain laps against and partly overlies (1) the Gulf slope of a composite province in which the Colorado plateau, the Rocky mountains, and the plains merge, and which is composed chiefly of Cenozoic and Mesozoic formations; (2) the province of the interior basin with the subprovince of the Ozarks, in which the formations are Paleozoic; (3) the Appalachian and Cumberland provinces, in which the formations are uplifted or deformed Paleozoic; and (4) the Piedmont province, which is made up of ancient crystallines.

Through recent progress in geomorphy, or the interpretation of land forms, it has become feasible to interpret geologic history from land forms as well as from deposits and to restore partially degraded surfaces from the study of their remnants. In this way the land forms in the provinces against which the Coastal plain laps have been found to yield more or less definite records of continental attitude during various periods in the term extending from the present as far back as the beginning of the Cretaceous—i. e., the term represented by the deposits of the Coastal plain. Now certain unconformities (i. e., ancient surfaces) in the Coastal plain series of deposits are sometimes continuous with, sometimes closely similar to, remaining or restored surfaces in the adjacent provinces; and in many cases the continuity is so clear or the homology so close as to warrant identification of the ancient surfaces in the contiguous provinces. Thus the provinces which are distinct in the present stage of continental development, and which have been distinct during various earlier stages, may be correlated by means of phenomena produced during stages in which they were more closely related. Moreover, since correlation effected through interpretation prepares the way for further interpretation, and hence eventually for more exact correlation, the identification of stages in continental development in adjacent provinces gives definite pictures of the local topography and general geography during these successive stages. The form and steepness of slopes of ancient surfaces represented by remnauts in the land province and by unconformities in the Coastal plain indicate relation to baselevel, and, in connection with the amounts of degradation and coeval deposition, the relative duration of the degradation period; the relation of the ancient surface to the present surface gives an indication and sometimes a measure of subsequent deformation; and the relation of ancient surface to tide level suggests and sometimes clearly indicates the sources of materials deposited in the Coastal plain formations during the periods of submergence in the Coastal plain province. Accordingly the comparative study of ancient surfaces permits correlation of deposits in contiguous provinces, and at the same time affords a means of determining the attitude of a considerable part of the continent during the intervals of high level and consequent interruption of deposition; and the high-level record, coupled with the record of the formations, gives a more or less complete history of geographic stages in the evolution of the continent, or of geologic growth.

In some cases the correlation of ancient surfaces is indefinite (1) by reason of limited exposures of unconformities, (2) by reason of extensive degradation in the contiguous land province, or for some other reason; but in most such cases the formation record supplements the geomorphic record. Thus when a continent stands near base level, degradation is chemical rather than mechanical, the shores are low and sweeping, the land contours are soft, and the relief is low and well modulated, while the formations laid down in the correlative deposition area, or on the former land surface immediately after submergence, are made up chiefly of precipitates and chemically degraded as well as mechanically divided sediments. If on the other hand the land stands high above baselevel, degradation is chiefly mechanical, the shores are rugged, the rivers are swift and soon carve deep, steep-sided gorges, giving angular land contours and an ill modulated surface, while the coeval formations are made up largely of coarse mechanical sediments in which the pebbles and smaller grains remain in part undecomposed. Thus the record of the formation and the record of the unconformity and the contiguous land surface are supplementary, and give independent yet consistent and cumulative testimony concerning continental condition, and the juxtaposition of the two commonly serves to bridge the gaps in both.

GENERAL CORRELATION.

The correlation by visible continuity and in general the correlation by lithologic similarity and measurably the correlation by similarity of sequence are objective and empiric, i. e., these modes represent the juxtaposition of objects and comparison of their extrinsic characters; while the correlation of formations and unconformities viewed as records of geologic agencies and continental conditions is subjective and philosophic and, pursued to its logical end, becomes a juxtaposition of episodes, or a correlation by historical similarity. The application of this mode of correlation involves such a study of agencies and conditions of geologic action as to enable the geologist to determine provisionally the origin of each phenomenon examined, whether deposit or topographic feature, formation or land form; and the subsequent comparisons involved in the correlation are comparisons of genetic records, which may be made in such manner as to eliminate the incongruous and preserve the congruous, and thereby develop a consistent history for the entire province under examination. This method has already been characterized as homogenic, i. e., correlation by homogeny, or identification by origin.* The method has been more or less closely approached elsewhere by different geologists, but so far as known it has not been definitely formulated by other investigators; certainly it has been independently developed in the course of the surveys in the coastal plain, and, it is believed, constitutes an important addition to the modes of correlation commonly employed.

In the practical application of the method, the deposits of given sections and circumscribed areas are first correlated empirically by visible continuity and lithologic similarity, and to some extent by similarity of sequence, in order that their relations may be generalized; next, the agencies of genesis are inferred from the materials of the deposits viewed individually and collectively; then the unconformities and pebble beds, with other aberrant phenomena, are generalized, and from them, in connection with the normal deposits, the conditions of genesis (i. e., the attitude of land, proximity of rivers, etc.) are inferred. By these means a tangible and definite picture of topography, geography and geologic agencies of the area is produced; and the various inferred features are tested by their consistency and the inconsistent eliminated or withheld for more extended comparison. Then the history of the contiguous area is wrought out in similar fashion and the episodes are compared severally and jointly, and the deposits and uncomformities are interpreted in the light of this comparison. The comparison is eventually extended to other portions of the province and to the contiguous provinces, and in each area the significance of the sum of phenomena is sought and the inferred histories are generalized by combination of the congruous and elimination of the incongruous until finally the history of a given period throughout the entire province is interpreted in terms of episodes, each inferred from the sum of phenomena representing the period.

---

* American Journal of Science, 3d series, vol. XL, 1890, p. 6; "The Lafayette Formation," U. S. Geological Survey, 12th Annual Report, 1893, p. 381.

The method of correlation by homogeny has not yet been extended to all of the formations of the province; but it has been applied successfully in the surveys of the unfossiliferous littoral formations (the Pleistocene Columbia and the late Neocene Lafayette as well as the early Cretaceous Potomac) and so far extended to the fossiliferous formations, in part by collaborators, as to establish its applicability to deeper water deposits. Though at first sight apparently complex the method is really simple, and some of its results are of exceptional interest. Thus, largely through the application of homogenic correlation in the Coastal plain and contiguous provinces, the general law of continental movement in southeastern United States has been made out with considerable clearness. It may be stated briefly thus: (1) In general the post-Jurassic history of the subcontinent represented by southeastern United States is one of progressive uplift along the Appalachian axis, progressive downthrow about the periphery (extending from the Atlantic coast through the Gulf of Mexico and some distance northward in the Mississippi valley), with concomitant seaward tilting of the Piedmont areas. (2) The movement was not uniform but spasmodic. (3) Each throe was complex, including an initial warping and a series of oscillations. (4) The first movement in each throe was (a) relative uplift of the Appalachian axis and downthrow of the periphery, coupled with (b) depression of the entire area perhaps to such an extent as to counterbalance the axial uplift, thereby (c) submerging a considerable part of the subcontinent and (d) stimulating degradation over the interior. (5) The second movement in each throe was general elevation, without recognized tilting, followed by gradual subsidence, and in some cases at least a diminishing series of minor oscillations. (6) The subsidence and elevation of each throe were coördinate, profound subsidence being followed by high elevation, limited subsidence by slight elevation. This law of continental movement, which is singularly simple and rythmic, is apparently exemplified in all of the oscillations in which the Coastal plain and the contiguous land area have participated.

The method of homogeny indicates with greater or less accuracy the relation of land and water, rivers and sea-bottoms, during different periods; and through its application the formations with the unconformities and the ancient land surfaces of contiguous provinces give data for the construction of geotectonic maps showing, in panoramic succession if desired, the successive stages in continental evolution; and some such maps have already been constructed. In this way the method gives opportunity for the development and presentation of much more definite and tangible illustrations of geologic growth than can be obtained from any method of correlation which is confined to formations and does not include the coeval land surfaces.

While the method by homogeny has thus far been practically confined to the Coastal plain and contiguous provinces, it is believed to be capable of considerable expansion in this and other countries. Of course the utility of the method culminates in broad provinces of simple structure and limited age; but it will doubtless be found measurably useful in many other cases. It is pursued in the Coastal plain chiefly because of its special applicability thereto, and with the hope that it will permit so clear a definition of the various episodes in continent growth as to indicate eventually the geographic distribution of organisms during each episode, whereby paleontology will be raised to a higher plane. For this purpose the extended province found in the Coastal plain of southeastern United States is exceptionably favorable.

RÉSUMÉ.

In brief the methods of correlation which have been developed and found applicable in the Coastal plain are as follows:

For local discrimination and correlation.. { Visible continuity, Lithologic similarity, Similarity of sequence.

For correlation in the province .... { Physical breaks viewed as indices of ancient topography and geography.

For correlation with contiguous provinces.. { Geomorphy and stratigraphy interpreted in historical events, including continent movements, transportation of materials, land sculpture, etc.

For general correlation .... { Homogeny, or identity of origin.

Mr. DAVIS showed that it was possible to decipher geological history not only through the records of deposition, but also by processes of degradation. As an example of this method he explained a topographical section from the city of New York westward. In this we have evidence of the existence of an ancient "*peneplain*," or base level lowland of Cretaceous age. This surface was subsequently elevated (more toward the west than toward the east) at the end of Cretaceous, or at the beginning of Tertiary time. It has since been dissected by the excavation of more recent valleys. The Hudson valley lowland was cited as an example of this recent dissection.

Mr. CLAYPOLE considered that the different methods of geologic correlation differed very greatly in their value. It is improbable that the plant or mammalian record will ever equal in its perfection that of the marine invertebrate fauna. The marine fauna is to the geologist what a primary triangulation is to the geodesist. It marks out the main divisions, which are subsequently further subdivided through the aid of other fossils, such as plants and vertebrates.

Mr. VAN HISE spoke of the methods of correlation employed for pre-Cambrian rocks, which occur in widely separated areas in the United States, and, so far as present knowledge goes, are nearly devoid of fossils. Physical data only are available for correlating these rocks. Experience has shown that among all physical criteria, unconformity is by far the most important. Other physical criteria, such as the degree of induration, metamorphism, and relations to eruptives, are valuable for the subdivision of the rocks of individual regions, but can not be safely used in identifying synchronous formations in widely separated areas. The idea that lithological character is any direct proof of geological age has retarded the scientific subdivision of pre-Cambrian rocks. The researches of Pumpelly and others in the eastern United States have demonstrated that Silurian, Devonian, and even Carboniferous rocks have become, under certain physical conditions, as highly crystalline as far more ancient rocks in the West. Since biological evidence for correlating the pre-Cambrian rocks is lacking, and since physical evidence does not warrant correlations of rock series in regions separated by great distances, it has been found necessary to abandon such names as *Huronian* and *Keweenawan* as universal terms. Evidences of life are not lacking in pre-Cambrian rocks, and it is to be hoped that the paleontologists will succeed in differentiating several periods below the Cambrian, as the Cambrian itself was differentiated from the Silurian.

Mr. HILGARD laid stress upon the importance of the abundance or scarcity of species in the correlation of strata. He thinks some quantitative estimation of the species should be made. He is of the opinion, also, that, as compared with marine fauna, plants have but little value for purposes of correlation owing to their local distribution, their accidental proximity to water, transportation, and preservation. Plants can be so used only after large areas are worked over.

Mr. WARD continued the discussion, presenting a brief summary of his communication given on preceding pages (p. 97).

The two more general principles of correlation by means of fossil plants may be stated as follows:

I. That the great types of vegetation are characteristic of the great epochs in geology.

This principle is applicable in comparing deposits of widely different age when the stratigraphy is indecisive. For example, even a small fragment of a Carboniferous plant proves conclusively that the rocks in which it occurs are Paleozoic, or a single dicotyledonous leaf proves that they must be as late as the Cretaceous.

II. That for deposits not thus widely different in age, as for example, within the same geologic system or series, ample material is necessary to fix their position by means of fossil plants.

Neglect of this principle has led to the greater part of the mistakes of paleobotanists, and has done most to bring paleobotany into disrepute. Geologists have expected too much of them, and they, in turn, have done violence to the truth in attempting to satisfy extravagant demands. On the other hand, where the material is ample, fossil plants have often corrected the mistakes of stratigraphical geologists, and solved problems concerning geologic age, which seemed impossible of settlement by any other class of evidence.

Mr. WALCOTT spoke upon the correlation of the Cambrian rocks of North America. He stated that the principles of correlation were essentially the same as those used by the New York survey prior to 1847, and that the only essential modification made in them has been through the influence of the theory of evolution.

Both physical and biotic data are available in the correlation of the Cambrian rocks within each of the four principal provinces; but biotic data are alone available, with slight exceptions, in correlating the strata of different provinces with each other. Within the Atlantic Coast basin the Cambrian rocks of Newfoundland, Massachusetts and New Brunswick may be correlated by their physical characters and the presence of similar organic remains. The same is also true as to the Rocky mountain and great Interior continental provinces. Throughout the Appalachian province the physical data suffice to correlate the Lower Cambrian from Vermont to Alabama, but such data are not sufficient to correlate it with that of the St. Lawrence valley. It is only by the presence of the Cambrian fauna that correlations can be made. The correlation of the deposits of the Appalachian and Rocky Mountain provinces are based almost entirely upon biotic data.

A study of the sediments of the Cambrian rocks establishes the fact that in Cambrian times there were geographic areas in which different types of sedimentation prevailed, and that great variation existed in what are considered the same geographic provinces. The sediments of the northeastern Atlantic coast province are almost entirely shales, with a small proportion of sandstone and a trace of limestone. Tracing the long Appalachian province from the Gulf of St. Lawrence to the southwest and south, we find an immense accumulation of shale, with some interbedded sandstone and limestone. This extends to the Lake Champlain region of New York and Vermont, where a great limestone of Lower Cambrian age is subjacent to several thousand feet of shale, in which lenticular masses of sandstone and limestone occur at irregular intervals. Further to the southwest, in southern Vermont and eastern New York, great thicknesses of argillaceous sediment were deposited. These now form the series of shales in which beds of finely laminated roofing-slates occur in massive strata. Toward the east, near the pre-Cambrian shore line, the Lower Cambrian sandstones are followed by arenaceous, dolomitic, and purely calcareous limestones of later Cambrian time. This condition of sedimentation continues far to the south, varied more or less by the presence of great thicknesses of shale above the lower quartzite. In the latter case the limestone of the Cambrian forms only a thin belt at the summit of the section.

In the Rocky Mountain province the silicious sediments, sandstones, and quartzites, are followed by limestones, and nearer the shore line the sandstones are subjacent to shale. Over the Interior province the record is sandstone, followed on the west and southwest by alternating limestones and sandstones.

The study of the fauna has shown that, while there is a general resemblance in the faunas of the Lower Cambrian in the Atlantic Coast, Appalachian, and Rocky Mountain provinces, there is sufficient differentiation to mark off distinct faunal areas in the early Cambrian sea.

The fauna of the Upper Cambrian is less specialized over great areas than that of the Lower and Middle zones. The opportunities for comunication between the provinces were greater, and it is only in the northeastern or Atlantic coast province that a marked difference is shown.

The Middle Cambrian fauna of the Atlantic coast province is so strongly differentiated from that of the Appalachian and western Rocky Mountain provinces that it was not until they were found to occupy the same relative stratigraphic position that it was suspected they lived in approximately the same epoch.

The evidence of sedimentation and the evidence of organic remains unite to prove that in Cambrian time the geographic and faunal provinces were differentiated and established over the area of the present North American continent. In fact, the sedimentation of the Cambrian was as varied as that of many of the later geologic periods, and more so than that of the immediately succeeding Silurian (Ordovician.)

The correlations, made mostly by biotic data, between the various areas of Cambrian rocks indicate that in Lower and Middle Cambrian time a great continental area existed over the interior of what is now the North American continent, and that the Cambrian sediments were accumulated in troughs, one west of the Rocky Mountains, extending from Nevada northward into British Columbia; another westward of the Appalachian Mountains, extending from Alabama northeastward to the St. Lawrence valley, and some minor troughs on the Atlantic side of the continent in which Cambrian sediments were deposited. In Upper or late Middle Cambrian time the interior of the continent sank beneath the ocean, and the sandstones of the Upper Cambrian were deposited over that area contemporaneously with the Upper Cambrian limestones of the Appalachian and Rocky Mountain provinces.

The results of these various correlations add a chapter to the history of the evolution of the North American continent.*

Mr. JAMES HALL spoke of the difficulties encountered in the earliest attempts at correlation of the rocks, even in the State of New York. He urged the importance of taking into consideration both physical and faunal characters of the rocks. In some cases, however, the physical characters of the rocks change greatly in passing from one region to another—sandstones grading into limestones, and limestones into shales—and these beds may also vary greatly in thickness. Fossils are of unequal value in such correlations; lamellibranchs are near-shore forms and fail in deep water; they are not, therefore, so valuable for purposes of correlation as brachiopods, which have a wider distribution.

Mr. H. S. WILLIAMS urged that in using fossils as a means of correlation, it is of the greatest importance to take into consideration the relation of organisms to conditions of environment.

In the case of living organisms the fact is well established that species, and groups of species constituting faunas, are closely adjusted to the varying conditions of bottom, distance from shore, depth of water, temperature and purity of the medium. These different conditions are expressed in the rocks of sedimentation in terms of fineness or coarseness of the component particles of the rock, and their mineral characters, as conglomerates, sandstones, argillaceous or calcareous rocks. We know also that along the same shore there are often rapid transitions from one kind of sediment to another in passing from place to place.

It is therefore unsafe to presume that two faunas which may present unlike species, preserved in localities separated only a few miles from each other, are necessarily successive faunas; they may have existed at the same time, but under different conditions. It results, therefore, that along a given coast line we should expect the faunas to vary more rapidly on passing from the beach outward to deep water than in a direction parallel to the coast.

The same principle of adjustment of species to particular conditions of environment would be expressed in the fossil-bearing rocks by

---

* The maps used by Mr. Walcott in explanation of the correlations made and the deductions drawn from the results thus obtained will be found on Plates II (p. 264) and III (p. 368) of Bulletin 81 (Cambrian correlation paper) of the U. S. Geological Survey, which has been sent to all the members of the Congress. In this bulletin, which was prepared for the Congress by Mr. Walcott, will be found also a full description of the methods of correlation used in the correlating of the Lower Paleozoic formations of North America.

decreasing abundance of a particular species on passing out of the center of its best conditions of environment, so that, unless great care were taken, the differences of faunas due entirely to change of locality might be interpreted as a change of geological horizon or chronological succession.

Again, in the case of gradually sinking land, the faunas which would be preserved as successive faunas in the deposit of a single rock section would represent the faunas actually living at the same time at varying distances from the shore line.

For the above reasons, in order to obtain accurate correlation by means of fossils, it is essential to observe the nature of the deposits in which each species is most abundant as expressive of its natural environment.

In comparing successive beds, the relative age of the deposits is not to be determined by the change of genera or the appearance of entirely distinct species, but by the comparison of species closely allied.

A complete change of genera in two successive fauna may indicate a very slight change of time, and a considerable shifting of the conditions of environment, hence the only certain indication of geological succession, to be read in terms of fossils, is to be found in a modification of greater or less degree in the specific characters of closely allied species found in the same character of deposits in successive beds.

Hence the age of beds ought to be determined by a comparison of the species of the same genera, rather than by contrast of genera or even of species.

There are also centers of abundance of particular species where the species exhibit considerable variability. Among such variable forms are found the typical characteristics of the species. Outside such centers are often found varieties, which are called extra-limital forms, but in some cases such extra-limital varieties, having been first found and described, stand in descriptive literature as the types of the species.

To determine the true chronological stage of differentiation of the species at a particular horizon, it is essential to learn the typical characteristics at the center of abundance, and for this reason it is unsafe to depend upon what has accidentally been taken as the type of the described species.

Mr. FRECH said that in comparing the Middle Paleozoic fauna of Europe with that of North America, there were two principal points of especial interest:

A. The identity of some comparatively small horizons.

B. The far greater differences that exist in these same beds.

The similar faunas are—

(1) That of the Niagara and of the Wenlock shales.

(2) In the Upper Devonian, the Rhynchonella of the Tully limestones, and the Goniatites of the so-called Naples beds.

(3) The Goniatites at the base of the Carboniferous, in Iowa, in Spain, and in middle Germany.

The Hamilton fossils are of especial interest because we have on the Rhine, in the so-called Lenneschiefer, a fauna of the same facies. But while these rocks were deposited under similar physical conditions, the number of identical species in the two countries is very small, and there are many genera in each country not found in the other. All the Lower Devonian is wanting in European Russia, and part of it is wanting in middle Germany, but the great physical change which followed is quite sufficient explanation for the differences which characterize the junction of the Devonian and Silurian.

M. BARROIS ne croit pas qu'il soit possible de comparer en détail les formations paléozoïques de l'Europe et de l'Amérique. Toutefois, malgré les différences climatériques et orographiques qui distinguaient ces régions à l'époque paléozoïque et expliquent leurs différences, il y a un frappant parallelisme dans l'évolution des faunes de ces régions éloignées. Certaines faunes, probablement synchroniques, présentent des analogies d'autant plus frappantes de chaque côté de l'Atlantique, qu'elles sont séparées par des niveaux peu comparables.

J'ai cité, il y a quelques années, les rélations des Goniatites de Rockford, Indiana, (base du Carbonifère) avec celles du marbre griotte, qui occupe en Espagne le même niveau. M. de Verneuil signalait déjà les rélations de faune de Niagara et de Wenlock. M. Frech cite avec raison le faciès européen des couches à Goniatites des Naples-beds. Citons enfin le calcaire dévonien moyen à Stringocéphales de l'Ardenne, découvert récemment au Canada par M. Whiteaves.

Mr. VAN HISE spoke of the distribution, character, and succession of the pre-Cambrian sedimentary rocks of North America. All rocks are regarded as pre-Cambrian which are earlier than the Olenellus fauna. These rocks contain the evidence of abundant life as shown by thick beds of carbonaceous shales, by various distinct fossils, and in many other ways. When a less highly developed fauna is found, as different from the Cambrian fauna as the Cambrian is from the Silurian, it is best to give this fauna a new name.

There are in many areas in North America great thicknesses of little altered pre-Cambrian sedimentary rocks. In many regions these rocks have been separated into series by widespread unconformities, and these series have been further divided into formations. Some of the more important pre-Cambrian regions are Lake Superior and Lake Huron, Central Arizona, New Brunswick, Newfoundland, southwestern Montana. As an illustrative example of the successions may be cited the first. In this area the descending order is Lake Superior sandstone

(Potsdam), unconformity; Keweenawan, unconformity; Upper Huronian, unconformity; Lower Huronian, unconformity; basement, crystalline complex. Each of these series is divided into several formations.

In individual regions it is possible to correlate series and formations upon a physical basis. In different regions the series have variable lithological characters and unlike successions. Because of the absence of a well-known pre-Cambrian fauna it is impracticable at present to make correlations in far-distant regions. Hence the term Algonkian has been proposed by the U. S. Geological Survey to cover the whole of the pre-Cambrian clastics. No working geologist in America now holds the indivisibilty of the pre-Cambrian in all regions.

If the foregoing conclusions are correct, the invariable succession advocated by Hunt, evolved almost wholly within the laboratory, is valueless. It is shown to be untrue at one or more fundamental points by the observed order of the rocks in every region in which there are tolerably full successions.

Mr. PUMPELLY corroborated the conclusions of Prof. Van Hise from his own familiarity with the regions referred to. The divisibility of the pre-Cambrian is clearly established on this continent in several widely separated regions. In the Lake Superior region the Potsdam transgression, with its basal conglomerate, has beneath it three sedimentary series of great importance, and well-observed relative ages—Keweenawan, Upper and Lower Huronian. The independent age of each of these is marked by enormous transverse erosion intervals, by basal conglomerates, or by discordant stratification; usually by all three.

The oldest—the Lower Huronian—has a basal conglomerate carrying pebbles of the various rocks of the complex on which it rests.

In the two lower series the cyclical deposition is marked—basal conglomerates and sand-rocks of the transgression, overlain by the carbonate and chert deposited in deeper waters.

These cyclical depositions presuppose rhythmic movements of the ocean—positive and negative movements—which must have been accompanied by corresponding sedimentation throughout the whole communicating oceanic area. The limestone of the Lower Huronian of Lake Superior is thicker than the Cambro-Silurian Stockbridge limestone in Vermont and Massachusetts, which, by its fossils, Walcott, Dale and Wolff have shown to span the whole period of the earth's history, from the Olenellus zone of the Lower Cambrian to the top of the Lower Silurian, inclusive.

There can be little doubt that the Upper and Lower Huronian represent two very great periods of the sedimentary time scale. It is probable that these same periods are represented elsewhere at many points on the earth. But in the absence of determinable fossils their identifiation in unconnected areas is impossible. There may be little doubt that the so-called Laurentian limestones of Quebec, our own pre-

Cambrian limestones of the Green Mountains, the Adirondacks and New Jersey, represent one of these periods or, among them, both periods; or one of the Huronian periods and the Keweenawan, but actual correlation is not yet possible.

Lithological similarity is a phantom. Even on Lake Superior, where the rocks have suffered relatively little from metamorphism, the identification, when the great Lower Huronian limestone is absent, is dependent wholly upon observable relations. Both series are equally rich in iron ores. The representatives of the period may be shore deposits in one place, those of off-shore in another. Again, as a rule, outside of Lake Superior the exposures of pre-Cambrian rocks occur in regions of great folding. And here, through dynamic action and the metamorphosing processes that accompany it, the rock has not only often wholly lost its original character, but the same formation varies in the most protean manner with the varying conditions of metamorphism to which it was subjected. An excellent illustration of this is the Lower Cambrian quartzite of the Green Mountains. A simple quartzite west of the central range, it shows step by step its transition into a wholly crystalline gneiss in the folds of the range. Nor is this all; the basal conglomerate of this quartzite preserves perfectly its heterogeneous pebbles on the top of the arch of the main fold; but on the eastern side, from whence the folding force acted, the pebbles are wholly obliterated and the slipping movement has produced a finely foliated biotite gneiss with parallel lamination, and on the westward, or over and in-folded side, a white granulite.

If this could happen to Cambrian beds the same must be more generally true of the older sediments. Excepting the limestones, the petrographic character, and partly the composition, of metamorphosed sediments are more dependent upon the action of folding and of the processes genetically connected with pegmatization than upon the original character of the sediment.

In the eastern part of the continent the equivalents of the Lake Superior pre-Cambrians, where they exist, are highly crystalline schists whose clastic origin is detected thus far only by their relation to occasionally associated pre-Cambrian limestones. And it is probable that extensive areas of pre-Cambrian sediments exist on other continents in the form of crystalline schists. It is impossible to correlate them. They represent a large part of the sedimentary time-scale, and also, doubtless, of the life scale. And while each of them represents a long period, the formations of one region may represent the time-breaks of another.

In view of the impossibility of correlating its members in separated regions, and of its importance in the time-scale, it seems very proper as well as convenient to give to this long period a name coordinate with Paleozoic, Mesozoic, etc., or at least with Silurian, Devonian, etc.

M. Barrois, à propos des observations du Prof. Van Hise, dit qu'il n'existe pas de base générale, soit biologique ou lithologique, pour la corrélation des roches pré-cambriennes de l'Europe avec celles de l'Amérique du Nord, et que même les termes appliqués à ces roches sont sujets à des malentendus. Il est certain que les divisions employées en France ne peuvent pas être raccordées avec celles qui sont aujourd'hui en usage aux États-Unis. La corrélation générale ne peut, jusqu'à présent, être basée sur les discordances de stratification; l'autopsie est la seule base sur laquelle il soit possible d'établir des comparisons.

Il indique certains parallelismes entre l'histoire des schistes cristallins d'Amérique, telle que l'a fait connaître M. Pumpelly, et les roches gneissiques de Brest, où les schistes argileux cambriens sont transformés en des gneiss d'un aspect archéen, tandis que les quartzites fossilifères, qui alternent avec ces gneiss, sont devenus des quartzites cristallins. Pour arriver à un accord commun au sujet des roches cristallins, les géologues doivent examiner le terrain collectivement.

Mr. Cope discussed the question from a general point of view with especial reference to the value of vertebrates for purposes of correlation, particularly for intercontinental correlation. He pointed out that there is a marked difference in the present vertebrate faunas of continents, and that the variation of such forms must be sought in vertical rather than in horizontal ranges. Such study shows that we have had invasions of a given region by a fauna from without; for example, a South American fauna invaded North America at one time and then retreated, while a North American fauna once invaded South America and traces of it still remain in that country. He is inclined to believe that certain vertebrate forms did not spread over the earth from a single place of origin, but that they originated at different places upon the earth. We have parallelism in separate places, but the parallelism is defective in the Laramie.

Mr. Gilbert was of the opinion that many methods of correlation must be used. He doubted the trustworthiness of the correlation of non-fossiliferous rocks by comparative change, even locally. He thought the abundance and scarcity of fossil forms comparable with lithologic differences, and considered the simple occurrence of a species as valuable for purposes of correlation as its abundance.

Mr. Van Hise explained that the distinction between the Algonkian and the Archean has not been widely made in Europe, because there, as in the Appalachian region, later and powerful dynamic movements have repeatedly occurred.

Mr. Cope added that life in its progress on the earth differed from minerals and rocks in that it has its own laws, which give it an independent element.

C.—DISCUSSION SUR LA CLASSIFICATION GÉNÉTIQUE DES DÉPÔTS PLÉISTOCÈNES.

Mr. CHAMBERLIN opened the discussion with the following words:

Pleistocene formations that are not, either directly or indirectly, of glacial origin are not embraced in the classification herewith proposed.

There are at least three appropriate modes of classification of the Pleistocene glacial formations:

(1) A classification based upon the structural characteristics of the deposits.

(2) A classification based upon the origin of the formations.

(3) A classification based upon the time relations of the deposits.

Only the second will be here considered. A purely structural classification has an indispensable value and is a fit subject for more and more critical consideration as our knowledge of the glacial formations increases. The classifications of the past will doubtless continue to need extension and more precise definition as glaciology advances. Nevertheless, our structural classifications do not seem to be so lacking in exhaustiveness nor so defective in discrimination as our genetic classifications, and, as time will not permit an adequate treatment of the three forms of classification, it has seemed best to pass the first by with a mere mention.

A chronological classification of the Pleistocene formations possesses the highest interest and constitutes one of the two great goals sought by glaciology. One of these is to ascertain how the formations were produced; the other, the times and sequences of their production. But it is too early yet to fix upon a satisfactory chronological classification. The data are not yet sufficiently gathered nor have they been tested with sufficient severity to admit of satisfactory correlation of the successive glacial stages, even within the limits of a single province. How much less then those of different continents. While recognizing, therefore, the supreme interest that attaches to a chronological classification, I am impressed with the feeling that it is best to postpone a formal attempt to establish such a classification until the data shall be more adequately developed. It is believed that the development of a

176

more satisfactory genetic classification will be a step toward a more satisfactory chronological classification.

The following outline is submitted for discussion: Six general classes are proposed.

A. GENERAL CLASSES.

I.—*Formations produced by the direct action of Pleistocene glaciers.*—As very much of that which is commonly embraced, for convenience, under the general phrase " glacial formations" is not the direct and simple product of glaciers, but springs in part or in whole from accessory agencies, it is thought serviceable to discriminate the simple from the complex formations. In this classification glaciers are assumed to be the primary and chief agency in the production of the formations classified, but the secondary and associated agencies are very important, and often the final expression of the deposits is due chiefly and sometimes wholly to these auxiliary agencies.

II.—*Formations produced by the combined action of Pleistocene glaciers and accompanying glacial drainage.*—All of the ice of the glaciers, except that portion which was transformed into vapor, passed away in the form of glacial waters, and to this there was added the rain precipitated upon the glacial expanse. The combination of the work of this large volume of water with that of the ice gave rise to results which neither the ice alone nor the water alone could accomplish. These constitute a distinct class of deposits.

III.—*Formations produced by glacial waters after their issuance from Pleistocene glaciers.*—While the glacial waters were acting on the glacier, or more especially in tunnels within or under the glacier, they were constrained by the presence of the ice walls and forced into modes of action that they would not have assumed of themselves on free and open surfaces of the land. The formations of free glacial waters after their issuance from the ice are thus distinguishable from the confined glacial streams of the ice-body itself. The products of glacial waters after their issuance differ from the products of ordinary surface waters in the fact that they were overloaded with detritus in an extraordinary and peculiar way, and in the fact that this detritus differs from ordinary land waste. This difference furnishes one of the most valuable criteria in the discrimination of Pleistocene formations.

IV.—*Formations produced by floating ice derived from Pleistocene glaciers.*—Assuming that the greater mass of the glacial deposits of Pleistocene age were formed by the primary agency of glaciers, it nevertheless remains important to distinguish the secondary products that were formed through the agency of floating ice derived from these glaciers. Just as the waters from the melting glaciers bore away and deposited a certain constituent of the material that had been wrought

451 GE——12

at first hand by the glaciers, so icebergs bore off and deposited in a fashion of their own another portion. The two classes are secondary and dependent upon the original glacial action and some of the characteristics of the material must be sought in the original glacial action, but in their final stages the deposits take on characters of their own.

V. *Formations produced by shore ice and ice-floes due to low Pleistocene temperature but independent of glacial action.*—The presence of the great ice sheets and the glacial conditions that produced them appear to have given rise to a class of independent ice deposits that may fittingly be put in this classification. These are believed by some writers to attain great importance. Whatever their extent, they need to be scrupulously discriminated from the products of both glaciers and glacier-derived icebergs.

VI. *Formations produced by winds acting on Pleistocene glacial and glacio-fluvial deposits under the peculiar conditions of glaciation.*—It is not unreasonable to suppose that the tracts bordering the great glaciers, especially those recently abandoned by them, were exposed to very effective wind action. The differences of temperature between the ice-clad fields and the adjacent ice-free lands may be supposed to have induced strong wind currents. Between their strength and the facilities offered for their action by bare surfaces of recently-formed incoherent material, there is ground for the belief that æolian deposits of more than usual magnitude and of peculiar characteristics were formed. So, also, it is highly probable, if not certain, that some of the rivers of glacial times were accompanied by broad flats which they had themselves built up and which they alternately flooded and left exposed, and that here the winds found a special field of effective action. Such æolian formations as resulted from these agencies are probably not so important in themselves as in the erroneous inferences to which they are likely to lead, if unrecognized. Such wind-blown sands and silts might easily be borne up to high levels and lodge there. If these are attributed to water action (which seems the natural alternative) they lead to hypotheses of flood-heights or of submersion of much magnitude, and these lead on to other conclusions of much consequence. The recognition of the class, especially in framing working hypotheses and interpretations, is thought to have some importance.

With the foregoing general statements, we may turn to the more special consideration of the several classes indicated.

B. SUBCLASSIFICATIONS.

1. FORMATIONS PRODUCED BY THE DIRECT ACTION OF PLEISTOCENE GLACIERS.

Under this head may be recognized three subclasses:

1. Formations that gathered at the bottom of the glaciers;

2. Formations derived from material borne on the glaciers and within them (but not at their basal contact), and deposited at their margins, or let directly down by their melting when stagnant;

3. Formations produced by the mechanical action of the edge of the ice.

1. It is believed that the base of the glaciers was the great seat of action; that here took place the disrupting of the material and the larger part of the rubbing, grinding, and crushing to which it was subjected in transportation. All of this material, however, did not come to final rest beneath the ice. A portion of it was borne away by the glacial drainage, a portion was thrust up into the ice and borne along to its edge and there deposited as superglacial material, and a portion may have once been uncovered by the retreat of ice and have been subsequently plowed up by a re-advance and so have taken on a new form of aggregation. There is, therefore, a discrimination to be observed between material that was *produced* at the base of the glaciers and that which was finally *deposited* at their base. It is only the latter class that is included here. Of deposits originating under the ice the following subclasses are distinguished:

(1) *Subglacial sheets of till.*—These constitute one form of ground moraines; the form which is perhaps most commonly recognized. There are to be embraced here those broad sheets of till which were spread out under the ice and left, on its retreat, as a blanket mantling the surface of the land. These sheets are not uniform in thickness nor universal in their presence. In this classification it has been thought best to separate all distinct and special forms of aggregation from the more nearly uniform sheets of till and to place them in the following subclass. This is done under the belief that the causes of these special aggregations were somewhat special and peculiar, and that these forms are worthy of distinction for working purposes until their final significance and classificatory value shall be determined.

(2) *Subglacial aggregations of till.*—These admit of subdivision into two varieties between which there is no sharp dividing line and which are perhaps separated from each other genetically only by the degree of their development. These are—

(A) *Drumlins.* (B) *Aggregations not strictly drumloidal in form.*

(A) Of the drumlins, four subvarieties may be recognized, and it may prove serviceable to distinguish these and treat them as distinct varieties until the mystery of the drumlin formation shall be solved and the importance, or otherwise, of these distinctions be determined.

(*a*) *Lenticular or elliptical hills.*—These are the typical variety of drumlins and consist of very remarkable aggregations of till in hills of dolphin-back form whose longer axes are two or three, or at most a few times longer than their transverse diameters. The longer axis lies in the direction of glacial movement. This is the most familiar form.

(*b*) *Elongated ridges.*—These have the same constitution as the preceding and have similar terminal contours. The body of the hill is, however, elongated to the extent of two or three or occasionally several miles. These elongated ridges commonly lie parallel to each other, giving a markedly fluted character to the surface. They are thought worthy of being distinguished for the present because the elongation of their forms may prove a significant feature and lead to the recognition of some of the essential conditions of drumlin formation.

(*c*) *Mammillary hills.*—These have the same constitution as the previous types, but differ from them in the extreme shortness of the axis. This, in some instances, is scarcely longer than the transverse diameter. These are thought worthy of being distinguished because they emphasize, more than either of the preceding varieties, the vertical element of the constructive process. I know of nothing more extraordinary in glacial formations than the building up of these domes to the height of 50 to 60 or more feet with such steep sides and on so circumscribed and so nearly circular a base. There are no cases, so far as I am aware, in which the base is strictly circular. There seems always to be an element of elongation in the direction of glacial movement.

(*d*) *Till tumuli.*—These are low mounds of more than usually stony material (so far as I have observed). They have not generally assumed the drumloidal curves of contour and profile, but their nature is such as to have suggested that they are the immature nuclei of drumlins. Further investigation is proposed and they are here introduced tentatively.

(*B*) There are several classes of aggregations of till that are not strictly drumloidal in form but which are thought to deserve recognition for the present, as varieties, for their possible suggestiveness respecting the physical processes of subglacial accumulation.

(*a*) *Crag and tail.*—These embrace the well-known accumulations of till in the lee of rocky crags or embossments.

(*b*) *Pre-crag.*—These embrace the less well-recognized accumulations of till in front of crags or embossments of rock. These two forms may coexist in connection with the same protruberance of rock and may coalesce. From this has arisen the suggestion that their coalescence might initiate a drumlin. In support of this it is cited that many drumlins have a core or pedestal of rock. Against this is cited the fact that many drumlins have no such nucleus of rock, so far as observation can discover.

(c) *Veneered hills.*—These are hills of rock coated somewhat uniformly with till, the surface conforming approximately to that of the underlying rock. These differ from the crag and tail, and pre-crag accumulations in the genetically significant fact of a much more uniform distribution of the till over the rock embossment, and in the subordination of the veneering to the preexistent contour rather than the formation of a new contour.

(d) *Till billows.*—There is a class of drift accumulations which take on a billowy surface. They differ from the drumlins in their want of conformity to axes lying in the direction of the drift movement. The drumlins are also usually separated from each other by low flat ground. The till billows, on the other hand, are arranged more closely together, are disposed more irregularly, and are connected with each other by saddles or cols. Between these billows are frequent undrained basins. The type, it will be observed, graduates into, if it does not strictly belong to, the class designated below as submarginal moraines. Tracts of these till billows are usually distributed parallel to the margin of the ice, and to that extent conform to the habit of terminal accumulations. I have thought that they might be an intermediate form between submarginal moraines and drumlins, but while they unquestionably graduate into the former I have never observed their graduation into the latter. It seems, therefore, probable that they should be removed from this division and placed below.

(e) *Irregular till hills.*—Besides the above forms which show a tendency to some definite law of development, there are a considerable class of aggregates of till that seem to pay no respect to laws of symmetry or to systematic principles of growth. At present no classification seems tenable except one based upon their very irregularity.

(3) *Submarginal ridges of till parallel with the ice border.*—Both the till sheets (1) and the subglacial aggregates (2) that have been described above occupy territory extending for considerable distances back from the border of the ice; indeed, ideally, the first class may be regarded as covering the entire territory occupied by the ancient glaciers. On the contrary, the ridges of till here considered lie along what was the immediate border of the ice at certain of its stages. They are thought to have been formed under the edge of the ice, but it remains to be determined to what an extent they were accumulated under the immediate border of the ice and to what an extent they were deposited at the distance of one, two, or three miles from the precise edge of the glacier. It does not seem at present possible to determine, or at least it does not seem to have been determined, whether the whole of the accumulation

was built up simultaneously throughout its entire breadth, or whether the outer portion was accumulated under the immediate edge of the ice, and the inner portion built up a little later in like manner under the edge of the ice when it had withdrawn somewhat. These ridges are from one to a few miles wide, are composed essentially of till (though assorted material may form a greater or less constituent), possess a gently flowing contour in the main, which distinguishes them from the rougher ridgings and sharper contours of frontal moraines formed by the mechanical thrust of the ice or by the dropping of superglacial material at its edge.

(a) *Submarginal or lodge moraines (a variety of terminal moraines).*— The most important form falling under the above head may be designated submarginal or lodge moraines. They are designated submarginal, not so much because they are believed to be formed near the edge of the ice and not absolutely at its edge, as because they are believed to be formed under the margin of the ice. Three varieties of moraines, all of which may be called terminal, are recognized, being produced in three distinct ways. The first are formed from material borne on or in the ice (the latter being brought to the surface by ablation before reaching the edge) which is dropped at the terminus of the ice and which, when the ice remains stationary for a sufficient period, grows into a bordering ridge. These may be given the rather homely but expressive name *dump moraines.* The second is formed by the mechanical thrust of the ice when it advances against any incoherent material that lies in its path. These may be designated *push moraines.* The third variety consists of that under consideration, and which may be designated *lodge moraines,* from the conviction that the material, instead of being carried or pushed or dragged forward to the extreme edge of the ice, is permitted to lodge under its thin border and constitute a submarginal accumulation. The lodge moraine is not in its nature or material radically different from the ground moraine, which lodged farther back from the edge of the ice and constitutes the subglacial till sheet. It differs from it, perhaps, only in the fact that the thinned and weakened edge of the ice presented conditions specially favorable to deposition and that, as a result, a thickened belt of drift formed under the border of the ice when it remained approximately stationary for a sufficient period and took on the special billowy contours above described. The submarginal moraines were doubtless subjected to more or less mechanical action of the ice as it oscillated forward and backward. This action is thought to have been of the nature of an overriding or an oversliding of the ice rather than of a pushing or plowing up by the edge of the ice. It is my growing conviction that this form of the terminal moraine is the predominant one in the great American glacial field. I incline to the opinion that the broad complex tracts of thickened drift that mark the border of the ice sheet at several of its stages were chiefly formed in this submarginal way, and that, while there is usually present a constituent dropped from the surface of the ice and a constituent formed by the mechanical thrust of the ice, the great mass of these moraines, in general, accumulated by lodgement under the border of the ice but near its edge.

(b) As only those ridges which have some measure of persistency and which mark notable stages of the ice action should be formally designated terminal moraines, it seems advisable to recognize under a different head local ridges of till arranged transversely to glacial movement. These are to be contrasted with the drumlins, which are elongated ridges of till whose axes lie in the line of glacial movement. These transverse local ridges are in some cases perhaps of the same nature as the submarginal moraines, except that the action was limited. But for the greater part they may be presumed to have sprung either from exceptional conditions of accumulation which lead to exceptional deposition when the force of the ice was weakened, or to exceptional conditions favorable to deposition, the conditions in both cases determined by local agencies. They are thus distinguished from the products of those general agencies that produced the persistent submarginal moraines.

2. *Formations derived from material borne on the glaciers and within them (but not at their basal contact) and deposited at their margins or let directly down by their melting when stagnant.*

(1) *Dump moraines (a variety of terminal moraines).*—In various well-known ways a certain amount of material finds lodgment on the surface of a glacier and a certain additional amount becomes incorporated within its body. Leaving out of consideration such part of this as finds its way to the bottom, both the englacial and superglacial material is carried forward, and by ablation is at length brought to the surface of the ice, and is carried on to its edge and dropped there. If the material is considerable and the border of the ice remains stationary for sometime, notable accumulations may result in the form of border ridges, constituting the variety of terminal moraines designated as dump moraines. When the englacial and superglacial material is inconsiderable in quantity the deposit may not amount to a ridge, but may yet constitute a very definite and distinctive belt of material. The boulder belts of several of the interior States are classed here. They can not be said to be moraines, in so far as that term implies ridging, but they are terminal border deposits that have much the same significance as terminal moraines and belong to the same general genetic class.

(2) *Englacial or superglacial till ("upper till").*—When this englacial and superglacial material is let down over the whole territory of the

ice, either during its successive stages of retreat or by being let down
directly through the melting of the glacier when it becomes stagnant, it forms a superficial sheet quite analogous to the subglacial
sheet already considered. This was some years ago designated "upper
till" by Torell, Hitchcock, Upham and others, but because the term
upper till was also used to designate a reduplication of the subglacial
till sheet by many other geologists it was thought best to propose the
term englacial or superglacial till. There still exist differences of
opinion as to how much of existing deposits is to be referred to englacial and how much to subglacial till, and the criteria for discriminating
between these are still under discussion, but the importance of the
classification and the significance of its bearing upon the interpretation of glacial action and of glacial history seems beyond question.

(3) *Medial moraines.*—These familiar forms of glacial deposits do not
call for remark, further than to note that they merge into dump moraines
at the frontal edge of the ice, and into superglacial till in cases in
which they are let directly down by melting without being carried on
to the terminus of the glacier, and to observe further that they are
subordinate elements in the great Pleistocene glacial deposits.

3. *Products of the mechanical action of the edge of the ice.*

(1) *Push moraines (a variety of terminal moraines).*—The distinctness of this familiar variety will doubtless be sufficiently recognized
without further remark. It is perhaps, however, worthy of note that
two subvarieties may be recognized, based upon the glacial or non-
glacial character of the material involved, since the latter variety is
sometimes overlooked.

(a) The first and common variety is formed of glacial material which
may belong either to the subglacial, englacial, or superglacial variety,
or it may be any form of glacio-fluvial or glacio-natant material. It
must, however, be presumed to have been previously brought forward
to the edge of the ice or beyond it, and to be thus subject to be plowed
up or pushed up into ridgings by the thrust of the advancing ice, which
is the essential factor in the formation.

(b) The second variety embraces local material of any kind that lay
in the path of the advancing edge of the ice and was pushed into
ridges by it. It is nonglacial material except in the simple fact that
it was ridged by ice action. It is entitled to be regarded as a moraine
so far as its origin as a topographic form and a rearranged formation
are concerned. In other senses it is not.

(2) *Lateral moraines.*—These familiar forms need no discussion.

N. B.—Interlobate moraines form a variety of terminal moraines and
not, as is quite often stated erroneously, a variety of lateral moraines,
They are produced along the line of contact of adjacent glacial lobes.

but the direction of ice movement is vertical or approximately vertical to the moraines and not parallel with them. The mode of ice action and of morainic aggregation is that of the terminal and not that of the lateral moraine. These interlobate moraines may belong to either of the three classes above designated—the dump, the push, or the lodge moraine, or they may be formed of the three combined.

II. FORMATIONS PRODUCED BY THE COMBINED ACTION OF PLEISTOCENE GLACIERS AND GLACIAL DRAINAGE (ASSORTED DRIFT).

The large amount of water derived from the melting of the ice and the added amount contributed by rains constituted a very important auxiliary agency and gave rise to much assortment and rearrangement of the drift.*

Three classes of deposits may be discriminated, those that are the products of (1) subglacial streams; (2) superglacial streams, and (3) marginal waters.

1. *Deposits of subglacial streams.*—Subglacial waters may be classified in two groups, the first embracing those which flowed in well-established tracts or tunnels formed in the base of the ice, the second embracing those which flowed in a more diffuse and irregular way under the ice. These appear to have given rise to corresponding deposits.

(1) *Osars* (*åsar*) or *eskers* (*kames of many authors*).—It is thought to be now beyond question that trains of gravel accumulated in tunnels formed by subglacial streams and that these, on the disappearance of the ice, formed ridges. The course of these seems to have been conditioned partly by the slope of the land and partly by the direction of glacial movement. They are best developed where these approximately coincide. It can scarcely be said, however, that osars are limited to such coincidence. The more extensive branching and typical forms were probably also conditioned by stagnant or approximately stagnant states of the ice in its vanishing stages. The Scandinavian term *åsar* seems entitled to precedence both because these remarkable forms are typically developed there and because they first received notable attention there. The term *eskers* is coming to be used by many American writers somewhat freely as a synonym and is preferred by some for phonic reasons, while the term *kames* is being used for a cognate variety of gravel accumulations, to be mentioned below.

---

* It is to be noted that only that assortment of the drift which was *contemporaneous* with the ice epoch and connected with the ice action is here taken into consideration. Modifications of the drift that took place subsequent to the disappearance of the ice, or independent of it, belong to a class quite distinct from that under discussion here.

Whatever the terminology, it seems important to distinguish the long-branching gravel ridges that represent the longitudinal drainage of the ancient glaciers from those gravel accumulations that are associated with terminal moraines, and that appear to be the débouchure deposits of glacial waters. A variety of osars or eskers may have been produced by superglacial streams, as will be recognized below.

(2) *Simple tracts or patches of drift formed by subglacial drainage.*—Thin sheets and lenses of sand and gravel in the midst of subglacial till are common phenomena and, while they may in many cases be produced by streams running in tunnels which afterward shifted their position and left no other mark than these patchy deposits, it seems probable that many of these detached sheets and lenses were produced by a diffuse and local drainage developed by any one of several combinations of conditions while the ice was still present and continuing its deposition of till. Similar patches on the surface were perhaps produced in a similar way.

2. *Deposits of superglacial streams.*—For the most part superglacial streams, after short courses, descend through crevasses or moulins and become englacial or subglacial. This was doubtless true of the Pleistocene glacial streams, and the material which they bore along found its final deposition either in connection with subglacial streams or with the glacial waters after they had emerged from the ice. Nevertheless, there are two forms of deposition by superglacial streams that probably find some representatives among Pleistocene formations. The first embrace those which were carried along by superglacial streams that succeed in reaching the edge of the ice or a channel which they cut back into the margin of the ice. In the one case it probably took the form of delta cones, in the other of narrow ridges formed through the restraining aid of the channel walls. The other variety are presumed to have been formed from deposits which gathered along the course of superglacial streams and were let down by the melting of the ice after the glacier became stagnant, retaining essentially the form of their original accumulation in the superglacial channel except so far as they were disturbed in the process of descent. Of these there are perhaps two varieties.

(1) *Superglacial osars (åsar) or eskers (kames of some authors).*—Under this head are to be classed such channel-deposits as retained their elongated form and became ridges, and hence fall under the Scandinavian type. In the earlier studies of the subject a considerable number of glacialists would have inclined to classify most osars under this head, but opinion appears to be inclining toward the reference of the greater part of the osar ridges to subglacial agencies.

(2) *Superglacial kames.*—Sheets or pockets of assorted material gathered on the surface of the ice were doubtless subjected to much dis-

turbance and rearrangement in the process of descent. The resulting deposits would constitute undulatory tracts of drift or groups of hillocks for which there is perhaps no specific name, but which may be thrown under the general class kames for the present.

3. *Marginal deposits.*—Under this class are embraced all those deposits of glacial streams that were made at the margin of the mer de glace and whose forms were dependent upon the conditions that obtained at the margin. The following classes are recognized:

(1) *Kames.*—To this class I refer irregular heapings of assorted drift, generally arranged in lines or tracts transverse to the glacial movement. They are often closely associated with and merge into terminal moraines of till. Sometimes they largely constitute the terminal moraines, the action of glacial waters being so great as to assort nearly all of the material brought forward to the edge of the ice. In such cases the morainic factor (in the genetic sense) is found in the mechanical action of the ice in restraining the action of the waters, in controlling the nature of the heapings, and in pushing off the accumulations, distorting and modifying them. Irregular heapings of assorted drift of this variety are not, however, wholly confined to border tracts, but occur irregularly distributed over the area abandoned by the ice. They differ from the osar type in their marginal relation to the ice and their transverse arrangement. There is, however, a gradation between the two and no sharp line of demarkation has yet been found; probably none exists. But there appears to be this important genetic difference; the typical kames appear to be the products of relatively active vigorous glaciers, while the typical osars appear to be the products of extremely inactive, if not stagnant, glaciers. This important difference of significance is thought to amply justify the recognition of the two classes, whatever may be the difficulties in sharply separating them.

The difficulties of sharply discriminating between the osar type and the kame type are increased by the fact that there is a class of gravel ridges having forms precisely like the typical osars that lie more or less transverse to the glacial movement. These yet await thoroughgoing investigation, but I incline to the belief that they are to be regarded as true osars, and that they were formed during stagnant conditions of the ice just before its disappearance, their transverse course being due to the control of the underlying topography. It has been stated that the course of osars was conditioned partly upon the direction of the ice movement and partly upon the topography of the land surface. In this instance, it appears that the topography dominated the osar formation, and that the movement of the ice became uninfluential. In support of this view it may be added that some osars on reaching the border of the ice turn to a course nearly parallel with it.[*]

(2) *Osar (esker) deltas or fans.*—It appears that when the glacial streams reached the border of the ice sheet and were free from bounding ice walls, they spread themselves out widely and dropped a large portion of their burden in the form of deltas or fans. These are not uncommon in the Interior as well as in Maine, where the osar phenomenon has its most remarkable development in America. These deltas are very significant, both in respect to the method of formation of the osar and in respect to the position and restraining functions of the ice sheet at the time of their formation.

(3) *Overwash aprons.*—When terminal moraines grew to constitute notable barriers along the border of the mer de glace, and when the ice pressed against these moraines so as to obstruct the transverse flow of the glacial drainage along their inner border, the waters derived from the ice crept over the moraines in numerous small streams, which deposited gravel, sand, and silt on the outer flank of the moraine. These deposits are often distributed along the moraines for great distances and constitute a fringe of assorted material to which Shaler has given the apt name "apron." These constitute one of the most satisfactory demonstrations of the marginal character of the moraines and of the relations of the ice to them. The material varies widely in coarseness, according to the conditions of formation, and a structural subclassification may be based upon it, embracing (a) gravel, (b) sand, (c) silt aprons. Immediately next to the moraine the material is sometimes exceedingly coarse, constituting little less than a boulder belt. At the other extreme of the series the silt sometimes forms a clay deposit and sometimes it takes on that peculiar assortment which constitutes loess.

The class of aprons here described are dependencies of definite terminal moraines. There were, however, tracts of assorted material formed by waters outflowing from the ice where no definite terminal ridging took place. Such forms may be designated *outwash* aprons in distinction from *overwash* aprons. This class is usually made up of sand or silt. In the latter case there are gradations into the great flanking tracts of loess which appear to have arisen in the manner indicated on very low slopes with prevailing slack drainage.

(4) *Pitted plains (in part).*—Both the osar deltas and the overwash aprons are characterized in certain regions by a surface marked with numerous depressions, sometimes symmetrical (kettles), sometimes irregular, with undulatory bottoms and embracing knobs and subbasins,

---

[*] Dr. Lundbohm has directed my attention recently to the fact that this is a not uncommon phenomenon in Sweden, as shown by the geological maps of the Swedish Survey.

giving the surface an expression resembling the kames. A part of these pitted plains seem to be intimately connected in origin with the ice edge and to be due to marginal conditions, among which it has been thought that the incorporation of ice fragments, the grounding of ice blocks, the movement of the ice edge, and the development of underground ice sheets were among the special agencies, but the full explanation of the pitted plains can scarcely be claimed to have been reached.

III. FORMATIONS PRODUCED BY GLACIAL WATERS AFTER THEIR ISSUANCE FROM THE PLEISTOCENE GLACIERS.

1. *By glacial rivers.*—One of the most familiar facts of glaciology is the detritus-laden condition of the icy streams as they issue from the body of an active glacier. A portion of this material is thrown down immediately at the margin under the special conditions there presented and constitutes the formations classified above, but the larger portion is borne onward to varying distances and deposited quite independently of the agency of the ice. Two varieties of this class of deposits are worthy of being specially recognized.

(1) *Valley drift.*—In those cases in which the previous surface agencies had developed a definite drainage topography and in which the gradient was favorable the detritus was borne down the valleys leading away from the ice border in trains of gravel and sand. As the glacial streams were usually greatly overloaded with detritus at the outset, they built up their valley bottoms by depositing material from bluff to bluff, constituting a valley plain, out of which subsequently beautiful systems of terraces were often cut. The most notable class of this type consists of those gravel valley-plains which head in a terminal moraine or, more strictly speaking, in the overwash apron of a terminal moraine. Sometimes the apron gathers in for many miles on either side, giving a very broad expanded head to the valley tract. As these valley tracts may head on successive moraines and may be traceable far down their valleys, they afford most admirable means of working out glacial history by determining the intervals between the successive moraines and topographic conditions under which they were formed. As the gradient of the streams that formed these valley tracts may be estimated, they afford valuable criteria for determining the altitude and the attitude of the land at the time of their formation.

(2) *Loess sheets.*—The above valley tracts grade from the coarsest gravel through sand to the finest silt. So long as they are confined definitely to the valleys a subclassification of them on the basis of coarseness or fineness of material would be rather structural than

genetic, though carrying a genetic significance. But when the waters spread widely beyond the immediate vicinity of the valley and the material took on a peculiar and distinctive assortment there seems to be sufficient ground for recognizing a second genetic class. The great tracts of fluvial loess (not all varieties of loess) are placed here. These are valley phenomena, in part. They have for their axes the great valleys of the region and their thickest deposits are along the valleys. They, however, spread widely over the adjacent country, so that conjointly they mantle the whole region. They also coalesce with the great fringing sheets of loess that are classed above as outwash deposits. The class graduates into typical valley deposits on one side, into typical fringing deposits on another, and, apparently, into wind deposits of loess on a third side.

2. *By fringing lakes.*—This class obviously embraces deposits of suspended material brought out from the ice into bordering lakes by glacial streams and spread over their bottoms, being generally of the clayey type, sometimes bearing lacustrine fossils, sometimes not; sometimes commingled with stony material dropped by floating ice derived from the edge of the glacier, sometimes not, at least not in notable quantities, and always more or less commingled with wash from the adjacent land not covered by ice. Theoretically the type is characterized by the peculiar border of the deposit next the impounding ice. Practically this characteristic is not always readily demonstrable.

3. *By bordering seas.*—This class differs from the preceding in the fact that the waters were not impounded by the ice (as they usually but not always were in the preceding case), and in the fact that the deposits are commingled with oceanic sediments, marine fossils, and impregnated with saline waters, which may or may not have been wholly removed subsequently.

IV. FORMATIONS PRODUCED BY FLOATING ICE DERIVED FROM PLEISTOCENE GLACIERS.

The type of this class is glacio-natant till in which the constituents are identical with those of glacial till except when formed under the action of currents which induced secondary modifications. Two subclasses are to be recognized:

1. *Local.*—In general these are lacustrine but may be oceanic. Commonly the deposits took place in glacier-fringing (usually glacier-formed) lakes and constitute only a secondary phase of glacier formation. In cases in which the ice entered an arm or bay or even the border of the ocean and the deposit took place in the immediate vicinity, the deposit remained essentially a local one. If marine fossils or marine sediments were commingled, or if the marine factor is for

any reason regarded as important, a subclassification, distinguishing between lacustrine and marine glacio-natant local deposits, is justified.

2. *Foreign.*—These are essentially marine glacio-natant deposits and are due to icebergs derived from distant glaciers bearing to the point of deposit material wholly of foreign origin. Among the genetic conditions involved in this case are the submergence of the land beneath the ocean at the time of the deposit and its subsequent elevation. Local lacustrine glacio-natant deposits may be formed at various heights above the ocean level and subsequently exposed by the drainage of the lake without involving oscillations of the crust or the sea-level. In the case of the Great Lakes, iceberg material borne from one side and deposited on the opposite and distant side might constitute a recognizable subclass under this head. Such deposits have been described.

V. FORMATIONS PRODUCED BY SHORE ICE AND ICE FLOES DUE TO LOW PLEISTOCENE TEMPERATURE, BUT INDEPENDENT OF GLACIER ACTION.

1. *Shore ridges due to ice push.*—In northern latitudes the shore action of ice (not including icebergs) is very notably producing shore ridges of unusual strength, configuration, and importance. It is held by some writers that much of the phenomena placed in the above classes is referable to this. Without conceding this, it seems beyond question that this class of deposits need special recognition in the study of the Pleistocene formations.

2. *Littoral deposits.*—If we confine the above class to those ridges which were pushed upon the shore above the reach of the waters we need also to recognize a class which were deposited beneath the border of the body of the water. These differ from ordinary littoral deposits in the special contribution resulting from the ice action.

3. *Offshore deposits.*—These embrace the material of the ice action of the shore borne back in suspension or by ice floes into still waters and there deposited. They must in the nature of the case very closely simulate the formations produced by floating ice derived from glaciers.

VI. FORMATIONS PRODUCED BY WINDS ACTING ON PLEISTOCENE GLACIAL AND GLACIO-FLUVIAL DEPOSITS UNDER THE PECULIAR CONDITIONS OF GLACIATION.

Recalling what was said under this head near the opening of this discussion, it may suffice here to simply indicate two classes that may be recognized under this head.

1. *Dunes.*—These differ in no important respect from ordinary dunes, except that the material is made up in part of grains formed by glacial

grinding instead of disintegration and wave wear, and in their correlation with the ice border and the glacial waters that issued from it, rather than with the sandy shores of lakes and seas.

2. *Æolian loess.*—While the larger part of the loess found in the glaciated regions is believed to be the product of glacial waters, it still remains, in my view, probable that certain parts of it were deposited by winds. This part is believed, in general, to have been derived from the water-deposited portion, but perhaps this is not universally true. Along the leeward side of the Mississippi River, for instance, we find dunes of sand and dune-like accumulations of loess that seem in both instances to have been derived by winds from the flooded flats of the river below. In like manner there seems ground for the belief that in Pleistocene times the glacial floods alternately extended and withdrew themselves, leaving great silt-covered flats exposed to wind action, and that from these silt was swept up and deposited over adjacent and perhaps somewhat distant highlands. It seems also not improbable that the conditions of the surface may have been such as to permit the lodgment of this more uniformly over the surface than is the habit with dunes. There seems ground for this in the distinction between the formation of dunes and the supposed deposits. Dunes are formed from sand driven along the surface by winds, but not in any notable degree carried by the winds in full suspension. The supposed silt deposits, on the other hand, are presumed to have been formed by silt borne in free suspension until, by contact with the earth, it was lodged. Such contact might obviously be widespread and the lodgment product might have a wide and measurably uniform distribution. While, therefore, coinciding with what seems to be the majority opinion among American geologists that the loess deposits of the glaciated region are chiefly water-lain, it appears to me prudent, if not important, to recognize the æolian class, and to search diligently for criteria of discrimination between the two classes.

The foregoing classification is consciously incomplete. In some instances the bases of distinction border closely upon the structural rather than the genetic, but it is believed that there is involved in every case an important genetic factor, though it may sometimes be most conveniently expressed in structural terms.

M. GAUDRY s'exprime en ces termes:

Il y a plus de trente ans que les géologues de l'Europe se sont accordés à reconnaître, avec Boucher de Perthes, la coexistence de l'homme et des animaux aujourd'hui éteints de la période quaternaire. Mais, depuis trente ans aussi, les géologues n'ont pu se mettre d'accord sur la chronologie de cette période, et, cependant, sans chronologie il n'y a pas d'histoire; rien ne peut être plus important que de connaître les phases par lesquelles l'humanité primitive a passé. Les études de

M. Chamberlin sur ce qu'il appelle les questions génétiques sont assurément intéressantes; elles le seront encore plus quand cet habile géologue les aura fait servir à la chronologie.

Nous avons grand besoin en Europe que les Américains viennent à notre aide. Il semblerait que Paris doive être un des points où l'on arrivera le plus facilement à jeter quelque lumière sur la chronologie quaternaire, car nous y trouvons deux faunes très distinctes: l'une au sommet de Montreuil, qui est certainement une faune de climat froid, comprenant le renne en abondance, l'Elephas primogenius à lames serrées, le Rhinoceras tichorhinus; l'autre au bas de Montreuil et à Chelles, qui est une faune de climat chaud, ne renfermant jamais de renne, mais du Cervus elephas, l'Elephas antiquus, le Rhinoceras Merckii; mais parmi nos géologues les uns croient la faune froide plus ancienne que la faune chaude, d'autres la supposent plus récente.

En Angleterre, il y a les mêmes divergences de vue qu'en France. En Allemagne, auprès de Berlin, on voit des couches qui indiquent un grand froid et d'autres qui indiquent la fusion de glaces, mais on n'a pas une faune marquant nettement, comme à Paris, un climat chaud. En Italie, au contraire, et dans le sud de la France on n'a pas de faune révélant manifestement un climat froid. Il est donc encore plus difficile que dans le nord de la France et en Angleterre de préciser la succession des phases du Quaternaire.

Il faut espérer que, réunis tous ensemble, nous finirons par trouver quelques jalons au moyen desquels nous pourrons établir la chronologie des temps pendant lesquels nos ancêtres ont vécu.

Mr. CREDNER said: The north German plain contains deposits closely related to those of the Pleistocene in America. Mr. Chamberlin's classification is admirable and wholly applicable to Germany.

Mr. DE GEER expressed his great satisfaction at hearing that one of the most experienced of North American glacialists is of the opinion that a genetic classification of the Pleistocene is the preferable, and at present, in fact, the only practicable one. He had some years since come to the same conclusion with respect to Scandinavia, and had himself advocated a genetic classification for that country. Although time does not admit of even a discussion of that classification, he would remark that, as it was especially intended for mapping, it involved different colors for the different natural genetic classes, and different tones of each color for the different petrographical subclasses, such as gravel, sand, and clay; darker for the coarser and lighter for the finer materials. Where a chronologic determination was possible, it was to be indicated by a letter, by which means arbitrary correlations could be avoided and, as often as was necessary, alterations could be made in the always more or less artificial chronological subdivisions without changing the scheme of colors or the general appearance of the map.

To suit Scandinavian conditions it might be desirable to make some minor alterations in Mr. Chamberlin's scheme. Thus the marine deposits might be made a separate class in consideration of their importance for the study of changes of level and for comparison with non-glacial formations.

Formations belonging to Classes IV and V of Mr. Chamberlin could perhaps be referred to subclasses under III, as they are rather additions to or disturbances of regular deposits than such deposits themselves. He agreed with the distinction suggested between Osars and Kames, which have often been confounded, whereas the former are in the main radial and the latter peripheral with reference to the flow of the land ice.

Mr. HUGHES pointed out that the chronological and genetic classifications were more or less theoretical, and that their value depended upon the accuracy of the reading of the structural and other characters, including the paleontology, and all that was a matter of direct observation. The inferences of M. Gaudry led to a chronological classification not inconsistent with the inferences of Mr. Chamberlin, viz., that elsewhere deposits, as divergent in structure, may be synchronous and owe their difference to various conditions of genesis.

He expressed his own opinion that what was known as the Glacial Period, in England at least, was one and continuous, interrupted only by oscillations, similar *in kind* to those which are observed at the end of the Rhone glacier or any other similar body of ice at the present day, and differing *in degree* only in proportion to the much larger quantities and greater lengths of time with which we have to deal in studying the remains of former glacial conditions in temperate climes. Without appealing to particular cases, upon which he had commented elsewhere, he thought that the erroneous reference of certain deposits (e. g., some paleolithic deposits) to preglacial and interglacial times had arisen from mistakes in the genetic classification, *remanié* beds, often containing débris of older glacial deposits, being put down as themselves of direct glacial origin.

He explained, by reference to the genesis of the Boulder clay, a fact that had been mentioned, namely, that in one part of the glacial drift there was an abundance of striated boulders, while in another part of the same deposits there were none. If the ice moves on beyond the source of material (that is, of rock-bosses above the ice) then all the fragments which have fallen upon it will have found their way down through the crevasses and become glaciated at the bottom; but if the drift consists chiefly of material that has been carried on or in the ice, as in the case of short glaciers, then there will be few striated stones.

He thought that under the names Osars, Esgairs, Kames, etc., we had at least two distinct types of accumulation, often consisting of what was originally glacial material, but, as those names were only the equivalent general words in different languages for the very same class of phenomena, he preferred using a qualifying term, such as radial-kame, etc., to giving a new and restricted meaning to an old, well-worn word. He referred the hollows in the "pitted plains" to interruptions between hills and ridges of esgair drift.

Mr. WAHNSCHAFFE advocated the chronological classification, and considered such a one possible for the Quaternary deposits of north Germany. These deposits begin with preglacial sands and gravels containing *Paludina diluviana*, which is still a living form, and *Lithoglyphus naticoides*. Above these follows a typical ground moraine, which is overlaid by stratified sand and gravel, containing the well-known diluvial fauna; and to these again succeeds the upper till, considered now as the ground moraine of the second glacial epoch.

Mr. H. CREDNER remarked that the occurrence of the sand between two ground moraines indicates a retreat and second advance of the ice sheet. Such interpolated sands are, in Germany, always local and no proof of a real interglacial epoch. The sand layers between the moraines are not continuous, but local, and can not be given the significance attributed to them by Mr. Wahnschaffe.

M. PAVLOW : Je demande la permission de faire quelques remarques générales sur les principes de la classification que nous discutons.

Nous sommes déjà beaucoup avancés dans nos discussions. Nous avons entendu une série de communications extrêmement intéressantes, mais il me semble que la direction de nos discussions est, pour ainsi dire, accidentelle et que nous risquons de nous éloigner de la base sur laquelle la question est fondée. Il me semble qu'il faudrait, tout d'abord, être d'accord sur que ce que nous comprenons sous le nom de Pléistocène; est-ce seulement l'époque glaciale, l'époque du refroidissement du climat, ou faut-il y comprendre tous les dépôts qui se sont formés depuis la fin de la période tertiaire jusqu'à nos jours. Si ce sont les dépôts de l'époque glaciaire, il faudrait savoir si nous avons à faire aux régions autrefois récouvertes des glaces, ou à toute la surface du globe qui est accessible aux recherches. Il est évident que le schema de notre classification doit être différent dans ces deux cas. Autrement, si l'on veut donner au nom de Pléistocène des limites plus larges, en y comprenant tous les dépôts post-tertiaires, il serait utile d'établir quelques subdivisions principales foudamentales et de renoncer pour quelque temps aux détails, tant intéressants qu'il soient.

Quelles doivent être, donc, ces subdivisions principales ? Cela dépenderait, en premier lieu, du genre de la classification que nous choisissons. Si nous nous arrêtons à une classification chronologique, nous parlerons des grandes subdivisions chronologiques que l'on puisse établir pour les différents pays, nous discuterons la question sur l'une ou plusieures glaciations dans des pays autrefois récouverts des glaces et sur les dépôts interglaciaux, tâchant d'établir la correspondance de ces dépôts dans les différents pays. Nous passerons plus tard aux pays qui n'ont jamais été récouverts des glaces et nous tâcherons d'y établir les subdivisions se basant sur la structure, sur la position des différents dépôts et sur leur faune. Cela terminé, nous arriverons à éclaircir les principes de la classification chronologique des dépôts pléistocènes.

Mais, autant que je le comprends, nous sommes ici pour discuter la question sur la classification génétique des dépôts pléistocènes. Alors les subdivisions fondamentales doivent être autres. Nous avons devant nous le problème à resoudre; quels sont les agents principaux qui ont pris part à la formation de cette série extrêmement complexe que nous désignons sous le nom des dépôts pléistocènes?

La glace continentale n'était qu'un de ces agents, principal peut être; certainement nous avons devant nous beaucoup d'autres; les courants d'eau, dépendants ou indépendants des glaces, le vent, les eaux stagnantes et la végétation produisant des tourbières, les neiges couvrant périodiquement les montagnes et les plaines au delà de la limite de la glaciation, les pluies et les averses transportant les particules minérales des sommets des hauteurs et les déposant sur les pentes douces, etc.

Je voudrais bien exposer devant cette auditoire, dont la compétence dans ces questions dépasse tout ce qu'on pourrait imaginer, quelques résultats de mes études sur les dépôts quaternaires de notre pays, et faire-connaître la classification des dépôts en question, classification qui est en usage en Russie; mais je voudrais d'abord entendre les opinions de mes illustres collègues sur les questions générales que je viens d'exposer.

Mr. DE GEER agrees with Mr. Wahnschaffe in so far as that he admits that a chronological classification is locally possible. He also recognizes at least two glacial epochs, marked by two great oscillations of the ice, but in many places—in Russia for instance—he does not believe that it is possible to determine where the corresponding stratigraphical limit is to be drawn. The same is true of many other places and deposits, much of the alleged chronological classification being in fact genetic. The use of a chronologic system in mapping has often compelled the field geologist to make entirely artificial subdivisions, and it therefore seems to be without doubt the best and most natural way to use a genetic classification until we can get a basis for another.

Mr. WAHNSCHAFFE, referring to Mr. Credner's assertion that there is no proof of an interglacial period in northern Germany, thought that the existence of a diluvial fauna between the two tills is sufficient proof.

Mr. CREDNER replied that no complete skeleton had been found, but only single bones which might have been transported and deposited with the gravel.

Mr. WAHNSCHAFFE again replied that the bones occurring in these gravels are proportionately large, when compared with the gravels themselves, and therefore can not well have been transported from a distance.

Mr. SHALER: Organic deposits may possibly occur very near the ice sheet, which allows an interweaving of organic and glacial deposits.

Mr. GILBERT recalled the observation of I. C. Russell in Alaska, that where the movement of the ice is very sluggish, it may become covered with soil, or even with a growing forest in which such animals as bears still live.

Mr. DIENER remarked that the presence of fossil remains is no proof against the existence of glaciers in the vicinity. Many animals are known to live on the borders of glaciers or to cross the ice streams themselves. In New Zealand birds of a rather tropical, or at least subtropical type, like parrots, are common on the mountain spurs facing the great Tasman glacier: in the Austrian Alps moraines no more than twenty years old are covered with pasture, and in the Caucasus many glaciers end in the midst of a rich vegetation, the abode of many different species of animals.

Mr. HOLST, who felt convinced that neither in America nor in Europe has there been more than one ice age during the Quaternary period, and who intends at some future time to publish his arguments in favor of that conviction, denied that two ice ages have been proved to have existed in Sweden. The reasons which seem to support the hypothesis of two ice ages in that country have only a very local signification, and can be easily and much better explained in another way. It must also be remembered that this hypothesis is not one that was independently developed in Sweden, but was an idea imported from Germany.

The strongest argument in favor of two ice ages in Sweden is found in the two moraines of the southern portion of the country, of which, as a rule, the lower is blue, the upper one yellow. For his part, he considers that the blue moraine was formed under the ice sheet and by it protected from oxidation, while the yellow material, which was englacial drift, was rearranged when melted out from the ice and thereby oxidized. In some cases the pebbles in the two moraines are different and have a different origin, but this fact is only observed along the great Baltic ice stream. It is easy to understand that the land ice in the neighborhood of such a great ice stream must be influenced by it and may thus sometimes change its direction. Perhaps also, in some cases, the upper part of the ice moved in a different direction from the lower. Although the two moraines are not in all cases, very well separated, and indeed are not found everywhere, wherever they do exist they are, as a rule, the result of one land ice. In Greenland Mr. Holst has seen the blue moraine of characteristic appearance lying under the ice sheet, and the yellow coming out from it as englacial drift—both therefore contemporaneous. In the northern part of Sweden, where no one, so far as he knows, speaks of two ice ages, the same two moraines, the blue and the yellow, are also found.

Far from being an evidence of two ice ages, the two moraines are no more than one ice age demands. Every ice age has two moraines; two ice ages should, therefore, have four.

Mr. DE GEER thinks it would be difficult to explain the existence of extensive beds, sometimes 30 or 40 feet thick, of fine sand free from pebbles, between the two sheets of moraine mentioned, if both were deposited by the melting of one and the same glacier. The colors are in certain regions the reverse of those stated by Dr. Holst, the upper till being dark and the lower yellow, in accordance with those of the different rocks from which they have been derived. Even the numerous and characteristic guide boulders in the two tills indicate a derivation from sources so situated that their courses must have been at right angles to each other, thus showing that both beds must be the ground moraines of entirely different ice sheets.

Mr. CHRISTIE described the section of peat and silt between two layers of till, occurring on the river Clyde, above Glasgow.

M. CADELL said that in the buried river channel of the River Carron, a tributary of the Forth, no less than five distinct beds of till or boulder clay had been found, separated by beds of sand or mud, during the progress of boring and coal mining. This preglacial river bed had a depth of over 250 feet below sea level, and was described in Croll's "Climate and Time." He also described a small channel in the surface of a bed of trap at Bo'ness, on the Frith of Forth, the lower part of which was filled with gravel derived from ice-transported rocks from a distance, which was covered by true stiff till, showing that there must have been a mild interglacial period when the old till with foreign stones was washed away by streams, and that a subsequent return of the ice sheet covered and filled up the little stream bed with its *moraine profonde.*

Mr. McGEE called attention (1) to the importance of land forms as products and records of the geologic agencies operating in Pleistocene time, and (2) to the superiority of genetic classification for the deposits and land forms of the Pleistocene; and then (3) set forth in detail a genetic classification of Pleistocene formations and land forms based on (a) an extended survey of glacial, aqueo-glacial and aqueous products in the Mississippi valley (chiefly in northeastern Iowa); (b) an examination of lacustral and glacial deposits and land forms, with associated volcanic products, in the Great basin of western United States; and (c) detailed studies, extending over a decade, of the glacial, aqueo-glacial, estuarine, and marine deposits in the Coastal plain of the Atlantic and Gulf slopes of eastern and southeastern United States.

THE INTERPRETATION OF LAND FORMS.

Various geologic agencies operate on the land, modifying rocks and fashioning the surface, and in this way leaving a record of their operations. During recent years the significance of such phenomena has come to be perceived; and it is now recognized that land forms yield a record of continental history which, within certain limits, is quite as intelligible as, and sometimes more definite than, the record commonly read from coeval and correlative deposits. The systematic examination of land forms and their interpretation as records of geologic history introduces a new branch of geologic science, called "physical geography" or "physiography" by different writers, which has been designated "geomorphic geology" by Powell and the "new geology" or "geomorphology" by the writer; but the term "geomorphy," first employed in a somewhat different connection by Sir William Dawson, though never extensively used with this meaning, is preferable.

In general the utility of the record read from land form varies inversely with its antiquity. The newer characters representing the work of agencies now operating or recently discontinued are commonly clear and distinct, while the older characters are commonly blurred and indistinct, and the oldest may be all but obliterated and so overplaced by newer characters as to yield at the best but a fragmentary inscription. So the value of geomorphy as a means of interpreting geologic history culminates when applied to the later episodes and ages, and progressively diminishes when applied to successively earlier episodes and ages.

Land forms represent the operations of both constructive and destructive agencies. Alluvial fans at the embouchures of waterways are wholly constructive, while deltas, strands, and certain terraces formed at or below water level and laid bare by deformation of the land or withdrawal of the waters exemplify indirect constructive forms; and both these categories of forms represent the works of hydric agency. Craters, cinder cones, and other typical land forms produced by volcanic agency are in like manner constructive; and so too are drumlins, terminal moraines, till plains, and certain other forms produced by glacial agency as well as kames, paha, sand plains, and certain other forms produced by aqueo-glacial agency and eolic agency. Certainly the most extensive and perhaps the most interesting and instructive land forms, however, represent the destructive category of processes, and those found most significant thus far are the forms produced by the destructive action of running waters. These include baselevels, gorges, and canyons; pene-plains and plains; buttes or ridges of obdurate material which always lag behind the general surface in baselevel planing, and which may be called "catoctins" from a typical example, the Catoctin range of Virginia and Maryland; the residual remnants of divides which baselevel planing has not yet reached;[*] together with scarps, bluff lines, salients, and other forms; and from them much has been learned concerning the attitude of the eastern part of the United States, and so of the physical history of the continent as far into the past as the beginning of the Cretaceous. Other things equal, however, that geomorphic record is best which is inscribed in both constructive and destructive forms.

Now the characteristic phenomena of the Pleistocene are especially susceptible of geomorphic interpretation. In the first place the Pleistocene period represents the latest chapter in geologic history, and the characters of its deposits and land forms are commonly clear and seldom entirely obliterated, so that its record is exceptionally intelligible. In the second place the characteristic agency of the Pleistocene glaciers—ice—is both destructive and constructive, while the oscillations in level and the modifications in drainage attending the ice invasion facilitated production of constructive forms by hydric agency in such positions as to be little affected by subsequent changes—the glacial and aqueo-glacial deposits laid down in one place are made up of materials gathered elsewhere, and both the destructive process of collection and the constructive process of deposition resulted in the development of characteristic forms. Thus the land forms of the Pleistocene afford a more complete geomorphic record than do those of any other period. Moreover, since the record of Pleistocene land forms is more extended than the record of Pleistocene deposits (for the deposits represent only the areas of deposition while the land forms represent these areas and also the correlative areas of concurrent degradation) the method of geomorphy is essential to securing the best results in the examination and classification of Pleistocene phenomena. Accordingly the interpretation of land forms is of preeminent utility in Pleistocene geology.

THE SUPERIORITY OF GENETIC CLASSIFICATION.

The chief purpose of scientific research is the ascertainment of natural relations; and when the relations are ascertained they are expressed in taxonomies or systems of classification. Now, the successive

---

[*] Recently called "monadnocks," by Davis, after the typical New England example, Mount Monadnock.

stages in the progress of knowledge are marked by degrees of development in classific arrangements. When knowledge is limited or superficial, classification is commonly based on one or a few extrinsic or adventive characters, and usually the process of classification is the Aristotelian or bifurcate one of successive division by difference; with more extended knowledge the classification is commonly based on limited groups of adventive or essential characters, and the arrangement usually partakes of the Aristotelian method of successive division by difference and of the Baconian one of successive grouping by resemblance; with still more extended and intimate knowledge the classification is commonly based on aggregates of characters, essential as well as adventive, and the method of arrangement is usually the Baconian one of grouping the like rather than separating the unlike; and with profound knowledge the classification is based on essential characters, the adventive characters being eliminated, and the process of arrangement approaches more or less closely to the Powellian method of integrating attributes, grouping characters, and seriating stages in such manner as to express the most comprehensive and essential relations, i. e., genetic relations. The first three of these stages in the development of classification are well exemplified in biology and more or less perfectly in other branches of science; but geology, which may be said to have passed from the primitive stage under the inspiration of the genius of Lyell, is yet hardly beyond the second stage, though rapidly passing into the third and in some of its branches entering the fourth under the inspiration of the genius of living leaders.

As pointed out by Spencer, the practical purpose of classification is twofold: (1) to facilitate finding or identification, and (2) to systematize knowledge. The first of these ends may be attained without research and with limited knowledge, e. g., rocks may be arranged in a museum by color, macroscopic texture, or other superficial elements of difference—indeed a convenient mode of arrangement of rocks for architectural purposes is effected by separating granites, marbles, freestones, etc., into primary groups and then successively separating the specimens in each group by differences in color or coarseness of grain. The second end is attained only when the arrangement expresses the sum of knowledge concerning the objects, so that the arrangement involves the results of research, e. g., in the attainment of this end rocks are arranged in a museum by the agencies and conditions of genesis inferred from texture, structure, and other characters; yet, in so far as the arrangement expresses systemized knowledge, it indicates essential relations and approaches more or less closely to the expression of genetic relations. In the first case the arrangement is primarily artificial, and being based on adventitious or nonessential attributes and relations may be characterized as *adventive;* while in the second case the arrangement represents an effort to imitate or express natural relations, and since it is based on essential attributes and relations (real or ideal) may be called *essential*. Now, since knowledge grows only by research, its progress is retarded by the lower classification; while, since the solution of one problem always opens other problems, so that research progresses cumulatively, knowledge is promoted by the higher arrangement. And the highest classification is that which most fully expresses essential natural relation or genetic relation—indeed the highest standard of classification attained by science (the arrangement through integration, grouping, and seriation) is preeminately an expression of origin and development, i. e., of genesis.

The history of the growth of every branch of natural science is a record of development in classification from the adventive toward the essential; and in no field of knowledge is this course of development better displayed than in Pleistocene geology in this country. The terminal moraines, the drift sheets, the far-traveled boulders have long been recognized by geologists and laymen alike, but knowledge of them yielded no profit until they were grouped as products of a continental glacier. The varied topographic forms of the glaciated region were recognized by tyros in glacial geology, but they remained of little interest until, largely through the discriminating genius of Chamberlin, they were interpreted as records of local operations in ice work, and thereby made to reveal their own genealogy. The structure of glacial deposits was long perceived, yet it remained a dead language until verified by interpretation as a record of genesis; and the value and living interest of glacial geology to-day depend directly on the completeness of interpretation of its facts and on the perfection of the arrangement of these facts as genetic records. Certain temporary purposes are indeed subserved by adventive classification of glacial phenomena, e. g., by form or structure, and useful results are attained by chronologic arrangement; but the genetic arrangement involves appreciation of relations in time and space, and is far more comprehensive and useful than either.

GENETIC CLASSIFICATION OF PLEISTOCENE FORMATIONS AND LAND FORMS.

Viewing the earth as a unit, geologic history has flowed on continuously in a series of confluent stages; but when a restricted part of the globe is viewed as a unit, the elsewhere confluent stages are sometimes found clearly separated, and these separations are of use in fixing chronologic divisions.

Conceived as a natural series, the various geologic agencies have operated and cooperated throughout the term of geologic history with increasing or diminishing energy, and few agencies have been introduced or extinguished during that term; but when a given agency is viewed by itself, its energy or activity is commonly found to be variable, weak at the beginning, gradually strengthening to a culmination, and afterward weakening again; and the method of research by isolating agencies in this manner is of great utility in developing genetic classification and also in interpreting general history.

Now the stage in earth history represented by the Pleistocene period grades imperceptibly into earlier stages, and it is only in poleward lands that records of discontinuity can be found; yet these records are useful and give a basis for chronologic division which it is desirable to recognize even in those regions in which the record is not clear. At the same time the Pleistocene was characterized by a remarkable increase in the efficiency of one of the geologic agencies, i. e., glacier ice; and while glacier ice undoubtedly operated (albeit with less energy and over smaller areas) during earlier ages, it is useful and probably not misleading to characterize the period by its principal agency in poleward lands as the *glacial period*. It is to be remembered, however, that the several individual agencies of the Pleistocene were in operation before that period, and that the characteristic agency in poleward lands did not operate over equatorial lands or broad oceans; and accordingly when classified genetically the Pleistocene phenomena grade into the phenomena of earlier periods.

The accompanying classification of formations and land forms of the Pleistocene represents in most particulars individual work in several fields. It differs from the elaborate and admirable classification proposed by Chamberlin chiefly (1) in that it includes land forms as well as formations, and (2) in that greater importance is given to the products of hydric and volcanic agency. A few of Chamberlin's subclasses (taken from the leaflet circulated in the Congress, and marked by asterisks) are introduced, sometimes with minor modification, to secure logical completeness and symmetry; for this classification is designed chiefly to supplement that proposed by Chamberlin,

A. Aqueous:
  1. Constructive—
    a. Below baselevel.
      I. Marine deposits and land forms produced by marine currents (commonly revealed by emergence).
      II. Estuarine deposits and concomitant land forms (commonly revealed by emergence).
      III. Lacustral deposits and concomitant land forms (commonly revealed by withdrawal of the waters).
      IV. Beds and crags of tufa and other precipitates (commonly revealed by withdrawal of the waters).

    b. At baselevel (commonly revealed by emergence).
        I. Sea-cliffs, wave-rounded islets.[1]
        II. Littoral deposits, shingle beaches, wave-built strands, spits, bars, shore terraces.
        III. Marsh deposits, horizontal plains of silt or mud.
        IV. Alluvial deposits, flood plains, natural levées, terraces of varigradation.[2]
    c. Above baselevel.
        I. Torrential or overplacement deposits and land forms characteristic of mountainous regions; grading into—
        II. Talus deposits and land forms, including ordinary cliff-talus, playas, etc.
        III. Spring deposits of tufa and sinter, etc; geyser products.
        IV. Cave earth, stalactites and stalagmites.[3]
  2. Destructive—
    (Chiefly derivative topographic forms produced by the modification or partial destruction of the foregoing through the action of running water; together with surface sinks and the characteristic topography produced by subterranean drainage.)

B. Glacial:
  1. Direct—
    a. Constructive.
        I. Ground products of glaciers.
          i. Till sheets or ground moraines with their characteristic configuration.
          ii. Till accumulations (due to the cumulative effect of arrested activity); including drumlins, veneered hills, crags-and-tails, irregular till-hills; including also greatly elongated and parallel or slightly divergent and sometimes loess-crowned ridges, thus forming the nucleii of paha.[4]
          iii. * Lodge moraines, transverse till-ridges.
        II. Surface products of glaciers.
          i. * Dump moraines, certain boulder belts.
          ii. Englacial and superglacial drift (sometimes, but with doubtful propriety, called "upper till"), including occasionally rock-flour, loess, and sand beds, and rarely boulders, associated with assorted drift matter.

---

[1] During the submergence of the early Pleistocene Columbia period, wave-rounded islets of two classes were formed: (1) Some eminences were completely submerged at the maxium depression, though the waters persisted much longer at a lower level, so that the knobs are rounded in plan and profile (examples, Maulden mountain, Bull mountain, and neighboring eminences in Maryland); (2) certain isolated eminences were never submerged, though the waters washed their bases and built broad terraces on their flanks, so that they are wave-rounded in plan but not in profile (examples, the two Lumpkins mountains and Duck hill in Mississippi, Gordons mountain in Tennessee, and neighboring eminences).

[2] Pleistocene history of Northeastern Iowa, 11th Ann. Rep. U. S. Geological Survey, 1892, p. 273.

[3] A relation between cave formation, stalagmitic growth, etc., is suggested by the similarity of the cave fauna and the Quaternary fauna of this and other countries. Cf. op. cit., p. 564.

[4] Op. cit., pp. 220, 229, 405, 545.

III. Marginal products of glaciers.
    i. Glacial talus, or local and sublocal drift usually confined to transverse valleys about the ice-margin.[1]
    ii. *Push moraines.
    iii. * Lateral moraines.
  b. Destructive.
    1. Ground products of glaciers.
      i. Tors, parallel ice-molded ridges, smoothed and striated plains, billowy rock-surfaces (e. g., the Rainy lake region).
      ii. U-canyons cirques.
      iii. Rock basins, transverse troughs (e. g., Sardine lake in Bloody canyon, Cal.; the basins of the Great lakes).
    II. Marginal products of glaciers.
      i. Lateral cascades of U-canyons.
      ii. Ice-molding of summits without modification of valleys in regions of low relief; ice-molding of valleys without modification of summits in regions of high relief.[1]
2. Indirect—
  a. Constructive.
    I. Products of englacial streams.
      i. Aasar, gravel ridges (including most kames and eskers) with zones and patches of assorted drift, all produced by subglacial streams.
      ii. Paha, with some aasar, kames and eskers, and zones or patches of loess and sand, including most of the high-level loess, all deposited by superglacial streams generally at or after the time of cutting through the ice sheet.
      iii. Sand-plains, boulder and gravel fans, overwash aprons, zones of boulders and assorted drift, all produced by streams issuing from the ice or flowing along the ice margin.
    II. Products of glacial streams after leaving the ice.
      i. Certain sand-plains and overwash aprons heading in terminal moraines; the more extended bodies of loess, including most of the low-level phase; all produced by glacial rivers heading in the ice.
      ii. Fossiliferous loess, both high-level and low-level, grading into paha; produced by deposition in ice-bound or marginal lakes.
      iii. Silt sheets (and perhaps some sand-plains in regions of high-relief), including "gumbo" tracts and some "crawfish flats;" the Albany clays and their equivalents in part; all produced by deposition of glacial débris about baselevel in swamps, alluvial marshes, shallow bordering lakes, or deeper fresh-water lakes due to ice-dams or deformation.
      iv. Silt sheets, loam beds and other accumulations of glacial débris, perhaps containing marine fossils or giving origin to salt springs; produced by deposition of glacial débris about base-level in coast-wise marshes, shallow estuaries, or bordering ocean.

[1] Op. cit., pp. 510, 514.

III. Products of floating ice.
     i. * Glacio-natant till.
     ii. Beds of boulders, cobbles, and gravel, with scattered boulders more or less definitely stratified and intermixed with loam and silt (examples: the basal bed of the Columbia formation in New Jersey and the "loess-base" in northeastern Iowa).
IV. Products of water and ice affected by glacial climate.
     i. Shore ridges produced by the grounding of floes and structural features of like origin.[1]
     ii. Beds of boulders and gravel transported in ice-floes.
     iii. Estuarine, marine and alluvial beds of loam formed during the glacial depression, the collection of the material being facilitated by increased precipitation and vernal thaws and freshets (examples: most of the Columbia loam).

   b. Destructive.
     I. Products of englacial streams.
         Certain pot-holes, channels and driftless patches, chiefly produced by subglacial streams and moulins.

     II. Derivative products (chiefly representing post-glacial work).
         i. Gorges carved in sand-plains and over-wash aprons, with the terraces and other topographic forms produced thereby.
         ii. Certain terraces.[2]
         iii. Incised plains, i. e., plains in which the prevailing profiles are horizontal lines, with narrow V-shaped notches marking waterways.
         iv. "Pinnacly" or dentate topography of the loess and the low-level Columbia formation.

C. Eolic:
   1. Constructive—
     a. Dunes and drifted sand-plains.
     b. *Eolic (?) loess.
   2. Destructive—
     a. Sand-pits and ponds.[3]
     b. General wind carving or deflation.[4]

---

[1] Vide 7th Ann. Rep. U. S. Geol. Survey, 1888, pl. LXIV, p. 583.

[2] Classified in detail in Pleistocene History of northeastern Iowa, op. cit., p. 273.

[3] In northeastern Iowa there are certain tracts in which the basal sands of the loess are exposed to eolic action and have acquired a peculiar chopped-sea topography, the dunes alternating with sand-pits, sometimes of such depth that for a considerable part of the year their bottoms are below "ground-water" level, when they are transformed into ponds. Near the divide between Elkhorn and Platte rivers in Nebraska, on the head waters of Clearwater creek, there is a tract of sand-dunes of which the largest is hollowed out on the west-southwest into the form of a crescentic shell, the sand-pit cut into the dune being of such depth as to form a pond during the greater part of the year; and this dune (known as the "Devil's Basin") is only an extreme type for that region, in which nearly all large dunes are broken by sand-pits on their windward slopes.

[4] Cf. Johannes Walther, Nat. Geog. Mag., Vol. IV, 1892, p. 173.

D. Volcanic: [1]
  1. Direct—
    a. Lava sheets.
    b. Craters and cinder-cones.
    c. Tuff beds, lapilli sheets.
  2. Indirect—
    a. Tuff beds and lapilli sheets laid down in neighboring water bodies, more or less intermixed with other sediments.
    b. Derivative land forms produced by running waters acting on the foregoing deposits.

Mr. CHAMBERLIN, in closing the discussion, said that there was great difficulty in applying a chronological classification, and that such a classification might even act as a barrier to observation and to the recognition of the truth. Chronological classification is the ultimate goal of glacial studies, but it is something for which we are not as yet prepared. Red, oxidized subsoils are not developed in northern latitudes. Organic deposits between glacial layers are abundant in the West, but do not belong to a single horizon. Many facts of erosion and physical geology indicate that the glacial epoch in America was widely differentiated and of long duration. How many distinct periods it embraced we do not as yet know.

Mr. COPE added: An abundant tropical fauna is found in the "Equus beds," which, if they be of interglacial age, indicates at this time a very warm climate. This fauna is succeeded by a truly boreal fauna. In this is contained material for a chronological subdivision of Pleistocene deposits.

---

[1] While vulcanism is not a characteristic agency of the Pleistocene, there is some reason for believing that the agency was stimulated by the strains and deformations produced by shifting loads of ice and water during this period. Perhaps the most significant evidence of association is that afforded by the extinct miniature volcanoes whose craters are now occupied by the Soda lakes, Nevada. The larger of these suffered two and only two distinct eruptions, respectively coinciding exactly with the terminations of the high-water stages of the Pleistocene Lake Lahontan.

D.—DISCUSSION SUR LE COLORIAGE DES CARTES GÉOLO-GIQUES.

Mr. Powell, opened the discussion by exhibiting charts displaying the color schemes recently adopted for the general geological maps about to be published by the U. S. geological survey, and said:

There are four great classes of rocks which are more or less diverse in external appearance, internal characteristics, and mode of genesis, and which are in the main readily distinguishable even by casual observation. These are (a) fossiliferous clastic rocks, (b) superficial deposits, (c) ancient crystalline rocks, and (d) volcanic rocks.

For the distinction of these four classes of rocks and of the numerous divisions within each class there are available three kinds of devices, each of which already has the sanction of general usage. These are (1) color, (2) pattern, and (3) name or name symbol. They are sometimes used singly, but they are most advantageously used in combination.

Colors vary through a wide range in tone, and they likewise vary through a wide range in tint. When colors are used in combination with patterns, the patterns have the effect of producing variations in tint, and for this reason it is advantageous to restrict the distinctions made by colors to distinctions of tone only. Distinctions through tint may, however, be used to strengthen distinctions through pattern.

The possible variety of patterns is infinite, and the number of simple patterns readily distinguishable from one another is large.

The use of names, or of initial letters or other abbreviations representing names, fails to express area and boundary, and is therefore not coordinate in efficiency with the use of colors and of patterns, but it has an important subordinate function.

In view of these considerations, and of certain other considerations arising from general usage, the general plan of geologic cartography of the U. S. geological survey has been arranged as follows: To each of the four great classes of rocks there is assigned a group of patterns. Within each class, subdivisions are represented by colors and by specific patterns, the two being ordinarily used in combination. The identification of each indication through color and pattern is secured by means of letter symbols.

208

The fossiliferous clastic rocks are indicated by patterns in parallel lines. Superficial deposits are indicated by patterns in circular figures. Ancient crystalline rocks are indicated by hachure patterns. Volcanic rocks are indicated by patterns in angular figures.

Color is used in two ways: First, solid, or completely covering the surface, and in pale tints; second, in patterns. The first use has a chronologic function; the second has various functions in the several rock classes. In the notation for the fossiliferous clastics, the chronologic element is accented by combining an underprint in solid color with an overprint in pattern of the same color, differing only in tint. In the notation for the other classes of rocks, overprinted patterns are made to differ from underprints in tone as well as tint. In the notation for the superficial deposits, all patterns are given paler tints than the patterns for the fossiliferous clastics. In the notation for volcanic rocks, all patterns are given deeper tints than the patterns for the fossiliferous clastics. In the notation for ancient crystalline rocks, one pattern is printed over another pattern of different tone, each being of medium tint.

The following table shows the color scheme for fossiliferous clastic rocks, in which only patterns in parallel lines are used:

Period.	Letter symbol.	Color.
Neocene	N.	Orange.
Eocene	E.	Yellow.
Cretaceous	K.	Yellow-green.
Jura-Trias	J.	Blue-green.
Carboniferous	C.	Blue.
Devonian	D.	Violet.
Silurian	S.	Purple.
Cambrian	C.	Pink.
Algonkian	A.	Red.

The arrangement of this color series is essentially prismatic, but the chromatic circle is completed by the addition of the extra-prismatic colors, purple and pink, and the completed circle is redivided between the prismatic colors, red and orange, for the sake of securing in the serial arrangement the largest possible number of correspondences with existing widely prevalent usage.

The prismatic colors are never used with their full prismatic value, but are modified with white and black, so as to greatly reduce their brilliancy. The colors for volcanic rocks are made most brilliant; those for the superficial deposits least brilliant. The clastic rocks are chiefly modified with white; the ancient crystalline rocks with black.

The period colors are printed solid in pale tint. For other pur-

451 GE——14

poses the same series of colors are also used in pattern, but in darker tint.

Among the fossiliferous clastic rocks the period color is used alone when it is impracticable or undesirable to indicate the component formations. When the formations are discriminated, they are indicated by overprinted patterns of parallel straight lines in darker tint of the same color. Each formation is further indicated by a letter symbol, composed of the capital initial (or monogram) of the name of the period and the lower case initial (and when necessary other letters) of the name of the formation. When the formation is defined and the period is not certainly known, the ground color may be omitted, and the overprint constituting the formation pattern will be used alone. In such case the associated letter symbol should include only the formation symbol in lower case. When boundaries of the formations are not accurately known, the overprints are allowed to blend if the uncertainty is due to intergraduation, or to fade out without meeting if the uncertainty is due to concealments of contacts.

The parallel lines used for the distinction of formation terranes among the fossiliferous clastics may be vertical, horizontal, or oblique; narrow, broad, or alternately narrow and broad; and separated by spaces which are narrow, broad, or alternately narrow and broad. Combination and permutation of these differences will afford twenty-seven patterns for the distinction of the formations of each period, a number of distinctions which will undoubtedly suffice for all practical purposes. The lines of oblique patterns may descend from left to right or from right to left, but the distinction thus afforded will not be used on any atlas sheet to discriminate patterns otherwise identical.

Mr. VAN HISE, in further explanation of the color scheme of Mr. Powell, explained that the Archæan rocks are designated by an underprint of light brown, with an overprint of hachure patterns in discrete or broken lines, in two colors of medium tint. For metamorphic crystalline rocks of known age a similar hachure pattern in one color is printed over the ground color adopted for the period. Thus metamorphic Cambrian rocks are designated by a hachure pattern of broken lines, over an underprint of pink. For crystalline rocks of unknown age the hachure pattern is used without underprint.

Mr. WILLCOX objected to the scheme of coloring proposed by Mr. Powell, that the best use had not been made of the chromatic scale. It did not use the nine colors that most forcibly contrasted with each other. In this scheme, three of the primary colors are made up of red and blue, and one only of yellow and blue; a better distribution would be to have two of red and blue, and two of yellow and blue.

Mr. POWELL replied to the remarks of Mr. Willcox that, in choosing primary colors for his scheme, he had aimed not so much at selecting colors evenly distributed through the chromatic scale, but those that would be most readily distinguished from each other when printed on a map.

Mr. CADELL asked, why, among the colors used by the U. S. geological survey, no use was made of various shades of black or gray for the Carboniferous formation which was used in the British survey with advantage? So long as the formation was called Carboniferous, he thought that gray or black was the most suitable color that could be used, and was surprised that, when formations were so numerous and distinct colors to represent them were comparatively scarce, no use should be made of such a distinctive pigment as gray in any comprehensive scheme of map coloring proposed to be adopted.

Mr. POWELL replied that blue was used in place of the dark shades for the Carboniferous; that dark colors are misleading in regard to the occurrence of coal, which occurs in the Cretaceous and Tertiary as well as in the Carboniferous.

Mr. CHRISTIE found the black color very inconvenient because it often made the details of the map covered by such colors illegible.

Mr. CADELL said that cross-hatching and stippling, as a means of distinguishing between different geological groups or zones, could only be adopted when the maps were printed in color. Where hand coloring was employed, as in the British geological maps, it was impossible to carry out any such system properly.

Mr. POWELL explained that the U. S. Survey system is very economical when the color patterns are transferred to stones.

Mr. HUGHES thought it very difficult to devise a scheme that will meet the demands of every one. Some reference must be had to the permanence of the colors, the readiness with which they can be applied and the distinctness with which they show what is desired. He thinks the fittest scheme must survive. He pointed out that for maps on a small scale, on which a large number of different formations covered small areas, strong contrasts were necessary, but it would be rarely convenient in the case of large scale maps, where many details and subdivisions were given, to represent these by any modifications of the colors used on the small scale maps. He thought that as the Congress had after due notice and deliberation offered a tentative scheme of coloring, there was not much advantage to be gained by forcing the reconsideration of the question or putting small suggested alterations to the vote. He thought that the working out of the scheme must be left to individual map-makers, but that it would be practically decided by the geological surveys of the several countries. At the same time he would strongly advocate the appointment of a committee to watch the map-making of the world and report to each successive Congress upon it; as well as a committee to report from time to time upon the development of nomenclature and classification.

E.—COMPTE-RENDU DES EXCURSIONS GÉOLOGIQUES.

Avant que le Congrès se fut réuni, un certain nombre des membres se rendit dans l'État de New York, sur l'invitation qui leur fut faite par M. H. S. Williams, professeur de géologie à Cornell University, pour y étudier les terrains paléozoïques des États-Unis dans les localités classiques.

EXCURSION PALÉOZOÏQUE.

Directeur, H. S. WILLIAMS, aidé par MM. C. S. PROSSER et GILBERT VAN INGEN.

Les géologues qui prirent part à cette excursion se réunirent à Utica le 18 août.

Le 19 août ils allèrent à Little Falls où ils examinèrent, à la carrière de M. Hiram Boyer, les contacts entre les gneiss archéens et la base des séries paléozoïques. Le 20 août, excursion, sous la conduite de M. Ed. Hurlburt, aux chutes de la rivière Trenton, où se montre une belle coupe du calcaire Trenton (Silurien inférieur), riche en fossiles caractéristiques. Le 21 août, dans les environs de Oriskany Falls, on examina le grès Oriskany (Dévonien) et les calcaires à *Pentamerus*, inférieur et supérieur. Pendant le trajet de Oriskany Falls au lac Cayuga on traversa les séries Utica, Clinton, Salina, Corniferous et Marcellus.

A Union Springs, aux bords du lac Cayuga, les géologues purent voir le lendemain les carrières à gypse du série Salina et les calcaires du Lower Helderberg et Corniferous, et, plus au nord, suivre les couches en ascendant jusqu'au grès Oriskany. Après avoir un goûter, gracieusement offert par Mme. Anthony, ils traversèrent le lac dans un yacht à vapeur à Shelldrake Point. Là ils visitèrent un ravin où se montre une belle coupe de la série Hamilton, riche en fossiles, et se rendirent le soir à Ithaca.

Le 23 et 24 août à Ithaca. Retenus par la pluie le premier jour, les voyageurs examinèrent, le matin du 24, la coupe de Fall Creek, où se montre les couches depuis le Portage inférieur jusqu'au sommet du groupe Ithaca, y comprenant les zones à *Spirifer lævis*, à *Lingula complanata*, à *Cryptonella eudora* et à *Spirifer mesocostalis*. Ensuite ils visitèrent les belles collections du musée de l'Université de Cornell, où ils purent voir les fossiles caractéristiques des terrains qu'ils venaient d'examiner pendant l'excursion.

Ici l'excursion se termina et les géologues se séparèrent, quelques-uns allant à Rochester pour y examiner d'autres coupes paléozoïques, les autres passèrent par Mauch Chunk pour voir les terrains carbonifères, et quelques-uns enfin allèrent directement à Washington.

Pendant la durée du Congrès une seule excursion géologique eut lieu, qui eut pour but d'examiner les terrains mésozoïques et tertiaires au bord de la rivière Potomac.

212

EXCURSION AUX BORDS DE LA RIVIÈRE POTOMAC EN AVAL DE WASHINGTON.

Directeur, M. N. H. DARTON.

Partis de bonne heure, le 30 août, sur un petit vapeur, les excursionnistes ont d'abord visité les falaises de Fort Washington sur la rive gauche du Potomac (Mésozoïque moyen), sur lesquelles reposent en discordance de stratification les sables noirs de l'étage Severn (Crétacée supérieur). Au-dessus de ces couches, et également en discordance de stratification avec elles, se trouvent les sables à glauconie de la série Pamunkey (Éocène), avec *Cucullea gigantea* et autres fossiles, recouverts d'une couche de graviers de l'étage Columbia (Pléistocène).

A Cockspit Point, sur la rive droite, on a pu voir les couches arénacées de la série Potomac à plantes fossiles.

A Clifton Beach, sur la rive gauche, on vit les sables verts (greensands) fossilifères de la série Pamunkey, qui s'étendent assez loin le long de la rive.

A Pope's Creek, également sur la rive gauche, se trouvent les sables verts à *cardita planicosta*, etc., de la série Pamunkey, sur lesquels reposent les couches à *infusoria* de l'étage Chesapeake (Miocène) et les sables oranges de l'étage Lafayette (Pliocène?).

Les terraces de l'époque Columbia se montrèrent le long de la rivière pendant presque tout le trajet.

Deux excursions seulement s'effectuèrent à la suite de la réunion du Congrès. Une excursion de dix jours avait été projetée par M. Bailey Willis pour visiter la Pensylvanie occidentale, où se rencontrent tant de phénomènes d'une haut intérêt géologique que les travaux de Rogers et ensuite de Lesley ont si bien fait connaître. M. Frank Thompson, vice-président de la Compagnie des chemins de fer de la Pensylvanie, avait généreusement offert aux membres du Congrès le transport gratuit sur ses lignes, et d'autres préparatifs avait été faits, mais, en vue du très petit nombre de membres désireux de faire le voyage, on a dû en abandonner l'idée.

EXCURSION AUX MONTAGNES ROCHEUSES.

Directeurs, MM. S. F. EMMONS, ARNOLD HAGUE, G. K. GILBERT et J. W. POWELL, aidés par MM. G. H. WILLIAMS, I. C. WHITE, ED. ORTON, T. C. CHAMBERLIN, N. H. WINCHELL, J. P. IDDINGS, W. H. WEED, W. CROSS, M. E. JONES.

Celle-ci se trouva être l'excursion la plus nombreuse de toutes. Environ quatre-vingt-dix membres du Congrès, dont soixante-et-un de l'Ancien Continent et vingt-neuf du Nouveau, y prirent part, et dix dames honorèrent l'excursion de leur présence.

Partis de Washington à 9 heures 25 minutes du matin le 2 septembre, on suivit de point en point le programme précédemment publié (p. 12).

Le premier jour, pendant qu'on remontait la vallée de la rivière Potomac, M. G. H. Williams, professeur de géologie à l'Université John Hopkins, expliqua la structure géologique du pays. Ensuite en traversant les montagnes Appalaches, ce fut le Professeur I. C. White, chef géologue de l'État de West Virginia, qui expliqua les coupes de terrains paléozoïques qu'on passait en revue.

Le lendemain dans l'Ohio, Prof. Ed. Orton, chef géologue de cet État, accompagna les excursionnistes pour quelque distance, leur expliquant la structure des bassins renfermant le pétrole et le gaz, et, pendant de courts arrêts du train, leur montra plusieurs puits fonctionnants. Plus tard Prof. T. C. Chamberlin fit un exposé des phénomènes glaciaires qu'on pouvait distinguer du train.

Dans l'après-midi, arrivés à Chicago, on fit arrêt de quelques heures pour parcourir la ville, pendant que le train allait d'une gare à l'autre.

Le lendemain matin (troisième journée), pendant que le train remontait la vallée du Mississipi, des délégations des villes sœurs de St. Paul et de Minnéapolis vinrent inviter les voyageurs de visiter ces villes, où ils furent reçus de la manière la plus hospitalière. On conduisit les voyageurs en voiture par les deux villes pendant quelques heures, leur montrant tout ce qui pouvait les intéresser. En passant d'une ville à l'autre, on s'arrêta à Fort Snelling, où M. N. H. Winchell, chef géologue de l'État de Minnesota, attira l'attention sur les évidences de l'antiquité de l'époque glaciaire, démontré par l'érosion du fleuve.

A cinq heures du soir on remonta dans le train, et le lendemain on se trouva sur les Grandes Plaines du nord, que l'on traversa pendant toute la journée, faisant quelques courts arrêts pour voir de plus près les "Mauvaises Terres."

En arrivant au Yellowstone Park, le cinquième jour, on se divisa en deux sections, afin qu'on put accomplir avec plus de facilité la tournée projetée de six jours pour examiner les différents objets d'intérêt géologique. M. Arnold Hague guida une division et M. J. P. Iddings l'autre, pendant que M. W. H. Weed, qui était campé près des geysers, expliqua la croissance du sinter dans leur bassins, leur durée, etc., aux membres de chaque division.

On quitta le Yellowstone Park le douzième jour, et dans l'après-midi, à leur arrivée à Butte City, les excursionnistes furent reçu par les propriétaires des mines, et ceux qui en étaient désireux furent conduits dans les travaux souterrains. M. F. S. Van Zandt, propriétaire de la Bluebird Mine leur offrit un "lunch" avec vin de Champagne, et fit illuminer sa mine à la lumière électrique, afin qu'on put plus facilement examiner les galeries, dans lesquelles des dames mêmes sont descendues.

Le treizième jour, en entrant dans le bassin du Grand Lac Salé, on s'arrêta à plusieurs reprises pour examiner les terrasses et la voie d'écoulement de l'ancien lac Bonneville. Pendant l'arrêt de quelques jours qu'on fit à Salt Lake City, un banquet fut offert aux voyageurs par le professeur L. E. Holden de cette ville. La table fut mise avec 125 couverts, et les décorations, au lieu d'être de fleurs, consistèrent en fruits magnifiques, tels que l'irrigation seule peut produire.

Plusieurs excursions se firent pendant le séjour à Salt Lake City, dans des trains spéciaux, sous la conduite de M. K. G. Gilbert, aidé de M. E. Jones, pour examiner les terrasses du lac Bonneville, et les escarpements de failles qui les traversent le long de la base des montagnes. M. S. F. Emmons a aussi expliqué ce qu'on a pu voir, dans un si court délai, de la structure des montagnes Wasatch.

Arrivés dans le Colorado, la réception qu'y reçurent les voyageurs fut encore plus enthousiaste. On mit à leur disposition le magnifique établissement de bains à Glenwood Springs; en approchant de Leadville, Canyon City et Denver des députations vinrent au devant d'eux pour s'entendre avec eux sur les excursions en voiture, qu'ils pouvaient désirer faire dans ou autour de leurs villes respectives.

Pendant qu'on traversait le Colorado on fit quelques arrêts de courte durée pour examiner, ne fut-ce que rapidement, les phénomènes suivants: Les mines de houille dans les roches crétacées à Newcastle; les coulées toutes récentes de basalte dans la vallée de Eagle River; le mamelon (butte) de rhyolite près de Nathrop, qui contient des topaz; une série de roches algonkiennes récemment découverte près de Salida; les couches à poisson d'âge silurien à la carrière Harding, près de Canyon City, que M. C. D. Walcott venait de décrire à la dernière réunion de la Geological Society of America. Pendant la nuit qu'on passa à Leadville, plusieurs membres sont descendus dans les fameuses mines d'argent de cette ville sous la conduite de M. Emmons. A Pueblo on a examiné avec intérêt les belles collections des minerais du Colorado dans le "Mineral Palace."

A Manitou un train spécial conduisit ceux des géologues, qui désirèrent faire l'ascension, au sommet de Pikes Peak.

À Denver, le 20 septembre, les excursionnistes furent reçus par les membres du Colorado Scientific Society, qui les conduirent en voiture autour de la ville, en pleine vue des Montagnes Rocheuses, et leur firent voir les belles collections de minéraux et roches du Colorado appartenant à la Société. Dans l'après-midi, M. Richard Pearce de cette Société donna une réception aux géologues, et dans la soirée, les citoyens d'origine allemande les invitèrent à un "Commers" dans le Turnhalle.

À Denver la compagnie se divisa. Environ trente-six des membres s'en allèrent à Flagstaff dans l'Arizona, d'où ils firent, sous la conduite du Maj. J. W. Powell, directeur du U. S. geological survey, l'excursion au fameux Grand Cañon du Colorado. Quelques autres continuèrent seuls leur explorations dans les montagnes. La plus grande partie retournèrent directement à Chicago, et de là une partie est allé au lac Supérieur, mais le plupart continuèrent leur course à l'Est en suivant le programme, passant par les chutes du Niagara et arrivait a New York le 26 septembre.

EXCURSION AU LAC SUPÉRIEUR.

Directeurs, C. R. VAN HISE et RAPHAEL PUMPELLY.

Grâce à la courtoisie de M. F. W. Rhinelander, fils du président des chemins de fer Milwaukee, Lake Shore et Western, les géologues qui prirent part à cette excursion purent voyager dans le wagon spécial à restaurant des directeurs de la compagnie: et ce wagon a été gracieusement continué à leur service sur les lignes des chemins de fer Milwaukee et Northern; Chicago et Northwestern; Duluth, South Shore et Atlantic; Canadian Pacific; et Wisconsin Central. Le chemin de fer à voie étroite de Hancock et Calumet leur a aussi fourni un train spécial pendant la durée du trajet sur sa ligne.

Parti de Milwaukee le 23 septembre au soir, on se trouva le lendemain matin à Iron Mountain, Michigan, où l'on put visiter les calcaires et les couches à minerai de fer du Huronien inférieur. Ici l'on examina aussi les roches ferrifères détritiques formant la base du Huronien supérieur.

Le 25 septembre on s'arrêta à Republic, Michigan. Ici on put voir, 1° la discordance de stratification entre le Huronien supérieur et inférieur, 2° le conglomérat à la base du Huronien supérieur, 3° les couches ferrifères et les schistes quartzeux du Huronien inférieur, et enfin 4° le contact entre le Huronien inférieur et le complex fondamental. De Republic on se dirigea sur Ishpeming, Michigan, et pendant le voyage on vit des belles coupes des granites-gneiss de l'Archéen. A Ishpeming on visita la mine de Goodrich, où la discordance entre le Huronien inférieur et supérieur est exceptionnellement apparente, et aussi les gisements de fer du Huronien inférieur.

Quittant Ishpeming, on fit dans la nuit du 25 au 26 septembre le trajet à Garden River dans le Canada, à l'est du Sault Ste. Marie. Ici se montra la discordance entre le conglomérat à schiste (slate-conglomerate) de Logan et le granite du complex fondamental, et, un peu plus vers l'est, la discordance entre le calcaire du Huronien supérieur et le conglomérat à schiste du Huronien supérieur. De cette manière on a pu étudier les roches caractéristiques de ces quatre séries dans cette région.

Pendant la nuit du 27 au 28 septembre on fit le trajet de Garden River à Michigamme, et l'on y examina les mica-schistes à staurolite et à grenat du Huronien supérieur.

Quittant Michigamme dans l'après-midi du 28 septembre on se rendit à Houghton et à Calumet, aussi dans le Michigan, et on put voir pendant le trajet de belles coupes des granite-gneiss du complex fondamental. Entre Houghton et Calumet on a pu voir, à plusieurs reprises le long du chemin de fer, des coulées de lave interstratifiées dans les grès du série Keweenaw.

Le 29 septembre on a visité à Calumet les fameuses mines de cuivre natif, où l'on a pu examiner le conglomérat cuprifère. De là on est allé en voiture aux ravines

Hungarian et Douglas-Houghton, où se montrent les contacts entre la série Keweenaw et les grès Cambriens.

Dans la nuit du 29 au 30 septembre on alla de Calumet au district de Penokee-Gogebic pour y examiner les gisements de minerai de fer, s'arrêtant plusieurs fois pendant le trajet entre Sunday Lake (Michigan) et Potato River (Wisconsin) pour examiner, 1° la discordance entre le Huronien supérieur et le complex fondamental; 2° les trois membres du Huronien supérieur; 3° les relations entre les gisements de fer, ces trois membres et les roches éruptives; et enfin 4° les roches éruptives et détritiques, interstratifiées dans la série Keweenaw.

Partant de ce point nos géologues furent reconduits à Milwaukee, où ils arrivèrent le 2 octobre, et ici l'excursion se termina.

QUATRIÈME PARTIE

EXPLICATION DES EXCURSIONS

A—LA GÉOLOGIE DES ENVIRONS DE WASHINGTON
B—EXCURSION AUX MONTAGNES ROCHEUSES
C—EXCURSION AU LAC SUPÉRIEUR.

A

GEOLOGY

OF

WASHINGTON AND VICINITY

BY

W J McGEE

With the collaboration of

PROF. G. H. WILLIAMS, BAILEY WILLIS, AND N. H. DARTON.

CONTENTS.

GEOLOGY OF WASHINGTON AND VICINITY.

THE GENERAL PHYSIOGRAPHY.

There are in Eastern United States three distinct physiographic provinces. Most conspicuous of these is the Appalachian zone, an area of long, low mountain chains of wonderful parallelism. At the eastern base of the mountains lies the Piedmont plateau, an undulating plain standing 500 to 1,000 feet above sea level. Between this plateau and the ocean lies the Coastal plain, a generally smooth lowland rising gently from ocean waters to altitudes reaching about 300 feet.

The rocks of the Appalachian zone are Paleozoic, running from the Carboniferous down to the Cambrian and probably to the Algonkian, aggregating 25,000 to 40,000 feet in thickness. The entire series is nearly or quite conformable; the materials range from coal seams toward the summit, and pure limestone at various horizons, to coarse sandstones, and in Pennsylvania to great beds of conglomerate. The strata, originally horizontal or slightly inclined westward, have been deformed and altered in a variety of ways. In the western and central portions of the province they have been flexed symmetrically and thrown into a series of anticlinal and synclinal corrugations, seldom more than a mile or two in width though often scores or even hundreds of miles in length—a series of mountain folds unparalleled elsewhere on the globe in length, symmetry, and concordance in direction. In the central part of the zone the symmetric flexing is combined with faulting, and in many cases the faulting is of that overthrust type which characterizes the Scottish Highlands and the Canadian Rocky mountains. In the eastern margin of the zone the symmetric flexing fails, faulting (both normal and overthrust) prevails, and the rocks are more or less profoundly metamorphosed—the limestones transformed into marbles, the shales into slates, the sands into quartzites. Throughout the province the distinctive structure and the rock composition are both reflected in topographic configuration; the prevailing forms are long narrow ridges, separated by long and generally narrow valleys; but these land forms represent respectively the outcropping edges of hard strata and soft beds rather than original flexures.

223

The rocks of the Piedmont belt are more or less crystalline, chiefly metamorphic schists and gneisses of considerable diversity in composition, but sometimes including ancient eruptives, as well as quartz veins and dikes. The structure of the province is obscure and diverse, and has not yet been fully investigated. It is known, however, that, in the latitude of Washington at least, the Piedmont belt is separable into two distinct parts. Of these the western is composed of semicrystalline slates, phyllites and schists having a constant inclination toward the east; while the eastern part is made up, except for a few included folds of the less crystalline rocks, of highly crystalline gneisses and a variety of foliated eruptives, all of which have a prevailing dip toward the west. The nearly vertical position of strata intermediate between these extremes gives a pseudo fan-structure to a section of the Piedmont plateau in Maryland. The line between the western semicrystalline area and the eastern gneissic area is indefinite; and there is an apparent progressive increase in the intensity of metamorphism from the western border to the eastern limit of the Piedmont belt, by which casual students have been misled. The surface of the zone is characterized by meandering stream channels and wandering divides, with moderately strong local relief; yet, while the harder rocks of the province find a certain expression in the topography, the general configuration is independent of rock structure but represents baselevel conditions during past eons.

The composition and configuration of the Piedmont zone are locally diversified by considerable areas of Mesozoic rocks, commonly referred to the Triassic. These rocks are red sandstones and red or purple shales, with occasional beds of conglomerate. They are characterized by strong dips toward the Appalachian zone; and they are frequently cut and sometimes interbedded with or overlain by contemporaneous or younger dikes and sheets of trap. In the northern part of the Piedmont zone, the trap occurs in considerable volume, and forms prominent ridges by which the topography of the entire Piedmont belt is dominated; but in general the sandstones and shales are soft and friable, and find topographic expression in low-lying plains and basins.

The rocks of the Coastal plain are clastic, ranging in age from Pleistocene to middle Mesozoic, probably reaching a total thickness of 2,500 to 3,500 feet. The entire series inclines gently seaward, the inclination increasing from the newer to the older formations. The strata are manifestly made up of the debris of the Appalachian and Piedmont provinces, are rarely lithified, and range from alluvium or alluvium-like silts along the rivers and toward the coast, and glauconitic marls and fine clays in the middle of the series, to coarse gravels and beds of arkose toward the base and near the old shore lines. Except for the gentle inclination of the strata, and except for a dislocation coinciding with the inland margin of the province, the strata are not visibly deformed, but retain substantially the attitudes as well as the composition of original deposition. The surface of the province is commonly characterized by meandering rivers, throughout the middle Atlantic slope by broad estuaries, and in general by broad, low divides, often terraciform—the configuration seldom expressing structure or localized earth movement, but representing simple erosion combined with wave action during several continental oscillations of general character.

The western boundary of the Appalachian zone is indefinite; the characteristic corrugations gradually die out and the flexed strata of the Appalachian pass into the undisturbed strata of the interior plain.

The common boundary of the Appalachians and the Piedmont zone is generally trenchant, consisting of a prominent ridge of quartzite—the Blue ridge. Somewhat south of the latitude of Washington the ridge is simple and single; where cut by the Potomac river west of Washington it is triple or quadruple; in Maryland and Pennsylvania it is frequently multiple; and in Virginia and the Carolinas it is sometimes interrupted and again divided; but in general it definitely marks a fairly decided transition from comparatively simple to comparatively complex structure, and from incipient metamorphism to pronounced alteration in the rocks.

Throughout the middle Atlantic slope the common boundary of the Piedmont zone and the Coastal plain is pronounced; along this line there is a sudden and decided transition in the rocks from highly altered crystallines to practically unaltered clastics; along this line the waterways change from narrow, rock-bound gorges of considerable declivity to broad tidal canals, and each river passes from the one province to the other in a cascade or rapid; along this line the rivers are diverted from courses cutting across the trend of structure and athwart the provinces to courses parallel with the line of cascades, thus peninsulating most of the Coastal plain; and along the line thus accentuated by the diverted drainage there is commonly a prominent scarp of Piedmont rocks overlooking the flat-lying rocks of the Coastal plain. This physiographic boundary is one of the most trenchant on the surface of the globe, and the natural line is emphasized by a prominent cultural line to which it gave origin; all the principal cities of the eastern United States from New York to the Carolinas are located along this natural landmark.

The eastern boundary of the Coastal plain may be drawn at the shore of the Atlantic; but it may more properly be drawn 100 miles offshore at the great submarine escarpment, 3,000 to 10,000 feet high, hugged by the Gulf stream. In general configuration, in inclination of the surface, and unquestionably in structure and composition. the subaerial and the submarine portions of the Coastal plain are essentially a unit, and the present coast line is but an accident of present relation between sea and land.

Despite the diversity in rocks, structure, and configuration in the three provinces, the principal rivers of the middle Atlantic slope traverse all alike. The Mohawk and the Hudson run around the northeastern extremity of the typical Appalachian zone, separating the three distinctive provinces from the analogous (but probably not homologous) physiographic tract of New England; the Delaware, with its great secondary the Lehigh, the Susquehanna, the Potomac, and the James, rise well within the Appalachian zone, cut through the successive ridges in a series of clefts, cross directly the Piedmont plateau, and, although diverted at the fall line, thence intersect the Coastal plain to the Atlantic; and except at the fall line their courses are essentially independent of structural conditions. Yet even along the great rivers the boundaries of the physiographic divisions find expression. The Appalachian-Piedmont boundary is marked by narrow notches in the Blue ridge, forming the far-famed "water gaps" of the Delaware, of the Lehigh, of the Susquehanna near Harrisburg, of the Potomac at Harpers Ferry, and of the James at Balcony falls; the Piedmont-Coastal boundary is still more strongly marked by the line of cascades on every river, large and small, from the Raritan in New Jersey to the Roanoke in North Carolina, and by the deflection of the waterways which peninsulate the lowland plain from New York to Richmond.

THE LOCAL PHYSIOGRAPHY.

The city of Washington, like the other metropoles of the middle Atlantic slope, is located at the common boundary of the Piedmont and Coastal zones. The western part of the city is built on the ancient crystallines, the eastern on the nonlithified clastics; though outliers of the clastic formations occasionally occur on the upland some miles farther westward. Located like neighboring metropoles at the head of navigation, the city marks the position of the fall line. At Washington the Potomac river is tidal, and perhaps half a mile wide; within 4 miles upstream the channel contracts at ordinary stages to barely 100 feet, changing meantime from a slack-water canal into a rushing torrent. This is the "Little Falls of the Potomac." Then follow 12 miles of nearly continuous rapids to the "Great Falls of the Potomac," where at ordinary stages the river contracts to about 50 feet and descends 40 feet in a succession of plunges, of which the highest is about 15 feet. Between Washington and Great Falls the river occupies a narrow gorge excavated in a broader one, whose bottom averages 150 feet above tide; above Great Falls the river wanders over the bottom of the older gorge. Each tributary between Washington and Great Falls has a rapid or fall at a distance from its mouth varying directly with its volume and inversely with its distance below the cascade of the Potomac; that of Rock creek is some four miles from the mouth, and is a picturesque rapid in which the contracted stream rushes over a rugged rocky bed cumbered by huge boulders. Midway between Washington and Great Falls the river displays an anomalous feature (of which the Susquehanna is the type) in that, although of high declivity, lateral corrosion exceeds the vertical cutting, especially at freshet stages, so that even the channel remains broad and shallow.

Just west of the city the embouchure of the gorge expands, and its walls merge into the general Piedmont scarp overlooking the Coastal lowland. Just east of the city lies Anacostia river, a goodly mill-stream only, clear and rapid in its headwaters among the Piedmont hills, but sluggish and marsh-bordered for the last five miles of its course. A century ago it was navigable, and transatlantic shipping embarked and debarked at Bladensburg; but now it is clogged with alluvium and barely navigable above the Washington navy-yard. Between the rivers lies a triangular amphitheater, bounded on the west by the Piedmont scarp, on the north by a terraciform upland, on the east and southeast by low bluffs carved out of Coastal plain deposits, and opening southward through the Potomac estuary. Most of this amphitheater, together with the upland borders toward the north and west, is occupied by the city.

Southwest of the city there are extensive terraces, evidently wave-fashioned, but deeply invaded by erosion. North of the city the upland is similarly terraced, though broad and deep ravines interrupt the continuity of the plains; and beyond the Anacostia most of the surface represents two or more wave-fashioned plains which, although deeply scored by erosion, sometimes maintain their integrity quite to the verge of the river bluffs. The Fort Myer upland, southwest of the city, is simply the scarp of a broad terrace. Kalorama heights and Columbia heights, toward the northwest, are the salients of a similar terrace. Good Hope hill, on the southeast, is a remnant of another terrace. The bluff on which the National Asylum of St. Elizabeth is located is the scarp of a lower terrace of wonderful horizontality and continuity. Farther westward and northward the surface rises in less regular divides, crests, knobs, and spurs; but here and there terrace remnants are found up to over 400 feet above tide, or nearly to the greatest altitudes of the region.

The terrace plains are built; the broad, low, wave-fashioned plains flooring the amphitheater are composed of the newest deposits of the region; the higher terraces carved on the walls of the amphitheater are of earlier, yet late Tertiary, origin. The smaller ravines as well as artificial excavations reveal the materials of the terraces in hundreds of exposures; the larger ravines as well as artificial excavations reveal the clastic formations beneath and east of the city in numberless exposures; Potomac river and its larger tributaries are bound between steep, often precipitous, walls of the crystalline rocks. The entire region is dissected by waterways and by a multitude of storm-cut ravines, and so the local relief is strong except toward the interiors of the broader terraces.

THE GENERAL GEOLOGY.

THE ROCKS OF THE PIEDMONT PLATEAU.

Present State of Knowledge.—Since the beginnings of American geology the prevailing crystalline character of the Piedmont terrane has been recognized, and the rocks have commonly been referred to the Archean and frequently correlated on petrographic ground with the Huronian, Laurentian, and other ancient rock systems of distant parts of the country. During the last decade Dr. George H. Williams began systematic work on the Piedmont rocks in the vicinity of Baltimore. More recently his studies have been extended westward across the entire zone along several lines in Maryland, Virginia, and North Carolina. The more important results of these researches have been published by the Geological Society of America.[1] By means of these studies the petrographic character, structure, and relations of the Piedmont rocks about the latitude of the National Capital have been made known.

The Rocks and their Relations.—The Piedmont plateau is divisible into an eastern highly crystalline and a western semicrystalline portion. The former consists of gneisses and holocrystalline mica schists, quartzites, and marble, containing an abundance of more or less dynamically metamorphosed eruptive masses. All of these rocks have a prevailing north-northeast strike and a westerly dip. The western portion, on the other hand, is composed of partially metamorphosed sedimentary strata (sericite and chlorite schists, ottrelite schist, phyllite, and limestone) and is nearly free from ancient eruptives. The strike of these rocks conforms to that of the eastern portion, but their dips are prevailingly toward the east. In spite of apparent conformity and even indications of transitions between these two portions of the Piedmont region, they are separated by a great time break and unconformity. The easterly dips on the west and the westerly dips on the east, together with the nearly vertical strata between, produce a radiating or fan structure, and the axis of this fan is not coincident with the contact between the crystalline and semicrystalline portions. The thickness of either series of rocks, as indicated by their present dips, would be so vast that we must assume that the same beds are repeated over and over again by tightly compressed folds or thrusts. In the absence of all paleontologic data it is impossible to assign a definite age to either of these series. In the light of what has been discovered elsewhere, however, it is not improbable that the western and semi-crystalline areas represent the older Paleozoic horizons, metamophosed by more intense dynamic action than has affected them farther west-ward, while the holocrystalline rocks on the east are a remnant of the pre-Cambrian continent, from which the Paleozoic sediments were de-rived. The apparent conformity between the two regions may be explained by supposing that the highly crystalline rocks also formed the floor upon which the now semicrystalline schists were deposited as sediments. These older rocks, already greatly altered and folded, underwent at the time of the Appalachian uplift one more final fold-ing, which gave them their now prevailing trend and carried the over-lying Paleozoic sediments with them. This supposition is also in accord with the fact that several closed synclinals of slate and semi-crystalline schists are found pinched into the gneisses far east of the main contact.

THE FORMATIONS OF THE COASTAL PLAIN.

Present State of Knowledge.—Although geologic reconnoissance was extended over the portion of the Coastal plain lying in the middle Atlantic slope early in the present century, detailed surveys were not made until long after. So, while the composition, structure, and age of the deposits were known in general terms, little was known of the precise limits of the several formations, or of the geologic history recorded within them (particularly about the National Capital), until the middle of the last decade. Soon after the organization of the present Geo-logical Survey, systematic study was initiated; within the next three years certain formations were discriminated and classified, and the methods of investigation applicable in this distinctive, if not unique, geologic province were developed. Subsequently detailed surveys were undertaken, under the auspices of the Geological Survey, by Mr. Nelson H. Darton. Certain formations were by him discriminated and classi-fied, and the composition, attitude, and precise areal distribution of the formations lying between the Potomac river and Chesapeake bay (the "western shore" of Maryland) as well as in much of "tide-water Virginia," were ascertained. The areal distribution of the clastic for-mations developed about Washington, as determined by Mr. Darton, is represented on the Washington atlas sheet of the U. S. Geological Survey; maps of other portions of the Coastal plain are not yet pub-lished, although the surveys are well advanced.

The surveys north and south of the Potomac-Chesapeake peninsula, and of the peninsula lying east of Chesapeake Bay (the "eastern shore" of Maryland and Delaware) are not yet completed. Accordingly, while the formations enumerated below are probably representative of the Coastal plain throughout much of the middle Atlantic slope, they are, in general, accurately known only in the immediate vicinity of Washington.

In the researches within the Coastal plain certain methods, developed as the work progressed, have been constantly used; and since these methods are distinctive, and since, moreover, they affect materially the results of the work, they may briefly be stated:

(1) Reconnoissance and preliminary surveys showed that the Coastal plain deposits are commonly thin but extensive, and each composed of distinctive materials, only a part of the series being fossiliferous. Moreover the Coastal plain is vast, extending over fully 15° of latitude and 25° of longitude, and includes the deposits of the greatest river of the continent, of many variously conditioned rivers of less size, and of coasts receiving little terrestrial drainage; from which it was inferred that the distribution of organisms during past eons was affected by diverse conditions of environment, much as the fauna and flora of the present are affected. Accordingly it was deemed feasible to define the formations by composition, attitude, and physical relations, and to trace formations from place to place throughout the province by means of stratigraphic continuity, independently of fossil remains presumptively varying from place to place with the varying environmental conditions of the periods of deposition. Thus the formations discriminated in the Coastal plain are essentially physical units.

(2) As research progressed, it was found that, in many cases, the materials of the successive Coastal plain deposits may be traced to their sources, and that their character and distribution indicate the proximity of shores, the depths of waters, the positions and characteristics of sediment-bearing rivers, etc. Thus it was found that each formation represents a certain general relation between sea and land, the recognition of which easily explained local variations in the physical condition of the deposits; and thus the tracing of the formations by stratigraphic continuity was faciliated and extended. So each Coastal plain formation is a physical unit, and at the same time an expression of the general physiography of the continent during the period of its deposition.

(3) As researches into the relations of land and sea during the several eons progressed, it was found that in many cases the character and distribution of deposits composing the formations indicate not only the position and size of sediment-bearing rivers, but the declivities and other conditions of those rivers, which in turn indicate the attitude,

altitude, and general configuration of the land surface during the period of deposition. It was also found that in many cases the land-forms themselves record geologic history as definitely and intelligibly as the deposits from which history is commonly read; and accordingly the deposition-record was in many cases supplemented by the degradation-record. So, many of the Coastal plain formations not only represent general physiographic conditions, but yield detailed records of geography and topography during the periods of deposition.

(4) As the discrimination of successive deposits of the sea and of the variously superimposed topographies of the land in the Coastal plain and Piedmont provinces progressed, it became evident that any local tract gives a record of a certain series of physical episodes, each of definite character, and that recognition of the conditions of each episode facilitates the tracing of deposits from place to place, even throughout the entire Coastal plain and far within the contiguous provinces of concurrent degradation. Thus it was found feasible not only to correlate formations with aspects of the land in each tract but to correlate the tracts of a vast area by means of genetic identity, or by homogeny.[2] So, certain of the Coastal plain formations discriminated in the Atlantic slope represent not simple records of local physiographic conditions, but exact indices of geographic and topographic conditions extending over a considerable fraction of the continent.

For these reasons the taxonomy of the Coastal plain formations is largely independent of the paleontologic scale. Accordingly, while each formation is known to record a definite episode of continental history, its paleontologic position can seldom be indicated with accuracy in the present state of knowledge, and perhaps can not be ascertained until researches have extended over the entire Coastal plain, and until the distribution of organisms during each episode in Coastal plain development is determined with precision.

The Formations and their Relations.—The clastic formations found in the middle Atlantic slope, the geologic groups to which they are provisionally assigned, the thickness, attitude, and certain other characters of each, the history indicated by their physical relations, together with the approximate paleontologic position of each episode (whether of deposition or degradation), are indicated in the accompanying table, the distribution being shown in the Washington atlas sheet issued by the United States Geological Survey.

FORMATION.	CHARACTERS.	PALEONTOLOGIC POSITION.
Alluvium .. Thickness variable; chiefly below tide; undisturbed*	{Late Pleistocene and modern.
Erosion interval; dissection of Columbia...............	Pleistocene.
Columbia .. Thickness 5-40 feet; altitude 200 feet; undisturbed*.	Early Pleistocene
Erosion interval; extensive removal of Lafayette	Pliocene (?)
Lafayette .. Thickness 5-50 feet; altitude 500 feet; undisturbed*	Pliocene (?)
Erosion interval; extensive planing of Chesapeake..........	Miocene (?)
Chesapeake. Thickness 10-125 feet; tilted slightly; fossiliferous..	Miocene.
Erosion interval; extensive planing of Pamunkey and Severn.	?
Pamunkey .Thickness 3-100 feet; tilted slightly; fossiliferous..	Eocene.
Erosion interval; extensive planing of Severn and Potomac..	?
Severn ..... Thickness 2-25 feet; tilted seaward; fossiliferous .	Cretaceous.
Erosion interval; profound dissection of Potomac..........	Cretaceous.
Potomac ... Thickness 5-500 feet; considerably tilted; fossiliferous	Early Cretaceous.

Long interval of extensive and profound erosion........... Jurassic (?)

There is a notable dearth of alluvium throughout the middle Atlantic slope; west of the "fall line," which is not only the common boundary of two strongly distinguished provinces, but a line of modern dislocation as well, the land is rising so rapidly that the rivers, albeit rapid and generally rushing torrents, are unable to cut their channels down to baselevel; east of the "fall line" the land is sinking so rapidly that deposition in the estuaries, albeit localized and rapid, does not keep pace with the sinking.

Anterior to the vaguely limited period which may be assigned to alluvium deposition the land stood higher than now, for the antecedent formations are deeply and broadly trenched by the Potomac, the Anacostia, and other Coastal plain rivers; but whether it was the entire region or only the now sinking Coastal plain that formerly stood higher is not certainly known. It seems probable, however, that both Piedmont and Coastal provinces were elevated after Columbia deposition, that both were subsequently depressed to some extent, and that while the downward movement of the Coastal plain continues, the movement of the Piedmont plateau was long since reversed.

The Columbia formation[3] commonly consists of brown loam or brick clay, grading downward into a bed of gravel or boulders. Toward the embouchures of the larger rivers from their Piedmont gorges the loam commonly thins and the boulder bed thickens; in the remoter parts of the estuarine valleys the loam thickens, the boulder bed thins, the materials become finer, and a sand bed often separates loam and gravel; farther down the estuaries the gravel bed commonly disappears, and the loam becomes interstratified and sometimes intermixed with silt. Between the rivers the deposit extends over divides up to altitudes of about 200 feet in the latitude of Washington, increasing northward and decreasing southward; and in such interstream

* Except by a late Neocene displacement which is yet in progress.[7]

areas the deposit is more heterogeneous than along the rivers, and contains a considerable element of materials corresponding with those of the immediate subterrane. As a whole the deposit evidently represents littoral and chiefly estuarine deposition. The materials differ from those of the modern alluvium in (1) greater dimensions of the boulders, (2) greater coarseness of sediments in general, and (3) less complete trituration and lixiviation of the several elements. These differences are indicative of long cold winters, heavy snow fall, and thick ice, but not of glaciation (in this latitude) during the Columbia period.

The Columbia formation has been traced throughout the greater part of the Coastal plain from the mouth of the Hudson to beyond the Mississippi, or over an area of more than 200,000 square miles, its thickness and composition varying from place to place with the volumes of rivers and with the character of sediments transported by them; and the altitudes of occurrence indicate submergence decreasing from about 400 feet in the latitude of New York to 200 feet at Washington, and perhaps 75 feet in the latitude of Cape Hatteras, thence increasing to nearly or quite 700 feet on the Savannah, diminishing next to less than 50 feet at Mobile bay, and again increasing to variable maxima farther westward and northwestward.

Traced northward the formation is found to pass under the terminal moraine and the drift sheet it fringes; at the same time the size of boulders and other indications of contemporaneous cold multiply, and an element of ice-ground rock flour occurs in the upper member, from which it was long inferred to represent an early episode of glaciation; and during the present summer (1891) Salisbury has found it to pass into a premorainal drift sheet in northern New Jersey. From the relative extent of erosion and degree of oxidation, the Columbia formation and the corresponding drift sheet are inferred to be 5 to 50 times as old as the later glacial deposit, and a rude but useful measure of the duration of the Pleistocene is thus obtained.

During the post-Columbia period the inner gorge of Potomac river from Washington to Great Falls was excavated. Anterior to the Columbia period the land stood so high at Washington and northward that the antecedent Lafayette formation was profoundly eroded—indeed, north of Potomac river only isolated remnants of the Lafayette persist; but farther southward the high level diminished to such extent that the Lafayette formation maintains its continuity over wide areas. This period of erosion was long, yet not so long as to permit planation—deep and broad canyons were carved, to be subsequently converted into estuaries; ravines were deepened and slopes steepened, and much of the Lafayette formation was degraded; yet the interstream areas were not reduced to baselevel. Recent researches

indicate that it was during the transition, from the low level of Lafayette deposition to the high level of post-Lafayette degradation, that
the modern displacement of the middle Atlantic slope was initiated and
the rivers were diverted into their present anomalous courses.

The Lafayette formation[4] commonly consists of well-rounded, quartzitic gravel, more or less abundantly embedded in a matrix of red or
orange-tinted loam, the gravel element predominating in the north-
westernmost exposures and the loam predominating toward the interior
of the Coastal plain. The pebbles are evidently derived chiefly from
the Potomac formation; the loam is derived in part from the same formation, but in probably larger part from the residua of the Piedmont
crystallines. The deposits differ from those of the younger Columbia
formation in that the pebbles are finer, more completely water-worn,
and more largely quartzitic (the Columbia alone containing boulders
and abundant pebbles of the local and sublocal Piedmont crystallines);
and they may be discriminated from the older Potomac deposits by the
smaller size and better rounding of the pebbles, and by the dearth of
arkose (which is abundant in the earlier formation), as well as by a number of less striking characters.

The Lafayette formation, like the Columbia, has been recognized
throughout most of the Coastal plain except in the northern portion of
the middle Atlantic slope, in the Mississippi bottoms, and in a number
of more restricted areas from which it has been degraded. Its composition varies from place to place in such manner as to indicate the
local sources of material and conditions of deposition; yet despite this
local diversity it is marvelously uniform throughout the 200,000 square
miles over which it has been recognized—indeed, though the youngest
member of the clastic series, this formation is at the same time more
extensive and more constant in aspect than any other American formation.

The Lafayette formation overlaps unconformably all the older members of the Coastal plain series in such manner as to indicate that all were
extensively degraded anterior to its deposition; yet the floor on which
the formation rests is more uniform than its own upper surface, indicating that, while the antecedent erosion period was long, the land stood
low, so that it was planed nearly to baselevel and seldom deeply
trenched. During the post-Lafayette elevation, on the contrary, the
land was deeply trenched and not planed, indicating a higher altitude
than during the earlier eon, but a shorter period of stream work. This
record within the Coastal plain proper coincides with a geomorphic
record found in the Piedmont and Appalachian zones. Throughout
these zones the major and most of the minor rivers flow in broad and
deep yet steep-sided gorges excavated in a baseleveled plain. The Potomac gorge belonging to this category extends from Washington well

toward the sources of the river; it is within this gorge that the newer Washington-Great Falls canyon is excavated; the same ancient gorge is admirably displayed at Great Falls, and again at the confluence of the Shenandoah at Harper's Ferry. Moreover the ancient gorges of this category are best developed in the northern part of the middle Atlantic slope, where the Lafayette formation is most extensively degraded. Now, by the concordance of history thus recorded in plain and plateau, the degradation epochs of the adjacent provinces may be correlated and the ancient gorges of the Piedmont plateau and of the Appalachian zone as well may be referred to the period of high level immediately following Lafayette deposition. While the positive evidence for this correlation is hardly conclusive, the negative evidence is more decisive—the Coastal plain deposits yield no other record of continent movement of sufficient amplitude and extent to account for this widespread topographic feature.

Accepting the correlation, some conception of the relative antiquity of the Columbia and Lafayette periods may be formed. In general, post-Lafayette and pre-Columbia erosion was sufficient to remove fully half of the earlier formation throughout its vast extent, and to trench it and the older formations beneath, along the present shore lines of Atlantic and Gulf, to depths ranging from 150 or 200 up to 600 or 800 feet, or to effect from 50 to 5,000 times the degradation of the post-Columbia period. Again, the post-Lafayette gorges of the Piedmont and Appalachian zones exceed the post-Columbia gorges excavated by the same rivers in the crystalline rocks certainly not less than 500 times, and perhaps more than 5,000 times. Moreover, if the correlation be accepted, the immense canyons of the middle Atlantic slope which, albeit more than half filled by later deposits, yet accommodate great estuaries, must be referred to the post-Lafayette high level, and the pygmy submarine trenches of the Atlantic coast[6] must be referred chiefly, if not exclusively, to the post-Columbia high level; in which case the relative erosion measures are many thousands to one. It is indeed known from the steepness of wall of the Piedmont and Appalachian gorges that the excavation was effected rapidly, and hence that the land stood high above baselevel for a relatively limited period only— a period exceedingly short in comparison with the antecedent period of baseleveling; and accordingly that the post-Lafayette high level may not have persisted, and probably did not persist, to the beginning of the Columbia period. Yet however the several variables be evaluated, it is manifest that the pre-Columbia and post-Lafayette degradation interval must have been many times longer than the interval of degradation following the Columbia period. The relative antiquity of the Columbia and Lafayette formations thus indicated is shown graphically in the accompanying fig. 1.

Beneath the Lafayette formation lies the Chesapeake,[5] a heavy bed of fine sands and clay, sometimes containing more or less abundant glauconite and infusorial remains and characteristic Miocene fossils. This distinctive bed is the most extensively developed member of the Coastal plain series on the "western shore" of Maryland. Although the faunas of the inferior and superior portions are somewhat diverse, the materials of the formation are essentially alike from base to summit, and the faunas intergrade in such manner that it is impracticable to divide the deposits on this ground, at least in the latitude of the National Capital. Although the formation undoubtedly extends eastward to the ocean and both northward and southward for scores or hundreds of miles, the deposits have not been actually traced much beyond Delaware bay on the north and James river on the south; but a wide-spread paleontologic equivalent has been recognized in Georgia and Florida.

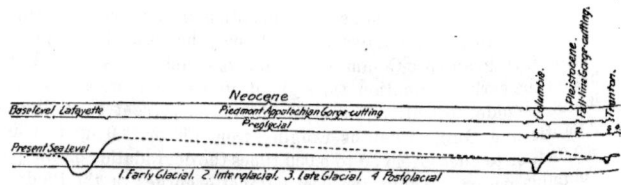

FIG. 1.—Diagram showing relative antiquity of Columbia and Lafayette formations.

Except as modified by the displacement coinciding with the fall line, the newer deposits sensibly maintain the attitudes of original deposition; yet the Chesapeake and older formations are slightly deformed. This deformation, best displayed by the surface configuration, is displayed also by the Chesapeake formation. It consists of a slight inclination toward the fall line from an axis approximately parallel with and 4 or 5 miles distant from that boundary, together with a somewhat more decided seaward inclination beyond.

The Chesapeake formation is separated from the Lafayette above and the Pamunkey below by strong unconformities, each recording considerable degradation of the underlying formations; but in both cases the inequalities in contact are comparatively gentle, indicating widespread planing rather than restricted trenching; from which it may be inferred that the degradation period was long but that the land stood near baselevel. This deposition record of the Coastal plain has been correlated only in a general way with the degradation record of the Piedmont province. In the latter province the extensive ancient baselevelling undoubtedly corresponds to several successive periods of Mesozoic and Cenozoic deposition and interruption of deposition in the Coastal plain, of which the Chesapeake period was one.

The Pamunkey formation[5] consists of a homogeneous sheet of sand (commonly glauconitic) and clay, with occasional calcareous layers; and it commonly abounds in characteristic Eocene fossils. Like the Chesapeake it lies in a gentle anticlinal, its western margin inclining landward, and the great body inclining seaward.

Although it has not been actually traced on the ground beyond the limits of the "western shore" in Maryland, and "tide-water" Virginia there are good reasons for believing that the Pamunkey formation extends throughout nearly all of the Coastal plain in the middle Atlantic slope, and probably stretches thence southward with unbroken continuity until it merges with the calcareous Eocene series of the eastern Gulf slope.

The unconformity separating the Pamunkey from subjacent formations is of the planation type, and thus tell of a long degradation period, during which the land was little elevated above base level. In general terms this degradation period may be correlated with the baselevel period of the plateau and the mountains; but there are some indications that the lifting of the land was greater in the south than in the north.

The Severn formation[5] commonly consists of fine black, micaceous and carbonaceous sands, sometimes glauconitic, rather poorly fossiliferous, the organic remains being of characteristic Cretaceous facies. Southward from the National Capital the formation thins and soon fails; northward it thickens and expands, undoubtedly passing into the extensive glauconitic Cretaceous beds of New Jersey. Whether the attenuation southward is due to nondeposition, to extensive degradation in this direction, or to both combined, has not yet been determined. The formation inclines seaward gently yet more steeply than the Pamunkey; its extension beyond the gentle anticlinal axis parallel with the fall line is too slight to give decisive indication of the usual landward dip of this part of the province.

The floor upon which the Severn formation rests is more uneven than its newer homologues, indicating not only extensive planation but decided trenching, and therefore may be inferred to represent long-continued degradation of land standing considerably above baselevel; yet the land record of this episode is lost in the remoteness of the period and the faintness of the record.

The basal formation of the Coastal plain series (the Potomac)[7] outcrops along the fall line from the Delaware to the James as a heterogeneous mass of sand, clay, arkose, and quartzitic or quartzic gravel. The arkose unquestionably represents the neighboring Piedmont crystallines; the quartzite is evidently derived from the extensive Paleozoic beds forming the Blue ridge; the quartz represents the veins by which the Piedmont crystallines are frequently intersected. The more obdurate materials are not, however, confined to the Potomac formation in which they were originally deposited; they have been rearranged and incorporated with the Lafayette, the Columbia, and probably the Chesapeake formations, and have been accumulated in modern taluses and torrential deposits. Moreover, since the advent of the white man the pebbles and cobbles have been collected for paving and guttering; and before his era they were extensively used by the aborigines for the manufacture of rude implements. Although poorly fossiliferous in the District of Columbia, so far as known, the Potomac formation has yielded a remarkable fauna and a wonderfully rich and interesting flora. The faunal remains, collected principally between Baltimore and Washington comprise dinosaurian bones of unique species but, according to Marsh, strong Jurassic affinities; the flora, obtained chiefly from Virginia, has been monographed by Fontaine, by whom it is regarded of Cretaceous facies and probably equivalent to the Cenomanian of Europe, though Ward deems it somewhat older.

The Potomac formation has been traced southward along the fall line in isolated exposures across the Carolinas and Georgia, to reappear in considerable volume in Alabama, where it is designated the Tuscaloosa formation.⁹ It has also been traced northward through Maryland and Delaware, and has been recognized in New Jersey.

The Potomac formation rests unconformably on the Piedmont crystallines, filling steep sided and narrow gorges at low levels, overspreading the moderately undulating plains at high levels. The ancient configuration revealed by this unconformity comprises an extensive Piedmont peneplain, half reduced to baselevel and afterward deeply trenched by the waterways, much as the smoother baselevel surface of later times was trenched during the post-Lafayette high-level. The duration of the pre-Potomac degradation period was vast. At the close of the Paleozoic the eastern United States was extensively deformed, uplifted, and eroded, until many thousand feet of the surface was carried into the sea; then came the Newark or Triassic period of local deposition, which was followed in turn by extensive deformation, the faulting amounting probably to many thousands of feet; and then followed comparative quietude until not only the channels of the waterways, but the entire surface over some hundred thousand square miles was approximately baseleveled, undoubtedly by the degradation of thousands of feet of rock beds.

This sub-Potomac unconformity gives some indication of the relative position of the Potomac formation in the Mesozoic period as well as of the relative duration of the several coastal plain periods of deposition and degradation. Let post-Columbia erosion represent unity; then post-Lafayette degradation may be represented by 1,000, and the post-Potomac and pre-Lafayette baselevel period may be represented by

100,000; then, using the same scale, the post-Newark and pre-Potomac erosion must be measured by something like 10,000,000, and the post-Carboniferous and pre-Newark degradation by 20,000,000 or 50,000,000. These figures are but rude approximations; they moreover, in one sense misleading, since degradation undoubtedly proceeded much more rapidly during the earlier eons; yet they give some conception of the relative importance of a long series of episodes in continent growth, and indicate definitively the wide separation of the Newark and Potomac periods.

The time relations between the post-Potomac formations are represented graphically in Fig. 2. The intervals are of course only rudely approximate, yet they stand for estimates, not guesses.

THE GEOLOGY OF THE APPALACHIAN ZONE.

Present State of Knowledge.—The general features of this province were long ago made known by the classic work of the Rogers Brothers in Pennsylvania and Virginia; but since the expansion of the field of geologic science during recent years it has been found necessary to survey in greater detail much of the area already once or twice traversed. The federal surveys of the southern and central Appalachians were for some years in charge of Mr. G. K. Gilbert, and more recently have been carried on by Mr. Bailey Willis. One of the results of this work has been to raise questions as to the validity of the early correlation of the central Appalachian series with that of New York, except in a general way. The great groups of New York are indeed known to occur throughout the Appalachian province, yet the minor subdivisions with their distinctive faunas are found to undergo modification of such character and extent as to indicate that identity in each particular case can be determined only by more extended and detailed studies than have thus far been made. Another result has been the discrimination and delimitation of certain well-defined formations in Virginia, West Virginia, Maryland, Tennessee, North Carolina, Georgia, and Alabama; but the rocks of a considerable part of the province remain to be classified in accordance with the modern method. For the present it will suffice to say that an essentially complete American Paleozoic series of rocks is represented in the province.

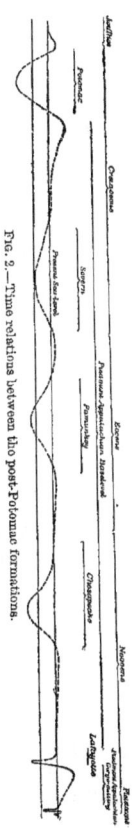

FIG. 2.—Time relations between the post-Potomac formations.

The Origin and Relations of the Rocks.—The Appalachian Paleozoic province is characterized by the occurrence of sediments deposited in the Mediterranean sea of North America, which existed during the lapse of time from the early Cambrian to the close of the Carboniferous period. It is bounded on the north and east by ancient crystalline rocks, the bases of a great mountain system, now deeply eroded, and the remains of a continent whose former extent is only to be inferred from the enormous volume of sediments it yielded to the Paleozoic sea; and on the south and west, Mesozoic and Cenozoic deposits limit our observation of the older strata.

The history of subsidence and uplift, of erosion and sedimentation, may be summarized as follows:

Cambrian: The invasion of the sea, which began the known deposits of Cambrian strata along the Appalachian crystalline area, found a continent mantled in the products of rock disintegration.[9] These materials, easily swept away, produced a mass of fine sandstones and shales, and near the source they retained fragments of feldspar, hornblende, and other minerals, which gave rise to transition beds between the clearly crystalline and the clearly sedimentary rocks. Limestones formed where the mechanical débris was not too abundant, and the result is a complex of deposits measuring 7,000 feet and more in thickness. The uppermost member is the Potsdam, a sandstone in its typical locality, elsewhere a shale or a limestone carrying the characteristic Upper Cambrian fossils.[10]

Lower Silurian: This period is divided into two epochs, separated by an interval of erosion of the earlier member. The conditions of deposition continue generally unchanged from Cambrian into Silurian time, the principal result being a great thickness of chert-bearing dolomite. This formation is the most widespread, the most uniform, and the most massive of all the Paleozoic series. From Massachusetts and New York to Alabama, and westward under the Mississippi Valley, it is everywhere the great limestone member of the stratigraphic column. It is usually 3,000 to 4,000 feet thick. This phase of deposition was closed by an uplift, which permitted the formation of wave-wrought conglomerates and sea-cliff débris from the limestone along the coast line in Tennessee and in Massachusetts, and probably throughout the entire interval where detailed search has not been made. This brings us about to the close of the Trenton period of New York.

The second epoch began with the transgression of the sea, and continued until the coast line of the Cambrian ocean had been submerged. The conditions of the source of sediments were precisely like those that existed during the Cambrian, and a very similar series of conglomerates and sandstones were formed. The submergence of the land was deeper than any that preceded or followed it; sediments to a depth of 1,200 feet accumulated locally and thinned out westward to a few hundred feet. About Cincinnati they are represented by the highly fossiliferous shales and limestones of that name.

Upper Silurian: The preceding period closed with an uplift, which is possibly contemporaneous with the unconformity locally evident in the northeastern province. The first deposit of the Upper Silurian is a widespread sandstone, of peculiarly clean character, followed by the ferruginous shales of the Clinton formation, which contain the important fossil iron ores. The later history of the period is recorded in limestones, the Niagara, Salina, and Helderberg, which are best represented in New York, Pennsylvania, and Ohio, and thin out or disappear southward.

Devonian: In the Oriskany calcareous sandstone, followed by the Corniferous limestone in New York, we have a lithologically variable horizon, which contains fossils of both Upper Silurian and Devonian types, and marks the transition from conditions favoring the deposit of impure limestones of the Silurian to the great subsidence under the load of mud and sand deposited over New York, Pennsylvania, and Virginia during the Devonian. The lowest member of this series is a highly bituminous shale, the most persistent of all Paleozoic formations, except the great limestone, although in Tennessee and Alabama it is often not over 20 feet thick. In Pennsylvania it exceeds 500 feet, and in New York the formation reaches 1,200 feet. Above these dark shales follow greenish argillaceous sandstones, succeeded by red shales and sandstones. The total thickness of these mechanical deposits exceeds 8,000 feet in northern Virginia, but they thin out rapidly southward, and are not clearly recognized in Tennessee.

Carboniferous: The mechanical sediments of the Devonian are overlain by beds of limestone, which are sometimes shaly, sometimes massive and chert-bearing. Above these are the sandstones and conglomerates at the base of the Coal Measures, deposits of coarse materials spread over a vast area during a single epoch. Then ensued the conditions of alternating sea and marsh, which built up to a thickness of 3,000 to 4,000 feet the mass of sandy shales, shales, limestones, and coal beds of the Appalachian coal field.

The Appalachian Structure.—It has long been the assumption that the deformation of Paleozoic sediments in the Appalachian province took place at the close of the Carboniferous period. That certainly was the time of greatest development of folds and faults, but there is good reason to believe that there were initial disturbances as far back as the Trenton period. The forms of structure called "Appalachian," and often referred to as a single type, differ greatly in different regions. But they are all manifestations of one phase of deformation, namely, compression. A belt of strata extending along the old shore line from Canada to Alabama has been narrowed in a direction perpendicular to that shore by a reduction to five-sixths or four-fifths of its undisturbed width. This compression, which probably went on at several epochs during the Paleozoic age, raised long narrow arches with intermediate troughs (anticlines and synclines), and in some localities pressed these folds till they closed upon themselves. The force also produced movements (faults) along planes of weakness developed in the folding mass, movements which sheared across strata opposed to them in such a way as to slide older and deeply buried formations over the edges of younger deposits. Thus a geologic map of the Appalachian province usually represents many narrow parallel belts of strata in some regions, such as Pennsylvania and Virginia, winding around alternating anticlinal and synclinal axes; in other districts, such as Tennessee, extending for scores of miles adjacent to a continuous fault line.

The history of Mesozoic and Cenozoic time is recorded in the Paleozoic province in geographic forms, in mountains, baselevel plains, and river systems. What we have thus far read of this history is explained elsewhere.

THE LOCAL GEOLOGY.

CRYSTALLINE ROCKS OF WASHINGTON.

General Features.—The entire area covered by the Washington atlas sheet is composed of the crystalline rocks of the Piedmont plateau. These are, however, concealed in the eastern and southern portions of this area by the comparatively thin covering of coastal plain deposits, from whose irregular and sinuous western edge they emerge to form the surface. Satisfactory exposures of these rocks are to be found only in the deep ravines cut by the streams (e. g. the Potomac and Rock creek or their tributaries), since at the surface of the plateau their character has been obscured or obliterated by extensive superficial decay and by cultivation.

The older rocks of the Washington sheet belong entirely to the eastern or holocrystalline portion of the plateau province, as already described. They are for the most part granitoid gneisses of varying composition, which grade into wholly massive varieties of probably eruptive origin on the one hand, while they retain occasional evidence of clastic origin (obscure conglomeratic layers) on the other. Toward the west, as displayed along the Potomac section, which is nearly transverse to their strike, these rocks become somewhat more foliated and schistose as they approach the boundary of the western or semicrystalline area which passes near Great Falls in the extreme northwestern corner of the sheet. There are also much farther east occasional bands of very schistose rock (notably those seen along Broad branch) which pass indiscriminately from one formation to another, and which owe their present character to unusual dynamic action.

The final period of orogenic disturbance which imparted to the entire Piedmont plateau, in common with the Appalachian system, its present structure, gave to the crystalline rocks within the Washington sheet a north-south strike. The occasional faint evidences of original bedding that have survived within this area seem now to accord closely with the foliation which has been developed in all the rocks, igneous and clastic alike, during the extreme metamorphism to which they have been subjected. This is a dip almost constantly to the west within the entire area, and growing more and more steep toward the west, in accordance with the general structure of the Piedmont plateau, as explained in a preceding section. Only in the extreme northwestern corner of the sheet, near Great Falls, do the rocks begin to incline toward the east.

Leading Rock Types.—A partial examination (still in progress) of the crystalline formation within the limits of the Washington sheet has brought to light the following easily distinguishable rock types, which are provisionally enumerated, although it is probable that further study will both modify and enlarge the list.

Granite, granite-gneiss, and gneiss; quartz-orthoclase-mica rocks. This is by far the most extensively developed of all the crystalline formations of the area in question. It embraces undoubtedly eruptive granite, secondary foliated (squeezed) granite, and typical gneiss, probably metamorphosed sediments. On account of their close lithological resemblance, decayed condition, and concealed contacts these rocks cannot, however, at present be accurately subdivided on the map. Hence they are represented by a single color. Toward the east, notably along Sligo and Piney branches, these rocks are very massive, often quite devoid of any foliation, and are not infrequently filled with inclusion of other rocks in which characteristic granite contact minerals are largely developed. All this points to an eruptive origin, and these characters persist even where a secondary foliation has been developed in accord with the prevailing strike and dip. Farther westward the rocks appear more like typical gneisses, being banded, more micaceous, and more schistose. Apparent beds of conglomerate have also been noticed in them along the south bank of the Potomac river and near the Klingle Ford bridge over Rock creek.

Diorite; massive, dark green amphibole-biotite-granite: These rocks present a marked contrast to the last type in their dark color. They always contain green hornblende, biotite, orthoclase, and plagioclase, sometimes one and sometimes the other in excess. Quartz is also usually present and not infrequently rutile, sphene, and epidote as well. Under the microscope they generally show evidence of profound dynamic action. In all probability they represent ancient eruptive masses which have been subsequently greatly changed and recrystallized by earth movements. They are most extensively developed around Georgetown and near Cabin John. In quarries at the former place clearly defined inclusions of other rocks have been noticed, which substantiate the theory of their eruptive origin.

Serpentine and steatite: A few small lenticular areas of serpentine and soapstone occur within the area under consideration. They are usually closely associated with the more basic hornblendic rocks, and are, probably, like these of eruptive origin, although this hypothesis can not as yet be considered as definitely proved.

Gabbro: Two small elongated exposures, presumably dikes, of trap-like rocks, which the microscope shows to be in all respects identical with the Baltimore hypersthene-gabbro,[11] occur near West Falls Church station, and a larger area surrounds Bethesda Park, which grades toward the south into the diorite.

Broad Branch schists: On the road leading northward from the Pierce Mill road, along Broad branch, a narrow band of thinly foliated sericitic, chloritic, and siliceous schists is exposed. These rocks differ considerably in character and appearance from those about them, but still they grade imperceptibly into the granite and gneiss which lie both on their eastern and western sides. The belt, although quite narrow, has a considerable extent from north to south in the direction of its strike. Under the microscope all of these schists show evidence of the most extreme dynamic action. Their distinguishing characters (mineralogical composition, foliation, etc.) are clearly secondary; and they may readily have been produced by an unusual amount of compression brought to bear on the normal material of the granite or gneiss. This schist belt is therefore probably the result of extraordinary pressure at the axis of a closed synclinal fold, rather than the product of metamorphism of beds orginially distinct from those around them.

Siliceous gneisses and schists of Great Falls: The barrier at the Great Falls of the Potomac is an unusually siliceous, and therefore unusually hard, band in the gneiss. In some places this rock is so siliceous that it contains hardly anything except quartz and mica, and thus becomes a quartz schist. It exhibits throughout definite microscopic evidence of having been subjected to great pressure.

In spite of the considerable variety shown by this list, the crystalline rocks near Washington are much more uniform and monotonous than those forming the eastern part of the Piedmont plateau farther northward. This is particularly the case with the eruptives. The gabbros and gabbro-diorites, so abundant near Baltimore,[11] in Harford county, Maryland, and in northern Delaware,[12] are represented by only two or three unimportant occurrences on the Washington sheet; while the various peridotites, pyroxenites, and derived rocks[13] are altogether absent.

Granitic rocks are largely developed near Washington, and many of them preserve, both in their massive character and included fragments, fair evidence of eruptive origin. Nevertheless, even these are far inferior in petrographic variety and interest to the undoubtedly intrusive granites, granite-porphyries, and felsites occurring farther northward in Maryland.

CLASTIC FORMATIONS OF WASHINGTON.

The general Structure.—In the vicinity of Washington the formations of the Coastal plain series are extensively displayed, and all the principal members are characteristically developed. The general structural relations are illustrated in the accompanying section Fig. 3.

The Potomac formation lies on the steeply-sloping surface of the crystalline rocks and is overlain by the gently eastward dipping Severn, Pamunkey, and Chesapeake formations in regular succession, upward and eastward. The Lafayette formation caps the higher plains and the Columbia formation occupies the lower terraces. The regular succession is interrupted locally in the ridge east of Washington, where the Pamunkey formation overlaps the Severn for a few miles, and is in turn overlapped by Chesapeake beds which lie directly on the Potomac formation in small outliers north and west of Washington. These formations also thicken seaward, each with slightly increased rate from below upward, and their separating planes of unconformity incline gently eastward. The Lafayette formation lies across the planed surfaces of the successive formations from the crystallines westward to the Chesapeake formation eastward. The Columbia formation lies on terraces cut in the crystallines and in the Potomac, Severn, Pamunkey, and Chesapeake formations from tide level to about 150 feet above.

A line of dislocation extends from northeast to southwest across the western half of the Washington atlas sheet. The downthrow is on the

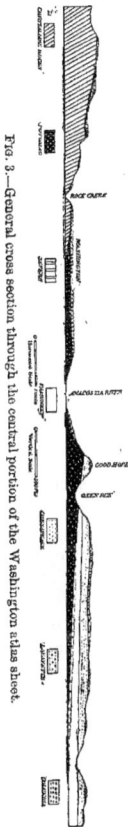

Fig. 3.—General cross section through the central portion of the Washington atlas sheet.

eastern side and amounts to from 50 to 100 feet. This fault traverses the Lafayette and Potomac formations and the crystalline rocks, and in the divides its presence is marked by an escarpment of crystalline rocks, usually capped by Potomac and Lafayette deposits. The date of the movement was mainly between the Lafayette and Columbia, but apparently some movement has taken place since the deposition of the Columbia.

The Columbia Formation.—The lower terraces of the Potomac valley and its larger branches and the valley of the Western branch of the Patuxent are occupied by the Columbia formation up to altitudes varying from 80 to 145 feet. About the city of Washington the more general Columbia terrace levels are at 40 and 80 feet, respectively, above tide, the Capitol being situated on the western edge of a prominent outlier of the 80-foot terrace. The formation exhibits its typical development in the District of Columbia, where it consists of two members— a lower series of gravels and an overlying brown or buff loam. The gravels are heterogeneous in character, comprising remains of the more obdurate material of preceding formations, in large part of local origin. The loams are often quite pure, but they are frequently intermixed with sand and pebbly streaks and disseminated pebbles. Southward from Washington the Columbia terraces border Potomac river to widths of from 1 to 2 miles, and the materials as a whole become finer. In the Anacostia valley the formation consists mainly of brown sands with pebbly streaks, but at Washington these sands merge into the loams and gravels of the typical phase. Along the northern side of the Washington-Great Falls gorge of the Potomac there is a narrow shelf at 145 feet above tide, which is capped at intervals by Columbia loams and gravels. The thickness of the Columbia formation about Washington averages from 20 to 30 feet.

The Lafayette Formation.—This formation occupies portions of the wide, high plains surrounding the Washington amphitheater, especially toward the south and southeast. Its materials are mainly gravels and loams. The basal and marginal beds are in larger part gravels, usually stained buff or orange superficially, and packed tightly in stiff loams and sharp sands. The upper beds are predominantly loamy, and farther eastward loams and fine sands with gravel streaks prevail. In the outliers north and west of Washington the formation consists of gravelly red loams. The plain on which the Lafayette formation was deposited is depressed by a wide, shallow basin, in which is excavated the present Potomac valley below Washington. The Lafayette formation extends for some distance west of the fall line fault in a series of outliers, usually with underlying remnants of the Potomac.

The Chesapeake Formation.—This formation underlies the high plains southeast of Washington, where it is overlain by a capping of the

Lafayette formation over the greater part of its area. The formation extends to the edge of the high bluffs east of Washington, and also is caught in small outliers of the higher terrace levels at Soldiers' Home Park and between West Washington and Tenleytown. The formation consists of very fine-grained materials, mainly sands, with a variable proportion of infusorial remains and clay. In their nonweathered condition the beds are usually very compact, dark gray to olive-green in color, and massively bedded. Surface outcrops consist of soft meal-like sands of light buff color. Some clay beds occur locally, notably in the eastern part of the District. Infusorial remains are nearly everywhere present, and faint casts of molluscan remains are generally abundant in the nonweathered material. The outliers in the ridges about Washington consist of buff-colored, meal-like beds, lying on an irregular surface of the Potomac sands, and in turn overlain by Lafayette gravels. The thickness of the formation increases gradually eastward and is about 125 feet in the Marlboro region.

The Pamunkey Formation.—The Pamunkey formation occupies a wide area east of Washington, and is a conspicuous member of the Coastal plain series in this region. In its nonweathered condition the formation is mainly a bluish or greenish-black marl, consisting of fine-grained quartz sands mixed with varying amounts of organic matter and clay, and usually containing a considerable proportion of the mineral glauconite. On weathering the glauconite is decomposed and its iron constituent oxidizes and stains the sands to a dull red brown or snuff color. The weathered phase is general on the surface in the regions in which the formation has long been bared of overlying formations. In the streams leading out of the high plains east of Washington the nonweathered marls are often exposed, and in the region to the northwestward the formation is bare for many square miles. Fossil shells frequently occur in great abundance in the marls, and there are many prolific fossil localities within a few miles of the city of Washington.

The thickness of the formation increases eastward, from 2 to 5 feet in the bluff just east of Washington to over 100 feet in the Marlboro region.

The Severn Formation.—In the vicinity of Washington this formation is a thin bed of black sands, lying between the Potomac and the Pamunkey formations east of Potomac and Anacostia rivers. It is the attenuated southern extension of the great Cretaceous green-sand formation of New Jersey and Delaware; but in this region it consists mainly of fine carbonaceous and more or less argillaceous sands, containing small scales of mica, but very little glauconite. It usually abounds in casts and impressions of distinct Cretaceous fossils; and fossil shells occur in abundance at several localities. In the bluff east of Washington it is locally cut out by an overlap of the Pamunkey formation, but it comes in again toward the northeast with a thickness of 20 or 30 feet, and is occasionally exposed in streams and road-cuts throughout the eastern portion of the Washington atlas-sheet.

The Potomac Formation.—The Potomac outcrops occupy a wide area in the vicinity of Washington, especially to the southwestward. In Washington and the Potomac estuary the formation is generally hid beneath the Columbia formation, and in the plateaus toward the southwest the Lafayette formation covers it extensively. The deposits consist mainly of clays and sands of light color, commonly most irregularly intermixed. The basal beds exposed along the western margin are mainly gray sandy arkose, with pebbles and boulders. In Virginia the sandy arkose and arkosic sands give place eastward to gray, greenish, brown, and buff sandy fissile clays. North of the Potomac they grade upward into a great series of fine quartz sands and clays, the argillaceous elements increasing in proportion eastward. Along the Baltimore and Potomac railroad, and thence eastward to the Severn formation, the clays are extensively developed, and the sands occur as locally indurated sheets and crusts, or more rarely intermixed with the clays. The formation attains a thickness of over 300 feet east of Washington, but it is eroded westward finally to a feather edge.

Post-Columbia Deposits.—The overwash deposits on slopes and along the smaller streams as well as the river muds and marshes, and the freshet deposits along the larger rivers are post-Columbia in age; but owing to their relative unimportance they are not represented on the present edition of the geologic map. As the rivers are drowned and sinking, and the present area of submergence was preceded by erosion, alluvial deposits are mainly under water in the Washington region and consist of river muds.

Artificial.—The tidal marshes adjoining the southern part of Washington have been built up above tide level with materials obtained by excavations from the adjoining channels. This area is represented on the map as artificial.

THE GEOMORPHY.

During recent years certain geologists have come to recognize that within certain limits earth history may be read from the land-forms developed by degradation as well as from the strata formed by concurrent deposition; and the Coastal plain and contiguous provinces of eastern United States are so conditioned that these lines of research may be successfully prosecuted within them.

Although the parallel mountain ranges are the most conspicuous features of the Appalachian province, the broad gently undulating intermontane plains are only less conspicuous and far more extensive; and only less conspicuous than the intermontane plains are the narrow steep-sided gorges of all water-ways incised within the plain and sometimes notching the mountains—indeed, the entire province is really an undulating baselevel plain with ranges embossed upon it, and with a series of wide-branching drainage systems sharply inscribed within it.

The most conspicuous and extensive feature of the Piedmont zone is the far-stretching peneplain or undulating baselevel plain comprising the greater part of its area; only less conspicuous are the narrow steep-sided gorges in which its waterways flow. The Piedmont plain thus homologizes the Appalachian province with respect to classes of features; but (excepting certain knobs and ridges of hard material due to lagging of erosion, or, "catoctins") the embossed mountains are lacking.

Although the most conspicuous configuration of the Coastal plain is that of the present surface, there are in this province a series of configurations characterizing a number of ancient surfaces, each of which is a great stratigraphic unconformity; and the researches in this region have progressed so far that the general character of each of these surfaces has been ascertained. The present surface is a terraced lowland, trenched by broad yet shallow estuaries and partly dissected by minor waterways flowing in narrow steep-sided channels, produced by rapid excavation; but portions of the lowlands are not yet invaded by the minor drainage. In general the surface is water-carved, and represents sluggish trenching along drainage lines. The next older surface (the contact surface between the Columbia and the Lafayette) is the most strongly accented of the province; it represents a peneplain strongly and deeply trenched yet nowhere planed to baselevel, save possibly in the deeper gorges far below the reach of observation. The next lower surface (the Lafayette-Chesapeake surface) is smoother than that of the present, and the configuration as well as the relations of structure to that configuration indicate widespread baselevel planation with little trenching along the waterways. The next surface (the Chesapeake-Pamunkey) is similar, but even smoother. The Pamunkey-Severn surface in like manner is smooth and so related to the structure as to tell of extensive planation without localized vertical cutting. The Severn-Potomac surface on the other hand is decidedly rugose, and its relations to structure are such as to indicate that it represents a peneplain extensively degraded, yet chiefly along the drainage lines. The base-level surface upon which the entire series of Coastal plain deposits rests—the sub-Potomac floor—is much like the present Piedmont surface, i. e., a rather strongly undulating peneplain, trenched by deep-cut gorges.

In addition to these general features of the three provinces there are a multitude of minor features, of which a portion have been studied and interpreted. Thus Chamberlin and Gilbert as well as White in the western part of the mountain province, and McGee in the eastern part of the same province and also in the plateau, have ascertained that the early Pleistocene deposits rest on the great Appalachian-Piedmont peneplain; Willis has traced the same or a remarkably similar peneplain into the southern Appalachians in North Carolina; Davis [14] has recognized and admirably described an ill-defined pre-Triassic and well defined pre-Cretaceous peneplain in New England and the northern Appalachians; Emerson has incidentally developed certain features of a pre-Triassic land surface in New England; Hayes and Campbell in the southern Appalachians, and Keith in the central area of the range, have traced other baselevels and deduced their dates; and by these and other researches several important features in the geomorphic history of eastern United States have been elucidated. It is known that the drainage and the topographic forms resulting therefrom in the Appalachian zone were developed by orogenic movement and are therefore tectonic, and it is believed that they are certainly consequent in the western part of the province, and probably antecedent in the eastern part; it is known that much of the drainage and configuration of the Piedmont plateau is of the subsequent type, depending upon planation and measurably reflecting rock composition, and also that another part is superimposed; and it is known that while the principal drainage lines of the Coastal plain, are affected by relatively recent deformation the greater number of the streams and of their land-formed progeny belong to a series of autogenetic systems, repeatedly yet concordantly superimposed.

The episodes thus recognized blend as a consistent and essentially complete series of continent movements indelibly recorded in the land forms of the mountains, the plateau, and the lowland. The series begins with the faintly recorded incomplete baselevelling of the pre-Triassic time; this shadowy record is followed by the more definite one (at least in the latitude of Washington) of a long baselevel period, followed by a brief high-level period, during which the land first tilted seaward and then sank until the Potomac deposits were laid down; next follows the extensive record of that long baselevel period which Davis styles "pre-Cretaceous," though it may be questioned whether this record does not merge with that of the pre-Potomac episode on the one hand and that of the long post-Cretaceous baselevel period on the other; then follow the series of alternating episodes of sluggish deposition and indolent degradation recorded in the Severn-Pamunkey and Chesapeake formations, with their intervening unconformities—a series of episodes which may not be discriminated in the faintly inscribed record of the ancient Piedmont and Appalachian baselevels; afterward follows the well defined episode of high level recorded in the Piedmont-Appalachian gorges and in the broad and deep trenches through which half the volume of the Lafayette formation was carried into the sea; and then follows the inconspicuous but easily legible record of the Columbia submergence and the post-Columbia emergence—the former certified by the semifilling of the ancient canyons, the latter by the shallow submarine channels and the pygmy fall-line gorges; finally, in the northern part of the Coastal plain comes the record of submergence and subsequent lifting during the later ice invasion. The long series of generally consistent land movements is complicated in the middle Atlantic slope by the displacement, probably beginning in the Lafayette period and certainly continuing to-day; but properly interpreted this complication only affords a check on the accuracy of the general reading.

BIBLIOGRAPHY.

1. "Petrography and structure of the Piedmont plateau in Maryland," by George H. Williams. Bull. Geol. Soc. Am., Vol. II, 1890, pp. 301–322.
2. "Southern extension of Appomattox formation," by W J McGee. Amer. Jour. Sci. (3d ser.), Vol. XL, 1890, p. 36.
3. "Three formations of the middle Atlantic slope," by W J McGee. Amer. Jour. Sci. (3d ser.), Vol. XXXV, 1888, p. 125.
4. Described by Safford in 1856 (Geological reconnaissance of Tennessee, pp. 148, 162) and by Hilgard in 1860 (Geology and agriculture in Mississippi. p. 3) under the name of Orange sand; described by McGee in 1888 (Amer. Jour. of Sci. (3d ser.), Vol. XXXV, p. 328) under the name Appomattox; formerly named Lafayette from original records (of 1855–'56) by Hilgard in 1891 (Amer. Geol., Vol. VIII, p. 129).
5. "Later formations of Virginia and Maryland," by N. H. Darton. Bull. Geol. Soc. Amer., Vol. II, pp. 443, 439, 438.
6. "Post-glacial subsidence of the middle Atlantic coast," by A. Lindenkohl. Amer. Jour. Sci., Vol. XLI, 1891, pp. 489–499.
7. "Geology of the head of Chesapeake Bay," by W J McGee. U. S. Geol. Survey, 7th Ann Rep., 1888, pp. 546, 616–634.
8. "Tertiary and Cretaceous strata, etc.," by E. A. Smith and L. C. Johnson. U. S. Geol. Survey, Bull. 43, 1887.
9. "Secular rock disintegration, etc.," by R. Pumpelly. Bull. Geol. Soc. Amer., Vol. II, p. 210.
10. "Middle Cambrian faunas," by C. D. Walcott. U. S. Geol. Survey, Bull. 30, 1886.
11. "Gabbros and associated rocks near Baltimore," by George H. Williams. U. S. Geol. Survey, Bull. 28, 1886.
12. "Gabbros and associated rocks in Delaware," by F. D. Chester. U. S. Geol. Survey, Bull. 59, 1890.
13. "Non-feldspathic intrusive rocks of Maryland," by G. H. Williams. Am. Geol., Vol. VI, 1890, p. 35.
14. "Geological dates of topographic forms, etc.," by W. M. Davis. Bull. Geol. Soc. Amer., Vol. II, 1890, p. 549.

B

GEOLOGICAL GUIDE BOOK

OF THE

ROCKY MOUNTAIN EXCURSION

EDITED BY

SAMUEL FRANKLIN EMMONS

253

INTRODUCTION.

The undersigned was called upon, a few weeks before the opening of the Congress, to prepare a geological guide book for the use of those members of the Congress who should take part in the proposed excursion to the Rocky Mountains.

Parts of the vast region to be passed over during the excursion had never been systematically examined; of other parts the results of recent examinations had not yet been published, and the published surveys of still other parts were scattered through many bulky Government, State, and other reports, all of which it would have been impossible to consult in the time given, and many of which had been more or less superseded by later and as yet unpublished observations. Under these circumstances the only feasible plan for preparing such a guide book was to call upon geologists within reach, who were most familiar with different parts of these regions, to contribute descriptions of the geology of such parts. Those thus called upon responded most promptly and generously, but their contributions necessarily varied somewhat in detail and method of treatment, according to the varying conceptions of the authors. It was not possible, in the limited time given, to return their manuscripts to the authors for revision so as to produce the necessary uniformity, and the duty therefore fell upon the editor of hastily rewriting a considerable portion of the material contributed, and of filling in any gaps to the best of his personal knowledge. The resulting guide book was necessarily somewhat unequal in the amount of detailed description given of different parts of the region, and also incomplete in illustration and bibliographic reference. In spite of these imperfections, it so well subserved its purpose that a unanimous request was made by the other secretaries, who had taken part in the excursion, that it should be published in the Compte-rendu, after revision by the various contributors of the parts for which they were to be held responsible, and the addition of as many illustrations and bibliographic references as possible. This request has been complied with, as far as has proved practicable, in the following pages. In a few cases the original contributors have been too busy to revise their notes, and the editor has been obliged to present the description of their regions much as they originally appeared. In other cases, where the descriptions contained in the original guide

255

book were a compilation of the notes of several individuals, it has been difficult to divide the parts so as not to interfere with the continuity of the description. Whatever has been lost in this respect, however, has been more than made up by the greater fullness of treatment. The lists, prepared by each contributor, of publications containing the most recent and important geological information in regard to the respective areas described have, for convenience of reference, been combined into a single list placed at the end of the sketch, references being made by means of numbers in the text corresponding to the numbers affixed to each title.

Some of the visiting geologists have kindly contributed notes and sketches made by them during the journey, which form a valuable addition, and it is only to be regretted that they are not more numerous.

S. F. EMMONS,
Editor.

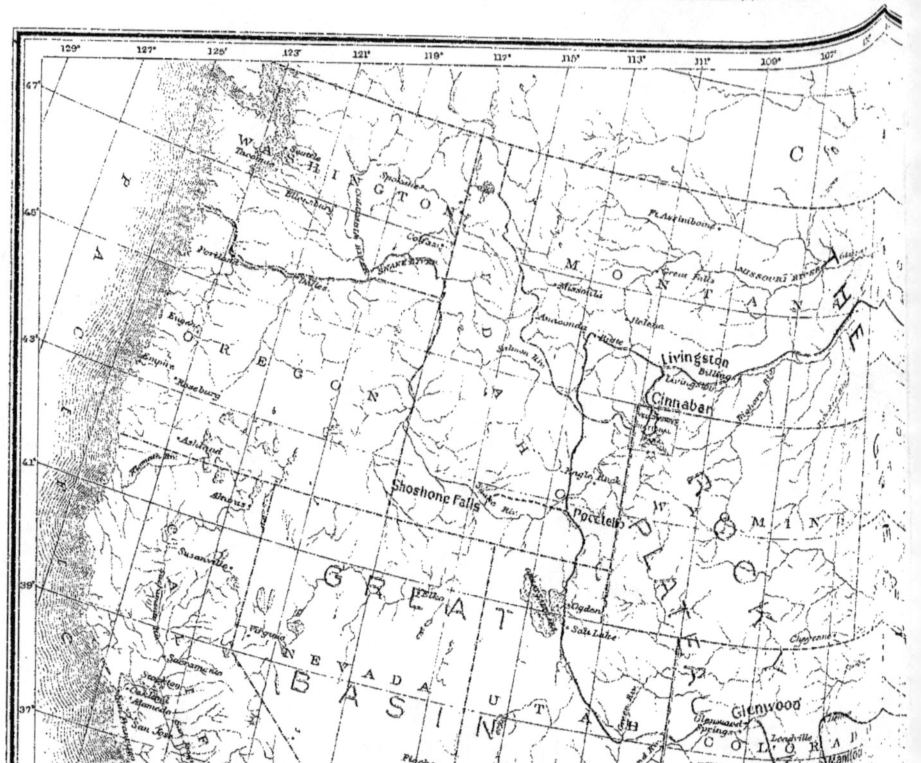

WASHINGTON
OREGON
IDAHO
MONTANA
WYOMING
NEVADA
UTAH
COLORADO
ARIZONA
NEW MEXICO
GREAT BASIN
LOWER CALIFORNIA
Seattle
Tacoma
Spokane
Ellensburg
Colfax
Portland
Dalles
Snake River
Eugene
Empire
Roseburg
Ashland
Susanville
Virginia
Sacramento
Stockton
Oakland
Alameda
San Jose
Bakersfield
Santa Barbara
Los Angeles
San Bernardino
San Diego
Ft. Assiniboine
Missouri River
Great Falls
Missoula
Helena
Anaconda
Butte
Livingston
Billings
Cinnabar
Shoshone Falls
Eagle Rock
Pocatello
Elko
Ogden
Salt Lake
Pioche
Glenwood
Glenwood Springs
Leadville
Manitou
Canyon City
Del Norte
Cheyenne
Grand Canyon
Eldorado
Flagstaff
Holbrook
Albuquerque
Rio Gila
Rio Grande
Deming
El Paso
Eddy
Presidio
129° 127° 125° 123° 121° 119° 117° 115° 113° 111° 109° 107°
47° 45° 43° 41° 39° 37° 35° 33° 31° 29° 27° 25°
119° 117° 115° 113° 111° 109° 107° 105°

WEST

INTERNATIONAL

(Base map p
Main R

Scale

100   200   300   400   500   600 Kilometres.

Page

Niagara Falls, by G. K. Gilbert 455
   Itinerary—Niagara Falls to New York City, by C. D. Walcott 459
Excursion to the Grand canyon of the Colorado...................... 461
   Itinerary—Denver to Albuquerque, by S. F. Emmons 465
   Itinerary—Albuquerque to Flagstaff, by G. K. Gilbert 468
   Itinerary—Flagstaff to the Grand Canyon, by G. K. Gilbert 472
Notes and sketches by visiting geologists.......................... 475
   Note on Walnut canyon, by Prof. T. McK. Hughes.................. 476
   Geological section in Congress canyon, by Dr. Fritz Frech 481
   Grand Canyon sketches, by H. M. Cadell 481
Bibliography 482

LIST OF ILLUSTRATIONS.

Map of United States showing route............................ 256
Pl. I. Geology of St. Paul, Minneapolis, and vicinity.............. 308
 II. Old Faithful in action 356
 III. Yellowstone canyon...................................... 359
 IV. Panorama from Mount Washburne 359
 V. Outlet of lake Bonneville at Red Rock pass............... 377
 VI. Gates of Bear river 379
 VII. Fault-trough at Little Cottonwood canyon 394
 VIII. San Rafael swell.. 400
 IX. Map of Florence oil field 428
 X. Map of Niagara river 455
 XI. Bird's eye view of Niagara river 456
 XII. Congress canyon below Hance's cabin 481
 XIII. Grand Canyon above mouth of Congress canyon 481
Fig. 4. Section through the Piedmont plateau in Maryland 272
 5. Section near Hancock, Md 277
 6. Section near Cumberland, Md 278
 7. Section across the Alleghany mountains 279
 8.
 9. Sections across the oil field of Northwestern Ohio........ 294
 10.
 11. Section from Vermillion lake to Duluth 304
 12. Section across Mississippi river above Fort Snelling...... 308
 13. Section across Mississippi river below Fort Snelling...... 309
 14. Snowy range sections A, B, and C 330
 15. Glacial boulder near Yellowstone canyon................. 342
 16. Map of Yellowstone park................................ 347
 17. Shore lines and fault scarps near Farmington, Utah...... 380
 18. Geological map of Wasatch mountains................... 386
 19. Geological section between Cottonwood canyons 389
 20. Profile of South Moraine 394
 21. Solitude station 401
 22. Section, Book Cliffs to White river plateau 408
 23. Section, Coal seams at Newcastle, Colo 408
 24. Section, reversed fold and fault on Roaring Fork 413
 25. Pikes Peak railway...................................... 432
 26. Section at Bear Creek 439
 27. Section at Van Bibber creek 441
 28. The whirlpool at Niagara 458
 29. The Nutria fold 471
 30. Cliff dwellings in Walnut canyon........................ 475
 31. Section in Walnut Canyon 476
 32. Section in Congress canyon............................. 477

LIST OF RAILROAD LINES FOLLOWED BY THE EXCURSION.

Day.	Main Route.	Distance Miles.	Kilo- meters.
First	Baltimore and Ohio R. R.:		
	Metropolitan Branch—Washington to Washington Junction.	43	69
	Main Line—Washington Junction to Wheeling, W.Va......	310	499
Second.............	Central Ohio Division—Wheeling, W.Va., to Newark, Ohio..	101	162
	Lake Erie Division—Newark to Chicago Junction. Ohio	88	142
	Chicago Division—to Chicago, Ill	271	436
	Chicago, Milwaukee and St. Paul R. R.:		
	Chicago Division—Chicago to Milwaukee, Wis.............	85	137
Third.............	La Crosse Division—Milwaukee to La Crosse, Wis..........	198	318
	River Division—La Crosse to St. Paul, Minn	127	204
	River Division—St. Paul to Minneapolis. Minn	10	16
Fourth	Northern Pacific R. R.:		
	St. Paul, Minneapolis and Pacific Coast Line—Minneapolis,		
Fifth	Minn., to Livingston, Mont.........................	996	1,603
Twelfth	Yellowstone Park Branch—Livingston to Cinnabar..........	51	82
	Yellowstone Park Branch—Cinnabar to Livingston..........	51	82
	Pacific Coast Line—Livingston to Logan, Mont.............	49	79
	Butte Branch—Logan to Butte, Mont	71	114
	Union Pacific R. R.:		
Thirteenth	Utah Northern Line—Butte, Mont., to Pocatello, Idaho......	263	423
	Utah Northern Line—Pocatello to Ogden, Utah	134	216
	Rio Grande Western Railway:		
Sixteenth..........	Ogden to Salt Lake City	36	58
	Salt Lake City to Grand Junction, Colo......................	292	470
Seventeenth	Denver and Rio Grande R. R.:		
Eighteenth	Denver and Salt Lake Line—Grand Junction to Leadville ..	169	272
	Denver and Salt Lake Line—Leadville to Colorado Springs..	202	325
Nineteenth	Manitou Branch—Colorado Springs to Manitou.............	5	8
	Manitou Branch—Manitou to Colorado Springs.............	5	8
	Denver and Salt Lake Line—Colorado Springs to Denver	75	121
Twentieth	Union Pacific R. R.:		
	Kansas Division—Denver to Limon, Colo...................	90	145
Twenty-first......	Chicago, Rock Island and Pacific Railway:		
	Kansas City and Colorado Line—Limon. Colo., to Topeka,		
	Kans	476	766
	Union Pacific R. R.:		
	Kansas Division—Topeka, Kans., to Kansas City, Mo	67	108
	Hannibal and St. Joseph R. R.:		
	Kansas City to Cameron Junction, Mo......................	54	87
Twenty-second	Chicago, Rock Island and Pacific Railway:		
	Southwestern Division—Cameron Junction, Mo., to Chi-		
	cago, Ills	464	746
Twenty-third......	Chicago and Grand Trunk Railway:		
	Chicago, Ills., to Port Huron, Mich	335	539
Twenty-fourth	Grand Trunk R. R.:		
	Southern Division—Sarnia, Ontario, to Suspension Bridge,		
	N. Y	184	296
	New York Central and Hudson River R. R.:		
	Suspension Bridge to Niagara Falls....................	2	3
	Niagara Falls to Buffalo....................	22	35

LIST OF RAILROAD LINES FOLLOWED BY THE EXCURSION—Continued.

Day.	Main route.	Distance.	
		Miles.	Kilometers.
Twenty-fifth.......	West Shore R. R.:		
	Buffalo, N. Y., to Weehawken, N. J	428	689
	BRANCH ROUTE TO SHOSHONE, IDAHO.		
	Union Pacific R. R. :		
	Oregon Short Line—Pocatello, Idaho, to Shoshone, Idaho...	106	171
	BRANCH ROUTE TO FLAGSTAFF, ARIZ.		
	Atchison, Topeka and Santa Fé R. R.:		
	Denver, Colo., to Albuquerque, N. Mex......................	529	871
	Atlantic and Pacific Division—Albuquerque to Flagstaff, Ariz..	344	554
		6,733	10,854

N. B.—Figures above the line in the text refer to titles in Bibliographic list, P. 482, et seq.

EXCURSION TO THE ROCKY MOUNTAINS.

PHYSICAL GEOGRAPHY OF THE REGION.

By G. K. Gilbert.

In journeying three-fourths of the distance from the Atlantic to the Pacific, the excursion crosses a number of distinct topographic provinces, as well as districts characterized by a wide range of climatic conditions. By way of introduction to the details described in subsequent chapters, a few paragraphs will be devoted to the characterization of the general physiographic and climatic types of the country to be traversed.

The most general classification of the United States into physiographic provinces indicates an Appalachian region at the east, a Cordilleran region at the west, and a broad plain between drained by the Mississippi and the St. Lawrence rivers. In the Appalachian region six belts are distinguished—the Coastal plain, the Piedmont plain, the Blue ridge, the Appalachian valley, the Appalachian mountains, the Cumberland plateau. They all trend approximately northeast and southwest, and, as is the case with other physiographic provinces, they are clearly differentiated in some districts and difficult of discrimination in others. We are concerned chiefly with their aspects on the line of journey.

Washington City stands at the boundary between the Coastal plain and the Piedmont belt. If any river on the Atlantic coast, from New Jersey to the Carolinas, be followed from its mouth, a cataract is sooner or later reached and navigation interrupted. A line across the land connecting these cataracts is known as the "fall line," and gives the northwestern limit of the Coastal plain. The Coastal plain is characterized by level strata, little indurated, having essentially the attitude of deposition. These strata are Cretaceous and newer. The general elevation is not great, and the northern portion of the plain is interrupted by great tidewater bays, due to the sinking of the land and the drowning of the river valleys.

The Piedmont belt is in general a plain somewhat higher than the Coastal plain, and bearing small hills or even small mountains. Its surface is generally undulating, and the greater lines of drainage are abruptly incised. The plain is carved from indurated and folded rocks of various ages, partly Jura-Trias, but chiefly metamorphic. On the outward journey the belt is crossed between Washington and

Weverton, and on the return it is theoretically crossed between the Highlands of the Hudson river and New York City, but its manifestation there is not characteristic.

The Blue ridge overlooks the Piedmont district from the northwest, and is a nearly continuous upland from New England to Georgia. At most points it is a single ridge rising 2,000 feet above adjacent lowlands, but in North Carolina it is expanded into a mountain chain, with greater altitude. The outward journey intersects it in approaching Harpers Ferry; the return journey, at the Highlands of the Hudson.

The Appalachian valley is a wonderfully persistent belt of lowland separating the Blue ridge from the Appalachian mountains. For several geologic periods the Appalachian region has stood at so low a level that its streams have had small declivity, and mechanical erosion has been slow. The chemical factor in erosion has thus acquired relative prominence, and the broad outcrop of lower Paleozoic limestone which occupies the valley area has been degraded by solution until its surface is far below the contiguous outcrops of sandstone and shale. It is an undulating plain, sharply incised along principal lines of drainage, and otherwise characterized by "swallow holes" or "limestone sinks," and by caves. The outward journey traverses it from Harpers Ferry to North Mountain Station; on the line of the return journey it is not well distinguished.

The Appalachian mountains*[1] consist of Paleozoic strata, from Cambrian to Carboniferous, which have been acutely folded and faulted. The steeper limbs of the fold are usually on the northwest side. Anticlines are often pushed northwestward over synclines, and in numerous instances this process has culminated in thrust faults. The principal epoch of folding ended early in Mesozoic time, and during the latter part of the Mesozoic the district, which then stood several thousand feet lower than now, was degraded to the condition of a peneplain. Subsequent uplift renewed the activity of the streams, and the district was carved into a grand cameo, in which the topographic features express the rock texture and rock structure in a peculiarly effective manner. The outcrop of each series of soft rocks is recorded in a system of valleys; each great bed of hard rock caps a ridge, but none of the ridges rise above the level of the old peneplain, and all the greater ridges have even tops expressing that factor of their history. On the outward journey this belt is traversed from North Mountain to the Monongahela river; on the return journey it is entered at Schenectady, but its features are masked by Pleistocene deposits.

The Appalachian folding diminishes in intensity northwestward until finally the dips of the strata are gentle. Wherever a massive sandstone approaches the height of the old peneplain in the region of gentler folds it has been preserved in the form of a high table, usually

* Figures above the line refer to titles in Bibliographic list at end of Guide Book.

with an abrupt escarpment toward the southeast. The Cumberland plateau is a belt of such tables. It is thus a connecting link between the Appalachian mountains and the interior plain, but it is not everywhere to be recognized. On the line of the outward journey it is not discriminated. On the line of the return journey it is represented by the Catskill mountains, which are visible in the distance from the west (right) side of the train as it follows the Hudson River.

The central plain of the continent does not yield a simple and definite classification into physiographic provinces, but the portion traversed by the excursion may be imperfectly characterized by describing three topographic types which it illustrates: the Lake region, the Prairie region, and the Great plains. The Lake region[2] is a rather uneven plain, having for its foundation, at the north, crystalline rocks of complex structure which are so degraded that all mountains are obliterated, and at the south, Paleozoic rocks, level or gently folded, and likewise degraded to approximate evenness. Over these are deposits of glacial drift, irregularly disposed so as to break the surface into a large number of basins, holding lakes, lakelets, ponds, and swamps.[3] On the outward journey this region is traversed in the States of Ohio, Indiana, Illinois, Wisconsin, Minnesota, and a part of North Dakota; on the return journey, in Illinois, Indiana, Michigan, Canada, and western New York.

The Prairie region is less susceptible of mapping than most other divisions, because it depends on complex conditions involving climate as well as surface configuration. Humid lands are normally characterized by forest and are suited for agriculture; arid lands are normally characterized by the absence of forest, and are unsuited for agriculture without irrigation. The normal distribution of forest is modified by fire, as annual fires destroy forests but permit the growth of grasses. In regions of great humidity a forest fire does not spread with facility; in regions of rugged surface it is limited by topographic accidents; in regions of moderate humidity and smooth surface a forest is destroyed by fire, and thus a fertile region becomes bare of trees and is clothed by grasses. The outward journey traverses prairie in western Minnesota and adjacent North Dakota. A typical portion of the prairie belt is traversed on the return journey in Missouri, Iowa, and Illinois.

The Great plains slope eastward from the eastern foot of the Rocky mountains, descending from altitudes of 5,000, 6,000, and 7,000 feet (1,500, 1,800, 2,100 m.) nearly to sea level. In detail the surface undulates gently and is devoid of commanding eminences, except that the Ozark uplift has produced low mountains in Texas, Arkansas, Indian Territory, and Missouri; the Black Hills uplift has produced another group at the western edge of South Dakota; and a few buttes, mostly of volcanic origin, stand near the Canadian boundary. The underlying strata are chiefly Cretaceous and Tertiary, and in general these have been reduced to an even surface. In a few districts, however, local causes have accelerated corrasion, causing the sculpture of shaly beds into the picturesque "Bad lands." On the outward journey the Great plains are traversed in North Dakota and the eastern half of Montana; on the return journey, from Denver to the Missouri river. The branch route to the Colorado canyon traverses the western edge of the plains from Denver to Las Vegas.

The Cordilleran region includes the Rocky mountains, the Plateau region, the Great basin, the Snake River plain, the Sierra Nevada, and a number of more westerly divisions beyond the limits of the journey.

The name Rocky mountains is applied in Colorado and northern New Mexico to the complex mountain chain lying between the Great plains and the Plateau region, and in Montana and northern Wyoming to the mountain chain between the Great plains on the east and the Snake River plain on the west. In central Wyoming there is a discontinuity, the mountain chain being interrupted by an arm of the Plateau region known as the Laramie plain. The Rocky mountains of Colorado constitute the greatest mountain mass of the United States, although a few peaks in the Sierra Nevada exceed them in height. Their principal uplift took place at a later date than that of the Appalachian Mountains, and they are far more rugged. The type of structure involves less of closely compressed folding, and faulting has played a more prominent part in producing their present relief. The nucleal rocks range from crystalline schists through the Paleozoic. Mesozoic strata appear about the flanks, sharing in the uplift, and lacustrine beds of Cenozoic date rise against the bases on all sides and are to be discovered in most of the mountain valleys. In various places, and especially toward the south, there are great masses of volcanic rocks, and dikes abound in all parts. The Rocky mountains of the northern group are of similar type, and also consist of lofty and rugged ranges. From these two masses flow the longest rivers of the country. The Mississippi-Missouri rises in the northern chain, receives many affluents from the southern, and flows southward to the Gulf of Mexico. The length of its main line is 4,900 miles (7,080 km.). The Columbia, rising in the northern chain, reaches the Pacific Ocean after flowing 1,400 miles (2,200 km.). The southern chain likewise sends the Rio Grande 2,000 miles (3,200 km.) to the Gulf of Mexico, and is drained on the west by the tributaries of the Colorado, whose waters flow 1,300 miles (2,100 km.) to the Gulf of California. On the outward journey the route lies along the mountains of the northern chain from Cinnabar to Pocatello; on the return journey it lies among the Rocky mountains of Colorado, from Grand Junction to Canyon city, and thence to Denver skirts the base of the Front range. The branch route to the Colorado canyon crosses the southern extension of the chain in New Mexico between Las Vegas and Albuquerque.

The Plateau region lies between the southern Rocky mountains on the east, and the Great basin on the west. Structurally, it consists of stratified rocks of all ages, in chief part conformable, and usually lying nearly level, but dislocated in great orogenic blocks which stand at different altitudes. With them are associated eruptive rocks, in part injected as laccolites, and in part heaped in conical mountains. Its physiographic character is further determined by the fact that the region stands at great altitude above the sea, and has been profoundly dissected by the streams constituting its drainage system. It thus consists of a number of tables, standing at various heights, and separated partly by cliffs and partly by deep canyons. The return route traverses the region from Soldiers Fork in Central Utah to Grand Junction in western Colorado. On the branch route to the Colorado canyon it is traversed from Albuquerque westward.

The Great basin is a district without drainage to the ocean. It is bounded on the west by the Sierra Nevada, and on the north, east, and south by the basins of the Columbia and Colorado. At the north its lowlands have a general elevation of 4,000 to 6,000 feet (1,200 to 1,800 m.), and they descend southward to the level of the sea. Upon this sloping plateau are set a multitude of mountain ranges, for the most part of moderate height and of moderate length.[4] They are in general parallel with one another, the prevailing trend at the north being north and south, and at the south, northwest and southeast. The intervening valleys are flooded by alluvium derived from the erosion of the ranges, and are usually 15 to 20 miles (24 to 32 km.) broad. The ranges are party volcanic, but consist chiefly of Paleozoic strata, sometimes with folds, but nearly always profoundly faulted. The typical structure is in contrast with that of the Appalachians. In the Appalachian mountains are thrust faults, indicating compression of the strata; in the Great basin the faults exhibit hade to the downthrow. The ridges of the Appalachian mountains, as exhibited to-day, are due to differential erosion; the mountain ridges of the Great basin are due directly to uplift. The excursion enters the Great basin at Red Rock pass, in Idaho, and continues within it to the summit at the head of Soldiers fork, in eastern Utah.

The type of mountain structure of the Great basin, which has been called the "Basin Range type," is not restricted to the region of interior drainage, but prevails throughout a belt extending southward and eastward, about the margin of the plateau region, in Arizona and New Mexico.

CLIMATIC FEATURES.

The entire route falls within the climatic province of variable rains. In some localities there is ordinarily greater precipitation at some seasons than at others, but there is no district where at any season precipitation is regular or continuous. Rainfall and snowfall are associated with cyclonic or other disturbances of atmospheric equilibrium. The temperatures are those normal to the temperate zone; except on mountains and loftly plateaus, wheat and Indian corn are grown and the apple is the leading fruit.

The most conspicuous climatic contrasts are afforded by the local differences in the annual amount of precipitation and in the annual and diurnal range of temperature. The outward route from Washington to Minneapolis, and the return route from the Missouri river to New York traverse a region of essentially the same climatic type. The annual temperature mean ranges from 40° to 55° F. (4° to 13° C.), the precipitation ranges from 30 to 45 inches (90 to 135 cm.). The surface is normally timbered and deciduous trees prevail. Occasional patches of prairie interrupt the forest throughout, but they increase toward the west and finally predominate. The whole country is fertile without irrigation. The annual range of temperature (the difference between the coldest month and the warmest) is about 50° F. (27° C.). The length of the summer suffices for the growth of the leading food staples.

Westward to the base of the Rocky mountains the rainfall gradually diminishes to less than 15 inches (45 cm.); trees disappear except along the water courses, where the cottonwood (*Populus monilifera*) flourishes, and they are replaced by grasses. Farther westward the grasses become scant, and are partially replaced by an open growth of low bushes. Farming without irrigation becomes at first precarious and finally impossible, and grazing supersedes agriculture as the leading industry. At the same time the annual temperature range increases to about 60° F. (33° C.), and the diurnal range is likewise greater, as the excursionists will readily discover.

Throughout the more westerly portion of the route, in the Rocky mountains, the Great basin, and the Plateau region, the most important local climatic condition is altitude, and the native floras and faunas are arranged in belts which follow contours, but these contours run somewhat lower at the north than at the south. Precipitation varies with altitude, and temperature inversely with altitude. The plain of western Utah, with an altitude of 4,500 feet (1,375 m.), has an annual temperature of 75° F. (24° C.) and a rainfall of about 7 inches (21 cm.). Yellowstone Park, with an altitude of 6,500 feet (1,980 m.), has an annual temperature of 40° F. (4° C.) and a precipitation of 20 inches (60 cm.). At Leadville, 10,200 feet (3,100 m.) above the sea, the annual temperature is 35° F. (2° C.) and the precipitation 13 inches (40 cm.).

In western Utah the winter is 80° F. (44° C.) cooler than the summer; in Yellowstone Park, 50° F. (27° C.); at Leadville, 43° F. (24° C.).

The principal zones of vegetation are known as the desert, the pinon, the pine, the balsam fir, the spruce, and the subalpine.[5] In the desert zone there are no trees, except that the cottonwood occasionally follows the lines of streams. Bushes, of which the sage is the most important, have an open growth, usually offering no impediment to the progress of pedestrian or horseman. With these are grasses, invariably growing in discrete bunches so as not to constitute a turf. The prevailing color of earth, grass, and bushes is ashen. Near springs and perennial streams grow other grasses and bushes with more chlorophyl, so that bright green is to the desert wanderer the welcome sign of water.

The pinon zone is characterized by two species of evergreens, the nut pine (*Pinus edulis*) and the "cedar" (*Juniperus occidentalis monosperma*). The pine is ordinarily from 20 to 30 feet (6 to 9 m.) high, and the cedar from 15 to 25 feet (4 to 7 m.).

The pine zone is characterized by the yellow pine (*Pinus ponderosa*), a beautiful tree, 70 to 100 feet (20 to 30 m.) high, the groves of which stand in open order, without underbrush and without low branches. This tree is rarely associated with others, but waterways traversing its zone sometimes nourish a few individuals of various deciduous species, including the maple, the ash, and the box elder.

The characteristic tree of the balsam fir zone is the Douglas fir (*Pseudotsuga douglasii*). With it are associated the Rocky Mountain pine (*Pinus flexilis macrocarpa*) and the aspen (*Populus tremuloides*). The Douglas fir is a commanding tree, the rival of the yellow pine.

The spruce zone is characterized by the Engelmann's spruce (*Picea engelmanni*) and the foxtail pine (*Pinus aristata*). The spruce, which ordinarily predominates, is a beautiful tapering cone, its lower branches resting upon the ground.

In the subalpine zone Engelmann's spruce and the foxtail pine become gnarled and procumbent.

In the western region the precipitation of the winter is ordinarily greater than that of the summer, and on the mountains and uplands it takes the form of snow. In the spring and early summer this snow is melted and streams are nourished which flow to the lowlands, where the temperature is favorable to agriculture. Thus, despite the aridity of the lowlands, man is enabled to cultivate a limited portion of the land. The land thus cultivated yields a much greater return than can be obtained in the humid districts farther east. The great power of solar rays transmitted through a clear atmosphere of low humidity, combined with the rapid evaporation of moisture from the leaves of plants, gives a wonderful stimulus to vegetation, and, where the water for irrigation is abundant and skilfully applied, the yield is large.

THE APPALACHIAN REGION.[*]

By G. H. WILLIAMS.

The first day's ride of the excursion is well calculated to give a clear idea of the character of that regular mountain belt which bounds the entire North American continent on its southeast side. The general type of Appalachian structure is well known to all geologists, either through the classic work of the brothers Rogers,[1] or from the summary of their results given by Dana and others. The railroad course to be followed during the first day traverses the entire belt from east to west along one of its narrowest, deepest, and most characteristic sections, i. e., that which has been excavated by the Potomac River.

The recent work of several geologists, but notably that of Mr. W J McGee,[6] has clearly demonstrated the divisibility of the mountain and coastal portion of the eastern United States south of New York into three topographically and geologically distinct provinces or zones. Together they embrace nearly the entire sequence of geological formations, while at the same time the age, petrographical character, and structure of each is widely different from that of the others.

The most recent, as well as the most easterly, of these three zones is called the Coastal plain. It varies greatly in width between New York and Florida, but is throughout composed of nearly unconsolidated Mesozoic and Tertiary strata (clays, sands, and gravels), which dip very gently seaward, and in this direction grow steadily younger. Over all, however, is spread a capping of Pleistocene gravel (Columbia). The western edge of this Coastal plain may be regarded as approximately coincident with a line drawn from New York to Washington, and it is not improbable that a flexure or fault, still in process of development along this line, separates it from the crystalline region on the west.[7]

The next of the three provinces or zones in point of geological age, although not the one geographically contiguous to the Coastal plain, is the Appalachian Mountain belt. These two are separated by the third, and much the most ancient zone, composed of highly crystalline rocks, or semicrystalline rocks, and known as the Piedmont plateau.

The Appalachian Mountain Belt embraces nearly the entire sequence of Paleozoic strata. In the section to be traversed during daylight,

---

[*] A more extended account of the Maryland section through the Appalachians accompanied by a geological map of the strata in colors, has been published in the World's Fair book "Maryland," 1893, and reissued by the Joh Hopkins University Press as "The Geology and Physical Features of Maryland."

however, no beds lower than the Chazy-Trenton horizon have as yet been certainly identified upon paleontological evidence, while at the top of the series the Permian strata are wanting. Between these limits the series is quite complete, although many of the members are, in comparison with their Pennsylvanian equivalents, considerably attenuated. The formations distinguishable along the Potomac section are for the most part the same as those recognized by the Pennsylvania geological survey. The names and numerals by which these are usually designated are given, for convenience of reference, in the following table:

	No.	New York and Pennsylvania names.	Maryland and Virginia equivalents.
Carboniferous.	XV XIV XIII	Productive Coal Measures	Pittsburg series. Barren Coal Measures. Alleghany River series.
	XII	Pottsville conglomerate (Millstone grit).	Great conglomerate.
	XI	Mauch Chunk red shales.......... Mountain limestone	Greenbrier shales.
	X	Pocono sandstone (Vespertine)	Montgomery grits.
Devonian.	IX	Catskill sandstone................	Hampshire.
	VIII	Chemung...................... Hamilton (Marcellus) shales	Jennings-Romney.
	VII	Oriskany sandstone	Monterey.
Silurian.	VI	Lower Helderberg limestone.... Salina group (sandstone)	Cement rock.
	V	Niagara limestone. Clinton shales	Rockwood.
	IV	Medina sandstone .... Oneida conglomerate	Massanutten sandstone.
Siluro-Cambrian.	III	Hudson River shales	Martinsburg shale.
	II	Trenton-Chazy limestone	Shenandoah limestone. "Valley limestone."
	I	Potsdam sandstone...............	

This thick succession of conglomerates, sandstones, shales, and limestones accumulated as a part of the deposits of the vast sea which, during Paleozoic times, occupied the interior of the North American continent. Since these formations are so much thicker in the Appalachian belt than in the Mississippi basin they must, as has been shown by Hall[8] and Dana,[9] have been deposited in a trough which was undergoing a gradual depression. Their aggregate thickness in Pennsylvania is estimated at 40,000 feet. There is every evidence that this vast deposition took place from the east toward the west, and we are obliged to assume as the source of supply for so great an amount of material, a continental mass over and beyond what is now the Coastal plain. It is not improbable that the more crystalline portion of the Piedmont plateau may represent a remainder of this eroded and sunken continent. The conditions of accumulations through Paleozoic times were probably quite similar to what they have been and still are for the Coastal plain, if we imagine the direction of the drainage and position of the sea to be reversed, i. e., on the west instead of on the east.

At or near the end of the Carboniferous period, when the accumulations of sediments must have transformed the deep Appalachian trough into shallow marshes or estuaries, the vast thickness of strata were folded into the remarkably regular series of wrinkles which give to this mountain range at present its peculiar character. A cross-section of the Appalachians is nowhere symmetrical. On the contrary, it presents only one side or half of a symmetrical range, for it is composed of anticlinal and synclinal folds, all more or less overturned toward the west and also becoming steadily more and more abrupt toward the east. Along the section cut through the Appalachians by the Potomac River for instance, which the line of the railway follows very closely, one finds on the west side, between Cumberland and Grafton (in Garrett County, Md., and in West Virginia), only the low, flat folds of upper Paleozoic strata inclosing the nearly horizontal coal beds. Further east these gently sloping basins are replaced by others whose sides are steeper and which also display older rocks. It is then observed that each anticlinal has its western side more steeply inclined than its eastern, and still farther east both sides may dip toward the ocean so as to make the fold overturned. Last of all, the fold may become too sharp for the strength of the materials to stand, when the flexure becomes a thrust with the same general dip and strike. This latter may be seen in the isolated sandstone mountain, "Sugarloaf," 40 miles west of Washington. The perfect regularity, with which the folds become more and more abrupt toward the east, is at some points interrupted. On this section it is notably the case at Cumberland and Hancock, where a much sharper fold than the surrounding structure would lead one to expect discloses much lower horizons than are to be seen in the adjoining ridges (see the two sections beyond, figs. 5 and 6). Nevertheless, the regularity is so great as to have led so good a geologist as Rogers to the idea that the wave-like folds of the Appalachian system had been actually produced by undulations in a flexible crust due to horizontal pulsations or waves in the earth's liquid interior.[10]

Eruptive rocks in the gentler folds of the Appalachian system are noticeably absent. Along this section they are only to be found at all in the eastern portion, where the folds become abrupt or are replaced by faults and thrusts. Thus near Harpers Ferry and Weverton some ancient dikes occur, and the Blue Ridge sandstone (Lower Cambrian) is underlain somewhat farther north by large masses of ancient quartz-porphyries and rhyolites. More basic greenstones also occur in this region.*

*Am. Jour. Sci., 3d series. Vol. XLIV. Dec. 1892, p. 482.

The relation between the topography and drainage of the country and the geological structure is nowhere more apparent than in the Appalachian region, as was long ago pointed out by Lesley.[11] The drainage is consequent to the post-Carboniferous folding. The rivers in all probability flowed in the synclinal valleys. Many of the smaller streams follow the direction of the folds, while the larger ones, like the Potomac, Schuylkill, and Susquehanna, owe their transverse course to mutual reaction and adjustment through repeated cycles of elevation, tilting, and depression. This subject has been recently treated by Prof. William M. Davis in his studies of the rivers and valleys of Pennsylvania.[12]

The Piedmont Plateau.—For a distance of 43 miles after leaving Washington the railroad traverses, in a northwest direction, a rather low and rolling country, before entering the mountain belt proper at the station called Point of Rocks. It is composed of gneisses and mica-schists, sericite and chloritic schists, marbles and quartzites, whose strike follows in the main the general Appalachian trend. The relief of this country is given it by rapid streams or torrents, which are still excavating deep, rocky channels. The section of this belt affords a good idea of its general character, for although it increases greatly in width farther south, it everywhere retains a constant character at the eastern base of the Appalachians, and is for this reason appropriately designated the Piedmont plateau.

Topographically the Piedmont Plateau in Maryland and Virginia begins at the Catoctin Mountain, which meets the Potomac at Point of Rocks and pursues a straight course, across the former State, nearly northward from that point; geologically, however, the peculiar formations of still undetermined age, which are most characteristic of this region, begin farther to the east. At the base of Catoctin Mountain stretches a broad transgression of Triassic (Newark) red sandstone, which is crossed by the railroad at its narrowest point. From beneath the eastern border of this emerge the upturned edges of the Frederick Valley limestone, which has recently been found from its fossils to be the same as the Trenton-Chazy limestone which forms the valley farther west. East of this, with constant easterly dip, succeed overlying slates and another ridge of sandstone, which, however, only assumes topographical importance, as a high ridge, in the isolated mass of Sugarloaf. This mountain, seen just north of the railroad, is a thick monocline of easterly dipping beds, 1,350 feet in height. Geologically this mass forms the western boundary of the Piedmont plateau in Maryland. To it succeeds that vast complex of semicrystalline and holocrystalline rocks whose origin, age, and structure repeated earth movements have rendered most obscure. Along the section these extend eastward, becoming more and more crystalline, until they are buried beneath the overlying deposits of the Coastal plain.

The formations of this great Piedmont belt, and their equivalents farther north in New England, have, until recently, been usually classified as Archean. Very careful and detailed mapping is now, however, resulting in their subdivision, and in the identification of some as metamorphic Paleozoic strata, others as foliated eruptives, while still others belong really to the pre-Cambrian ages (Algonkian or Archean). It is to the detailed study of this field in Maryland and northern Virginia that the attention of the writer has of late years been directed. Some of his more general conclusions in regard to the structure of the Piedmont plateau have been communicated in a paper to the Geological Society of America.[13] The more important of these, as far as they relate to the region to be passed over, may be summarized as follows:

The Piedmont plateau is divisible into an eastern highly crystalline and a western semicrystalline portion. The former consists of gneisses and holocrystalline mica-schists, quartzites and marbles, containing an abundance of more or less dynamically metamorphosed eruptive masses. All of these rocks have a prevailing NNE. strike and a westerly dip. The western portion, on the other hand, is composed of partially metamorphosed sedimentary strata (sericite and chlorite schists, ottrelite schist, phyllite and limestone) and is nearly free from ancient eruptives. The strike of these rocks conforms to that of the eastern portion, but their dips are prevailingly to the east. In spite of apparent conformity, and even indications of transitions between these two portions of the Piedmont region, they are separated by a great time-break and unconformity.

The easterly dips on the west, and the westerly dips on the east, together with the nearly vertical strata between, produce a radiating or fan structure, and the axis of this fan is *not* coincident with the contact between the crystalline and semicrystalline portions. The thickness of either series of rocks, as indicated by their present dips, would be so vast that one must assume that the same beds are repeated over and over again by tightly compressed folds or thrusts.

The adjoining section (Fig. 4) made along the line of railroad between

FIG. 4.—Section through the Piedmont Plateau in Maryland.

C = Catoctin sandstone; N = Newark (Triassic) red sandstone, with northwesterly dips; L = Frederick valley limestone; Sh = Hydromicaceous and chloritic schists; Sl = Slates; D = Mesozoic diabase dike; D' = Buck Lodge diabase; D'' = diabase dike; S = Serpentine (two belts); A = Axis, near Dorwood Station; Gn = Gneiss; G, G = Granitoid gneiss, with abundant inclusions; G'' = Granite, in part hornblendic; P = Potomac clays and Coastal plain deposites.

Washington and Point of Rocks will illustrate the general character of the Piedmont plateau about the latitude of Washington.

In the absence of all paleontological data, it is impossible to assign a definite age to either of these series. In the light of what has been discovered elsewhere, however, it is not improbable that the western and semicrystalline areas represent the older Paleozoic horizons, metamorphosed by more intense dynamic action than has affected them farther west, while the holocrystalline rocks on the east are a remnant of the pre-Cambrian continent, from which the Paleozoic sediments were derived.

The apparent conformity between the two regions may be explained by supposing that the highly crystalline rocks also formed the floor upon which the now semicrystalline schists were deposited as sediments. These older rocks, already greatly altered and folded, underwent at the time of the Appalachian uplift one more final folding, which gave them their now prevailing trend, and carried the overlying Paleozoic sediments with them. This supposition is also in accord with the fact that several closed synclinals of slate and semicrystalline schists are found pinched into the gneisses, far to the east of the main contact.

WASHINGTON, D. C., TO CUMBERLAND, MD.

ITINERARY.

By G. H. WILLIAMS.

Station.	Distance.		Elevation.		Station.	Distance.		Elevation.	
	Miles.	Kilometers.	Feet.	Meters.		Miles.	Kilometers.	Feet.	Meters.
Washington *	0	.....	10	3	Washington Junction .	43	69	229	70
Terra Cotta	4	6	.....	.....	Point of Rocks	43	69	.....	.....
Silver Spring	7	11	.....	.....	Weverton	52	84	249	76
Garrett Park	12	19	.....	.....	Harpers Ferry	55	89	272	83
Rockville	16	26	.....	.....	Martinsburg	74	119	634	193
Derwood	19	31	.....	.....	North Mountain	81	130	547	167
Gaithersburg	22	35	.....	.....	Cherry Run	87	140	398	121
Boyd	30	48	.....	.....	Hancock	96	154	474	143
Dickerson	36	58	.....	.....	Cumberland †	152	245	639	195
Tuscarora	39	63	.....	.....					

* Population, 230,392.　　　　　　　† Population, 12,729.

Washington. On leaving Washington, the railway line at first passes over the unconsolidated deposits of the Coastal plain. The lowest of these is the Potomac series (of early Cretaceous or late Jurassic age), consisting of coarse littoral conglomerates at the base, formed of well-rounded pebbles of quartzite. Above the conglomerates are a series of variegated clays, often valuable for pottery, and sometimes containing deposits of limonite.

Terra Cotta. At this station are works where drain pipes, etc., are made from the clays of the Potomac series.

Silver Springs, on the northern boundary of the District of Columbia. The rock here is a very granitoid, though much rotted, gneiss. It is not visible from the cars. The road then passes over typical holocrystalline gneiss, with steep westerly dips, to

Garrett Park, where these are cut by serpentine and a still later hornblende granite, filled with included fragments of the surrounding gneiss.

Rockville, a thriving town, recently much developed as a suburban residence part of Washington. The division between the crystalline and semicrystalline portions of the Piedmont plateau passes just west of here. The rock near the station is a nearly vertical feldspathic gneiss, but just beyond the sericite schists are seen.

This place is directly north of the Great Falls of the Potomac, the source of Washington's water supply, near which the gold mines of Montgomery County are situated.

Derwood is where the schists are quite vertical, and where the axis of the Piedmont fan crosses the road (see section Fig. 4).

Gaithersburg. Here are several lenses of serpentine in the schists; the latter are fully exposed in the deep ravine of Great Seneca Creek.

Boyd. Here a large mass of intrusive trap occurs, which is petrographically identical with the Triassic dike of diabase just west of it. The age of both is in all probability the same.

Dickerson. The isolated mass of Sugarloaf Mountain, which has for some time been visible, is now well seen on the right (north). It is an east dipping monocline of sandstone, divided into two parts by a longitudinal valley, and is probably due to a double thrust of the sandstone from the east. Near this point, also, a broad transgression of Triassic red sandstone is entered, and a little beyond a great dike of diabase is cut through. This latter traverses the whole State of Maryland from north to south, and passes into Virginia. Its hardness makes it a topographical feature, which is known in the neighborhood as "Ironstone Ridge." A mile or so farther, the Monocacy River is crossed by a high bridge; a short break in the red sandstone exposes the shales and blue limestone of Frederick Valley (Hudson and Trenton), which at the railroad are mostly covered by alluvium deposited by the Potomac River, which is here for the first time approached.

At **Tuscarora** red sandstone appears again and continues to **Washington Junction.** For a mile or more before reaching the latter station the upper member of the Trias is well exposed along the line of the railroad. This is a coarse conglomerate of rounded, or sometimes angular and broken, pebbles of variegated Paleozoic limestone, embedded in a red argillaceous matrix. It is a striking rock and enjoys quite a wide celebrity as a decorative stone. It is known as "Potomac Marble," or "Calico Rock," and is the material from which the great columns in the rotunda of the Capitol at Washington were made.

At this point (**Washington Junction**) connection is made for the city of Frederick, some 15 miles to the north, which occupies the center of the broad, fertile limestone valley. Before reaching this place, however, the main line of the Baltimore and Ohio railroad diverges and pursues a course down the Patapsco River direct to Baltimore. Near this point of divergence, on the Monocacy River, 12 miles NE. of **Washington Junction**, an important battle of the war was fought in 1864, and the northern extension of the same valley in Pennsylvania was the scene of the great battle of Gettysburg.

Point of Rocks, a short distance beyond **Washington Junction**, marks the entrance of the Appalachian Mountain belt proper. The river here cuts through a high ridge capped by a monocline of Cambrian sandstone, known as Catoctin Mountain. This runs nearly due north and forms the western boundary of the Frederick Valley. The road passes for 8 miles across a valley of slate, drained by Catoctin Creek, and called, from its principal town on the Maryland side of the river, the Middletown Valley. It then intersects another abrupt ridge nearly parallel to Catoctin Mountain at

Weverton. This is the junction for a branch road northward to Hagerstown, one of the largest cities of Maryland. The road now crosses a slight depression occupied by crystalline rocks (granite-gneiss with some basic dikes) and soon reaches

Harpers Ferry, a place which, both geologically and historically, is of more than usual interest. From the railway bridge over the Potomac may be seen on the right (north) a lofty ridge of contorted sandstone with underlying shales and slates, known as Maryland Heights. Upon the opposite side of the river rises the continuation of this ridge in Virginia, known as Loudoun Heights. The course of the Potomac is for some distance above this point nearly south, and here it is joined by one of its most important tributaries, the Shenandoah River. In the triangular space between these two streams lies the town of Harpers Ferry, which, from its surroundings, possesses great strategic importance.

The two ridges cut through at **Weverton** and **Harpers Ferry** well illustrate a characteristic feature in Appalachian topography. The former originates a short distance south of the river and continues its course across Maryland as the "Blue Ridge," and, after its junction with the Catoctin Mountain, as "South Mountain," in Pennsylvania. The Harpers Ferry elevation, on the other hand, soon dies out toward the north, but continues its course hundreds of miles southward, across Virginia and North Carolina, as the "Blue Ridge." Thus, near the Potomac, one important fold dies out and is continued by another, *en echelon*, or offset somewhat to one side, as is a frequent occurrence among the long parallel ridges of the Appalachian system.

The geology at **Harpers Ferry** is complex, and has given rise in former times to different interpretations by different investigators. To the west are the contorted layers of the blue Valley limestone. known from its fossils to be of Trenton-Chazy age (II). On the east of these, and apparently overlying them, succeeds a thick mass of shales, slates, and the contorted sandstones seen in the front face of Maryland Heights which Walcott has recently shown to be Lower Cambrian. Then follows toward the east, occupying the space between here and

Weverton, the axis or base of granite-gneiss, whose cleavage dip (undoubtedly a secondary feature) is constantly toward the east.

Beyond **Harpers Ferry** the railroad temporarily leaves the river and crosses the broad Shenandoah Valley. This is composed of the same Trenton-Chazy limestone as the Frederick Valley, and is possessed of a like fertility. Its continuation southward is the great "Valley of Virginia," notable for its caves (Luray and Wiers), its "natural bridge," and for the fact that it is just now the scene of remarkable industrial activity and development, by which the richness of its natural resources is being rapidly brought to light.

At **Martinsburg**, the largest place passed by the railroad in this valley, there are shales of Hudson River age (III), developed by a fold in the limestone.

At **North Mountain** the great valley is left for another intersection of a sandstone ridge, while to the north of the next station,

Cherry Run, there occurs a great fault by which rocks of all the formations, from the Niagara (V) to the Hamilton (VIII), inclusive, are brought successively in contact with the Trenton-Chazy.

Hancock is a town of some importance, situated at the narrowest part of Maryland, where the State is not over 4 miles wide. It was formerly a prominent station of the National turnpike; its chief industry now is the manufacture of cement from the Helderberg limestone (VI). Here occurs one of the more abrupt anticlinal folds mentioned above, whereby the whole sequence of Upper Silurian and Devonian formations is exposed on either side of a wide compound arch (fig. 5). At the railway station fossiliferous shales are exposed.

Within the next 3 miles westward, Oriskany sandstone (VI), Helderberg limestone (cement rock), and the red Salina sandstone band at its base, are traversed. At 3 miles from the station the cement mills are situated, and near them are some small folds of the Salina rock of remarkable perfection. The center of one of these was used one hundred years ago as a blacksmith's forge, and it is not ill-suited for this purpose.

Between **Hancock** and **Cumberland** the road follows the remarkable sinuosities of the river for 56 miles to accomplish a distance which, in a direct line, is only 32. Within

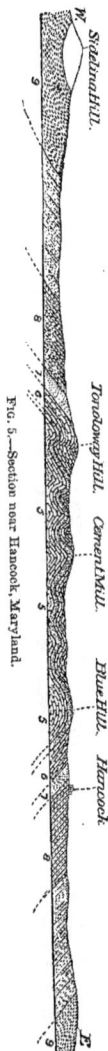

this distance the track crosses, more or less obliquely, several rather low and flat folds of Upper Silurian and Devonian strata, and finally reaches

Cumberland. Although it contains but 13,000 inhabitants, this is the largest place in Maryland, next to Baltimore. It has considerable importance as the center of the coal industry of western Maryland and from the manufacture of Portland cement of excellent quality. Geologically, Cumberland is important because it affords one of the most complete sections of the entire Paleozoic series to be found anywhere within the Appalachian system. This is not seen to the best advantage along the railroad line, which here enters upon a southwesterly course, but by following up Wills Creek and Jennings Run, which head in the Frostburg coal basin. On the west side of the town rises an abrupt N. to S. ridge, Wills Mountain (fig. 6). This is composed of red layers of the lower Medina (IV), capped by the massive white bed of the same horizon, which forms an anticlinal fold whose eastern flank dips gently east while its western plunges downward nearly vertically. On the east side of this axis, in the city itself, may be seen in succession Clinton, Niagara (V), Salina, Helderberg (VI), Oriskany (VII), and Hamilton beds (VIII), all filled with their characteristic fossils. On the west side of the mountain, owing to its abrupt downward plunge, the series is still more complete and extends, within a short horizontal distance (8 miles), to the top of the Coal Measures.

Above Cumberland the road follows the river in a southwest direction as far as Keyser, parallel with the strike of the beds, and along the west base of Knobby Mountain. This mountain is made of Salina and Helderberg beds, capped by Oriskany sandstones, which are visible in the cliffs. On the right (west) can be seen the first range of the Alleghany Mountains, which is capped by the Pottsville conglomerate (millstone grit), the Mountain limestone, Pocono sandstone and Chemung beds forming the intervening slopes, while the valley at the base is in Hamilton shale.

At Bradys, in limestone concretions of the Helderberg limestone, have been found beautifully and curiously developed crystals of celestite.[14]

On the left, before entering the town of Keyser, are quarries in Helderberg limestone, and on the right, cliffs of

FIG. 6.—Section near Cumberland, Maryland.

Oriskany sandstone. Beyond **Keyser** the road crosses the Potomac into West Virginia, and fine exposures of Hamilton shales are seen in the cuts. The road now bends to the northwest and crosses, at right angles to the strike, steeply dipping beds of the Devonian and of the Carboniferous up to the Coal Measures, a thickness of about 13,000 feet (4,000 m.) of rock strata.

FROM CUMBERLAND (MD.) TO THE OHIO RIVER.

By I. C. WHITE.

The area across which the Baltimore and Ohio Railroad passes from Cumberland westward to the Ohio River includes all of the Alleghany Mountain country proper, and also about three-fifths of the breadth of the great Appalachian coal field, the eastern line of which is found only five miles northwest from Cumberland, and the western margin of the same near Newark, Ohio. The air-line distance across the Appalachian field along the line of the Baltimore and Ohio Railroad is about 164 miles, but the distance by rail between Cumberland and Newark, the two margins of this coal field, is 315 miles.

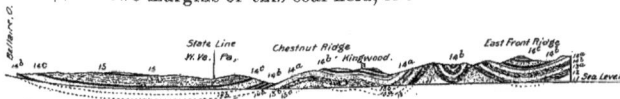

FIG. 7.—Section across the Alleghany Mountains to the Ohio River.

15 Permian. 14ᶜ Upper Coal Measures. 14ᵇ Lower Coal Measures. 14ᵃ Millstone grit. 13ᵇ Mountain limestone and Mauch Chunk shales. 13ᵃ Pocono Sandstone. 12 Catskill. 11 Chemung.

As shown by the above illustration of the geological structure, the rocks of the Alleghany Mountain country are crumpled up into a series of large folds, and these, together with the great erosion to which the region has been subjected, have produced the wild and rugged scenery between Piedmont and Rowlesburg.

The Pottsville Conglomerate (XII), and the Pocono sandstone (X) are the mountain-makers of the Alleghanies, while the great anticlinal ridge (Wills Creek Mountain), through which the Baltimore and Ohio Railroad leaves Cumberland for Pittsburg, is made by the White Medina sandstone (IV) at the base of the Upper Silurian.

Westward from the Alleghanies proper the folds in the strata become gradually more gentle, and finally die away, before reaching the Ohio River, into low undulations of the beds, which are so insignificant as to be almost imperceptible to the eye.

The great arches already referred to bring up the lowest rocks of the region in the vicinity of Cumberland, so that from there westward we pass upward through the geological scale from the middle of the Medina series in the Wills Creek gap at Cumberland to the summit of the Permo-Carboniferous, or Permian, beds in the middle of the great plateau between the Monongahela and Ohio rivers.

The character of these several terranes and their thicknesses, etc., along this line, will now be briefly described in ascending order.

Medina series (IV of Rogers, base of Upper Silurian), thickness 2,055 feet.

	Feet.
(a) Oneida Conglomerate	355
(b) Red Medina	1,200
(c) White Medina	500

The thickness of the Red Medina and the Oneida beds in the above measurement is based upon data obtained from a deep boring in Wills Creek gap, one mile east from Cumberland. The well starts 790 feet above the base of the Red Medina, penetrates the Hudson River dark shales at 1,145, and stops in them at 2,010, probably about 1,000 feet from the top of the Trenton limestone.

The White Medina is finely exposed in the great arch of Wills Creek Mountain, where its top rises to 1,300 feet, almost vertically, above the bottom of the gorge through which Wills Creek finds an exit to the North Potomac River at Cumberland.

This great arch dies rapidly away to the northwest and the Medina passes below the surface where the North Potomac passes across its trend 15 miles southwest from Cumberland, but 10 miles farther southwest the arch swells up again, and the Medina coming above the surface makes the summit of New Creek Mountain. This rock is the great mountain-maker of the Appalachian belt east from the Alleghanies and west from the Blue Ridge, and on account of its hardness is highly valued for ballasting railroad tracks.

Clinton series, (V) thickness 721 feet.

	Feet.
(a) Lower olive shales	100
(b) Iron sandstone	20
(c) Middle shales and limy beds	300
(d) Fossil iron ore	1
(e) Upper limestones and shales	300

The Niagara limestone proper has never been differentiated from the great mass of limy shales and impure limestones which occur in the interval between the Salina series and the Medina sandstone, anywhere along the Appalachian system. It is possibly absent, but until the richly fossiliferous beds which are here grouped under the Clinton are thoroughly studied, no one can say positively that the Niagara limestone of the New York column is not represented in the series of rocks given above and all classed as Clinton.

This series is beautifully exposed at Cumberland, along the Pennsylvania Railroad, near the southern entrance to Wills Creek Gap. The "fossil ore" (d) was once mined here, and has the same appearance and fossils as in New York and Pennsylvania.

Salina series, (*VI*) thickness 680 feet.

	Feet.
(a) Red shales and thin, light-colored limestones with marly shales	300
(b) Water lime, a bed of dark, magnesian limestone, from which the famous "Cumberland cement" has long been manufactured; finely exposed at the quarry along the West Virginia Central and Pennsylvania railroads in Cumberland, thickness	30
(c) Gray and yellowish, thin-bedded, sparingly fossiliferous, impure limestones.	350

Lower Helderberg series (*VI*), thickness about 350 feet.

Massive gray and dark-colored limestones, many of the layers quite pure, and a splendid flux for iron ores, richly fossiliferous; finely exposed at Cumberland, and from there southwestward to **Keyser**, a distance of 24 miles.

Oriskany sandstone (*VII*), thickness 75 to 100 feet.

A coarse, dirty yellow, calcareo-siliceous rock, highly fossiliferous (mostly casts); makes Knobby Mountain, which starts at Cumberland and trends away to the southwest; often forms great cliffs along the Baltimore and Ohio Railroad, between **Cumberland** and **Keyser**; one of these just east from **Keyser** is known as Bull Neck, and here the railroad passes across the Oriskany in a cut through a sharp syncline, making a fine exposure of the rock and its fossils. The high cliff just opposite **Keyser**, across the Potomac River from the Baltimore and Ohio station, is made by this rock, and is known as Queens Point. From **Cumberland** eastward this stratum is quite massive and often forms ridges 1,000 to 1,200 feet high.

At **Cumberland** the Baltimore and Ohio Railroad, main line, turns southwestward, following up the North Potomac River, and hence runs along the strike of the rocks to **Keyser**, 24 miles from **Cumberland**, so that not only the Oriskany sandstones, but all of the other beds below it, down to the Medina white sandstones, are frequently seen between the two points.

The Corniferous limestone appears to be absent entirely from the Alleghany Mountain region, since although a thin, earthy limestone (Selinsgrove) is present a few feet above the Oriskany at many points, it evidently belongs to the Marcellus epoch of the Hamilton.

Hamilton shales, (*VIII*). A series of black slates, olive shales, and dark gray sandy beds, with an earthy limestone near the base, all quite fossiliferous. These beds underlie the station site in **Cumberland**, and are partially exposed just east from it in a cutting; also exposed in the vicinity of **Keyser**.

These rocks make valleys wherever they extend, and hence are usually covered up and concealed from view by soil and detrital matter, so that the exact thickness can not be determined, but estimating this from the usual breadth of the valleys, and dip of the hard rocks above (Portage) and below (Oriskany), if no rolls are present, the entire thickness of the Hamilton rocks, including Marcellus, Hamilton proper, and Genesee, in the region of **Cumberland** and **Keyser**, can not be less than 3,000 feet.

Chemung Series, (VIII).   Total thickness 5,000 feet.

The Portage beds consist of olive gray shales and flaggy layers, sparingly fossiliferous; thickness probably about 1,000 feet.

The Chemung beds consist of olive shales, flaggy sandstones, and one massive conglomerate (Allegrippus) near the top; thickness about 2,500 feet.

The Catskill beds are a series of reddish shales, green and red sandstones, with conglomerate layers in lower half; thickness 1,500 feet.

The Chemung* rocks have a fine development along the line of the Baltimore and Ohio Railroad, east as well as west from **Cumberland**.

West from **Cumberland** the first exposure of these beds is about two miles west from **Keyser**, where the Chemung proper is fairly well exposed, dipping down under the great trough of the Cumberland coal basin.   Then follow the red shales and sandstones of the Catskill, but the latter reappear along the Savage River after the railroad has crossed the Potomac coal basin and begins the long climb up the "17-mile" grade to the summit of the Alleghany Mountains at **Altamont**.   The Catskill beds are covered up by the Pocono (Lower Carboniferous) sandstones in the summit cut at **Altamont**, but reappear just west from it and extend along the Baltimore and Ohio to **Deer Park** and a mile beyond, where the Negro Mountain anticlinal, which crosses the Baltimore and Ohio at **Mountain Lake Park**, 3 miles west of **Deer Park**, brings up the soft shales and conglomerates of the Chemung proper.   Then these beds turn over and descend under the Youghiogheny River coal basin, to be brought to the surface again on the crest of another great arch at **Terra Alta**, and are constantly in sight from there down the western slope of the Alleghanies to Cheat River, at **Rowlesburg**, and two miles beyond, where the red Catskill beds pass under water level, to be seen no more to the west.

Pocono sandstone (X), Lower Carboniferous.

	Feet.
Thickness near Piedmont about	1,000
Thickness on Cheat River, near Rowlesburg	600
Thickness at Mannington	400

A series of hard, gray, current-bedded sandstones, usually of rather fine grain, but occasionally containing flat pebble conglomerates, inter-

---

* The term *Chemung* as here used includes the three epochs of Portage, Chemung, and Catskill, as proposed by Dr. Jno. J. Stevenson in his vice-presidential address before Section E, at the Washington meeting of the "A. A. A. S.," August 1891.

stratified with sandy shales, comes next above the Catskill red beds. The series is first seen on the Baltimore and Ohio between **Keyser** and **Piedmont**, where it has a thickness of 1,000 feet, and makes the east front ridge of the Alleghany Mountain, on the south side of the Potomac, and Dan's Mountain, on the north side, but soon dips rapidly down under the Cumberland coal basin.

These beds come up again a few miles west from **Piedmont** and are seen making the steep upper escarpment of Big and Little Savage mountains, westward to the top of the "17-mile grade" at the summit of Back Bone Mountain, near **Altamont**, where the highest cut, at 2,620 feet elevation, is through these sandstones.

It is seen again at **Oakland**, making a fine cliff near the Baltimore and Ohio hotel, and dipping down under the Youghiogheny coal basin, to come up and pass into the air just east from **Terra Alta**, and then, coming down again, makes the summit of Briery Mountain, the most western ridge of the Alleghany range, and, passing under the Baltimore and Ohio Railroad for the last time just west of **Rowlesburg**, assists in shaping the grand scenery along the Cheat River gorge between **Buckhorn Wall** and **Rowlesburg**.

The upper portion of this sandstone series is the great oil and gas depository of West Virginia; at **Mannington**, on the Baltimore and Ohio Railroad, where this stratum lies 1,800 feet below the surface, a large oil field, extending northeast to the Pennsylvania line and southwest for an unknown distance, has been developed in these beds.[15]

Mountain limestone and Mauch Chunk red shale (XI).—This subdivision of the Lower Carboniferous, like the Pocono beds below, is more closely allied to the Catskill series, both in its lithological aspects and in its fossils, than to the Coal Measure series above, which always rests unconformably upon it.

The bottom member of the series, the Mountain limestone, where crossed by the Baltimore and Ohio, just east of **Piedmont**, is 340 feet thick, quite fossiliferous, and consists of gray limestone interstratified with marly gray shales, and some red beds, the lowest 30 feet being a silicious limestone merging gradually into the Pocono sandstone below. On Cheat River, west from **Rowlesburg**, where the Baltimore and Ohio grade cuts through this series, it is 273 feet thick, of which the silicious member at its base measures 105 feet. West from **Rowlesburg** the Mountain limestone is seen no more, but at **Mannington** the drill passes through it at 1,600 feet below the surface, where it is 90 feet thick, while at **Bellaire** it is still 30 to 50 feet thick, as determined by drill holes. As is well known, this Mountain limestone thins away entirely, to the northeast along the Alleghanies, and is not found anywhere in northeastern Pennsylvania, but southwest from the Baltimore and Ohio Railroad it thickens greatly, attaining 900 feet in Randolph and Pocahontas counties, W. Va., 1,100 feet in Greenbrier County, and on New River, in Summers County, a deep boring, 12 miles above Hinton, finds it 1,415 feet thick.

The Mauch Chunk Red Shale division of No. XI consists of bright red shales, impure limestones, and massive, green sandstones, and frequently contains iron ore near the top in irregular layers and nuggets. The thickness of this series on the Baltimore & Ohio near **Piedmont** is 850 feet, but on Cheat river, near **Rowlesburg**, it is reduced to only half that amount, or 440 feet, while westward from this latter locality the Mauch Chunk continues to dwindle in thickness, being only 140 feet in the oil borings at **Mannington**, and probably not more than half of this at **Bellaire** on the Ohio river, while further northwest in Ohio it seems to disappear completely.

These beds are quite variable in thickness along the Appalachian Mountain region, being 3,000 feet at Mauch Chunk, Pa., and about 2,500 in Greenbrier county, West Virginia. Like all of the Paleozoic deposits, they attain their maximum thickness to the east, and diminish rapidly westward. The flora of these beds seems to be more closely related to that of the Catskill below than to that of the Coal Measures above.

The Pottsville Conglomerate, Millstone Grit, etc. (XII).[16]—The basal member of the Coal Measures, rests unconformably (not in dip but by erosion) upon the underlying Lower Carboniferous beds, and consists of white, hard, and often conglomeritic sandstones, interstratified with dark shales which contain coal beds (New River series), thin and insignificant along the Baltimore & Ohio Railroad and elsewhere to the northeast, but thickening up to the southwest, and becoming the famous coking and fuel coals of New River, West Virginia, and the Pocahontas region of Virginia and West Virginia.

The sandrocks of this series, being composed of nearly pure quartz grains and therefore almost indestructible by atmospheric agencies, have played a very important part in forming the wild scenery along the Alleghanies; in fact, this series conspires with the Pocono beds below to make the Alleghanies, the Pottsville beds topping out the highest summits, like Eagle Rock near **Deer Park**, on the "backbone" of the Allgehanies (3,350 feet). The wild scenery where the Baltimore and Ohio passes out of the Cheat river canyon, west from **Rowlesburg**, is made principally by these beds. Wherever these rocks come to the surface, they form "rock cities," waterfalls, canyons, mountains, and a rugged, picturesque country generally. The series varies greatly in thickness, as will be seen by the following:

Thickness on the B. and O. R. R.:	Feet.
Near Piedmont	473
Near Oakland	700
Near Rowlesburg	362
At Mannington	255
At Bellaire	200
Thickness at Pottsville, Pa	1,000
Thickness on New river, W. Va	1,400

Lower Coal Measures, Alleghany River Series (XIII).[16]—The next higher series of rocks, called the Alleghany River Coal Series, contains the greater portion of the coal in the Appalachian field. These measures consist of alternate beds of coal, shale, limestone, and sandstone, and where the Baltimore and Ohio Railroad first crosses them at **Piedmont**, West Virginia, have a total thickness of 308 feet; 310 feet near **Rowlesburg**; 225 feet in the deep shaft at **Newburg**; 262 feet at Nuzum's, 10 miles west of **Grafton**; and 220 feet at the Ohio river near **Bellaire**.

In Pennsylvania this series has a thickness of 200 to 370 feet, while in Ohio the same series varies between 175 and 250 feet. In northern West Virginia these measures are 200 to 300 feet thick, but they thicken rapidly to the southwest, and on the Big Kanawha, at Powellton, are 1,000 feet, and maintain that southwest from there to the Tug river in Logan county.

The principal coal beds of this series are the following in ascending order:

(1) Clarion. (2) Lower Kittanning. (3) Middle Kittanning. (4) Upper Kittanning. (5) Lower Freeport. (6) Upper Freeport.

The Barren Measures, Elk River Series (XIV).—The Barren Measures occupy the interval between the Upper Freeport coal and the next great coal bed about 600 feet above, viz, the Pittsburg seam. This Elk River Series contains several (5) coal beds, but the most of them are too thin and impure to be commercially valuable, with the exception of the two lower ones (Mahoning and Masontown). The rocks consist of massive sandstones, red shales, and thin limestones. Marine fossils occur for the last time about the middle of this series at the horizon of the Crinoidal limestone, since above that stratum nothing but brackish or fresh water forms are found. These beds furnish valuable building stone (Mahoning, Morgantown, and Connellsville sandstones), and make a great band of red soils from western Pennsylvania clear across West Virginia into Kentucky, and back around through southern Ohio to western Pennsylvania.

This series is, in the hills at **Piedmont**, about 500 feet thick. Where next found on the Baltimore and Ohio Railroad, at **Newburg**, these rocks are 650 feet thick, and the railroad runs in them from there to **Fairmont**, except for a short distance on the crown of the Chestnut Ridge axis, 10 miles west of **Grafton**.

The fauna of this series belongs to extreme Upper Carboniferous types, and some forms are closely related to Permian species, while the flora above the middle of the series is also of Permian aspect.

	Feet.
Maximum thickness (on the Great Kanawha in West Virginia)	800
Minimum (in Ohio)	400
Average thickness	600

The Upper Coal Measures, Monongahela River Series (XV).—These beds crown the summit of the mountain in the bottom of the syncline at **Piedmont**, the principal coal (the Pittsburg) being known there as the "Big" bed, since along George's Creek, between **Piedmont** and Frostburg, it is often 20 feet thick, and is mined and shipped under the name of Cumberland coal. This great bed of valuable fuel is the basal member of the Upper Coal Measures, and is the most important stratum in the Appalachian field. It is the celebrated Connellsville coking coal, the Westmoreland and Penn gas coal, the shipping coal all along the Monongahela River, and extends under a vast area between that stream and the Ohio River.

A small remnant of it is found in the summits of the hills at **Newburg**, and the Baltimore and Ohio Railroad does not cross its outcrop again until we come to the town of **Fairmont**, where, at Monongah, Gaston, Montana, and West Fairmont, it is extensively mined for coking, gas, and general fuel purposes.

Westward from **Fairmont** it dips down below water level, being 400 feet below the valley at **Mannington** and 800 feet at **Board Tree**, where it is 1,500 feet under the highest summits. Rising westward from the center of the great Appalachian trough at **Board Tree**, it comes to the surface again at **Bellaire** and furnishes valuable fuel for that city and **Wheeling**.

The rocks above the Pittsburg coal in Pennsylvania, northern West Virginia, and Ohio are mostly limestones and calcareous shales, with some sandstone, and four other coal beds, viz, in ascending order:

Redstone. Sewickley. Uniontown. Waynesburg; this last topping out the series.

Southwestward through West Virginia and southern Ohio, the limestones are replaced by red shales and the last four coals practically disappear, while much massive sandstone comes into the column.

This series has an average thickness of 360 feet along the Monongahela river, but it thins to 250 feet in Ohio and thickens to 413 feet in the central portion of the Appalachian trough.

The Permian, Dunkard Creek Series (XVI).—This series of beds tops out the Carboniferous system in the Appalachian field, and has its greatest development along the Baltimore and Ohio Railroad between the Monongahela and Ohio rivers. The series begins with the roof shales of the Waynesburg coal, and includes all of the beds above, about 1,100 feet in all. It consists of massive sandstones, red and variegated shales, with thin limestones and impure, mostly worthless, coal beds.

The fauna has never been studied, but appears to be of fresh-water origin, and is largely composed of bivalve crustaceans, while the flora has a distinctly Permian facies, containing *Callipteris conferta, Saportæa, Tæniopteris, Baiera*, and other Permian types, while the species that resemble Coal Measure forms also occur in the Wichita, or undisputed Permian of Texas, so that the evidence is now complete that this Dunkard Creek series, formerly called Permo-Carboniferous, is the American equivalent of the European Permian.

Tabular view of the geology between Cumberland and the Ohio River.

[By I. C. White.]

Station.	Distance from Washington.		Elevation.		Geological formations.
	Miles.	Kilometers.	Feet.	Meters.	
Cumberland	152	245	639	195	Medina to Hamilton.
Bradys Mill	159	256	642	196	Lower Helderberg.
Potomac Bridge	172	277	786	240	Hamilton.
Keyser	175	282	800	244	Lower Helderberg to Hamilton.
Piedmont	180	290	925	282	Pottsville to Dunkard Creek series.
Bloomington	182	295	1,024	312	Pottsville to Elk River series.
Fraukville	188	303	1,699	518	Catskill to Pottsville.
Altamont	197	317	2,620	799	Pocono to Pottsville.
Deer Park	200	322	2,442	744	Chemung to Catskill.
Mountain Lake Park	203	327	2,400	731	Chemung.
Oakland	206	332	2,372	723	Catskill to Pocono.
Little Youghiogheny Bridge.	206½	333	2,371	723	Mauch Chunk.
Great Youghiogheny Bridge.	207	333	2,372	723	Pottsville conglomerate.
Huttons	212	541	2,477	755	Lower Coal Measures.
Snowy Creek	214	344	2,469	758	Mauch Chunk.
Terra Alta	216	348	2,549	774	Chemung, western ridge of Alleghanies.
Rodemer's tunnel	220	354	2,083	635	Catskill.
Cheat River Bridge	227	367	1,392	424	Chemung to Catskill.
Rowlesburg	227	365	1,392	424	Do.
Buckeye Run	228	367	1,515	462	Catskill.
Buckhorn Wall	231	370	1,720	524	Mauch Chunk and Pottsville.
Cassidys Summit	233	374	1,855	565	Top Lower Coal Measures.
Tunnelton	234	375	1,820	555	Do.
East portal, Kingwood tunnel.	234	375	1,819	554	50 feet under Upper Freeport coal.
West portal, Kingwood tunnel.	235	378	1,779	542	Upper Freeport limestone and coal.
Murray's tunnel	238	383	1,554	474	Upper Freeport coal at track level.

Tabular view of the geology between Cumberland and the Ohio River—Continued.

Station.	Distance from Washington.		Elevation.		Geological formations.
	Miles.	Kilo-meters.	Feet.	Meters.	
Newburg	241	388	1,215	370	Elk River series; Pittsburg coal in hill-top; shaft through Lower Coal Measures.
Independence..........	242	389	1,156	352	Elk River series (No. XIV).
Raccoon Creek	244	392	1,105	337	Do.
Thornton	248	399	1,038	316	Do.
Grafton...............	254	409	987	301	Do.
Fetterman.............	255	410	984	300	Do.
Valley River falls	261	420	969	295	Pottsville to Elk River series.
Texas	268	431	883	269	Elk River series (No. XIV).
Fairmont..............	276	444	877	267	Do.
Barnesville...........	277	444	871	265	Upper Coal Measures.
Dunkard Mill run	284	458	922	281	Waynesburg coal, 150 feet above track; Pittsburg coal, 230 feet below track.
Mannington	293	476	957	292	Base of Permian; Pittsburg coal, 400 feet below valley; oil field here in Pocono S. S., 1,800 feet below valley.
Glovers Gap..........	300	483	1,050	320	Permian.
Glovers Gap tunnel ...	302	486	1,146	349	Permian, 900 feet above Pittsburg coal.
Burton	304	489	1,060	323	Permian.
Hundred	306	492	1,025	312	Permian, 700 feet above Pittsburg coal.
Littleton	311	500	936	285	Permian.
East portal, Board Tree tunnel.	314	505	1,104	336	Permian; center of Appalachian trough; 1,100 feet Permian beds.
West portal, tunnel....	314½	506	1,077	328	Permian.
Bellton	318	512	886	270	Permian; Belton coal group.
East portal, Welling tunnel.	323	520	1,202	366	Permian
Cameron..............	325	524	1,049	320	Permian; natural gas here in Gordon oil sand, 2,700 feet below valley.
Easton................	330	531	967	294	Permian.
Rosbys Rock	336	543	773	286	Do.
Moundsville...........	342	550	640	195	Upper Coal Measures, Pittsburg bed, 125 feet below track level.
Benwood	349	562	648	197	Pittsburg coal at track level.
Bellaire...............	350	573	657	200	Do.
Wheeling..............	353	568	645	197	Pittsburg coal 60 feet above track level.

FROM THE OHIO TO THE MISSISSIPPI RIVER.

GENERAL SKETCH.

By G. K. GILBERT.

This is for the most part an area of Paleozoic strata lying nearly horizontal. The Coal Measures constitute the underlying rocks from Bellaire on the Ohio River to Newark, Ohio. The lower members of the Carboniferous constitute the underlying beds from Newark to a point just west of Chicago Junction, Ohio. From this point to Tiffin, Ohio, the Devonian rocks underlie. From Tiffin to a point west of New Baltimore the region is underlain by Upper Silurian rocks. From New Baltimore, Ohio, to a point west of Union Mills, Ind., the Devonian rocks again constitute the substructure; and from the last-named point to Watertown, Wis., the Upper Silurian terrane underlies the drift. From Watertown, Wis., for a few miles the Lower Silurian rocks come to the surface, except where concealed by drift, and thence to La Crosse the Cambrian rocks appear. The disturbance of these strata throughout the region of the route is so slight that it nowhere expresses itself in well-marked surface features.

The greater part of the route is through a region overspread with glacial drift, the southern part of the great North American glaciated area. From near Newark, Ohio, to Kilbourn City, Wis., the older formations are generally concealed by the drift deposits. The surface features by which this region is characterized and by which it is distinguished from the territory south of the limit of glaciation include an imperfect drainage system abundant in lakes, lakelets, ponds and marsh meadows, and the replacement of the topographic forms developed by subaërial erosion by the less regular but equally characteristic forms of glacial deposit.

The layer of drift ranges from a few feet to several hundred feet in thickness, and its most important element is till or boulder-clay. It is exhibited in a general sheet of ground moraine, traversed by relatively thick bands of the nature of terminal moraines. The moraines are usually several miles broad and their ensemble is not commanded by the tourist, who is able only to observe changes in the topographic character of the country as he enters and leaves their belts. The surfaces of terminal and ground moraines are alike undulatory, but the slopes of the terminal moraines are relatively steep. Hillocks, often somewhat elongated, alternate rapidly with hollows for the most part

289

undrained. The more elongate hills of the terminal moraines are apt to lie parallel to the course of the moraine; the hills of the ground moraine usually trend parallel to the direction of ice movement. Associated with the terminal moraines are many hillocks and abrupt ridges of water-worn and assorted gravel and sand, and sometimes these water-wrought materials sheathe parts of the moraine where its core is of boulder-clay.

In low and level tracts the *till* proper is frequently covered with sand or silt. Many such deposits were made in temporary lakes, accumulated between the ice on the one hand and a terminal moraine or higher land on the other. Many such temporary lakes accompanied the recession of the ice sheet, and have left small lacustrine plains to mark their former existence. Lacustrine plains of much greater extent, but due to the same causes, lie in the basins of the Great Lakes, and are traversed by the train in western Ohio and in the neighborhood of Chicago.

The moraine, first fully and accurately described in America, and which received the designation "kettle moraine,"[25] delimits a phase of the great ice sheet in which its margin was divided into lobes, each one a large glacier. Between Newark and Kilbourn City the route crosses the area overspread by the Scioto, the Maumee, the Saginaw Bay, the Lake Michigan, and the Green Bay lobes.

At Kilbourn City, Wis., the line of travel leaves glaciated territory and enters the driftless area[26] of the Upper Mississippi Basin. Thence to La Crosse the topography and the constitution of the surface material stand in sharp contrast to the corresponding features of the region farther east. Rock exposures, which have been rare eastward, and altogether wanting over considerable areas, are here of almost constant occurrence wherever the surface has any considerable relief. Frequently, too, butte-like hills or fantastically carved, castellated towers of sandstone give some indications of the extent of the subaerial erosion the region has suffered. From the presence of these bold eminences within the driftless area, rising 200 or 300 feet above the more or less completely base-leveled plain on which they rest, and from their absence in the area covered by ice, instructive inferences may be drawn as to the work effected by the ice in the country over which it passed. Before La Crosse is reached, the bluffs in the immediate vicinity of the Mississippi are capped with a thin sheet of loess, but this does not attain great thickness east of the Mississippi river in this latitude.

FROM THE OHIO RIVER TO CHICAGO.

ITINERARY.

By Edward Orton.

Station.	Dis-tance.		Elevation.		Popula-tion.	Stations.	Dis-tance.		Elevation.		Popula-tion.
	Miles.	Kilometers.	Feet.	Meters.			Miles.	Kilometers.	Feet.	Meters.	
Bellaire............	0	0	635	194	9,934	St. Joe	304	480	815	248	....
Barnesville........	24	39	1,276	389		Auburn Junction..	314	505	874	266	....
Cambridge.........	49	79	842	257	....	Garrett............	317	510	891	272	....
Zanesville	75	121	742	226	21,009	Avilla	322	518	961	293	....
Newark	101	163	868	265	14,270	Albion............	332	534	926	282	....
Mount Vernon......	126	203	991	302		Cromwell	342	550	937	286	....
Mansfield..........	163	262	1,154	352	13.473	Syracuse	350	563	869	265	....
Chicago Junction...	189	304	930	283	....	Milford Junction ..	355	571	842	257	....
Tiffin	213	343	758	230	10,801	Bremen............	371	597	820	250	....
Fostoria	226	364	777	237		La Paz Junction...	379	610	857	261	....
Bloomdale	233	375	755	231	....	Walkerton Junc-tion...............	388	624	718	219	....
Walker	238	383	....	...		Wellsboro	403	649	753	230	....
North Baltimore....	239	385	740	226	....	Allda	410	660	....	...	....
Deshler............	251	404	720	219	....	Suman	416	609	746	227	....
Holgate...........	264	424	721	220	....	Willow Creek	426	686	....	...	....
Defiance...........	277	446	711	217	....	South Chicago.....	428	690	589	180	....
Sherwood..........	284	457	....	...		Chicago	460	740	589	180	1,099,850
Hicksville	297	478	761	232	....						

From **Bellaire**[19.20] for about 35 miles (55 km.) northwestward, the road traverses the northern edge of the great Pittsburg coal basin, the beds of which are rising to the northwest, 10 to 15 feet to the mile (2 to 3 m. to the km.). The coal seam has deteriorated in quality as it has been followed westward into Ohio. It still has good thickness (6 feet) and it preserves the coking quality which gives to it such great value in the Connellsville district of western Pennsylvania, but the percentage of sulphur has been so greatly increased that it no longer furnishes an iron-making fuel and for the same reason it is not acceptable as a gas-making coal. The seam lies 100 feet (30 m.) above the river near **Bellaire**. The geological section furnished by the river hills at this point covers the entire Upper Productive Coal Measures and reaches well up into the Upper Barren Measures. At no point in Ohio is there a better exposed natural section of this part of the geological column.

291

There are many horizons of limestone in these two divisions of the Coal Measures. The limestones are of fresh-water origin in the main, are readily soluble in atmospheric waters, and give rise by their decay to soils of unusual fertility. It thus happens that the hills of this portion of the State rank as high in agricultural wealth as the plains and valleys of the western half of the State. Some of the limestones of this series yield natural hydraulic cements of fair quality.

In the vicinity of **Barnesville**[21,22] the highest rocks of the Ohio scale are found. They have been proved by Fontaine and White, by means of their flora,[24] to have a Permian aspect, but having much in common with the underlying division they are generally recognized as a transition series under the name Permo-Carboniferous.

At **Cambridge** the railroad reaches the uppermost seam of the Lower Coal Measures of the State, the Upper Freeport Seam. A coal field of great importance begins at this point and extends for many miles to the southward and southeastward. Within the limits of this field the deepest-lying coal of the State is likely to be found. The seam has been followed by the drill to a depth of at least 500 feet (152 m.) below the valley levels to the southeastward. This coal has an excellent reputation as a steam coal. It has a maximum thickness of 7 feet, but, more than any other Ohio coal, it is liable to disastrous "wants" or "cutouts," due to its invasion by the great Mahoning sandstone, which is the next higher element in the normal section. A low anticline traverses the series near **Cambridge** which has proved effective in the accumulation of gas and oil to a small extent. The reservoir rock is the Berea grit, which here lies about 1,000 feet (304 m.) below the surface. A very important oil field is found at Macksburg, 30 miles (48 km.) south of **Cambridge**.

At **Zanesville** [17,20,21] two of the most important coal seams of the lower measures, the Lower and Middle Kittanning seams, are extensively mined. They underlie much of the town and are well developed in the surrounding country. A score of miles to the southwestward they constitute the Hocking Valley coal field which is now producing 5,000,000 tons annually and which is decidedly the most important coal field of the State.

Fire clays and shales are found associated with the coal seams in large amounts. They are proving far more valuable than the coals themselves, constituting as they do an excellent basis for the manufacture of building brick of all qualities, of paving brick (vitrified), of earthenware, and finally of encaustic tiles of the finest quality. The American Tile Works established here are the largest of their kind in the world, employing a thousand men and furnishing a product that competes successfully in the New York markets with the best grades of imported floor and ornamental tiles. **Zanesville** is the most important clay manufacturing center in the State in the several lines above noted.

To the west of **Zanesville** and extending to the north and west as far as **Chicago Junction**[21] the Sub-Carboniferous series, consisting of conglomerates, sandstones, and shales, and containing an abundant fauna, admitting of several distinct subdivisions, occupy the country traversed by the railroad. The conglomerate portion of the series has been cut through by the Licking River between **Zanesville** and **Newark** and is well exposed in the picturesque gorge traversed by the railroad between these two points.

Near **Newark** the great terminal moraine[20] which marks the southern extension of the Drift is crossed and the drift-covered area is entered.

The sections exposed between **Newark** and **Chicago Junction** agree with those last named, but the rocks appear at comparatively few points.

At **Chicago Junction** the basal rocks beneath the drift belong to the black shale division of the Devonian. The drift beds are here 100 or more feet in thickness.

Near **Tiffin**[17,21] the first exposures of underlying Devonian and Silurian limestones are met with. They are found mainly in the valleys from which the mantle of the drift has been removed by erosion. Here begins the low anticlinal arch upon which the great accumulation of gas and petroleum in northwestern Ohio has been proved to depend. The central portion or summit of the arch is crossed between **Fostoria**[21] and **North Baltimore.**[22] Its structure is shown in the accompanying diagrammatic sections, (Figs. 8–10) the extremes of which are about 30 miles (48 km.) apart on the approximately north and south line of the axis of the arch.

The surface rock of the gas-producing portions of the arch is Niagara (Wenlock) limestone; of the oil-producing portions, the surface rocks are mainly Lower Helderberg (Ludlow) limestones.

The oil and gas districts are flat-lying regions, the extremes of the surface elevations of several hundred square miles not exceeding 50 feet (15 m.). The general level may be taken as 750 feet (225 m.) above tide. The reservoir of gas and oil lies 1,000 to 1,300 feet (300 to 400 m.) below the surface. At **Fostoria** and again at a point 6 miles beyond **North Baltimore**, the surface of the oil-bearing rock is nearly 500 feet (150 m.) below tide level. At **Bloomdale**, on the axis of the arch, it is 300 feet (90 m.) below tide. The measurements afforded by the drill indicate a gentle arch in the producing bed; when it is reached at 500 feet or more below tide, it is found charged with salt water. From 400 to 450 feet below tide, it carries oil; at higher levels gas is found.

The gas from this, and other adjacent and altogether similar fields in Ohio and Indiana, is transported by pipe lines to Toledo, Tiffin, Sandusky, Detroit, Fort Wayne, Indianapolis and scores of smaller towns,

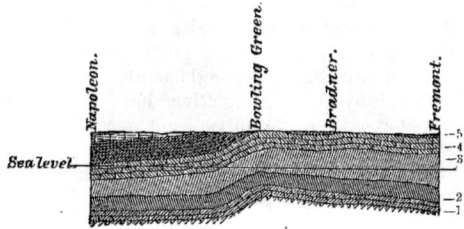

Fig. 8.

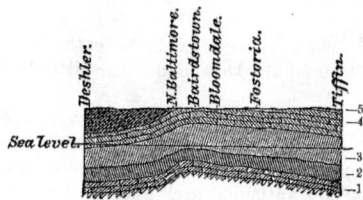

Fig. 9.

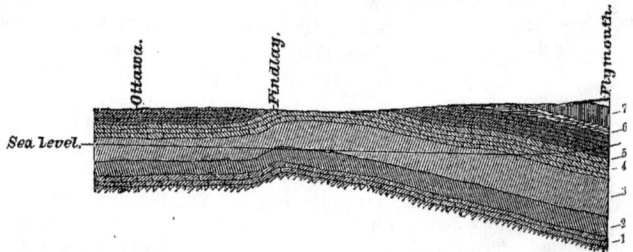

Fig. 10.

Sections across the oil field of Northwestern Ohio. (U. S. G. Survey. 8th Ann, Rep., p. 604.)

7. Ohio shale.
6. Upper Helderberg limestone.
5. Lower Helderberg limestone (Ludlow).
4. { Niagara { limestone. { shale. { Clinton limestone (Wenlock).

3. { Hudson River shale. { Medina shale.
2. Utica shale.
1. Trenton limestone (Bala).

where it is used on a very large scale for manufacturing purposes and especially for domestic fuel. It is sold to the consumer for the latter use at rates ranging from 5 to 25 cents per thousand cubic feet. It renders its most valuable service in the last-named capacity, introducing into the house the same convenience in heating and cooking that artificial gas does in lighting. The fire is always "laid," is kindled by a match, is regulated to any required point below its maximum by a cock, brings no soot or dust, and in short is the perfection of fuel. It is obviously the standard to which the fuel of the future will be obliged to conform. To this must be added that in the great majority of the favored towns, which it supplies, it is sold at a lower price than the coal which it displaces.

It brings equal advantage to the various lines of manufactures to which it is applied. In glass making, particularly, it introduces such economy and such improvement of product, that, other things being equal, competition with manufacturers that command it is very difficult on the part of those who depend on bituminous coal for fuel. It is a reproach to the intelligence of the communities that have obtained access to large supplies of gas to be obliged to add that, for several years after its discovery, it was generally furnished free to manufacturers and without measurement of any sort. This policy was entered upon in order to induce manufacturers to locate their plants in these towns. A single town has brought in establishments of this sort, mainly glass houses, enough to lead to the consumption of 30,000,000 cubic feet per diem (800,000 cu. m.) for the last three years.

But no gas field is able to bear such a strain as this amount of use brings; consequently there has been a rapid decline in the production of the wells, and it is now evident to all that the end of the large production is not far off. Five to eight years prove the limit for all sections that have found gas in large quantity and that have used it in the wanton way above described. The decline is most clearly revealed by the fall of pressure in the wells. The original pressure, due to the salt water contained in the same porous rock that holds the gas and oil, and falling under the laws of artesian wells, ranged from 400 to 500 pounds to the square inch (27 to 30 atmospheres).

Petroleum is also produced in large amounts from territory adjacent to the gas fields. The daily yield of the entire territory is between 40,000 and 50,000 barrels (of 42 gallons, or 158 liters each). The finest of illuminating and lubricating oils are made from the crude oil, and it also yields an unusually large percentage of paraffin. Both gas and oil contain a considerable percentage of sulphuretted hydrogen and probably other compounds of sulphur, and are consequently offensive in their natural state; but complete success has been attained in the deodorizing of the oil by several processes. Single wells, 5$\frac{5}{8}$ inches (13 cm.) in diameter, have produced oil at as high a rate as 8,000 barrels per diem, when first drilled. One well recently finished is even said to rival the great fountains of Baku.

Bloomdale. This village is in the center of the original dry gas territory. The gas rock is struck at 1,065 feet (320 m.) below the surface, or 310 feet (93 m.) below tide. The wells in this immediate neighborhood yielded, when the field was fresh, between 3,000,000 and 4,000,000 cubic feet (80,000 to 100,000 cu. m.) per diem. The original pressure was 440 pounds to the square inch (29.3 atmospheres). The largest gas wells of the field have produced not less than 30,000,000 cubic feet (800,000 m.) in twenty-four hours, but such flows are short lived. Wells of this character are invariably overtaken by oil or salt water within a few days or weeks, if allowed to flow unrestrained.

North Baltimore.[21.22] On the east side of the town dry gas was originally found, but on the west side a great oil field has recently been developed. Hundreds of derricks can be seen from the cars. In the oil wells the Trenton limestone is struck at 1,190 feet (360 m.) below the surface, or 450 feet (145 m.) below tide. Three miles further west, the Trenton limestone sinks to 500 feet (150 m.) and more below tide, and salt water occupies it everywhere, except in narrow ridges that are sometimes found in which a little gas or oil is contained.

These towns of the oil region all stand upon lacustrine deposits belonging to the period of the melting of the great ice sheet (Champlain). When the ice front occupied a position in the basin of Lake Erie a large body of water was held between it and the uplands to the south and west. At first this lake found outlet westward to Fort Wayne, Ind., but afterwards as the ice retreated further, other avenues of escape were opened and the water level fell, recording its position at different horizons by beaches. A beach line is crossed at Fostoria and another near Defiance, and still another near Hicksville. These old beaches, composed as they are largely of sand and limestone gravel, give rise to light and warm soils and sometimes to soils of extreme fertility.

Defiance.[18.21] The Devonian black shale is found on the western slope of the arch at this point, though generally concealed under heavy beds of boulder clay. The eastern outcrops of the shale were left at **Chicago Junction.** The drift beds seldom fall below 100 feet (30 m.) in thickness in this portion of the State, except in the river valleys, and they often rise to 200 and 300 feet (60 and 90 m.) in thickness. One section to the southward showed 530 feet (160 m.) of drift without reaching bottom.

Between **Hicksville,** Ohio, and **St. Joe,** Ind.,[23] a narrow moraine line is crossed.

At **Concord** is a morainic line, perhaps to be regarded as a part of that just east of **St. Joe,** separated from it by the valley of the St. Josephs River.

At **Auburn** the drill has revealed the presence of a slight fold in the deep-lying Trenton limestone, sufficient for the accumulation of gas and oil to a small extent. Despite the monotonous aspect of the present drift-covered surface, the underlying rocks have recently been found to be more broken by small folds and fractures than any other parts of the two states included in this review. Four or more wells, more than 2,000 feet (610 m.) in depth, have been drilled within the limits of **Auburn**. The gas found in the productive rock has a very high initial pressure. The first flow of the best wells showed more than 1,000,000 cubic feet per diem as the daily yield. The gas has been utilized in the town to the full extent.

The drift beds exceed 200 feet (60 m.) in thickness here. The first rock stratum reached beneath the drift is the basal portion of the black Ohio shale (Devonian). In the Lower Helderberg limestone (Upper Silurian), which has a thickness of about 600 feet (183 m.), considerable deposits of gypsum and anhydrite are found about midway in the series. At or about this same point in this series very important beds of rock salt are found in Ohio, Michigan, and Canada. These deposits are now being worked on a large scale in all three places. A valuable deposit of gypsum occurs in the same horizon in Ohio.

From **Auburn** to **Garrett** the basal rock is the black shale previously reported. The drift is nonmorainic.

At **Garrett** and just west of Garrett one line of the terminal moraine is crossed.

At **Albion** another belt of the terminal moraine is crossed. The region about **Albion** illustrates well the lack of drainage which is common along the terminal moraine. Numerous shallow lakes, ponds, and swamps are found here. A single lake will sometimes occupy several square miles of the surface.

Between **Teegarden** and **Walkerton** the line traverses the terminal moraine, the black slate of the Devonian still underlying the drift.

From **Walkerton** to **Coburg** is modified drift underlain by the Devonian to **Union Mills**, and by the Niagara limestone (Upper Silurian, Wenlock) from **Union Mills** to **Coburg**.

At **Suman** an inner morainic line of the Lake Michigan glacier is crossed. The morainic belt extends westward as far as **Woodville**.

From **Woodville** to **Chicago** the underlying rock is Niagara limestone, buried beneath modified drift, deposited in Lake Michigan during its expanded condition, in late glacial and postglacial times. Through a part of this region the sands of old beaches have been extensively modified by the action of the winds, forming in places conspicuous dunes. The road follows the southern and southwestern margin of Lake Michigan during the last 25 miles (40 km.).

The Niagara limestone which has been repeatedly named in the preceding account is extensively quarried at several points near Chicago. It is the Guelph division of the Niagara which is found here. In composition it is a typical dolomite of great purity. It was originally a highly fossiliferous limestone, and many species new to science have been described from the internal casts of the several divisions of the Mollusca that occur here.

South Chicago, a few miles beyond the Illinois State line and 11 from the Central Station, is already within the city limits. The road now bends to the westward, a mile or more away from the lake, to **Parkside,** and then runs north along the west side of Jackson Park, which occupies a mile and a half of the immediate shore line, and is the site of the Columbian Exposition. Beyond Jackson Park it again reaches the shore, which it follows for 6 miles to the Baltimore and Ohio station situated on the borders of Lake Park, which occupies a mile of the lake front opposite the central business portion of Chicago.

FROM CHICAGO TO THE MISSISSIPPI RIVER.

ITINERARY.

By S. F. EMMONS.[*]

Station.	Distance.		Elevation.		Station.	Distance.		Elevation.	
	Miles.	Kilometers.	Feet.	Metres.		Miles.	Kilometers.	Feet.	Metres.
Chicago..................	0	0	589	180	Portage City.............	188	303	813	248
Western Union Junction	62	100	....	....	Kilbourn City	205	330	897	273
Milwaukee a.............	85	137	583	178	New Lisbon	221	356	894	272
Brookfield	109	175	....	....	Tomah...................	240	386	906	294
Pewaukee	115	185	....	....	Sparta	257	414	798	243
Oconomowoc.............	128	206	....	....	La Crosse b..............	283	455	657	200

a Population, 204,468.   b Population, 25

Chicago is the second city in size and in commercial importance in the United States, and probably the greatest railroad center in the world. Its population in 1890 numbered about 1,100,000 souls, an increase of 118 per cent over that given by the census of 1880. Its streets are laid out on the rectangular system adopted by all modern American cities. The first eleven east and west streets from the Chicago River southward have names (Lake, Randolph, Washington, Madison, Monroe, Adams, Jackson, Van Buren, Harrison, Polk, and Taylor streets);

---

[*] Geological data furnished by Prof. T. C. Chamberlin and Prof. R. D. Salisbury.

beyond they are numbered from Twelfth upward to Ninety-fifth street, which runs through South Chicago. The area covered by the named streets and by the nine principal north and south streets or avenues, next westward from Lake Park, comprises the principal business portion of the city, in which are its most imposing buildings. It is inclosed on the north and west by the so-called Chicago River, an inlet of the lake, which has been artificially deepened and connected with the Illinois and Michigan Canal. This runs southwestward into the Illinois River, so that the overflow of Lake Michigan through it reaches the Mississippi River. The site of this portion of the city was originally only about 7 feet above the lake level, but was artificially raised in 1885 to 14 feet.

The great fire of 1871, which started among wooden buildings in the western portion of the city under the influence of a strong southwest wind, destroyed nearly every building over an area of 2,124 acres (858 hectares), in less than twenty-four hours. This area included a great part of the massive buildings of stone and brick in the business center of the city, so that it is estimated that the value of the buildings alone that were destroyed was $53,000,000, while the total loss was $196,000,-000. It was rapidly rebuilt, and the enormous blocks which have risen over the burnt district are among the finest specimens of commercial architecture in the country.

Chicago's water supply is mainly obtained from Lake Michigan, though artesian water is also used. Lake Michigan water is pumped through two tunnels 66 feet below the shore level and extending 2 miles out into the lake, where they connect with a vertical iron cylinder 64 feet (19 m.) high, inclosed in a crib of iron and wood loaded with stone. The water capacity of these tunnels is 150,000,000 gallons (550,000,000 liters) per 24 hours.

Chicago has spread rapidly, especially in the last decade, out onto the surrounding prairies, where enormous manufacturing and commercial establishments form small cities within themselves. Such are the great Union stock yards, the Pullman works, etc. It has also provided liberal breathing space for its rapidly-growing population in an elaborate system of parks. Six of these cover from 180 up to 590 acres (44 to 240 hectares) each, besides which are numerous open squares and smaller parks within the older portions of the city.

From Chicago to Milwaukee the route lies parallel to the shore of Lake Michigan, and within the area of lacustrine deposition of the Champlain epoch, when the boundaries of lake Michigan were much extended. Along this line the Champlain deposits reach westward 8 to 12 miles (13 to 19 km.) from the lake. The deposits immediately along the line of the railway are the "gray pebble clays" of the Wisconsin geological survey. The whole of this region is underlain by Niagara limestone.

At **Milwaukee** the surface formation is the gray pebble clay of Champlain age. Within the limits of the city Devonian (Hamilton) strata occur. This is the only area of Devonian rock within the boundaries of Wisconsin. The rock is an impure limestone, from which an excellent hydraulic cement is made. The Niagara limestone, upon and against which the Devonian lies, is here of such a character as to afford an excellent building stone, and is extensively quarried for that purpose.

At **Wauwatosa**, also, there are extensive quarries in the Niagara limestone, similar to those in Milwaukee. The gray pebble clay, of Champlain age, extends somewhat west of Wauwatosa.

From **Wauwatosa** to **Pewaukee** till (ground moraine) is well shown, underlain by Niagara limestone.

Pewaukee to **Oconomowoc.** About midway between these two stations (from Hartland to Nashotah) the Kettle moraine is crossed,[25] and here its characteristic topography, its accompaniment of lakes and its constitution, may be seen. Between these stations also the Hudson River shale appears from beneath the Niagara limestone, but is concealed by the drift. The Niagara limestone dips to the eastward, and the Hudson River shale is the next subjacent formation and appears at the surface, except for the drift covering, as the Niagara thins out to the westward. Before Oconomowoc is reached, the Galena (Lower Silurian) limestone appears from beneath the Hudson River shale, as the Hudson River shale appeared from beneath the Niagara. The region about Oconomowoc is famous for its lakes. The moraine crossed between **Pewaukee** and **Oconomowoc** is the joint product of the Lake Michigan and Green Bay glaciers, the ice from the Lake Michigan glacier moving westward and the ice from the Green Bay glacier moving eastward. The two ice movements met along the line of this moraine, which is therefore an interlobate moraine, terminal to two ice movements in opposite directions.

Between **Oconomowoc** and **Columbus** the ground moraine finds one of its typical phases of development. Particularly in the region about Watertown, and northwestward from Watertown to Columbus, drumlins are well developed. The drumlins have here a northeast-southwest trend. This corresponds to the direction of ice movement in this region, as shown by the striæ. Associated with the drumlins are undrained or ill-drained depressions which have an extension in the same direction as the drumlins themselves. Between **Watertown** and **Columbus** the underlying strata are the Trenton (Lower Silurian) limestone, the St. Peter's sandstone, and the Lower Magnesian limestone, in the order named. The strata all dip slightly to the southeastward, and each formation appears from beneath the one preceding.

From **Columbus** to **Rio** is ground moraine, underlain by Lower Magnesian limestone, and from **Rio** to **Portage**, drift, underlain by Potsdam (Cambrian) sandstone. At **Portage** the Wisconsin River is reached, and it is followed to **Kilbourn City**, through Potsdam sandstone, overlain by drift. Just before **Kilbourn City** is reached the moraine formed at the western margin of the Green Bay glacier is crossed. At **Kilbourn City** the railway crosses the river at about the center of that portion of its valley known as the Dalles of the Wisconsin. The river here flows through a canyon-like gorge, 50 or 60 feet deep, carved out of the Potsdam sandstone. The tributaries to the Wisconsin in this part of its course have cut for themselves deep gorges and glens leading down to the main stream. From the railway line near **Kilbourn City** may be seen isolated elevations of Potsdam sandstone, which represent erosion remnants of the once more extensive Potsdam formation.

Kilbourn City to **Camp Douglas**. Crossing the river at **Kilbourn City**, the driftless area [26] is entered. The sharp erosion forms, the absence of drift topography, and the frequency of rock exposures combine to place this region in sharp contrast to that farther east. In the vicinity of **Mauston**, **New Lisbon**, and **Camp Douglas** fine castellated outliers of Potsdam Sandstone may be seen resting on a lower and often nearly level plain of the same formation.

Camp Douglas to **Sparta**. Between these stations the Potsdam sandstone is the only formation which appears. The divide between the Wisconsin and the Mississippi is here crossed, and from the summit the topography of the driftless area may be well seen.

From **Sparta** to **La Crosse** the route lies in the valley of the La Crosse river. The valley is carved from the Potsdam sandstone, which forms the bluffs on either hand. In the lower part of the valley the summits of the bluffs are capped with lower Magnesian limestone. Loess also occurs on the summits of the bluffs fronting the Mississippi, and on the bluffs bounding the La Crosse valley, near its junction with the Mississippi.

THE STATE OF MINNESOTA.

By ULY S. GRANT.[*]

Topography.—The topographic features of Minnesota may be briefly summed up, for its western three-quarters, as being a moderately undulating, sometimes nearly flat, and occasionally hilly expanse. The only exceptions to this topographic character are the northeastern and southeastern corners of the State, the former being almost mountainous, and the latter deeply eroded valleys, often destitute of drift. The northwestern corner is a part of the valley of the Red River of the North, and is extremely flat and monotonous. It is made up superficially of deposits from the glacial lake, Agassiz, and well represents an area which is topographically young, the river courses having but just begun to cut shallow narrow channels in the unconsolidated materials. The extreme northeastern part of the State has little drift, and is very rough and broken in outline, the rugged summits of hills of crystalline rocks giving many bold features to the landscape. In the southeastern corner the Mississippi river and its tributaries are inclosed by bluffs from 200 to 600 feet (61 to 183 m.) in height; these bluffs consist of nearly horizontal Paleozoic rocks.

The average elevation of the State is somewhat over 1,200 feet (366 m.) above sea level. The only part that can be termed at all mountainous is the district between Lake Superior and the Canadian boundary. Here three parallel ranges of hills trend northeastwardly, and rise in some places to an altitude of 2,200 feet (671 m.). The most broken country is that immediately bordering on Lake Superior, where, in from 2 to 10 miles (3 to 16 km.) back from the shore, the land rises over 1,000 feet (305 m.). The region around the sources of the Mississippi has an altitude of 1,700 feet (518 m.), and in the northwest-central part of the State a gentle range of hills (the Leaf hills) rises to a height of 1,750 feet (533 m.). The Coteau des Prairies, an elevated region in the southwestern portion of Minnesota, is from 1,800 to 1,900 feet (549 to 579 m.) above sea level.

Lakes and rivers.—The water area of Minnesota is larger than that of any other State of the Union, being 5,637 square miles,[†] or nearly one-fifteenth of the whole area of the State. The total number of lakes

---

[*] Compiled largely from the reports of the Geological and Natural History Survey of Minnesota.

[†] This estimate does not include any part of Lake Superior.

is about ten thousand, and they are scattered over nearly the entire surface of the State, but the larger proportion are confined to the morainic areas. These lakes are of all sizes, from small ponds to bodies of water 12 miles (19 km.) or more in diameter. Their shores are mostly of drift, and rarely does the underlying rock appear anywhere in their basins. This is not true of those lakes in the district between Lake Superior and the Canadian boundary, which are for the most part rock bound.

There are three drainage systems. The most important is that whose waters find their way into the Gulf of Mexico through the Mississippi, whose largest tributary in the State is the Minnesota river. This basin includes fully two-thirds of the area of the State; the whole of the southern half, the northern-central and the central-eastern part of the State belong to it. The next largest is the Hudson Bay system, including the Red River of the North, which drains the northwestern portion, and the Rainy river which drains a small strip along the northern edge of the State. The third system is that of the Gulf of St. Lawrence; this is by far the smallest of the three systems and includes only the streams in the immediate neighborhood of Lake Superior.

Flora.[27]—The most important and conspicuous contrast presented by the vegetation covering different parts of Minnesota is its division into forest and prairie. Forest covers the northeastern two-thirds of the State approximately, while about one third, lying at the south and southwest and reaching in the Red River valley to the international boundary, is prairie. The line dividing these areas has an almost wholly timbered region on its northeastern side and on its southwestern side a region that is chiefly grass land, without trees or shrubs excepting in narrow belts along the larger streams and occasionally groves beside lakes. In the northern half of the State the timber consists largely of conifers, from which large amounts of lumber are annually taken. The deciduous forest exists for the greater part in the southern half of the State, and is composed chiefly of various species of oak, elm, linden, poplar, maple, and ash.

The geological column.—In the State of Minnesota there are to be seen rocks from the fundamental complex of gneiss and schist (Archean) up to those which are being formed at the present day. There are, however, several wide breaks in the geological column where formations of great importance are entirely lacking. In the Paleozoic, Carboniferous and Permian are without representatives. In the Mesozoic, the Triassic and Jurassic are completely wanting, the only Mesozoic rocks that are known being a few isolated areas of Cretaceous strata, very limited in extent. The Tertiary is also entirely lacking.

Below is given a brief account of those formations found within the limits of the State.

Group.	System.	Series.
Cenozoic............................	Pleistocene	Recent Glacial.
Mesozoic	Cretaceous	
	Devonian............................	Corniferous.
	Silurian	Hudson River.
		Galena.
		Trenton.
Paleozoic		St. Peter.
		Shakopee.
		Richmond.
	Cambrian (Ozark series)............	Lower Magnesian.
		Jordan.
		St. Lawrence.
		St. Croix.
Algonkian (Taconic)	Keweenawan (Nipigon)	
	Huronian............................	Animikie.
Archean	Laurentian?	Keewatin.
		Vermilion.
		Gneiss, etc.

The Archean is made up of semicrystalline and crystalline schists, gneisses and granites, and basic eruptives.   The semicrystalline schists are mostly sericitic and argillaceous in character and, together with a large number of graywackes and greenstone schists, compose the Keewatin series.   In this series are found extensive beds of very pure hematite, which supply an excellent ore for steel-making.[28]   The crystalline schists are micaceous and hornblendic and lie below the less crystalline; they belong to the Vermilion series, as described in the reports of the Minnesota survey.   Still lower are gneisses and granite. The entire Archean has been broken through by ancient eruptives, mostly granites and quartz-porphyries, and by more recent diabases. The Archean has its greatest development in the region along the international boundary north and east of Vermilion Lake.   Here the strata strike in a general east-northeast direction, and the dip varies only a few degrees from the vertical.

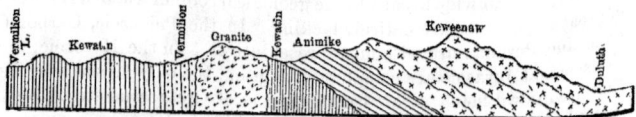

Fig. 11.—General section from Vermilion lake to Duluth.

This formation extends southwestwardly through the State, out-crops being occasionally seen, especially in the upper Minnesota valley, where, however, the three divisions described above are not recognizable. But, as a rule, west and southwest from Duluth these rocks are hidden by the drift.

The Algonkian (or Taconic of the Minnesota survey) is seen in its greatest development in the country lying just north of Lake Superior. It is separated into an upper and lower division—the Huronian and Keweenawan.

The Huronian in Minnesota is represented by dark carbonaceous slates and ferruginous quartzites. A belt of these rocks stretches from the extreme eastern end of the State west and southwest to the vicinity of Duluth and to the Mississippi river. They lie just south of the Archean complex and rest on its upturned edges. The dip is at low angles toward Lake Superior. After going a few miles northwest of Duluth the Huronian is seldom seen until reaching Pokegama Falls on the Upper Mississippi, and west of this these rocks are completely covered by glacial deposits. The Huronian in Minnesota is included under what has been described as the Animikie on the north side of Lake Superior.

The Keweenawan consists chiefly of interbedded sheets of intermediate and basic eruptives, with a few clastics. These rocks occupy the territory south of the strike of the Huronian and north of Lake Superior. They are seen in great development in the vicinity of Duluth and all along the lake shore to Pigeon Point.[29] The principal rock type is a very coarse-grained, gray gabbro, which has a geographical extent of 1,000 square miles (260,000 hectares). It is used somewhat for a building stone, and is well known as "the Duluth granite."

The Paleozoic.—To this belong the rocks of the entire southeastern and southern portions of the State. They are comparatively little exposed, except along the river courses.

The Cambrian extends along the Mississippi river and its tributaries from St. Paul to the southern limit of the State. It is especially well exposed in the bluffs along the Mississippi. The eastern part of the State, north of St. Paul and south of Duluth, is also occupied largely by Cambrian strata. A belt also extends from the last area southwestward along the southern border of the Archean to the extreme southwestern corner of the State. The rocks of this formation are mostly thick beds of magnesian limestone and of sandstone, but in the area lying in the southwestern corner of Minnesota an extensive quartzite occurs; associated with this is the famous Indian "pipestone" or catlinite.

The Silurian is seen overlying conformably the Cambrian in many places along the Mississippi river. It is represented only by a comparatively small thickness of strata, which are confined almost entirely to the Lower Silurian. Its chief member, the Trenton limestone, is seen in its best development along the Mississippi in the vicinity of

St. Paul and Minneapolis. The Silurian underlies the drift (occasionally appearing through it) of a considerable area in the southeastern portion of the State just west of the Mississippi.

The Devonian occupies a small portion of territory along the central part of the southern border of Minnesota. It is chiefly represented by the Corniferous (and Hamilton ?) limestones and is comparatively thin.

The Mesozoic.—The only Mesozoic strata found are small isolated areas of Cretaceous. These occur in basins in the older rocks and are exposed in a few places along river courses. The Cretaceous rocks contain many well-preserved fossil leaves, which have been described by Leo Lesquereux.[30]

The Cenozoic.—Nearly the whole of the State may be said to be drift-covered; the only exceptions are the extreme southeastern and the extreme northeastern portions. At any point on the northwestern boundary, as far east as Lake of the Woods, one may start southward and follow the Iowa boundary line without seeing any rocks *in situ,* except what he might encounter in crossing the valley of the Minnesota river, and the rare exposures of red quartzite in Rock, Pipestone and Cottonwood counties. East of this meridian he would encounter occasional exposures of such rocks along Rainy river, but southward from the northern boundary he would still have an almost equal scarcity of rock exposures, were he to set out again to the Iowa boundary line. The drift is so thick in the region of lakes Pemidgi and Winnibigoshish, and generally through the central portion of the State, that it does not afford rock exposures until reaching the vicinity of Motley. Rock is seen in scattered patches in Todd, Morrison, Mille Lacs, Kanabac, Stearns, Benton, and Sherburne counties, as well as at Pokegama Falls on the Upper Mississippi. But farther toward the south, except in the valleys of the Minnesota and Blue Earth rivers, the drift everywhere conceals the rock with an unbroken mantle from 100 to 200 feet, (30 to 60 m.) and sometimes 300 feet (91 m.) thick.

East of the meridian passing through the west end of Rainy Lake, the rock is more and more frequently seen projecting above the drift, both along the Iowa boundary and in the central and northern portions of the State, especially in the valleys of streams that flow eastward. There is a tract of the State heavily covered by drift east of Pokegama Falls, including the St. Louis Valley and its upper tributaries, in which many of the streams that enter Lake Superior in the State of Minnesota take their rise; but for the most part in the eastern half of the State the streams expose the rocks more and more frequently, indicating an attenuation of the drift sheet toward the east, so that at last they become continuously rock-bound. The drift fades out on the north toward the rock-bound shore of Lake Superior, as remarkably evinced along the international boundary, and on the south toward the equally ancient rocky valley formed by the St. Croix and the Mississippi.

Morainic accumulations are very extensively developed throughout the State. The most prominent are those of the eastern border of the glacial lobe which extended into central Iowa. This series of moraines enters the State about the center of the eastern boundary, runs south-westwardly to the vicinity of St. Paul and Minneapolis, and then turns and runs southward to the southern limit of the State. The moraines of the western side of this lobe cut across the southwestern corner of the State. Between these two major terminal moraines are many com-plicated net-works of minor moraines, formed as this great lobe of the glacier receded northward.

The "driftless area" is the only portion of the State that is entirely free from drift. This area, although confined mostly to Wisconsin, covers a considerable territory in the southeastern corner of Minnesota. Most of Houston and Winona counties and the eastern portions of Wabasha, Olmsted and Fillmore counties are included in this area. Here the topography is decidedly more rugged than in the adjoining drift-covered areas, and the rocks are exposed in bold outcrops.[26, 31]

The loess is well developed in the southeastern corner of Minnesota. It covers all of the driftless area in the State and extends a short dis-tance west of it, and also northward along the Mississippi nearly to Red Wing. Outside of the driftless area it is confined chiefly to the river valleys, extending along these much farther west than in the adjoining ground.

Lake Agassiz.[32]—This name has been applied by Mr. Warren Upham to a glacial lake which occupied the basin of the Red River of the North and extended over a considerable area in Manitoba. After the ice sheet had receded beyond the source of the Red River, the waters, formed from the melting of the ice, accumulated in a vast inland lake: the high ground to the south was the southern barrier of this lake, while on the north it was bounded by the edge of the retreating ice sheet. When the ice disappeared the present drainage system toward Hudson Bay was established. The area of this lake at the time of its greatest development was greater than that of Lake Supe-rior. In depth it varied much, reaching in some places over 400 feet (123 m.). The beaches of Lake Agassiz are well defined in Minnesota; they enter the western edge of the State at Lake Traverse, run east for a short distance, and then north, keeping parallel with the Red River of the North and usually within 20 (32 km.) miles of it; this northerly direction continues to almost the latitude of Red lake, where it sud-denly turns to the northeast and runs through Red lake to Rainy lake on the northern border of the State. The outlet of Lake Agassiz was the Minnesota River, which is now a small stream flowing in a gorge that could only have been cut by a volume of water many times the size of the present river.*

The recession of the Falls of St. Anthony.[33,34]—This subject is of interest, as it furnishes data for the calculation of the time which has elapsed since the glacial period. The results practically agree with those obtained from Niagara Falls and elsewhere, and furnish strong evidence that the disappearance of the ice sheet from North America was less than ten thousand years ago.

From the Falls of St. Anthony to Fort Snelling (see map), a distance of a little over 8 miles (13 km.), the Mississippi flows in a deep gorge about 1,000 feet (305 m.) wide. The rocks of this region are covered by drift averaging, where the river has cut through it, probably not more than 30 feet (9 m.) in thickness. Immediately below the drift occur the practically horizontal beds of the Trenton, with a thickness of from 20 to 30 feet (6–9 m). The upper and lower beds are of comparatively soft shales, but the greater part is of hard compact limestone. The Trenton lies conformably on the St. Peter sandstone, which is a pure sandstone of uniformly fine grain and, notwithstanding its age, has undergone very little induration. This very friable sandstone is nearly worn away and allows large blocks of the limestone to fall down, thus materially increasing the rate of recession of the Falls.

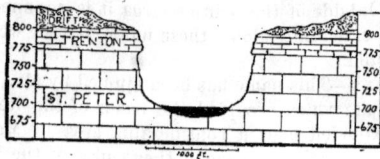

FIG. 12—Section across Mississippi River above Fort Snelling.

From the Falls to Fort Snelling the rock on the sides of the gorges has a freshly broken appearance, the large fragments thrown down by the action of the water on the easily crumbled sandrock, as the falls receded, still existing in the talus below the bluffs. Throughout this extent the strata are horizontal, the thickness of the drift sheet overlying them nearly uniform, and all other conditions, so far as they can be seen, that would affect the rate of recession, seem to have exerted an unvarying influence. The inference is that the rate of recession has been practically uniform between the two points named.

There is an aspect of age and long weathering presented by the rock in the bluffs of the Mississippi below Fort Snelling. It has a deeply changed color, a light yellow oxidized exterior, which marks all old

* Lake Agassiz is to be fully described by Mr. Warren Upham in a forthcoming monograph of the United States Geological Survey.

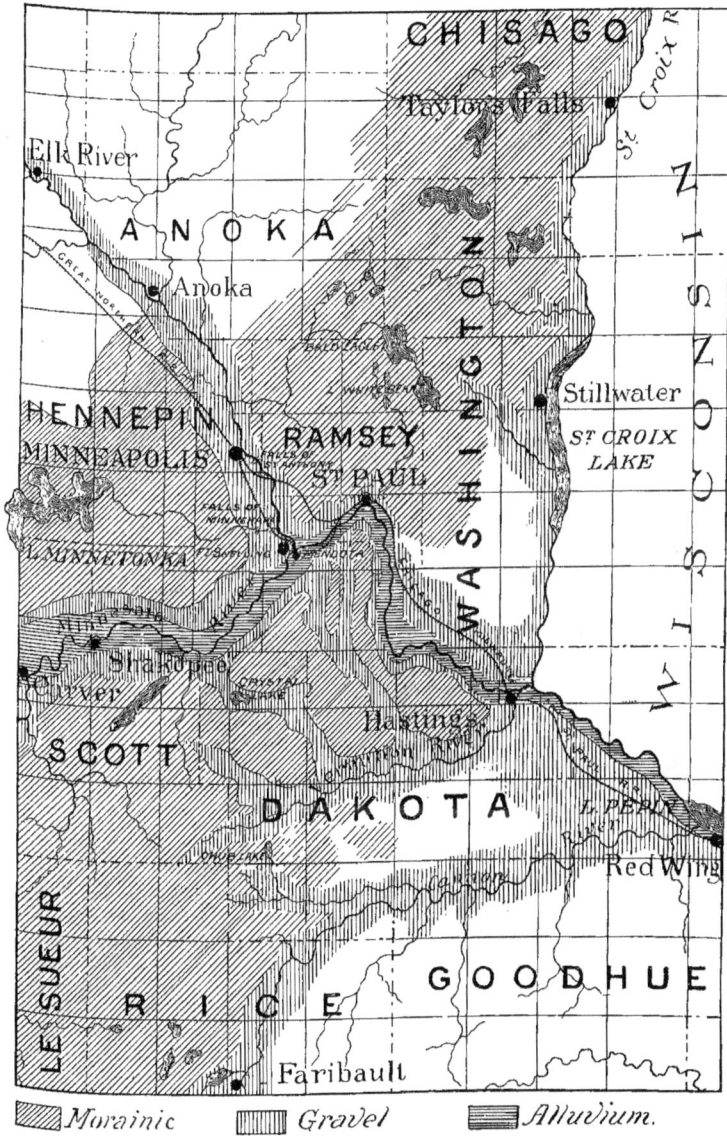

GEOLOGICAL MAP OF VICINITY OF ST. PAUL AND MINNEAPOLIS.

bluffs. This stained condition also pervades the rock at the mouth of Bassetts creek and at the quarries in the ancient river bluffs near the mouth of Shingle creek, on both sides of the river. (Both of these places are above the Falls.) Another notable difference between the bluffs above Fort Snelling and those below consists in the absence of caves and subterranean streams entering the river above Fort Snelling. Although the Trenton limestone exists in full force about St. Paul, in the bluffs east and north of the city, yet it has been cut through by some means prior to the drift so as to allow the entrance and exit of streams of water, at levels below its horizon, through the sandstone. None such are found above Fort Snelling, the surface drainage being shed by the limestone and precipitated over the brink of the gorge in several beautiful cascades. When such streams enter the river below Fort Snelling they either enter some subterranean passage and appear at the mouths of caverns in the sandstone, or as springs in the drift along the talus, or they find an ancient ravine down which they plunge by a series of rapids over boulders to the river level, rarely striking either the limestone or the underlying sandstone. Again, the rock

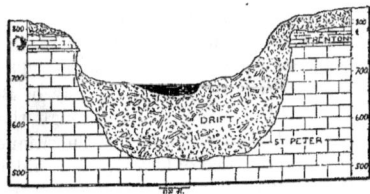

FIG. 13.—Section across the Mississippi River below Fort Snelling.

bluffs at St. Paul and everywhere below Fort Snelling are buried under the drift sheet. Their angles are sometimes seen jutting out from some wind-beaten corner, but nearly everywhere they are smoothed over by a mantle of drift and loam. Even the immediate river bank, where the lime rock should be intact, shows that it has been extensively disrupted, and its débris, often coarse and water-worn, in pieces from 4 to 10 feet long, is mixed with coarse boulders, gravel and drift at the height of 50 to 75 feet (15 to 23 m.) above the water level, the heterogeneous mass lying on the water-worn surface of the St. Peter sandstone.* But above Fort Snelling the upper edge of the limestone is intact all the way to the Falls and shows a fresh-cut section. It is surmounted by a continuous sheet of drift, which rises as a direct continuation of the bluff formed by the underlying rock. The individual strata of the drift show

---

*A good example of this is seen along the Chicago, Milwaukee and St. Paul Railroad tracks just west of the Union station at St. Paul.

that they were cut by the recession of the Falls in the same manner as the rock beds, and they do not conform in their undulations to the outline of the rock as if the gorge were present when they were formed, as in St. Paul. There is no spreading of loam over these cut edges, except what has fallen down from above at the time of their removal or subsequent to it. At Fort Snelling the direction of the Mississippi changes abruptly at a right angle, and the river enters the wider gorge in which the Minnesota flows, which gorge is entirely out of proportion with the amount of water that it carries. This gorge (of the Minnesota) continues in the same direction and with the same width beyond the confluence with the Mississippi, but takes the name of the latter stream; one mile below Fort Snelling this gorge is a mile and a half wide (2·4 km.).

These features of greater age, pertaining to the bluffs of the Mississippi below Fort Snelling, are seen in the old rock bluffs of the river above the mouth of Bassetts creek as far as Shingle creek. The rock there is deeply changed in color and is hid by the drift, and the bluffs, as left by the more ancient river, are far apart, the old gorge being three or four times as large as that between the Falls and Fort Snelling. These rock bluffs, consisting of the same limestone as that which at the Falls is below the water, here rise from 30 to 40 feet (9 to 12 m.) above the river and are buried under loam or drift and loam. This part of the old valley continues southwardly at first by way of Bassetts creek, across the western suburbs of Minneapolis, through the valleys occupied by lakes Calhoun and Harriet, and joins the Minnesota at some point above Fort Snelling, the precise locality being hid by a deposit of drift. It was cut down into the St. Peter sandstone over 100 feet (30 m.) at least, as shown by the well at the Sumner school house, and about 275 feet (84 m.), as shown by the deep well at the Lakewood cemetery. This would show that probably the ancient valley of the Minnesota, where it passes Fort Snelling, and all the way through Ramsey County and below, has been filled more than 200 feet (61 m.) by drift that originated since the excavation of the gorge. This supposition is borne out by all borings that have been made between the rock bluffs at lower points, as at West St. Paul and at Lake City on Lake Pepin. Such excavation is not found in the river gorge between Fort Snelling and the Falls of St. Anthony, the solid rock being found not more than 25 feet (7·8 m.) below the bottom of the water.

These facts warrant the conclusion that that part of the Mississippi gorge above Fort Snelling has been excavated by the recession of the Falls since the last general drift movement, and that prior to that event there was a gorge which passed from the present channel of the Mississippi, at the mouth of Bassetts creek, southward to the great gorge of the Minnesota at some place above Fort Snelling. It is probable that this old channel of the Mississippi was then occupied by waters that drained the northern part of the State, and had existed through many ages, dating back to pre-Cretaceous times. It seems to have remained open to the advent of the ice of the last glacial period, when morainic accumulations so choked it that the water of the river was driven out and compelled to seek another passage to the Minnesota. When this last event took place the Falls of St. Anthony probably began at Fort Snelling, the water being precipitated over the rock bluff of the pre-existing old gorge, unless, as was very likely the case, the whole valley was too deeply buried under water. Whether this was at the beginning or at the acme of cold, or at the recession of the ice, is a question to be considered, but at this time the only point that is claimed is that it was not earlier than the beginning of the last glacial epoch, and was very probably near the close of this epoch, for the following reason. It has been before stated that the Minnesota was the outlet of the glacial lake Agassiz, and as such must have carried away enormous quantities of water resulting from the rapid melting of the ice sheet toward the north. There is good reason to believe that this swelled the river to such a height that the level of its waters rose as high as, if not higher than, the limestone ledge at Fort Snelling. If such was the case there could have been no fall at that place. When Lake Agassiz began to be drained to the north, the Minnesota's volume of water was much decreased and the water level fell below the limestone ledge. Then the waters of the Mississippi began to be precipitated over the face of this ledge, and then, and not until then, the recession of the Falls began. If this reasoning is correct, the date of the beginning of the recession of the Falls could not have been until near the close of the glacial epoch.

From calculations based on the recession from 1680 to 1856 Prof. Winchell obtained from three sources the rates of 4.79, 6.73, and 5.08 feet (1.46, 2.05, and 1.54 m.) per year, and the periods necessary for the recession of the Falls from Fort Snelling (a distance of a trifle over 8 miles) (13 km.) 8,819, 6,276 and 8,315 years. The average of these results is 7,803 years.

LA CROSSE TO MINNEAPOLIS.

ITINERARY.

By ULY S. GRANT.

Station.	Distance.		Elevation.		Popu-lation, 1890.	Station.	Distance.		Elevation.		Popu-lation, 1890.
	Miles.	Kilometres.	Feet.	Mètres.			Miles.	Kilometres.	Feet.	Mètres.	
La Crosse	0	0	657	200	25,090	Lake City	69	111	703	214	2,128
Dresbach	4	6	674	205	....	Frontenac	76	122	718	219	....
Dakota	6	10	655	200	....	Wanconta	80	129	706	215	....
Richmond	10	16	672	205	....	Red Wing	86	139	685	208	6,294
Lamoille...........	15	24	658	200	....	Eggleston	94	151	689	210	....
Homer..............	19	31	661	201	....	Etter	99	159	689	210	....
Winona	24	39	660	201	18,208	Hastings	107	171	707	215	3,705
Minnesota City....	30	48	675	206	....	Langdon	113	182	811	247	....
Whitman..........	35	56	678	207	....	Newport..........	119	193	749	228	....
Minneiska.........	40	64	670	204	....	Daytons Bluff	126	203	707	215	....
Weaver............	43	69	672	205	....	St. Paul...........	127	204	703	214	133,156
Kellogg............	51	82	701	214	....	Mendota..........	......	......	759	236	....
Wabasha	57	92	711	217	2,487	Fort Snelling......	......	......	722	220	....
Reads Landing ...	59	95	686	209	446	Minnehaha	......	......	813	247	....
Kings Cooley.....	63	101	.....	.....	....	Minneapolis	137	220	826	252	164,738

At **La Crosse** the Mississippi river is first seen. At the point of crossing of the railroad the low-water and high-water marks are respectively 628 feet (191·4 m.) and 643·5 feet (196·1 m.) above the sea level. The river varies much in width from here to Lake Pepin, sometimes being not more than half a mile (0·8 km.) across, and again spreading out to nearly three times that distance. The railroad follows the river almost continuously until reaching **St. Paul**, and along this route is the finest scenery in the whole Mississippi valley. The river is confined between high bluffs, precipitous in many places, and its flood plain is as a rule comparatively narrow.

From **La Crosse** to **Winona** the road follows close to the river's edge. The high bluffs on either side are of Cambrian strata; St. Croix sandstone capped by Lower Magnesian limestone. In some of the shaly layers of the St. Croix, near **Winona**, are found many shells of *Lingula*; in fact one bed of 16 inches (0·41 m.) in width is made up almost entirely of linguloid shells. The bluffs rise in many places precipitously from the railroad, and their tops are often over 550 feet (168 m.) above the river and not more than a mile (1·6 km.) distant from it. Just before reaching **Winona** the valley widens, so that there are 2 to 3 miles of low level land between the bluffs and the water; this continues to about a mile (1·6 km.) north of **Minnesota City**, where the bluffs are again close to the river.

From **Minnesota City** to **Mount Vernon** this narrow valley continues. But beyond the latter point the bluffs rapidly recede from the river until they are 4 or 5 miles (6·4 to 8 km.) away from it. The railroad leaves the river proper and runs near the foot of the bluffs. The wide valley continues until reaching **Reads Landing** at the foot of Lake Pepin.

Lake Pepin is merely an expansion of the Mississippi. It is about 20 miles (32 km.) in length, and nowhere exceeds 3 miles (4·8 km.) in width. It is a most beautiful expanse of water, and many of the towns on its shores are summer resorts—as **Lake City**, **Florence**, and **Frontenac**.

At **Red Wing**, Barn Bluff (formerly known as La Grange mountain) is seen. It is a large butte of Cambrian rocks left within the river valley proper. Thus far the rock in the immediate vicinity of the railroad has been St. Croix sandstone, but before reaching **Hastings** the higher members of the Cambrian are entered.

At **Hastings** the river is crossed, and from this point to **St. Paul** the railroad keeps on the east side of the river, and usually at some distance from it. After crossing the river at **Hastings** the road runs through terrace gravel and sand, formed by washing of the drift material by the Mississippi, when it was flooded by the melting of the retreating ice sheet. Before coming to **St. Paul** several miles of recent river alluvium are crossed, and just before entering the Union railroad station at **St. Paul** the railroad skirts the base of a high precipitous bluff of St. Peter sandstone, which is capped by a bed of Lower Silurian (Trenton) limestone.

St. Paul and **Minneapolis**, the "twin cities" having an aggregate population of over 300,000, form an important business center. They stand on the highway between the great wheat producing fields of the Northwest and the markets of the East, and have tributary to them vast agricultural and lumber districts. The business centers of the two cities are about 10 miles (16 km.) apart, connected by half-hourly trains; the resident portions join. The cities are beautifully situated on the banks of the Mississippi, and abound in fine residences, pleasant drives, and picturesque little lakes. As an educational center **Minneapolis** and **St. Paul** are well in advance; within their limits are Hamlin University, under Methodist control; Macalaster College, under the charge of the Presbyterians; and the University of Minnesota, which is at the head of the State's system of public instruction and has 1,300 students.

St. Paul, the capital of Minnesota, is the older of the two cities, and is situated at the head of navigation on the Mississippi river. It forms an important commercial center and is well known throughout the Northwest on account of its wholesale trade. The western part of the city, called St. Anthony Hill, is the best resident district, and contains many beautiful homes, which overlook the river.

Minneapolis is best known for its lumber and flour interests. The new patent roller process of grinding wheat, together with the excellent quality of wheat of this region, have here built up the largest milling center in the world. The mills of the city have a daily capacity of over 44,000 barrels, and the Pillsbury "A" mill alone can produce 7,200 barrels of flour per day. The Falls of St. Anthony, famous as a great available water power, and Minnehaha falls are points of interest to the tourist.*

The geology of the vicinity of St. Paul has been partially explained in the preceding paragraph on the recession of the Falls of St. Anthony. Back from the river are large accumulations of drift materials in the form of moraines and intramorainic till. Along the river courses, at low altitudes, are small deposits of recent river alluvium, while higher up are larger and coarser deposits of modified drift, probably made by the rivers when flooded by glacial melting. Practically the same state of things exists around Minneapolis, with the exception of those peculiarities described in the pages treating of the recession of the Falls of St. Anthony.

From St. Paul to Minneapolis via Fort Snelling.—After leaving the Union Station at **St. Paul** the road passes along the foot of the bluff for some distance. Here the old, decayed and eroded condition of the St. Peter sandstone can easily be seen, together with some of its caves and subterranean water courses. In places, resting on this sandstone, are remarkable mixtures of soil, drift pebbles and boulders, and large angular fragments of Trenton limestone. These fragments are in all positions and are often over 20 feet (6 m.) long and 3 feet (0.9 m.) in thickness. It is probable that this composite mass was formed by the river when its level was much above the present level. Similar angular blocks of limestone, mingled with the glacial débris, can be seen in the bed of the Mississippi river above Fort Snelling; in fact, the whole river bed (above Fort Snelling) is more or less composed of such a mass of fragments which extend down to the solid rocks, 10 to 25 feet (3 to 7.6 m.) below the bottom of the river. After crossing the river the road runs along the southern bluff, of St. Peter sandstone capped by Trenton limestone, to **Mendota**.

---

* Guide books to each city can be procured from book dealers. The best one to Minneapolis is Hudson's Dictionary to Minneapolis, 1891, published by the Beard-Hudson Printing Company, 10 North Fourth street.

From **Mendota** to **Fort Snelling** the broad valley of the Minnesota is crossed, and beyond the latter point the narrow gorge of the Mississippi is entered. At **Fort Snelling** one can easily see the difference between the broad ancient gorge of the Minnesota and the Mississippi and that of the Mississippi above this point.* The new narrow gorge is now followed for a short distance, but the railroad soon rises above it and reaches Minnehaha falls after passing through a few shallow cuts in the Trenton limestone.

Minnehaha falls at present show a small volume of water; this is due to the damming up of the headwaters of the stream in order to raise the level of lake Minnetonka. The height of the fall is 56 feet (17 m.) from the brink to the surface of the water in the pool below the fall. Below this the stream flows eastward in a narrow gorge for about 700 feet (213 m.), when it enters a much wider gorge, and in this continues to the river. This wider gorge is of interest as being an old channel on one side of an ancient island in the Mississippi. The channel on the other side was larger and cut back to the head of the island more rapidly, thus tapping the supply of water for the smaller channel and leaving a deserted gorge which ends abruptly in a steep bluff some distance above where Minnehaha creek enters it.

In the vicinity of **St. Paul** and **Minneapolis** the Trenton formation contains some soft green shales, which are very rich in well preserved fossil remains; a few gasteropods, lamellibranchs, crinoids, and trilobites are found; also a considerable number of brachiopods and large quantities of bryozoans. In the limestone itself remains of gigantic Orthoceratites are often seen.

---

* For a more detailed description of these differences see the pages relating to the recession of the Falls of St. Anthony.

MINNEAPOLIS TO MOORHEAD.

BY ULY S. GRANT.

Station.	Distance.		Elevation.		Popula-tion.	Station.	Distance.		Elevation.		Popula-tion.
	Miles.	Kilometres.	Feet.	Metres.			Miles.	Kilometres.	Feet.	Metres.	
Minneapolis	0	0	813	248	164,738	Aldrich	138	222	1,327	404	69
Fridley Park	7	11	847	258	......	Verndale	142	228	1,349	411	635
Anoka	18	29	878	268	4,252	Wadena	148	238	1,350	411	95
Itaska	25	40	885	270	......	Wadena Junction	150	241	1,352	412	......
Elk River	30	48	893	272	679	Bluffton	153	246	1,323	403	......
Baileys	35	56	913	278	......	New York Mills	161	259	1,410	430	260
Big Lake	38	61	935	285	......	Richland	166	267	1,396	425	......
Becker	47	76	972	296	......	Perham	172	277	1,370	417	761
Clear Lake	53	85	991	302	......	Frazee	183	295	1,389	423	......
St. Cloud	65	105	1,025	312	7,686	McHugh	188	303	......	......	......
Sauk Rapids	66	106	1,005	306	1,185	Detroit	193	311	1,364	416	1,510
Rices	79	127	1,062	324	......	Audubon	200	322	1,310	399	156
Royalton	86	138	1,080	329	528	Lake Park	206	331	1,356	407	......
Gregory	92	148	1,100	335	......	Hillsdale	211	339	1,197	365	......
Little Falls	97	156	1,117	340	2,354	Winnipeg Junct'n	214	344	1,181	360	......
Darling	101	161	1,154	352	......	Hawley	217	349	1,151	351	270
Randall	107	170	1,178	359	......	Muskoda	221	356	1,090	322	......
Cushing	112	180	......	......	......	Glyndon	230	370	925	282	275
Lincoln	118	190	1,281	390	......	Tenny	233	375	922	281	......
Philbrook	125	201	......	......	......	Dilworth	235	378	909	207	......
Staples	131	211	1,274	338	......	Moorhead	239	384	905	276	2,088

Immediately after leaving the Union Station at **Minneapolis** the railroad passes through a small cut in the upper beds of the Trenton limestone, and a small area of this is also seen just after crossing the river. A mile or two distant on the right, low-lying morainic hills are distinctly seen. The road then enters a level sandy plain of modified drift, and no more Paleozoic strata are seen within the State. This flat land is more or less covered by small scraggly oaks, commonly known as "scrub oaks," and continues until after passing **Anoka.**

At **Anoka** the Rum river, a small but important lumber stream, is crossed; but no rock is visible, as the river bed is confined entirely to Pleistocene deposits. Between **Anoka** and East **St. Cloud** the road continues in an area of modified drift, more or less rolling; but at **Elk River, Baileys,** and **Becker** hills of morainic accumulations are seen at the right of the track.

The first of the Archean rocks are entered a few miles southeast of **St. Cloud.** This area is only a few miles in extent, but a much larger area is seen at and near **Sauk Rapids** and Watab. The rocks are entirely granites, syenites, and gneisses, and are quarried quite extensively, especially at the former town.

From **Sauk Rapids** to **Little Falls** an area of modified drift extends for some distance on each side of the road. At this place the Mississippi is again crossed and a small area of slate, mica-schist, and diorite is seen along the river; these rocks are all Archean, but their exact horizon has not been determined. They cause the rapids in the river. A few miles south of **Little Falls** the southwestern limit of the evergreen forest is crossed; scattered pines have been seen farther south, but here they appear in large amounts. The principal species is *Pinus banksiana*, black pine, more commonly known as the "jack pine." It rarely grows to be of sufficient size for good lumber.

At **Little Falls** the Mississippi is again crossed and the railroad now runs directly away from the Father of Waters, taking almost a northwesterly course to the western edge of the State. From **Little Falls** to **Staples** a region of rolling and often hilly morainic country is passed through.

From **Staples** to **Perham** a flat to undulating area of modified drift is crossed, excepting that there is a small hilly morainic area between **New York Mills** and **Perham**. Before reaching the latter place the southwestern limit of the evergreen is again crossed, and only a small area of deciduous forest is seen before entering the prairies of the western part of the State. Otter Tail county, in which are situated the two last named towns, is crossed by an elevated morainic range of hills, the highest in the western half of the State. This elevated region is known as the Leaf hills, and is thickly dotted with numerous crystal lakes. In Otter Tail county these lakes cover more than one-ninth of the surface of the county. This lake area is entered just beyond **Perham** and continues to and beyond **Lake Park** in Becker county.

In the vicinity of **Lake Park** the prairie has become fully established and no more timber is seen except along water courses. Beyond this place a more flat or undulating area of till is entered and the descent toward the Red River of the North is begun.

One-fourth of a mile (.4 km.) east of **Muskoda** the first of the ancient beaches of Lake Agassiz is crossed. This is a low, gracefully rounded beach-ridge of gravel and sand, called the Herman beach. Then a plain, two miles (3.2 km.) in width, representing an old delta deposit, is crossed before reaching the second or Norcross beach. A mile (1.6 km.) beyond this the Campbell beach is crossed, and about a quarter of a mile (.4 km.) further is the McCauleyville beach; the latter is seen just before coming to the Buffalo River. The low flat valley of the Red River of the North is now entered and the road runs in a straight line directly west to **Moorhead**, a distance of 13 miles (21 km). The ground here is almost as level as a table, and the streams have cut but shallow narrow troughs in it. Between **Moorhead** and **Fargo** the Red River, a sluggish stream, is crossed.

ARTESIAN WELLS OF EASTERN DAKOTA.

By Geo. H. Eldridge.

A most remarkable series of artesian wells is developed in two belts of country coinciding in a general way with the meridional valleys of the Red River, which the road crosses at Fargo and Moorhead, and of the James River, which is passed at Jamestown, about 100 miles farther west.

The geological horizons from which the wells derive their waters are: (1) The Pleistocene drift of the Red River valley. (2) The middle and upper portions of the Cretaceous. (3) The Dakota sandstone which forms the base of the Cretaceous system on the Great Plains. (4) The Cambrian sandstones.

The last of these affords but a single well, which is in the Red River valley, 100 miles (161 km.) north of the railroad. It has a total depth of 915 feet (279 m.), of which the last 12 feet are in granite. It obtains fresh water at 503 feet (151 m.) and salt water at 903 feet (271 m.)

The Pleistocene wells are confined to the Red River valley, and obtain their water from seams of sand and gravel in the glacial till at depths of 85 to 270 feet (26 to 82 m.) These wells are developed in a belt 375 miles long. Fresh water is obtained from the southern portion of the belt, brackish and alkaline waters from that to the north of Blanchford, in North Dakota. The source of these waters is not distant, and the supply depends on the rainfall and the configuration of the immediate region.

The Middle and Upper Cretaceous wells are limited in number, and, with one exception, are also located in the Red River valley. They are from 250 to 395 feet (76 to 120 m.) deep, and their water flow is copious but brackish.

The Dakota sandstones supply the most important series of wells, which occupy a belt, along or adjacent to the James River valley, extending from Yankton, on the Missouri River, to Devil's Lake, north of Jamestown, a distance of 400 miles (644 km.) The sandstones are 200 to 400 feet (180 m.) thick, and lie at depths of from 600 feet (60 to 120 m.) in the southern to 1,500 feet (450 m.) in the northern portion of the belt. The wells are from 4½ to 6 inches (11.4 to 15 cm.) in diameter and the flow of water in some of the wells is 8,000 gallons (36,000 l.) per minute, at a pressure of 25 to 28 pounds per square inch. Their waters are utilized for running the machinery of mills, for city fire systems, and for irrigating, to a limited extent.

The geological structure of the basin, so far as known, is that of a gently dipping series of Cretaceous rocks, resting unconformably upon the Cambrian and Archean of northwestern Minnesota, and rising slowly towards the base of the Rocky Mountains, where the first outcrops of Dakota sandstones are 250 to 400 miles (400 to 650 km.) away and at elevations of 3,000 to 5,000 feet (900 to 1,200 m.) above sea level.

THE GREAT PLAINS OF THE NORTH.

GENERAL SKETCH.

By Arnold Hague.

From Jamestown to the base of the Rocky Mountains the route crosses the Great Plains and affords an excellent opportunity to see this characteristic physical feature of the continent, which extends from British America to the southern boundary of the United States. The country traversed is singularly uniform in its general aspect, and to most travelers appears monotonous and dreary. It is gently undulating, and, save along the great drainage channels, presents but few rock exposures. With the exception of a few favored localities sheltered from the wind, the Great Plains are destitute of trees, but for the most part covered by a luxuriant growth of nutritious grasses.

One interesting feature of the route followed during the day is the gradual emergence of erosion topography from the distinctive drift topography which characterizes the ride across Wisconsin and Minnesota.

West of Jamestown the railway gradually ascends the Missouri coteau, a line of broad, but low, gently sloping hills of rounded ridges, extending along the east side of the Missouri River all the way from Bismarck southward to Pierre, the capital of South Dakota. Immediately bordering the Missouri the hills present a series of precipitous bluffs facing the valley. Taken together these undulating ridges form an elevated mass separating the broad, deep valley of the Missouri from the shallow parallel valley of the James.

The drift which near James Valley more or less completely masks the underlying sandstone ridges gradually thins out westward. It may be recognized by low lines of gravel stretching across the plains, resting unconformably upon the underlying sandstones. Frequently it takes on a morainal aspect in the form of wide belts of north and south trend. Beyond the Missouri River the drift expresses itself only in a thin, irregular mantle, from a few feet to a few inches in thickness, accompanied by quartz boulders. The drift finally disappears beyond Sims Station, and erosion topography characterizes the whole region, finding its most characteristic expression in the Bad Lands, or Mauvaises Terres, so-called by the early Canadian fur-trappers on account of the obstructions they offered to the traveler who attempted to cross them. The line of drift trends off to the northwest, and has not as yet been recognized along the Yellowstone valley.

Cretaceous strata form the greater part of the surface rocks all the way from the James Valley to the mountains. The divisions of the Cretaceous system recognized on the Great Plains and along the base of the mountains in Dakota and Montana are:

Series.	Subdivisions.	Prevailing rocks.
Livingston......	....	Conglomerates, sandstones, and volcanic agglomerates.
Laramie.........	....	White sandstones, clays, and shales.
Montana.........	Fox Hills......	
	Fort Pierre....	
	Niobrara......	Shales and limestones.
Colorado.........	Fort Benton...	
Dakota	....	Sandstones and conglomerates.

The Laramie Cretaceous, the coal-bearing series of the Great Plains and the Rocky Mountain region, is the prevailing horizon, and only in a few limited areas along the line of travel do the underlying rocks come to the surface east of the Yellowstone River; from Jamestown to Livingston, a distance of over 650 miles (1,046 km.), seldom do any but Cretaceous rocks appear at the surface from beneath the drift.

The Laramie consists of coarse sandstones and shales, the former more abundant at the base of the series, where also most of the coal seams are found. This important series has been traced north and south almost continuously along the Rocky Mountain front within the boundaries of the United States, and in British America has been recognized in the valley of the Yukon River, stretching northward nearly to the Arctic Circle. South of the United States it is less definitely known, but coal seams of Cretaceous age which probably belong to this horizon have been observed at various points in Mexico and also in Central America. Westward, the Laramie Cretaceous extends through the various breaks in the front ranges to the one hundred and twelfth meridian. The Northern Pacific Railway crosses it at one of its broadest expansions. Under varying conditions the coal obtained from the beds of this series range from a dry, porous, non-coking coal of very low specific gravity, through slightly caking coal, to a fairly dense coal affording an excellent quality of coke, and, in limited areas, to an excellent anthracite. Nearly 5,000,000 tons of coal were mined in these beds during the census year 1890.

Near the base of the mountains the Upper Cretaceous has been divided into two distinct groups which have been designated as the Laramie and the Livingston. The former is mainly a normal white sandstone composed of quartz grains; the latter is made up largely of coarse conglomerates, with intercalations of somber-colored volcanic material composed of fine and coarse basic agglomerates, and beds of fine sands and clays deposited in shallow water and derived from the denudation of land surfaces. The Livingston rests unconformably upon the Laramie and contains pebbles of Archean schists, Paleozoic limestones, and Cretaceous sandstones. Away from the mountains the two groups have not as yet been differentiated.

The railway, which for the greater part of the distance (save, for instance, when crossing the Bad Lands) traverses the top of the gently undulating surface, on reaching the Yellowstone Valley follows the river between high bluffs of Laramie sandstone. At a number of localities along the river, Cretaceous rocks older than the Laramie are exposed in the bluffs. Near the mountains the entire series of Cretaceous strata, from the base of the Dakota to the Livingston, are upturned and well shown in excellent cross-section along ravines cut by streams coming down from the mountains.

On a bright clear day the first glimpses of the Rocky Mountains may be had from Billings, snow-clad peaks standing out prominently both to the north and south, and three or four hours before reaching Livingston the Bear Tooth Mountains to the south present a grand panoramic view of rugged peaks stretching along the horizon as far as the eye can reach. These mountains, whose summits reach 12,000 feet (3,658 m.), and their western extension, the Boulder plateau, 10,000 feet (3,048 m.), consist of Archean gneiss and granite. The slopes and foothills are made up of Paleozoic strata dipping either to the north or east, and passing beneath Mesozoic beds, the uppermost members of the Laramie sandstone extending far out upon the plain. The Boulder plateau is intimately connected with the Snowy range, along the northern base of which the railroad runs from Big Timber to Livingston.

On the north side of the valley rise the Crazy Mountains, an isolated group standing out boldly from the main Rocky Mountain ranges. The central portion of these mountains consist in great part of igneous rocks that have broken through a broad synclinal trough of sandstones and shales, made up of a great thickness of sediments of the Livingston and overlying series of beds. From the central core of igneous rocks innumerable dikes have penetrated the sandstone, baking and hardening the sedimentary rocks, which have resisted erosion more than the easily friable beds.[35] The most southern peaks of the Crazies are situated about 15 miles (24 km.) north of the river, and are easily accessible to anyone desiring to study them.

451 GE——21

FROM JAMESTOWN TO LIVINGSTON.

ITINERARY.

BY ARNOLD HAGUE.

Station.	Distance.		Elevation.		Station.	Distance.		Elevation.	
	Miles.	Kilo-meters.	Feet.	Meters.		Miles.	Kilo-meters.	Feet.	Meters.
Jamestown	0	0	1,395	425	Little Missouri ...	256	412	......	....
Eldridge..........	7	11	1,541	470	Sentinel Butte....	272	438	2,707	825
Cleveland..........	20	47	1,794	547	Allard	312	502	2,246	685
Crystal Springs....	37	60	1,792	546	Glendive	322	518	2,069	631
Dawson............	50	80	1,748	533	Hoyt	337	542	......	....
Steele.............	58	93	1,859	567	Terry.............	361	581	2,242	683
Sterling	77	124	1,867	569	Ainslie	381	613	2,274	693
Bismarck	101	161	1,670	510	Miles City........	400	644	2,355	718
Mandan	106	172	1,646	502	Forsythe	446	718	2,514	766
Sweet Briar	122	196	1,806	551	Howard	456		......	....
Sims	142	229	1,962	598	Custer............	494	795	2,727	831
Kurtz	159	256	2,025	617	Billings	547	880	3,117	950
Gladstone	205	330	2,348	716	Park City	570		......	....
Dickinson	216	358	2,405	733	Big Timber.......	628	1,011	4,072	1,241
Belfield	236	380	2,579	786	Springdale........	643		......	....
Sully Springs......	247	397	2,575	785	Livingston	663	1,067	4,487	1,368

Jamestown is an attractive settlement and the center of an important wheat region, situated on the James River, a stream which running southward empties into the Missouri. From Jamestown west to Bismarck, the road for 100 miles (161 km.) passes over an undulating, grassy country, in every way characteristic of the Great Plains of the north. Leaving Jamestown the road gradually ascends, 450 feet (137 m.) in 20 miles (32 km.), to Cleveland. This is followed by gently rolling country for 50 miles (80 km.) to Sterling, from which point there is a gradual descent of 200 feet (60 m.) to the Missouri River. The glacial drift which everywhere covers the country east of the James Valley, gradually thins out to the west and south, and is lost as a continuous sheet before reaching the Missouri River. Drift in the form of small boulders may be easily recognized along the route, and is well shown near Crystal Springs. All the way from Jamestown to Bismarck the underlying rocks, as far as yet recognized, belong to the Laramie series, the beds becoming more and more marked as the Missouri is approached.

Bismarck, the capital of North Dakota, lies on the bluffs of the Missouri, which afford commanding views both up and down the broad valley.

Crossing the Missouri River by a fine bridge, good exposures of the Laramie may be seen in the bluffs on both sides of the river and back of the town of **Mandan**. Fox Hills strata occur along the river a few miles below **Bismarck**, but are not visible from the railroad; north of the railroad only Laramie strata occur.

Upon leaving the Missouri the train winds up the valley of Heart River, a small tributary of the main stream, coming out again upon broad plains much like those crossed on the east side of the Missouri. In many of the buttes and low hills one may observe a marked reddening of the strata. This is due to the heat which has resulted from the spontaneous combustion of the beds of liguite which are scattered throughout the Laramie. While some of these beds are now on fire, according to Dr. C. A. White many of them were burned out as long ago as Tertiary time, before the larger part of the great erosion occurred which produced the present configuration of the surface. Slag and ashes, the product of these early burnings, are now scattered upon the surface of even the highest of the hills, whence the softer material has been carried away.

At **Sims,** about 40 miles (64 km.) west of the main Missouri River, the first workable coal seams are seen. The mines here have been extensively opened and furnish a large amount of coal for locomotive purposes.

From the Missouri River there is a gradual rise of 450 feet (137 m.) in about 30 miles (48 km.), then a descent of 200 feet (61 m.) in 20 miles (32 km.) to **Curlew** Station. Rolling, monotonous country follows from Curlew to Fryburg, a distance of 90 miles (145 km.), and from **Fryburg** there is a descent of 500 feet (152 m.) in 15 miles (24 km.) to the Little Missouri River.

The Little Missouri River is the most important drainage channel between the Missouri and the Yellowstone. It takes its rise at the northwest extremity of the Black Hills, and follows a northerly course for 50 miles (80 km.) beyond the railroad, then bends eastward. In descending to the Little Missouri one passes through a small but characteristic area of *mauvaises terres* or bad lands. While these are by no means as extensive or impressive as those which characterize many of the great Tertiary basins of Wyoming, they present similar physical features and identical forms of erosion. For varied coloring and exquisite delicacy of tint they are most remarkable and unsurpassed by any similar country elsewhere.

Between the Little Missouri and the Yellowstone there is an ascent of 575 feet (175 m.), followed by a descent of over 800 feet (244 m.) to the latter river. All the country between these two streams is similar to that already passed over, but its surface is more broken, more diversified and picturesque, and dotted over by isolated buttes and long ridges of low hills, sculptured forms left by erosion.

The railroad crosses the boundary line between the States of Dakota and Montana just west of **Sentinel Butte** station. Sentinel Butte lies a few miles south of the railroad and is plainly visible to the tourist, standing out upon the plain a most impressive object, like a monument on the desert. Geologically, it presents much of interest, as its base is composed of the upper beds of the Laramie, and it is capped by conformable strata of fresh-water deposits regarded as of early Eocene age.

The railroad reaches the Yellowstone valley at **Glendive**, and from here it follows the river all the way to **Livingston**, a distance of 340 miles (547 km.), with a gradual ascent of over 2,200 feet. The valley between the bluffs varies from 1 to 6 miles in width, the river meandering from side to side. Several large tributaries to the Yellowstone which enter the river from the south are crossed, including Powder river, Big Horn, Clarks Fork, and Stillwater. The first named rises between the Black Hills and the Big Horn mountains; the second drains the Big Horn mountains; and Clarks Fork and the Stillwater have their source in the Absaroka range. The railroad crosses the river three times, and for the greater part of the distance follows along under the bluffs on the south side. The Yellowstone valley above the bluffs presents much the same physical features as seen in eastern Montana, and the geological features offer but little in the way of change to break the dull monotony. The bluffs and low-rolling hills are formed of yellow sandstone, for the most part horizontal. It is the prevailing color of these beds that has given the name to the river, and consequently to the now famous Yellowstone Park, where the river has its source.

From **Glendive**, bluffs of Laramie sandstone may be seen stretching far down the valley, which has here a northeasterly direction. About 10 miles (16 km.) above **Glendive**, at the foot of the bluff on the east side of the valley, occurs an exposure of Fox Hills beds. These beds are determined by their organic remains, but are difficult to recognize, as lithologically they are similar to the overlying Laramie, and both series are conformable. There is exposed here the base of the Laramie, whereas the top of the series occurs at Sentinel Butte. According to Dr. C. A. White the thickness of the Laramie in western Dakota and eastern Montana is nearly 3,000 feet (914 km.).

Shortly after leaving **Miles City** the lignite beds crop out at the base of the Laramie; and at **Howard**, Fox Hill strata are said to come to the surface from beneath the overlying Laramie, but they are by no means easily distinguished.

At **Billings** the first important exposure of Middle Cretaceous is encountered. Just east of the town the Fox Hills sandstones, together with the underlying Fort Pierre shales, are brought to the surface by an eroded anticline. The Pierre shales form the broad bottomland of the valley for several miles westward, and produce a marked contrast to the narrow valley and rugged inclosing bluffs of the Laramie. The town is built upon these Pierre shales that have been penetrated for 800 feet by an artesian-well boring. North of the town the shales form the smoothly sculptured base of a steep escarpment, being capped by a massive ledge, 75 feet in thickness, of Fox Hills sandstone, which is largely quarried and used as building stone in the town. Overlying the Fox Hills are the thinly-bedded gray sandstones of the Laramie, forming the highest cliffs bordering the river. Both series of sandstones present quite persistent bluffs on opposite sides of the valley, and about 25 miles (40 km.) west of **Billings** replace the Fort Pierre shales on the river bottom, the valley becoming narrow and canyon-like.

At **Laurel** a branch road crosses the river, running southward about 40 miles to the Red Lodge coal mines that supply the fuel used by the locomotives of the Northern Pacific Railway.[37]

West of **Park City** only Laramie rocks occur, but the country gradually becomes more rugged and the scenery diversified.

Big Timber, situated at the mouth of the Boulder, is built upon a broad terrace of Livingston deposits that extend back to the mountains, but have not as yet been traced more than a few miles to the east. The Livingston beds, however, stretch almost continuously as far west as the Bozeman Tunnel, through which the railway passes in crossing the Bridger range. The exposures of the interbedded volcanic agglomerates, largely made up of andesitic material, may be seen in the bluffs just west of **Springdale**, where the beds dip at an angle of 20° to the northeast. At one locality these agglomerates attain a thickness of 2,000 feet. Just beyond this place the railroad passes through a short ravine, showing an anticline of bedded clays and sandstones that form the base of the Livingston, the rocks being much disturbed, showing local folds and flexures.

Sheep Cliffs, just north of the river and midway between **Springdale** and **Livingston**, are especially interesting, as they show a massive sheet, of theralite intruded in the Livingston sandstone. This rare rock is here exceedingly fine-grained, almost black in color, with porphyritic crystals of augite and biotite; nepheline crystals only occur in the dense ground-mass of the rock. Much larger masses of this rock occur as intrusive sheets in the sandstones of the Crazy Mountains[35] to the north.

At **Livingston** the railroad crosses the Yellowstone for the last time, and immediately begins the ascent of the first range of the Rocky Mountains.

YELLOWSTONE VALLEY.

FROM LIVINGSTON TO CINNABAR.

ITINERARY.

Station.	Distance.		Elevation.		Station.	Distance.		Elevation.	
	Miles.	Kilo-meters.	Feet.	Meters.		Miles.	Kilo-meters.	Feet.	Meters.
Livingston.....	....	....	4,487	1,368	Daileys	31	50	4,915	1,498
Brisbin..........	10	16	4,682	1,427	Sphinx	41	66	5,065	1,544
Chicory	20	32	....	....	Cinnabar.......	51	82	5,179	1,579
Emigrant	23	37	....	....					

[By WALTER H. WEED.]

At **Livingston** the train leaves the main transcontinental route and passes over the Yellowstone Park branch of the Northern Pacific railroad to **Cinnabar**, a distance of 51 miles (82 km.). It travels up the Yellowstone river through a picturesque mountain valley, with high peaks on both sides. Those east of this valley and south of **Livingston** are known as the Snowy range.

That portion of the Snowy range seen from **Livingston** is really the front of the Rocky Mountains, which farther westward bend north and extend in a nearly continuous range to the Canadian line and to the east bend southeastward for 75 miles between the Yellowstone river and Clarks Fork. The peaks seen from the town show the folded Paleozoic limestones dipping at steep angles northward and passing beneath the less steeply inclined Mesozoic beds that extend outward into and form the valley. The front of the range is characterized by a fold, in general parallel to the Archean contact, the anticline being often faulted, and west of **Livingston** passing into several *en echelon* anticlines with steeply pitching axes. The highest point seen south of the town, locally known as Baldy, is a sharply defined mass of Archean schist brought into contact with Carboniferous limestone by a faulting of the anticlinal fold just alluded to. The general structure is shown in the section (Fig. 14–A, p. 330), which passes through this mass.

Livingston, situated on the north bank of the Yellowstone river, is one of the many towns of the West born on the advent of the railroad, but rapidly growing with the settlement of the surrounding country. The town is built upon an alluvial river terrace, cut in the upturned sandstones and grits of the Livingston series,[36] beds that consist of water-laid strata composed of andesitic material. These beds rest in apparent conformity upon the true Laramie, the horizon of the workable bituminous coal seams, but an uncomformity is proven by the variety of pebbles of Paleozoic and Mesozoic rocks found in the conglomerates, and is actually shown in the mountains to the westward. The greatest thickness of the Livingston yet measured is 7,000 feet, the overlying sandstones and clays being quite distinct lithologically, and carrying a purely fresh-water fauna and a flora of Fort Union type. Plant remains are abundant in the Livingston beds, the species being largely of Laramie types. Specimens may be collected from the rocks forming the hills immediately north of the town, where characteristic exposures of the series occur. The overlying strata can be seen from these hills, forming low sandstone ridges and a light gray bluff wall to the northward, and the beds form the high peaks of the Crazy Mountains lying to the northeast.

South of the river a gently sloping alluvial terrace rises to the foot of the mountains, effectually concealing all exposures of the Middle Cretaceous rock except along the river banks.

The branch railroad to the Yellowstone Park traverses the valley bottom toward the gap in the mountains through which the river has cut its way to the Great Plains. On the west, low combs of sandstone belonging to the Middle Cretaceous are occasionally seen, the Laramie, Montana, and Colorado groups being passed before reaching the "Gate of the Mountains." The first beds attaining prominence are those of the Dakota Cretaceous, whose conglomerates form the crest of a striking east and west ridge, or "hog-back," dipping at 20° away from Canyon Mountain and separated from the mountain slopes by a persistent depression eroded in the soft fossiliferous shales and limestones of the Jurassic. The red sandstones, elsewhere considered Triassic, are not definitely recognized in this section, the first great ledge of the mountain being a quartzite assigned to the Carboniferous.

The canyon now entered affords easy access to the beautiful intermontane valley of the Yellowstone, bringing the traveler at once into typical Rocky Mountain scenery. The gorge is cut across an anticlinal fold whose southern half is faulted and crushed. The walls show a perfect section of the entire stratigraphic series from Cambrian to Jurassic. Underlying the prominent quartzite mentioned above are the massive heavily bedded limestones of the Carboniferous, here used for burning lime, and characterized by abundant fossils. Beneath these massive beds are the fissile limestones of the Devonian, resting upon limestones of doubtful Silurian age, that are in turn underlaid by limestones and shales containing an abundant typical Cambrian fauna. The series from Cambrian to Laramie Cretaceous is throughout conformable, no gap or break appearing in this section. Near the south end of the canyon the beds of the west wall are repeated by local faulting, while those on the east side show considerable crumpling and pass beneath the Devonian and Carboniferous limestones that form the two arches at the south end of the wall. The general section may be briefly summarized as follows:

		Feet.
Carboniferous	{ Quartzites and interbedded limestones	300
	{ Limestones	2,000
Devonian	Limestones and shales	300
Cambrian	{ Limestones with interbedded shales	650
	{ Shales	200
	{ Quartzite	100

Emerging from the canyon these Paleozoic limestones may be seen resting at steep angles upon the gneisses and schists to the east, the valley of Deep Creek having been cut along the contact. The roof-like surfaces of the limestones and the square-cut bluffs and walls show in strong contrast to the spires and uniform slopes of the schists. The immediate valley bottom shows a couple of hills of Paleozoic limestone rising above the alluvial gravels. To the west the faulted anticlinal of Canyon Mountain brings the schists into view, but the nearer hill slopes are formed of Cambrian limestones replaced farther south by Devonian and Carboniferous beds, which an overthrust fault has superimposed upon the sandstones of Jurassic and Dakota Cretaceous. Near Brisbin an east and west ridge is formed of an anticline of Carboniferous limestones, the Mesozoic beds on its southern side passing beneath the dark-colored volcanic breccias and agglomerates that form Antelope Butte. From here southward the sedimentary formations are not seen until Cinnabar Mountain is reached near the railroad terminus.

The valley entered presents the most imposing scenery yet encountered; inclosed between the serrated crests of the Snowy Mountains upon the east and the more distant peaks of the Gallatin range upon the west, it stretches southward some 30 miles to the Archæan gorge of Yankee Jim canyon. Eroded along the line of a great fault the valley has long been the drainage way for the waters of the mountainous region to the south. The site of a Neocene lake whose sediments only remain where protected by a basaltic lava flow, it was filled by an ice sheet that flowed northward from the great plateau of the park, and its present most striking features were imposed upon it at this time. As the morainal deposits of this glacier, and the sculpturing effected by it, form a prominent part of the geology of this region, a somewhat full account is given of the glaciation of the valley at the end of this chapter.

West of the valley the peaks of the Gallatin range rise above retreating slopes, dotted with groves of pines and trenched by deep and narrow gorges. The range is eroded in a great accumulation of volcanic breccias, the rocks being basic andesites and lava flows similar to those described in the Snowy Mountains. The colors are usually dark, but rarely brilliant brick reds and purples prevail, as will be noticed on the slopes west of **Daileys**. The tufaceous beds often contain plant remains, and silicified tree trunks are not uncommon, together with agates, amethysts, and chalcedony. Hyalite is abundant and remarkably fine specimens have been obtained from the summits.

[By J. P. IDDINGS.]

The railroad from **Livingston** to **Cinnabar** passes in view of a transverse section across the end of the Snowy range, cut nearly at right angles to the strike of the beds. The general geological structure of the range is that of an anticlinal fold, the upper portion of which has been entirely removed by erosion, and which has been variously modified by faulting, especially at the southwestern end. The main body of the range consists of crystalline schists and granite, forming a high plateau and still higher peaks that reach 11,000 and 12,000 feet in altitude. Along the northern and southern flanks of the range the overlying Paleozoic strata dip away from the crystalline axis. While this is the structure of the greater part of the range it does not obtain for that portion of it passed in review by the railroad south of the main axis.

On the east, south of Deep Creek, a chain of rocky and precipitous peaks extends for about 12 miles to Mill Creek. These mountains consist of Archæan gneisses and schists, and constitute the end of the core of the great Bear Tooth range. They are extremely rugged, with narrow gorges or gulches cutting deeply into their mass. Their highest summits are 11,000 feet in altitude, or a little over 6,000 feet above the river level. Each of the great gulches has once been occupied by a glacier whose lateral moraines may be seen stretching far down into the open valley. The upper portions of the mountains are bare and forbidding and few of the summits have ever been ascended. Back of these peaks, to the east, the crystalline schists extend in a broad, flat-topped mass whose surface constitutes a high plateau 10,000 feet in altitude, but which has been dissected by canyons 3,000 or 4,000 feet deep. The surface of the plateau is finely glaciated and is covered with ponds and lakes, the rocks being almost completely destitute of soil or vegetation.

The most southern and highest peak of these gneissic mountains is Mount Cowen, 11,190 feet in altitude. It is immediately north of the valley of Mill Creek, opposite **Chicory** station. At the southern base of Mount Cowen is a double fault that has thrown down the country south in two displacements of over 3,000 feet each. These faults, and one at the south base of Sheep Mountain, have destroyed the anticlinal character of the main range in this vicinity. The Mill Creek faults disappear 9 miles east of Mount Cowen, from which point the anticlinal structure extends eastward.

A cross section of the region from near the mouth of Mission Creek on the Yellowstone River, 9 miles below Livingston, through Livingston Peak and Mount Cowen to Mill Creek (Fig. 14–A), will aid in the understanding of the geological structure of this part of the region. It exhibits the Paleozoic strata dipping steeply north and passing under less highly inclined Mesozoic beds that form the broad valley of the Yellowstone below Livingston. A slight fault crossing the peak south of Livingston Peak has thrown down the sedimentary strata against the crystalline schists for a short distance. It shows the great body of schist and granite forming the mountains along the east side of Yellowstone Valley to Mount Cowen and the faults at its southern base.

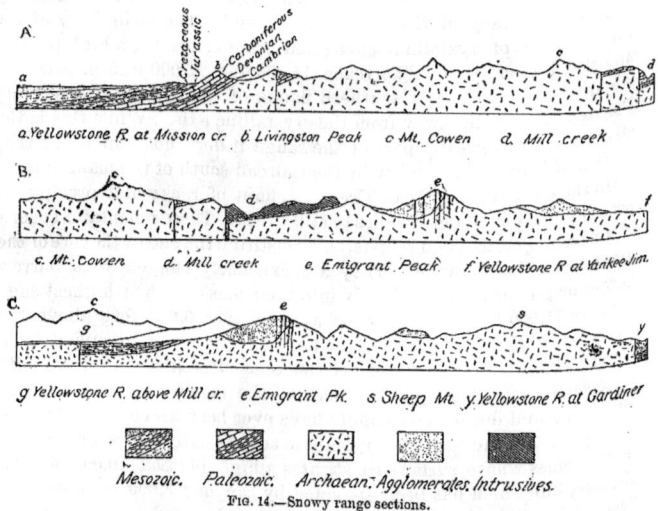

a. Yellowstone R. at Mission cr.　b. Livingston Peak　c. Mt. Cowen　d. Mill creek

c. Mt. Cowen　d. Mill creek　e. Emigrant Peak　f. Yellowstone R. at Yankee Jim.

g. Yellowstone R. above Mill cr.　e. Emigrant Pk.　s. Sheep Mt.　y. Yellowstone R. at Gardiner

Mesozoic.　Paleozoic.　Archaean. Agglomerates. Intrusives.

FIG. 14.—Snowy range sections.

The valley of Mill Creek opens a vista into the region of volcanic tuffs and breccia, mountains of which, 20 miles distant, may be seen from the railroad just north of Chicory station. The peaks a short distance back from the mouth of this valley consist of igneous rock that was intruded within sedimentary rocks and which undoubtedly rose along the fault plane at the base of Mount Cowen. Similar intrusive rock forms the high peak, Chico Mountain, between Mill Creek and Emigrant Gulch, the deep valley immediately north of Emigrant Peak

(10,960 feet—3340 m.) The limestone is exposed at the northwestern base of Chico Mountain, the main mass of igneous rock having been intruded in shale that lies near the base of the Cambrian deposits near the gneiss. From this point southward to Yankee Jim canyon, and beyond the great mass of the mountains east of the river, is gneiss and schist, upon the ancient surface of which volcanic ejectamenta have accumulated. The extremely uneven character of the gneissic country at the time of the volcanic outbreaks and the irregularities of subsequent erosion explain the present geological structure of this part of the region.

Emigrant Gulch, near its mouth, cuts 2,000 feet into crystalline schists that are exposed in a low belt around the northwestern base of Emigrant Peak. Above this andesitic tuff breccia or agglomerate rises for 5,000 feet to the summit of the mountain; while on the southern side gneiss rises to within 1,200 feet of the summit. The andesitic breccia at the summit is traversed by dikes of porphyrite with a general northwest and southeast trend. The more prominent may be seen from the railroad. The breccia has been indurated by the proximity of great bodies of intruded porphyrite which form mountains east of Emigrant Peak, which accounts for its having withstood erosion and having become one of the loftiest peaks in the region. (Fig. 14-B and C.)

South of Emigrant Peak, Six-Mile Creek has cut deeply into crystalline schists and granite. The mountain ridges surrounding its drainage basin consist of andesitic breccia, tuffs, and lava flows, which cap the granite at about 7,000 feet along the east side of the valley. The crystalline schists are well exposed in the narrow gorge through which the river runs at the southern end of its broad valley, and which is known as Yankee Jim canyon. The gneisses cross to the western side of the river, and are overlaid by volcanic lavas, as on the eastern side.

The cross section of the country through Mount Cowen and Emigrant Peak (Fig. 14-B) exhibits the geological structure of the region just described, which needs no further comment except to note that the section crosses the faults at the base of Mount Cowen about 3 miles west of where the first section crosses them, and intersects a large body of intruded porphyrite.

From Yankee Jim canyon south for several miles, knobs of chocolate-colored breccia may be seen on both sides of the river resting upon gneiss. The light-colored rocks east of the river, opposite Cinnabar Mountain, are mainly decomposed gneiss overlaid by some igneous rocks and cut by dikes of the same. The gneiss has been greatly fractured by joints representing the termination of a profound fault which exists farther south.

The high mass of Sheep Mountain opposite Cinnabar station is crystalline schist from base to summit, which is 10,628 feet in altitude.

These rocks continue to form high mountains along the northern side of the Yellowstone River east of Gardiner. The peaks surrounding the head of the valley of Bear Creek opposite Gardiner are volcanic ejectamenta and intruded bodies of porphyrite. The bench at the base of Sheep Mountain, about 600 or 800 feet above the river, consists of a lava flow of basalt resting on the upturned edge of faulted Cretaceous sandstones. Small wedges of these rocks have settled short distances and lowered parts of the basalt sheet to different levels, so that it appears from the south side of the river as though there were a number of superimposed sheets. The light-colored deposit upon the basalt is the remains of an ancient travertine, similar to that at Mammoth Hot Springs. A profound fault passes along the south base of Sheep Mountain in a southeast and northwest direction, the displacement being over 6,000 feet. This fault is shown in the cross section through Emigrant Peak and the summit of Sheep Mountain (Fig. 14–c). The geological structure is exceedingly simple and is largely a repetition of that in the second section (Fig. 14–b). The northern end of the section crosses the Yellowstone River two miles above Chickory Station, where it probably intersects the Mill Creek fault.

[By WALTER H. WEED.]

At Cinnabar Mountain the sedimentary rocks are again met with, a section being exposed from the Archean to the Laramie Cretaceous. The mountain received its name from a prominent band of bright red sandstone, the so-called Devils Slide, there being of course no mercury there. In the mountain the sedimentary strata are nearly vertical, being the sharply upturned end of a synclinal trough whose axis is the sag south of the mountain. It is the most convenient locality for the traveler to examine the stratigraphical section, as the rocks are well exposed and readily accessible. The northern part of the mountain is composed of Paleozoic limestones so closely compacted that the subdivisions are not easily recognized, but the quartzite at the summit of the Carboniferous, with its red magnesian limestones, is distinctly differentiated. Above these beds are the Triassic sandstones forming the Devils Slide, and the ripple-marked quartzite which overlies them forms the north wall of the most prominent of the gulches that seam the mountain side. The gray Jurassic shales are well exposed and contain an abundance of fossils characteristic of the Rocky Mountain Jura, such as *Myacites*, *Rhynchonella*, *Gryphæa*, *Camptonectes*, etc. It is in these rocks that the intrusives forming the south wall of the great gulch have been injected. Overlying these Jurassic beds are the grits and conglomerates of the Dakota Cretaceous, in which a limestone belt carrying fresh-water fossils may be seen. Above this the dark bituminous shales of the Fort Benton, with occasional arenaceous belts, are exposed, overlain by the lighter shales and shaly limestones of the Niobrara, while the leaden clays of the Fort Pierre form the southern part of the mountain. Fossils, though not abundant, may be found. The Fox Hills is shaly and is overlaid by the lithologically distinct sandstones of the Laramie, in which the coal seams [37] worked at Horr are located. The coal is of excellent quality, occurs in a number of seams, and makes a remarkably pure and firm coke. To the southward the sharp summit of Electric Peak is seen, Cinnabar Mountain being but the northern end of a long spur of that peak. Across the river the Coal Measures of the Laramie dip steeply eastward and are covered by a sheet of modern basalt. A few openings show the same coal seams here as at Horr.

Glaciation of the Yellowstone valley.—The local glaciers formerly abundant in the Rocky Mountain Cordillera attained an unusually extensive development on the plateaus and encircling ranges of the Yellowstone Park, and, as will be shown in the account of that region, sent glaciers down the valleys that drained the highlands in every direction. Along the northern border two streams of ice, pushing northward, found an outlet for their united flow down the valleys of the Yellowstone River, and have left impressive evidences of their power and magnitude that at once attract the attention of the observant traveler.[38] These memorials of the vanished glacier present several features of interest not common in this portion of the Cordillera. The mountain valleys of the Yellowstone were carved out before their occupancy by ice, and the ice stream expanded in its lower portion, producing morainal heapings closely resembling those of the continental type. True mountain moraines, both lateral and terminal, and formed of angular débris, occur in tributary gulches cut in the Snowy range to the east, but they are entirely lacking in the valley proper. Westward the ice sheet crowded upon the flanks of the Gallatin range and left Archean and Paleozoic boulders resting upon the volcanic rocks.

On emerging from the gorge south of Livingston the train skirts the margin of an alluvial bottom, inclosed by stream terraces cut in the old overwash plain of the glacier. The glacial gravels have been largely re-sorted and form benches that are susceptible of cultivation when irrigated. To the east, local moraines formed by small glaciers from lateral gorges may be seen flanking the valley. From **Brisbin** to **Chicory** the higher terrace is part of the original overwash plain, somewhat modified where the larger mountain streams debouch into the valley. Seen from the railroad, its characteristic gentle slope and gravelly surface can be distinguished, but just before reaching **Chicory** the steep hummocks of the moraine may be seen rising abruptly from the terrace. These hills are from 15 to 25 feet in height, their slopes covered with boulders and gravel; they mark the extreme northward termination of the Yellowstone glacier. Although several terminal moraines may be distinguished in places, there is too general an overlapping to permit of definite mapping of the successive positions of the ice front. In this part of the valley the sharply limited extent of the glaciation will be at once noted. To the east an extensive development of the overwash plain is seen, forming a flat dotted with farms between Mill creek and the river, with the hummocky surface of a strip of valley moraine at the base of the mountains.

East of the valley, the steep slopes of the Snowy Mountains show fine examples of mountain moraines upon their flanks, while their deeply incised canyons and polished slopes of gneiss show the abrading power of the tributary glaciers. West of the valley the rocks are wholly volcanic and readily distinguishable from the glacial drift. Near Fridleys (Emigrant station) the ice covered only the valley bottom, but as we ascend the valley toward the south the drift rises to higher altitudes, though nowhere does it creep far up the mountains. It is evident that the Yellowstone glacier once filling the valley was not reinforced by streams from the west.

Near Fridley's (Emigrant) the columnar cliffs of a recent basalt flow are seen. The lava caps Pliocene lake beds, and its upper surface is polished, planed, and striated by the ice sheet. Aross the river to the eastward a good example of a subglacial stream channel can be seen indenting the undulating morainal slopes. Approaching Daileys another exposure of basalt, capping lake beds, is seen to the east; boulders of basalt from the two localities are very abundant in the drift down the valley. To the west a remarkable series of terrace lines is seen, some thirteen being plainly distinguishable. They are found only within the glaciated area, and are due to diverted drainage, marginal to the ice. The valley presents very different features on its two sides. To the east the serrated summits of the Snowy mountains and the high point of Emigrant Peak rise abruptly from the valley. These mountains held tributary glaciers at one time confluent with the great glacier filling the valley, and at its decline pushing westward across the valley and depositing erratics of gneiss and limestone upon the slopes of volcanic agglomerate. The mountain slopes are strewn with scattered erratics, and striking examples of moraines of angular débris may be seen upon their flanks.

To the west of the valley, however, the limit of the drift is very sharply defined. The dark slopes of volcanic agglomerates rising gradually to the crest of the Gallatin range are free from drift and show no evidences of glacial sculpture. It is certain that the ice filling the valley extended but a short distance up the slopes. Within the drift-covered area the slopes are very generally terraced, and the evidence is clear that it is due to the diversion of the drainage from the mountain slopes to the west by the ice, the resulting marginal streams cutting the terraces. In certain cases these streams have cut remarkable canyons transversely across the slopes; in some instances, a mile or so long and 50 feet deep. Both canyons and terraces are confined to the drift-covered area, whose topography is in strong contrast to the unglaciated slope above.

The valley moraine is of the kame type, and does not extend up to the bounding slopes upon which scattered erratics alone are found. This valley moraine is formed of more or less waterworn or subangular material, and differs from the angular débris forming the usual types of lateral mountain moraines. Such a moraine is seen at the south end of the valley near the mouth of Yankee Jim canyon and was formed during the retreat of the ice.

Yankee Jim canyon is an ice-cut gorge whose polished walls are emphatic witness to the magnitude and power of the ancient glacier. Upon these glaciated surfaces *blocs perchés* can be seen at every favorable point. Not only did the ice fill this gorge to the brim, but the erratics on Dome mountain to the east and the higher slopes to the west show that it was here 3,000 feet thick and filled the adjacent valleys.

Cinnabar mountain was completely buried beneath the ice, and blocks of sedimentary rocks from its summit were carried southward and down the valley. Huge erratics of granite are abundant above the canyon and dot the slopes inclosing Cinnabar valley, but here the crowding of ice from Bear gulch, the Yellowstone and Gardiner rivers, and Cache creek has left a confused record. It is, strictly speaking, the beginning of the Yellowstone glacier, and from here southward glacial detritus and glacial sculpture will be observed throughout the park, an account of which is given in the chapter on that region.

THE YELLOWSTONE PARK.

By Arnold Hague.

The Yellowstone National Park is situated in the northwest corner of the State of Wyoming, with a narrow strip of country less than two miles in width in Montana on the north, and a still narrower strip extending westward into Idaho. Its boundaries, as determined by the act of Congress establishing the park, are ill-defined, as at the time of the enactment of the law the region had been little explored. Its relations to the physical features of the surrounding country were only slightly understood. It is probable that before many years Congress will read-just and clearly define the park boundaries, placing it entirely within the State of Wyoming. That portion of the Park most frequented by tourists lies south of the forty-fifth parallel of north latitude, and between the one hundred and tenth and one hundred and eleventh meridians of west longitude. For a long time the park country and the adjacent mountains remained an inaccessible land, which had defied all efforts of the early explorers to discover its secrets. The early fur trappers had been all around this unknown land, but they do not appear to have been attracted by it; the Indians never resorted there for permanent encampment, and as a dense forest growth covered the mountains it long remained an uninviting, trackless region. An occasional venturesome mountaineer entered the country, but not until 1870 was there any trustworthy account of a journey across its central portion. At that time the region was the largest tract of unexplored country in the Rocky Mountains. In 1871 Dr. F. V. Hayden, United States Geologist, visited the region, accompanied by a corps of skilled scientific assistants, including geologists, topographical engineers, and a photographer. His expedition was eminently successful, and immediately attracted the attention of the world. It must always redound to the credit of Dr. Hayden that he fully appreciated the exceptional character of the region and the advisability of its forever being held intact by the General Government. He laid the matter before the Congress of the United States, and upon his earnest solicitation the park was established.[39] In the organic act of 1872, defining the park, Congress declared that the reservation was "dedicated and set apart as a public park or pleasure ground for the benefit and enjoyment of the people." The wisdom and foresight of those who at that time urged the withdrawal from settlement of this tract of land from the public domain has never been questioned.

Since the setting aside of the region as a national park, the U. S. Geological Survey has done much towards investigating the natural phenomena found there.[40] The park is under the care of the Secretary of the Interior, who is authorized to carry out the provisions of the law, to make all rules and regulations for its protection and maintenance. The superintendent is a military officer, with headquarters at the Mammoth Hot Springs.

The area of the Yellowstone Park[41], as at present defined, is somewhat more than 3,300 square miles. The central portion is essentially a broad volcanic plateau between 7,000 and 8,500 feet above sea level, with an average elevation of 8,000 feet. Surrounding it on the south, east, north, and northwest are mountain ranges with culminating peaks and ridges rising from 2,000 to 4,000 feet above the general level of the inclosed table-land. Beyond the mountains the country falls away in all directions, the lowlands and valleys varying in altitude from 4,000 to 6,000 feet above sea level.

The Gallatin range incloses the park on the north and northwest. It lies directly west of the Snowy range, only separated by the broad valley of the Yellowstone River. It is a range of great beauty, of diversified forms, and varied geological problems. Electric Peak, in the extreme northwestern corner of the Park, is the culminating point in the range, and affords one of the most extended views to be found in this part of the country. Archean gneisses form a prominent body on the west side of the range, over which occur a series of sandstones, limestones, and shales of Paleozoic and Mesozoic age, representing Cambrian, Silurian, Devonian, Carboniferous, Trias, Jura, and Cretaceous strata. Immediately associated with these sedimentary beds are large masses of intrusive rocks which have played an important part in bringing about the present structural features of the range. They are all of the andesitic type, but show considerable range in mineral composition, including pyroxene, hornblende, and hornblende-mica varieties. These intrusive masses are found in narrow dykes, in immense interbedded sheets forced between the different strata, and as laccolites. The valleys are deeply scored by ice, and the rocks of the range may be found strewn all along the Gardiner River and well out over the valley of the Yellowstone.

South of the Park the Tetons stand out prominently above the surrounding country, the highest, grandest peaks in the northern Rocky Mountains. The eastern face of this mountain mass rises with unrivaled boldness for nearly 7,000 feet above Jackson Lake. Northward the ridges fall away abruptly beneath the plateau lavas of the park, only the outlying spurs coming within the limits of the reservation For the most part the mountains are made up of coarsely crystalline

451 GE——22

gneisses and schists, probably of Archean age, flanked on the northern spurs by uplifted Paleozoic rocks.

East of the Tetons, across the broad valley of the Upper Snake, generally known as Jackson Basin, lies the well-known Wind River range, famous from the earliest days of Rocky Mountain trappers. The northern end of this range is largely composed of Mesozoic strata, single ridges of upper Cretaceous sandstone penetrating still further north into the regions of the Park until buried beneath massive flows of lava.

Along the entire east side of the Park stretches the Absaroka range, so called from the Indian name of the Crow Nation. The range is intimately connected with the Wind River range, the two being so closely related that any line of separation must be drawn more or less arbitrarily, based more upon geological structure and forms of erosion than upon any clearly defined physical limitations. The Absarokas stretch for more than 80 miles, a rugged, unbroken mountain mass, without any good pass across them. They have always stood as a formidable barrier to all western progress, and to-day are only crossed by hunters and mountaineers by one or two dangerous trails known to but few. All the upper portion of the range is formed of eruptive rocks that have poured out in such enormous masses as to conceal an earlier range made up largely of Mesozoic and Paleozoic strata, extending from the Cambrian to the Upper Cretaceous. The latter are seen all along the east base of the range, and at the northern end in Clarks Fork valley, and in the Park at the junction of Soda Butte creek and the Lamar river.

At the northeast corner of the Park a confused mass of mountains connects the Absarokas with the Snowy range. This latter range shuts in the park on the north, and is an equally rough region of country, with elevated mountain masses covered with snow the greater part of the year, as the name would indicate. Only the southern slopes, which rim in the Park, bear upon the geology of the region. Here the rocks are mainly granites, gneisses and schists, with sedimentary beds, for the most part, referable to pre-Cambrian series. They are in great part overlain by Tertiary lavas.

The region has been one of profound dynamic action and a center of mountain building on a grand scale. So far as the age of these mountains is concerned, evidence goes to show that upheaval was contemporaneous in all of them, and coincident with powerful dynamic influences which uplifted the north and south ranges stretching across Colorado, Wyoming, and Montana. These dynamic movements blocked out for the most part the Rocky Mountains near the close of the Cretaceous, although there is good reason to believe that in the region of the Park profound faulting and displacement continued the work of mountain building into much later time. By the building up of these mountains a depressed basin was formed, everywhere inclosed by high land. Later the pouring out of vast masses of lava converted this depressed region into the Park plateau. Tertiary time in the Park was characterized by great volcanic activity, enormous volumes of erupted material being forced out. This activity extended through the Pliocene period and probably well on into the Pleistocene. Within recent time there is no evidence of any extensive outbursts; indeed, volcanic energy may be considered long since extinct.

The volcanic rocks present a wide range in chemical and mineral composition. They may all, however, be classed under three great groups—andesites, rhyolites, and basalts.

Andesites have played a most important part in bringing about the present configuration of the mountains surrounding the Park plateau. As already mentioned, in the Gallatin range they form large masses, and most of the culminating peaks in the Absarokas are composed of compact basic andesites or agglomerates accompanied by basaltic flows. Andesites, however, are not confined to the mountains, but also forced their way to the surface in the interior depressed basin.

That the duration of andesitic eruptions continued through a long period of time is made evident by plant remains embedded in volcanic ashes and mud associated with layers of breccia and more or less compact lavas, which accumulated to a depth of nearly 2,000 feet. Much of this plant material is in an excellent state of preservation and it is in these beds that the well-known fossil forests occur. In the grand escarpments along the Lamar valley the forest-bearing beds are admirably displayed, erosion having cut numerous lateral ravines and gorges in the lava beds, many trees still standing in upright position.

In late Cretaceous or early Tertiary times, a volcano burst forth in the northeast corner of the basin not far from the junction of the Absaroka and Snowy ranges. It rises from a base about 6,500 feet above sea level, the culminating peak attaining an elevation of 10,000 feet. This gives a height to the volcano of 3,500 feet from base to summit, measuring from the Archean rocks of the Yellowstone Valley to the top of Mount Washburne. The average height of the crater rim is about 9,000 feet above sea level, the volcano measuring 15 miles across the base. The eruptive origin of Mount Washburne has long been recognized, and it is frequently referred to as a volcano. It is, however, simply the highest peak among several others, and represents a later outburst which destroyed in a measure the original rim and form of an older crater. The eruptions for the most part were basic andesites and basalts. Erosion has so worn away the earlier rocks, and enormous masses of more recent lavas have so obscured the original form of lava flows, that it is not easy for an inexperienced eye to recognize a volcano, and that the surrounding peaks are the more elevated points in a grand crater wall. By following around the ancient andesitic rim, and studying the outline of the old crater, together with the composition of its lavas, its true origin and history may readily be made out. This older crater has, as yet, received no special designation, but when our maps and reports are finally published, this ancient geological ruin will receive an appropriate designation. This old volcano occupies a prominent place in the geological development of the park, and dates back to the earliest outbursts of lava which have in this region changed a depressed basin into an elevated plateau. We have here a volcano situated far inland, in an elevated region, in the heart of the Rocky Mountains. It lies on the eastern side of the continent, only a few miles from the great continental divide which sends its waters to both the Atlantic and Pacific.

After the dying out of the andesitic lavas, followed by a long period of erosion, immense volumes of rhyolite were erupted which not only threatened to fill up the crater but to bury the outer walls of the volcano. On all sides the basic lava slopes were submerged beneath the rhyolite to a height of from 8,000 to 8,500 feet. These great flows of rhyolite did more than anything else to bring about the present physical features of the Park table-land. But few large vents or centers of eruption for the rhyolite have been recognized, the two principal sources being the volcano to which reference has already been made, and Mount Sheridan, a volcano in the southern end of the park.

Mount Sheridan stands unsurpassed as a commanding peak, rising grandly above the general level of the plateau, with an elevation of 10,200 feet above sea level and 2,600 feet above Heart Lake at its eastern base. From the summit of this peak on a clear day one may overlook the entire plateau country and the mountains which shut it in, while almost at its base lie the magnificent lakes which add so much to the quiet beauty of the region, in contrast with the rugged scenery of the mountains. From no point is the magnitude and grandeur of the volcanic region so impressive.

Taking the bottom of the basin at 6,500 feet above sea level, these acidic lavas piled up until the accumulated mass measured 2,000 feet in thickness. In none of the deep gorges like the Yellowstone, Gibbon, and Madison canyons, are the underlying sedimentary rocks exposed.

The Park plateau, built up of rhyolite flows, embraces an area 50 by 40 miles, with a mean altitude of 8,000 feet. Strictly speaking, in the common acceptation of the word, it is not a plateau; at least, it is by no means a level country, but an undulating region characterized by bold escarpments and abrupt edges of mesa-like ridges. It is accidented by shallow basins of varied outline and scored by deep canyons and gorges. The rhyolites rest against the steep slopes of the Absarokas and bury the northerly spurs of the Wind River and Teton ranges. On both sides of the Gallatin range the rhyolites encircle sedimentary beds at about the same level. From Electric Peak to the Tetons, one may travel the entire distance of 60 miles without leaving the rhyolites. Nowhere, except in limited outbursts along the base of the Absarokas, do the rhyolites penetrate the mountains.

Although the rocks of the plateau for the most part belong to one group of acidic lavas, they by no means present the great uniformity and monotony in field appearance that might be expected. These 2,000 square miles offer as grand a field for the study of structural forms, development of crystallization, and mode of occurrence of acidic lavas as could well be found anywhere in the world. They vary from a nearly holocrystalline rock to one of pure volcanic glass. Obsidian, pumice, pitchstone, ash, breccia, and an endless development of transition forms alternate with the more compact lithoidal lavas which make up the great mass of the rhyolite, and which in color, texture, and structural development present an equally varied aspect. In mineral composition these rocks are simple enough. The essential minerals are orthoclase and quartz, with more or less plagioclase. Sanidine is the prevailing feldspar, although in many cases plagioclase forms occur nearly as abundantly as orthoclase. Chemical analyses, whether we consider the rocks from the crater of Mount Sheridan, the summit of the plateau, or the volcanic glass of the world-renowned Obsidian Cliff, present comparatively slight differences in ultimate composition.

Occasional thin sheets of basalt reached the surface before the completion of rhyolitic outbursts, flows of acidic lava overlying basic ones at several localities. In general, however, basaltic lava followed rhyolite. The basalt occurs near the outer edge of the rhyolitic mass, and in no single instance is there an extrusive flow of basalt in the central portion of the plateau.

After the close of the basaltic eruptions and the dying out of volcanic energy came the period of glaciation of the entire region. In the Teton range several characteristic glaciers still exist upon the slopes of Mount Hayden and Mount Moran, remnants of a much larger system of glaciers. Broad névé fields may be seen in the Bear Tooth mountains northwest of the Park. The Park region presents so broad a mass of elevated country that not only the surrounding mountains but the entire plateau was in glacial times covered with ice. Glacial lakes, kames, terminal moraines, and nearly all the phenomena of ice action usually seen in glaciated regions may be found here. A remarkable and exceptional feature, which is shown here on a grand scale, is the action of thermal siliceous waters on glacial drift.

Over the Absaroka range glaciers were forced down into the Lamar and Yellowstone valleys, thence westward over the top of Mount Evarts to the Mammoth Hot Springs basin. On the opposite side of the Park, the ice from the summit of the Gallatin range moved eastward across Swan valley and passing over the top of Terrace mountain joined the ice field coming from the east. The united ice sheet plowed its way northward down Gardiner valley to the Lower Yellowstone, where the broad valley may be seen strewn with the material transported from both the east and west rims of the Park. It has been named the Yellowstone glacier.

A second powerful glacier moved southward over the plateau and down the broad valley of the Snake, receiving a number of tributaries from both the Wind River and Teton ranges.

Since the building up of the Park plateau, glacial erosion has greatly modified all surface features of the Park. It has broadened and deepened preexisting drainage channels, opened new waterways, and cut

Fig. 15.—Glacial boulder near the Yellowstone canyon.

magnificent gorges in the rhyolite plateau. Such gorges as the Yellowstone, Gibbon, and Madison canyons, in the strictest sense of the word, have all been carved out in recent time. These canyons are several miles in length and from 700 to 1,400 feet in depth.

To the geologist one of the most impressive objects on the Park plateau is a transported boulder of granite which rests directly upon the rhyolite near the bank of the Grand Canyon, about three miles below the Falls of the Yellowstone. It stands alone in the forest, miles from the nearest glacial boulder. Glacial detritus carrying granitic material may be traced upon both sides of the canyon wall, but not a fragment of rock more than a few inches in diameter, older than rhyolites, has been recognized within a radius of many miles. This massive block, although irregular in shape and somewhat pointed towards the top, measures 24 feet in length by 20 feet in breadth and stands 18 feet above the base. The nearest point from which it could have been transported is distant 30 or 40 miles. Coming upon it in the solitude of the forest, with all its strange surroundings, it tells a most impressive story. In no place are the evidences of frost and fire brought so forcibly together as in the Yellowstone National Park.

Since the close of the glacial period no geological events have brought about any great changes in the physical features of the region other than those produced by the action of steam and thermal waters. Evidences of fresh lava flows within recent time are wholly wanting. Nevertheless, over the Park plateau the most unmistakable evidence of underground heat is everywhere to be seen in the waters of innumerable hot springs, geysers, and solfataras. A careful study of all the phenomena leads to the theory that the cause of high temperatures of these waters is to be found in the heated rocks below, and that the origin of the heat is in some way associated with the sources of volcanic energy. Surface waters in percolating downwards have become heated by relatively small quantities of steam rising through fissures from much greater depths. Geysers and springs return these meteoric waters to the surface. Thermal springs, geysers, and solfataras are in a sense volcanic phenomena, and remain as evidence of the gradual dying out of volcanic energy. If this theory is correct, proof of the long-continued action of thermal waters upon the rocks should be apparent, as it is fair to suppose that they must have been active forces ever since the cessation of volcanic eruptions. This is precisely what one may see all over the rhyolite area. Ascending currents of steam and acid waters have acted as powerful geological agents in rock decomposition and have left an indelible impression upon the surface of the country. Large areas of decomposed rhyolite and extinct solfataras show the former existence of still greater thermal activity. Rock decomposition and deposition of sediment from siliceous waters are extremely slow processes, if we may judge from what we see going on to-day in the different geyser basins. It is evident that to accomplish such changes a long period of time must have been required.

An evidence of the antiquity of the hot spring deposits is shown in an equally striking manner, and by a wholly different process of geological reasoning. Terrace mountain is an outlying ridge of the rhyolite plateau, just west of the Mammoth Hot springs. It is covered on the summit with thick beds of travertine, among the oldest portions of the Mammoth Hot springs deposits. It is the mode of occurrence of these calcareous deposits from the hot waters which have given the name to the mountain. Lying upon the surface of this travertine on the top of the mountain are found glacial boulders brought from the summit of the Gallatin range, 15 miles away, which have been transported on the ice sheet across Swan valley and deposited on the top of the mountain, 700 feet above the intervening valley. They offer the strongest possible evidence that the travertine is older than the glacier which has strewn the county with transported material. How much travertine was eroded by ice it is, of course, impossible to say, but so friable a material would yield readily to glacial movement.

The number of hot springs found in the Park exceeds 4,000. If to these be added the fumaroles and fissures, from which issue in the aggregate enormous volumes of steam and acid vapors, the number of active vents would be more than doubled. There are about 100 geysers in the Park. Between a geyser and a hot spring no sharp definition can be drawn, although a geyser may be defined as a hot spring throwing, with intermittent action, a column of water and steam into the air. A hot spring may boil incessantly without violent eruptive energy; a geyser may lie dormant for years without any explosive action and again break forth with renewed force

The thermal waters of the Park may be classed under three heads: first, calcareous waters carrying calcium carbonate in solution; second, siliceous acid waters usually carrying free acid in solution; third, siliceous alkaline waters rich in dissolved silica.

Calcareous waters are confined almost exclusively to the Mammoth Hot springs, where they have built up enormous deposits of travertine with only traces of salts of magnesia and alkalies. The travertine contains from 95 to 99 per cent of calcium carbonate. The Mammoth Hot springs lie just north of the northern escarpment of the Park plateau, and while they break out in close proximity to rhyolite bodies and undoubtedly receive their heat from volcanic sources, they reach the surface through Mesozoic strata, which here form the surface rocks. Jurassic and Cretaceous limestones have furnished the lime held in solution and precipitated as travertine. With a few insignificant exceptions, only siliceous waters are found issuing from fissures in the rhyolite rocks, from which they derive their mineral contents. Acid waters may usually be recognized by efflorescent deposits of alum and soluble salts of iron, and frequently by the presence of delicate crystals of sulphur. These acid waters possess an astringent taste. Although far less common than the alkaline waters, they occur scattered over the plateau at a number of localities and may be found at the Highland springs, on the west slopes of Mount Washburne, and in the Norris Basin. Alkaline springs present more of general interest than the acid waters, as it is only in connection with the former that the geysers are found. They are the principal waters of all the geyser basins and of most of the hot spring areas. Alkaline waters deposit mainly amorphous silica as siliceous sinter, but in an endless variety of forms, as shown in the geyser cones and incrustations upon the surface and edges of the hot pools. These sinters form the brilliant white deposits found over large areas in all the geyser basins. Scorodite, realgar, orpiment, oxide of iron, wad, and other minerals occur under favorable conditions as deposits from hot waters at certain springs, each adding something of interest to the marvels of the Park.

It is these unrivalled hydrothermal manifestations, and their geological relations to the earlier volcanic eruptions, that have made the Yellowstone Park famous throughout the world, and have justly gained for it the appellation of the Wonderland of America.

Another characteristic feature of the Park, and one that adds more than anything else to the scenic charms of the plateau, are the singularly beautiful, deeply eroded canyons that carry the waters from this elevated region to the broad valleys below. They are all of comparatively modern origin, presenting all the phenomena of recent canyon cutting. Of all these picturesque gorges, the canyon of the Yellowstone stands preeminent in grandeur and sublimity of its scenery. Nearly all these gorges carry waterfalls of great beauty, each adding some special attraction to the charm of the place. Nearly twenty picturesque falls may be found within the Park.

Across the plateau lies the continental divide, separating the waters of the Atlantic from those of the Pacific. Entering the Park from the southeast corner it runs with an irregular course in a northwest direction, following along the summit of Two Ocean Plateau, the waters of the plateau draining to both oceans. The watershed follows the undulating low ridge between Shoshone and Yellowstone lakes, and, with a broad sweeping curve around the streams running into the latter lake, crosses Madison Plateau and leaves the Park a short distance south of Madison Canyon. Several large bodies of water form such characteristic features on both sides of this divide that the country has deservedly received the designation of the lake region of the Park.

About 85 per cent of the Park is forest-clad. The timber is essentially coniferous, with here and there a few small growths of aspen. Two-thirds of the trees belong to the black pine, *Pinus murrayana*, and on many of the gravelly, rhyolite ridges no other species is seen. It rarely attains any great size, but for the purpose of water protection, one of the objects for which the reservation is maintained, it meets every economic requirement.

ITINERARY OF THE YELLOWSTONE PARK.

By Arnold Hague.

	Miles.
Cinnabar to Mammoth Hot Springs	7
Mammoth Hot Springs to Norris Geyser Basin	22
Norris Geyser Basin to Lower Geyser Basin	20
Lower Geyser Basin to Upper Geyser Basin	8
Upper Geyser Basin to Yellowstone Lake	39
Yellowstone Lake to Yellowstone Falls	18
Yellowstone Falls to Mammoth Hot Springs	33
Mammoth Hot Springs to Cinnabar	7

At Cinnabar station the travelers leave the railway and continue their journey for the next few days in stages. The road follows the Yellowstone river to Gardiner, thence up the Gardiner river to the Mammoth Hot springs. The northern boundary of the Yellowstone Park passes east and west through the junction of the Yellowstone and Gardiner rivers. About one and one-half miles beyond, the boundary line between the States of Montana and Wyoming is crossed. Entering the Park the road follows the river, with the long spurs of Mount Evarts on the left and those of Sepulchre mountain on the right. Mount Evarts, which rises 2,000 feet above the stream, affords an excellent exposure of Middle Cretaceous sandstones and shales dipping away from the river. Leaving the river the road crosses an ancient travertine deposit, and ascending a steep hill reaches the Mammoth Hot Springs hotel, situated on the finest of the travertine terraces. Travelers generally reach the hotel about noon, and the remainder of the day is spent in examining the hot springs and terraces, and the geological features in the neighborhood. Several days might be spent here with profit, visiting objects of interest within a radius of 10 miles of the springs.

MAMMOTH HOT SPRINGS.

The Mammoth Hot Springs deposits consist entirely of travertine, derived from waters heavily charged with carbonate of lime. The total area covered by travertine is about two square miles, occupying a narrow valley lying between Terrace and Sepulchre mountains. A continuous deposit extends from the Gardiner river to the top of Terrace mountain, a vertical distance of nearly 1,400 feet, the width and depth of travertine depending upon the form of the original valley. The top of Terrace mountain lies about two miles back from the river. On the north side of the valley, Cretaceous and Jurassic rocks

may be seen rising abruptly from beneath the travertine which rests
against them. A series of terraces extends all the way from the river
to the top of the mountain, eight of which are well-defined benches

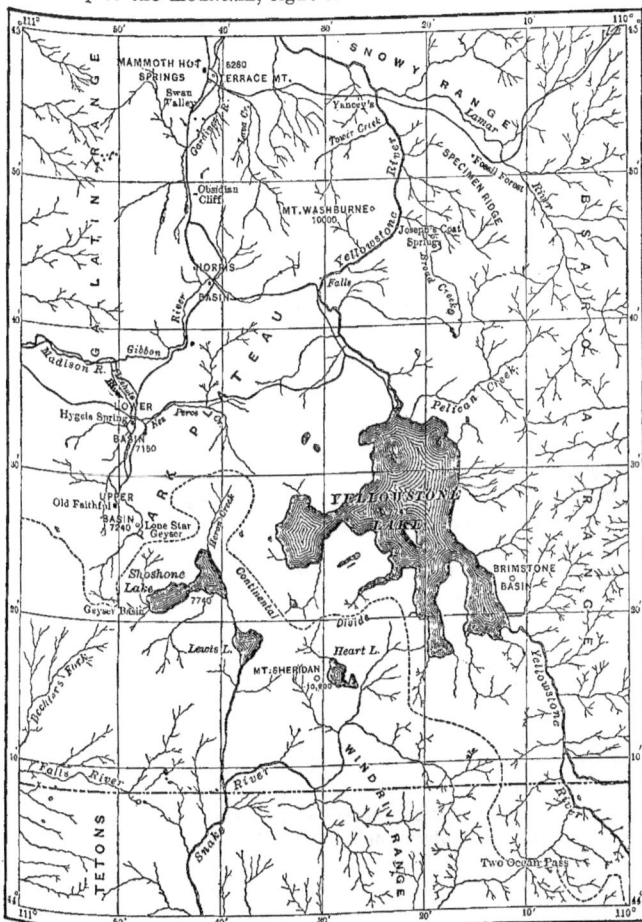

Scale: 1 Inch-12 miles.   **YELLOWSTONE NATIONAL PARK**

with more or less level floors and steep slopes facing the open valley.
Seen from any commanding point of view, this series of terraces,
inclosed within a mountain gorge suggests the terminal front of a

mountain glacier. The Hotel terrace is the broadest of all the terraces, with an area over 83 acres in extent. It is situated 500 feet above the river and from it most of the others are in full sight. All active springs are found either upon the Hotel terrace or upon those higher up the valley. The number of active springs may vary from year to year, some becoming extinct, while new vents are opened. The temperature of the springs ranges from 80° to 165° F., in all of which algæ have been found growing. This peculiar vegetation plays an important part in the formation of travertine by the secretion of lime, and much of the exquisite beauty of the springs and brilliancy of coloring is produced by these low forms of plant life.[43]

The principal objects of interest are the extinct springs, Liberty Cap and the Thumb, on the Hotel terrace, and such active springs as Pulpit Basins, Marble Basins, and the Blue spring, on Main terrace, and still higher Cleopatra's Bowl, Cupid's Cave, and the Orange spring. There are also innumerable small caves and fissures, each having special features of interest; some of the caves contain carbon dioxide in sufficient quantity to be dangerous to animal life. The largest active springs are centered on Main terrace, 8¾ acres in extent and 250 feet above Hotel terrace. Beautiful clear pools abound, the largest of which is nearly 100 feet in diameter, with a temperature of 136° F. Blue spring on this terrace is perhaps the most interesting of all in its phenomena of travertine deposition. The building up of travertine in a series of small basins, one above the other, the delicacy of coloring from algous growth, and the overflow of hot water are admirably shown. The spring has a temperature of 165° F.

Over the greater part of the travertine area hot springs have long since ceased to flow, although they may break out anew at any time. Where they have long lain dormant, the spring deposits are now covered by a coniferous forest. In some instances the trees have been killed by fresh outflows of hot water, the dead and bleached trunks still standing with their roots buried in travertine.

Sepulchre mountain, to the northwest of Mammoth Hot springs, stands out boldly as a volcanic peak, on the northern boundary of the Park. It consists of flows of compact andesite and breccias stretching in long gentle slopes toward the Yellowstone and Gardiner rivers. These spurs exhibit a succession of ice-carved benches, the surface being strewn with glacial débris from the Gallatin Mountains.

Mount Evarts on the east and Terrace mountain on the west shut in the Mammoth Hot springs on two sides, the former by a bold wall rising abruptly above Gardiner river and the latter by long gentle slopes inclined toward the river. Facing the basin on the south rises a grand escarpment of volcanic rocks, the northern edge of the Park plateau, stretching from the head of Lava creek westward as far as

Bunsen peak. Terrace mountain is an outlying mass of the plateau, formed of rhyolite and capped at its northern end by travertine. The cliff on the southern end of the summit of Mount Evarts is a thin extension of the great rhyolite lava sheet, separated from the main mass by the erosion of Lava creek.

Within a few miles of the Mammoth Hot springs are several beautiful waterfalls, formed by the waters of the plateau leaping over walls of compact basalt in their descent to the lower country. Among them may be mentioned Osprey falls on the Gardiner river, and Undine falls on Lava creek, both of which are well worth a visit.

MAMMOTH HOT SPRINGS TO NORRIS BASIN.

Leaving the Mammoth Hot springs (6,250 feet), the road gradually ascends over rhyolitic rocks more or less covered by alluvium and glacial débris. On the west side of the road, just before reaching the Golden Gate, immense blocks of travertine may be seen piled up one upon the other in a confused mass in a manner at first difficult to understand. They are best explained by supposing them to have been thrown forward from Terrace mountain by some sudden earthquake shock which thrust the easily displaced rock from its original position down the side of the mountain.

At Golden Gate the road enters the picturesque gorge of Glen creek, which separates the plateau escarpment from Terrace mountain. At the upper end of this gorge the road enters upon the open grassy plain of Swan Lake valley (7,200 feet), a northern extension of the plateau. A narrow ridge of andesitic rocks—a spur of Sepulchre mountain—shuts in the valley on the west. Opposite Swan lake andesite gives way to rhyolite, which in turn is replaced by basalt, forming the extreme southern end of the ridge and falling away gradually to the level of the plain. This basalt, in thin flows, stretches across the valley, and its southern limit is sharply defined by the course of Gardiner river. The valley is strewn with glacial drift from the Gallatin mountains, and evidence of ice movement is everywhere to be seen, especially in low morainal ridges trending across the valley, which are shown in cross section by the wagon road. From Swan Lake valley a fine view is obtained of the Gallatin range to the west, extending all the way from Electric Peak (11,125 feet) to Mount Holmes (10,578 feet) a distance of 13 miles in an air line. Snow lies upon the higher summits well into midsummer.

Crossing Gardiner river, the road passes up the valley of Obsidian creek, the lower part of which is known as Willow park, a long strip of meadow land the borders of which are in great part covered with a luxuriant growth of willow, beyond which rise the stately pines of the plateau. The scenery now assumes a monotonous appearance, due to the almost uniformly level character of the rhyolite ridges, through which the streams follow long straight valleys characteristic of this part of the Park. The rock is everywhere rhyolite, which fortunately exhibits a great variety of modifications. On the west side of the road, as it approaches Obsidian Cliff, there is a small exposure of columnar rhyolite.

Obsidian Cliff, on the east side of the stream, near the outlet of Beaver lake, is of more than ordinary interest to all tourists, but especially so to the geologist, on account of its peculiarities of structure and the development of spherulites and lithophysæ found in it. The cliff rises in nearly vertical walls from 150 to 200 feet above the stream. It has been formed by a surface flow breaking out and running over the rhyolite plateau, covering an area of about 10 square miles. The obsidian is a natural glass, the result of rapid cooling from a fused mass of highly acid lava, and has much the same chemical composition as the plateau rhyolite. On the surface, at the northern end of the cliff, it grades into pumice and lithoidal rhyolite. I am indebted to Mr. Joseph P. Iddings, who has made a careful study of Obsidian Cliff, for the following note:

The columnar structure at the southern end of the cliff is particularly well developed. The compact black obsidian passes into lithoidal rock northward along the face of the cliff, where various phases of crystallization may be studied in situ.

On the top of the plateau, to the east, the dense obsidian passes upward into white pumice, which has been more or less removed by glacial action. The fresh, unaltered condition of rock permits the mineralogical character of the crystallization to be observed in the utmost detail. The perfection of the lithophysæ and spherulites, and the richness of the microscopic spherulitic growths constitute its most important petrographical feature.

The lamination of the lithoidal portion of the mass and the general lamellar distribution of different kinds of structure, as well as of the gas bubbles in the vesicular and pumiceous parts of the sheet, furnish valuable evidence as to the agent most active in promoting the various kinds of crystallization in this and similar rhyolitic lavas. This agent was undoubtedly water-vapor.[42]

A short distance south of Obsidian Cliff lies Beaver lake, across which are several dams kept in repair by beavers who inhabit its waters. Since the rigid protection of game in the park the beaver are rapidly increasing in number, and several of their houses may be seen in the lake.

Shortly after passing Beaver lake the rhyolite begins to show the effects of hydrothermal action, and numerous areas of rock decomposition may be observed on both sides of the road.

Four miles south of Obsidian Cliff, Roaring mountain is passed. This is a bluff rising 500 feet above the road and one of the highest points on the lava ridge. It takes its name from the shrill, penetrating sound of the steam constantly escaping from one or more vents located near the summit, and on a calm day, or with a favorable wind, the rushing of the steam through the narrow orifices can be distinctly heard from the wagon road. The mountain is one mass of altered rhyolite, whitened by long action of steam and acid vapors. The entire ridge, lying between the road and Solfatara creek to the east, is largely formed of these highly altered rhyolites. It is evident that this region was at one time a center of long continued and energetic thermal activity.

Beyond Twin Lakes the water drains southward to Gibbon river, and from here to Norris basin along the roadside at the base of the cliff there is a succession of steam vents, hot springs, and mud pots.

NORRIS BASIN.

This geyser basin (7,350 feet) occupies a depressed area on the east side of the Gibbon river, the waters of the basin draining toward the stream. Rhyolite ridges surround the basin, gradually rising on all sides toward the summit of the plateau. In many respects this basin is the most instructive of all, as the varied phenomena of thermal action in all its phases are more clearly shown here than elsewhere. All stages of rock decomposition and of deposition of sediments from hot waters may be seen and are easily accessible to the pedestrian within a short walk of the hotel.

The basin contains 14 geysers, which, although neither so grand nor impressive as those in the Lower and Upper Geyser basins, are, on account of the varied phenomena exhibited, quite as interesting to the geologist. Several of them are of quite recent origin, as is shown by the freshness of the rocks through which the steam issues and by the absence of sinter deposits. The Monarch is one of the most interesting of the group. It breaks out through a narrow vertical fissure, 20 feet long, in the rhyolite. Eruptions take place every four hours, the water being thrown into the air for 50 feet. Other interesting geysers are the Arsenic, Constant, Congress, Fearless, and Pearl. Owing to the great number of steam vents and active orifices, each with some characteristic feature of its own, the basin presents one of the most weird and desolate regions in the Park.

NORRIS BASIN TO LOWER GEYSER BASIN.

Shortly after leaving Norris basin the road crosses Gibbon meadows, a broad open plain which in places is almost impassable owing to the wet marshy nature of the bottom. The Gibbon river runs through the Meadows. The ditches constructed for drainage purposes along the roadside expose, underlying the meadow, a fine white earth several feet in thickness, largely composed of diatoms developed in the siliceous waters. Similar diatomaceous ooze is found in nearly all the geyser basins, although seldom on so extensive a scale. Encircling the Meadows, at the base of the hills, are a number of local centers of thermal activity from which large quantities of hot siliceous waters drain into the basin.

A short distance off the road, on the east side of the Gibbon meadows, and easily reached by wagon, lie the Artists' Paint Pots. They consist of a number of small springs of hot water reaching the surface through brilliantly colored clays. These excessively fine clays are the products of rock decomposition by slow and long-continued processes, and the mineral matter held in suspension is the cause of the varied colors in the different pots. In the white pots the coloring matter has been leached out, leaving a pure white kaolin impalpable to the touch. The deep indian-red pots carry finely comminuted iron oxide, which, under favorable conditions, collects in certain springs. The hillsides are brilliant with decomposition products derived from rhyolite in various stages of alteration.

From the Meadows the road follows the Gibbon-river through Gibbon canyon, the imposing walls of which rise in sheer cliffs 1,000 feet above the stream. It is a grand exposure of rhyolite walls, exhibiting remarkable forms of rock erosion. The river makes a rapid descent, and at Gibbon falls, 80 feet high, flows over a fine example of obsidian worn smooth by the rushing waters. Along the base of the canyon walls, in close proximity to the river, steam vents and hot springs mark the course of thermal action. Beryl spring, on the west bank near the northern end of the canyon, is worthy of attention from the exquisite coloring of the constantly agitated water which has built up a delicate rim of sinter encircling the pool.

At Canyon creek the road leaves the Gibbon river and follows the ridge of rhyolite on the east side. A short distance below here, the Gibbon and Firehole unite to form the Madison, the latter stream being one of the main tributaries of the Missouri. From a commanding point on the road a distant view may be had of the Madison canyon, extending in an east and west line directly across the plateau. The precipitous walls on the north side of the canyon rise for 1,500 feet above the river. Madison plateau stretches as far as the eye can reach, without a break, beyond which both rhyolite and basaltic lavas extend westward over the great plains of Snake river.

From the same point of view, on a clear day, Mount Hayden, which is not only the culminating point in the Teton range, but the highest peak in this part of the Rocky Mountains, may be distinctly seen 60 miles to the south. Descending a steep ridge over rhyolitic gravels, the road comes out on the Firehole river, and after following the bank for two or three miles enters the Lower Geyser basin.

LOWER GEYSER BASIN.

The Lower Geyser basin is about three miles square and is the largest of the geyser areas. It is roughly rectangular in shape and at the junction of Firehole river with Nez Percé creek, the lowest point in the basin, has an altitude of 7,100 feet. The Firehole river coming in from the south runs in a northerly direction across the basin, receiving all the waters brought to the surface by the geysers and hot springs. Several tributaries of the river reach the basin from the plateau, and along these streams are a number of active hot-spring areas. Some of them situated along Sentinel creek and Nez Percé creek are among the most interesting in the basin. The Queen's Laundry, an immense hot spring on Sentinel creek, is well worth a visit.

The Lower Geyser basin contains innumerable hot springs, steam vents, paint pots, mud pots, and 17 geysers. The largest geyser in the basin and one of the finest in the park is the Great Fountain, situated in the extreme southern end of the basin, about one mile south of the hotel. It is in every way a typical geyser. The brilliant, deep blue pool of water measures over 100 feet in diameter, resting upon a broad circular sinter mound, which stands about three feet above the rhyolite base. The formation of siliceous sinter and the phenomena of geyser action are clearly shown here, while the exquisite beauty of the basin will ever place the geyser in the first rank. A large volume of water is thrown violently into the air to a height of 75 feet. The eruptive action lasts about 20 minutes.

The Fountain geyser is situated only 200 or 300 yards from the hotel, and plays fairly regularly three or four times a day. Instead of issuing from a cone or mound, the water is thrown out from a funnel-shaped basin.

A few steps from the Fountain are the famous Mammoth Paint Pots, similar to others seen throughout the Park, although on a much larger scale. The clay has a delicate blue color and the consistency of dilute porridge.

Other geysers of interest are the Surprise, Spray, and White Dome.

LOWER TO UPPER GEYSER BASIN.

The road, after leaving the hotel near the edge of the timber on the east side of the Lower Geyser basin, crosses the broad sinter plain, and, after reaching the Firehole, follows closely the bank of that stream to the extreme eastern end of the Upper Geyser basin.

From the junction of the Firehole and Nez Percé creeks, active and extinct hot springs may be seen all along the river bank or in close proximity to it, steam from fissures frequently rising within a few inches of running water and alongside of cold springs.

451 GE——23

On the west side of the river, between the two large geyser basins, lies a small but interesting geyser region known as Midway basin. It is a dazzling white sinter plain, without tree or meadow, the only relief to the eye being enormous volumes of steam rising from hot lakes and cauldrons. This basin contains the Excelsior, the grandest and most imposing geyser, and Prismatic lake, a singularly beautiful sheet of water unsurpassed for brilliancy of color and exquisite beauty of its rim. Excelsior geyser throws an enormous column of water, the more powerful eruptions emitting a stream 250 feet into the air, measuring 20 feet in diameter at the base and breaking into a fan-shaped body above. It rises from a circular cauldron of boiling, steaming water, the level of which lies about 15 feet below the surface of the sinter plain. The cauldron wall affords an excellent opportunity for a study of sinter deposition. The amount of water thrown out during any violent explosion reaches many thousand barrels, which, pouring over the edge of the cauldron, runs rapidly across the sloping plain and down the banks of the river into the stream below. The level of the Firehole river is frequently raised several inches, the water showing a marked increase in temperature for a long distance below the geyser. Frequently large blocks of sinter are hurled violently into the air by the force of the explosion.

The bluff on the east side of the Firehole river, opposite Excelsior geyser, is rhyolitic pearlite, and from the top of the cliff, on a clear day and the wind westward, Prismatic lake may be seen to great advantage.

From Midway basin the road passes a number of hot springs, but none of special interest until reaching the Sapphire group, situated on the west side of the Firehole. It contains a number of small geysers, each exhibiting some novel feature of thermal action peculiar to itself. Sapphire pool, a large circular basin raised slightly above the sinter plain, has scarcely any rival among the marvelous springs along the Firehole. It closely resembles the Great Fountain in the character of its sinter deposits and overflow basin. The water is of the deepest blue and the temperature always stands near the point of ebullition. Near by are the Jewel, Silver Globe, and Avoca.

For a mile along the river, before reaching the Upper Geyser basin proper, there is a line of hot springs and geysers, indicating a great amount of thermal action along the valley. Among the most important may be mentioned the Cauliflower, Gem, Artemisia, and Morning Glory. A good illustration of the difficulty which may sometime arise in distinguishing a hot spring from a geyser is seen in the case of the Artemisia, which for a long time was supposed to be a quiet pool, but in recent years has exhibited all the phenomena of explosive geyser action.

UPPER GEYSER BASIN.

At the Riverside geyser the road crosses the Firehole river by a bridge, which is generally regarded as marking the entrance to the Upper Geyser basin. The basin measures about two and one-half miles in length and one and one-quarter miles in width, and contains the greatest number and, with the exception of the Excelsior, the largest geysers in the park. At the Grand geyser, a central point, the elevation is 7,200 feet above sea level. Like the Norris and Lower Geyser basins it occupies a depressed area in the rhyolite lavas, with ridges rising more or less abruptly on all sides, the Madison plateau on the west presenting a bold escarpment. The Firehole river extends the entire length of the basin, and with its tributaries drains the region. The basin lies within 5 miles of the continental water-shed which separates the Atlantic from the Pacific drainage. The lowest passes between the two are not more than 800 feet above the basin, and the line of the Continental Divide makes a sharp bend and loop to the southeast in order to inclose the drainage area of the Firehole on the Atlantic side.

There are between 45 and 50 geysers in the Upper basin, and nine of them may be regarded as geysers of the first order. The Giant, Giantess, Grand, Splendid, Grotto, Castle, Beehive, Oblong, and Old Faithful are found within a few hundred yards of the river and within easy walking distance from the hotel, which is situated near Old Faithful. While the geysers present much in common, each offers distinctive features in the display of the water thrown out, the quantity discharged, the duration of explosive energy, and in the intervals between eruptions. The height of the column of water in these large geysers varies from 60 to 250 feet. Old Faithful is the most regular, with intervals averaging sixty-five minutes, throwing a column of water varying from 90 to 150 feet. The Giantess issues from a deep, funnel-shaped pool. The Giant has built up a sinter cone 10 feet in height. The Castle breaks out from the top of a series of sinter terraces. The Grand presents a quiet, shallow basin on the level with the sinter plain. The Splendid issues from an unpretending pool not unlike hundreds of hot springs in the park. Black Sand and Emerald springs afford excellent opportunities for a study of algous growths in hot waters. There is a most interesting group of springs and geysers, seldom seen by tourists but well worth a visit, on Iron creek, a tributary of the Firehole.

Plate II represents Old Faithful in action. It is a fine exhibition of the geyser, which is seen to the best advantage on a calm day, the absence of all wind permitting the water to fall freely in a perpendicular column broken up into graceful arrow-shaped bodies. The water column is surrounded on all sides by steam and dense vapor. The broad circular mound surrounding the orifice of the geyser is built up by a series of low sinter terraces, many of them holding pools into which the descending water falls. The great regularity of Old Faithful makes it an object of intense interest to tourists and one seen by all visitors to the Park.

LOWER GEYSER BASIN TO YELLOWSTONE LAKE.

Returning to the Lower Geyser basin the road turns off to the east and follows Nez Percé valley for nearly 10 miles. On entering the valley the Morning Mist group of springs is seen on the south side of the road, skirting the base of the hills and extending up on to the ridge. The hillsides are covered with small springs and steam vents, concealed by timber during the greater part of the day, but in early morning presenting a picturesque appearance by the numerous columns of steam which rise through the dark green forest. The Nez Percé is characteristic of the larger valleys of the region, and evidences of its former occupation by ice may be seen on the long monotonous lava ridges.

A steep ascent of 900 feet leads up to the top of the central plateau, and from an opening in the timber on the edge of a steep cliff a magnificent view may be had looking backward down the valley and out over the geyser basins and Madison plateau beyond. The road then crosses a treeless portion of the plateau over pearlite, obsidian, and various modifications of glassy rhyolite.

The Highland Springs area lies just south of the road on the very top of the plateau. It is a region desolate beyond description, but of great interest in a study of the action of acid waters upon siliceous lavas. The waters of Alum creek, which rise in the Highland springs, have a very obnoxious and astringent taste.

Leaving the plateau, a descent of 200 or 300 feet brings us out into Hayden valley, a broad, shallow depression in the rhyolite. The underlying rocks are modifications of glassy rhyolite, over which occur Tertiary lake beds and morainal material of the glacial period.

After skirting the southern end of the valley along the edge of the timber, the road turns southward, following Yellowstone river for eight miles to the Lake. Shortly after leaving the valley a hot spring area several acres in extent is reached, extending from the banks of the river up on to the slopes of the ridge. It is quite like other areas in most of its thermal manifestations, but the Mud geyser and the Mud volcano have attracted more than ordinary interest. The Mud geyser behaves like an ordinary geyser, except that instead of ejecting clear water it contains a mixture of clay and water, throwing the slimy material from 20 to 30 feet in the air. The Mud volcano is situated a short distance from the geyser on the steep side of a hill. A caldron 20 feet in depth

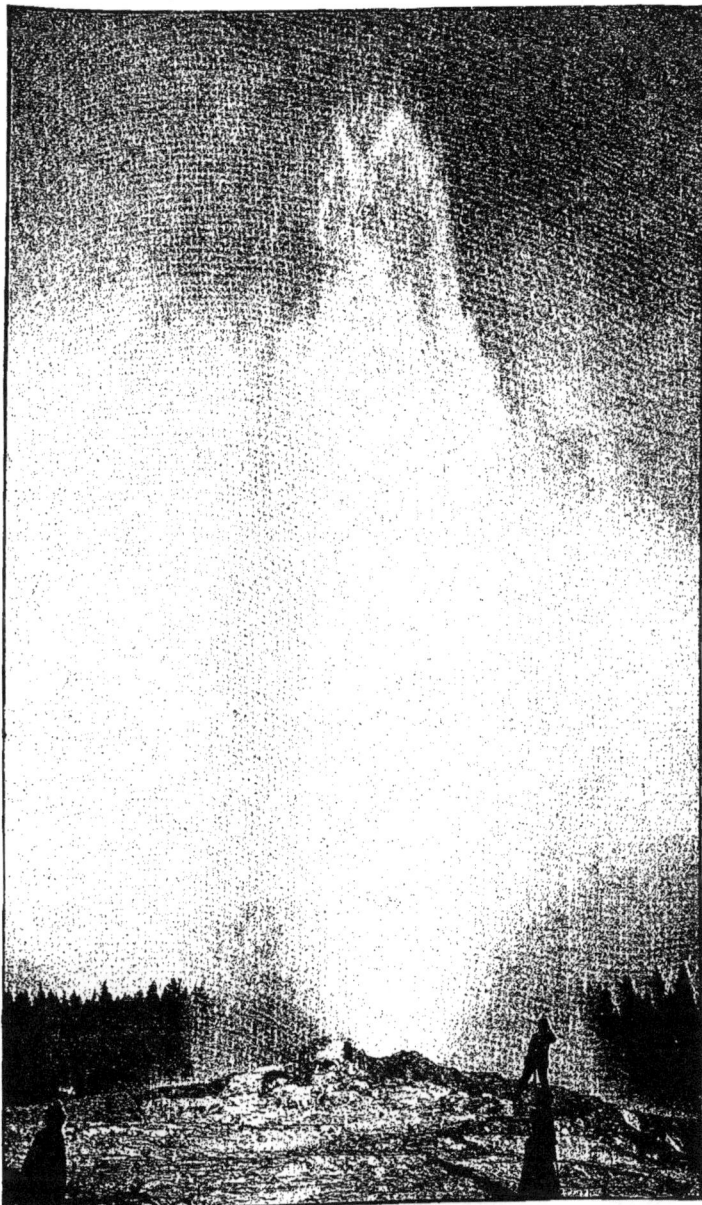

OLD FAITHFUL IN ACTION.

with steep slopes suggests a volcano, at the bottom of which are blackish gray slimes, not unlike the mud thrown out from many volcanoes. The country all about is strewn with clay pellets, showing that occasional explosive action throws out the mud for considerable distances, although in most instances the force is only sufficient to raise the clay to the rim of the caldron. Six miles from the Mud Volcano the hotel is reached, near the outlet of the lake.

YELLOWSTONE LAKE.

Yellowstone lake has an altitude of 7,741 feet above sea level, is over 20 miles in length and of very irregular width, reaching 15 miles across its broadest expansion. It embraces an area of about 140 square miles, and it requires a ride of 100 miles along its shore to complete the circuit. Rhyolites encircle the shores of the lake on all sides. On the opposite, or eastern side, the Absaroka range rises as a rugged mass of mountains, mainly composed of andesitic and basaltic rocks, the greater part of them being agglomerates and breccias. Similar rocks form the Promontory mountains projecting into the lake from the south. Flat Mountain to the south and west of the Promontory is a broad inclined table of rhyolite. Mount Sheridan, which stands out boldly to the southwest, is an extinct volcano from whose summits and sides have poured forth vast accumulations of rhyolitic lavas. Surrounding the lake on all sides and extending back from the shore are broad benches of sand and gravel, the highest of which is distinctly marked at 150 feet above the present level of the lake. This ancient lake bench may be traced along the cliffs in Hayden valley. On the beach along the west shore may be seen andesitic boulders from the Absaroka range that have been transported by ice across the lake in large quantities. On the arrival of tourists a small steamer makes a trip around the lake, giving an opportunity for obtaining fine views of the Absaroka range, Mount Sheridan, the islands, and the imposing valley of the Upper Yellowstone.

YELLOWSTONE LAKE TO YELLOWSTONE FALLS.

In order to visit the Canyon and Falls of the Yellowstone the route follows down the river and across Hayden Valley in a northwest direction over an undulating grassy plain. The first object of special interest to the visitor is Crater hills, where there is a remarkable grouping of steam vents and solfataras. These isolated hills, rising above the valley for 150 to 200 feet, are covered from base to summit with hundreds of steam vents, from which issue acid vapors depositing the most delicate sulphur crystals in innumerable cavities and fissures. Only a slight flow of water is visible at the surface, but at the base of the largest hill and close to the roadside lies Chrome spring, a constantly boiling pool of the most delicate sulphur-yellow tint. Two miles beyond Crater hills, Alum creek is again crossed just before it empties into the Yellowstone, and a short distance northward the road leaves Hayden valley and follows the west bank of the Yellowstone for three miles along picturesque rapids inclosed between walls of rhyolite.

YELLOWSTONE CANYON AND FALLS.

The Grand Canyon hotel, on the west side of the river, is situated on a glacial bench about one-quarter of a mile back from the canyon and the Lower falls of the Yellowstone. The Yellowstone canyon trends to the north and northeast, curving around au east spur of Mount Washburne, its length from the upper falls to Junction Butte being about 18 miles. At the Lower falls the walls of the canyon measure 700 feet, and five miles beyond, 1,000 feet, the width of the canyon varying from one-quarter to three-quarters of a mile. Rhyolite forms the canyon walls which are more or less decomposed by hydrothermal action, their color being due to various conditions of oxidation of the iron. Where the greatest decomposition has taken place the iron has been leached out, leaving a white kaolin-like material. A few hot springs may be seen steaming in different parts of the canyon, showing that thermal action is not yet extinct. The gaily colored walls extend about five miles beyond the Lower falls, but below this the rock is dark colored, still fresh and unaltered. The Lower falls are 310 feet in height, and may be seen to the greatest advantage from Inspiration Point, one of the cliffs which overlook the canyon below the Falls. Lookout Point furnishes another excellent point of view.

In following the bridle path along the brink of the canyon one should see the massive glacial granite boulder (Fig. 15), near where the side path leaves for Inspiration Point.

One-half mile above the Lower falls are the Upper falls, 110 feet in height. Here the rhyolite is undecomposed, mostly glassy, and more or less spherulitic, and is well exposed ou Cascade creek.

From the hotel an expedition may be made to the summit of Mount Washburne which on a clear day commands a view over the Park and the Absaroka range.

YELLOWSTONE FALLS TO MAMMOTH HOT SPRINGS.

Leaving the hotel at the Falls the road climbs up a steep ascent to the top of Solfatara plateau, an extension northward of Central plateau, which was crossed when going from the head of Nez Percé Creek to Hayden Valley. The thermal action seen on Solfatara plateau is in every way similar to that observed at Highland springs, but is by no

SKETCH OF YELLOWSTONE CANYON AND INSPIRATION POINT.

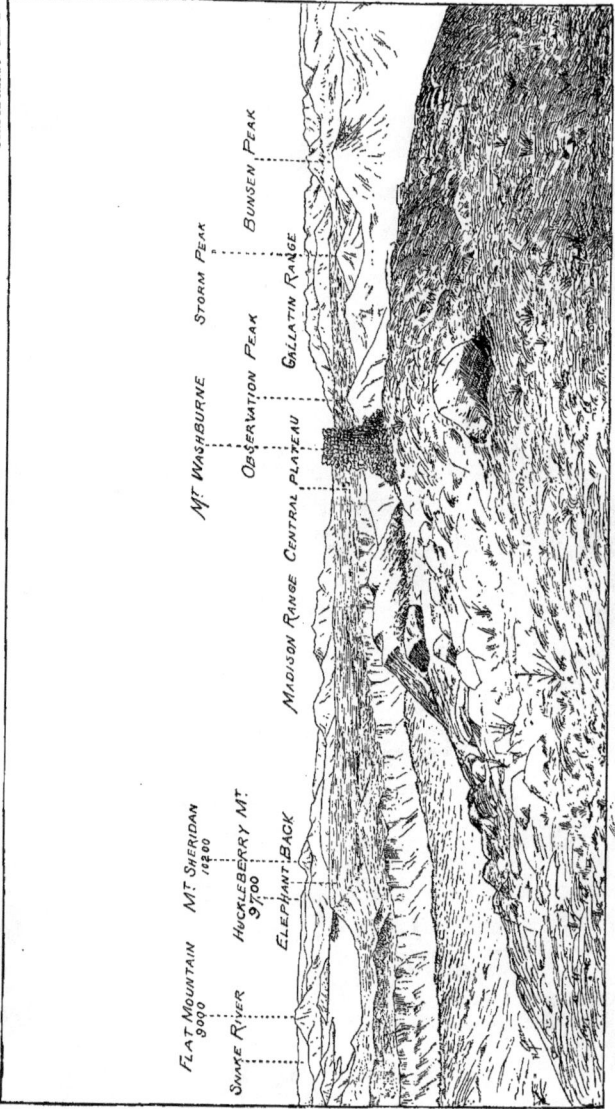

FLAT MOUNTAIN
9000

MT SHERIDAN
10200

HUCKLEBERRY MT
9700

SNAKE RIVER

ELEPHANTBACK

MT WASHBURNE

OBSERVATION PEAK

MADISON RANGE CENTRAL PLATEAU

GALLATIN RANGE

STORM PEAK

BUNSEN PEAK

PANORAMA FROM SUMMIT OF MOUNT WASHBURNE—LOOKING SOUTH.

means displayed on so grand a scale. The summit here has an elevation of 8,100 feet, and is for the most part formed of various modifications of glassy rhyolite. The road with a gradual descent follows
down the Gibbon River valley through a pine forest characteristic of
the Park. At the picturesque Virginia cascade, on the Gibbon, the
rock exhibits most interesting forms of erosion in the easily crumbling
rhyolite. Three miles beyond the Virginia cascade, Norris basin is
reached, and from here to Mammoth Hot springs the route follows the
same road traversed upon entering the Park.

Plates III and IV are from sketches made during the excursion by
MM. Eysséric and Golliez of the visiting geologists.

THE FORMATION OF HOT SPRING DEPOSITS.[43]

By Walter H. Weed.

Travertine.—The travertine deposits of the Mammoth Hot springs form one of the most interesting features of the region. Covering two square miles and attaining a thickness in places of 250 feet, the deposits have few equals in size, while the beauty of the terraced basins, the brightly tinted slopes covered by the steaming waters, and the varied views presented can not fail to impress every observer.

With the exception of the Hot River, all the active springs now issue from the terraces above the hotel; elsewhere the older deposits are quite generally covered by a forest growth of pines and spruces. In wandering about the springs one is sure to notice the brightly tinted basins surrounding the vents, with the red or orange colors of the slopes overflowed by the hot waters. These colors are due, not to mineral matter, but to the presence of algæ growing in the hot waters and frequently so covered by carbonate of lime as to be scarcely recognizable. These plants take a most important part in the formation of the travertine deposits, and in fact it is their presence that has caused the great beauty of the deposits. The varied tints are due to a different color of the algæ at varying temperatures, examples of which are seen in the beautiful mosaic of basins about the vents of the Blue springs on the Main terrace.

The fact that these deposits of travertine are mainly due to plant life has been fully proven by a careful study of the old deposits and of those now forming. This action of algæ was first observed by Ferd. Cohn at Carlsbad and other European localities, and the study of the Yellowstone deposits simply extends and confirms his theory. The proof is readily available at the Mammoth Hot Springs, and, though other causes cooperate to produce a separation and deposition of the carbonate of lime from the hot waters, the plant life is seen to be the chief factor in the production of the many varieties of calc sinter found about the springs. In the case of the fibrous tufa forming the fan-shaped masses found about many of the vents, an examination with a lens shows that the fibers are simply encrusted algæ threads. The rippled surface of the deposits covered by the overflow of the larger springs shows a furry covering of orange-colored algæ, the upright threads extending down into the mass. In both these cases the algæ filaments serve as a nucleus for encrustation, besides absorbing carbon dioxide and thus causing the separation of lime carbonate. The masses of gelatinous algæ, often several inches thick and forming mat-like coverings in the sluggish overflow channels, show the action of plant life most clearly. The successive layers of membrane-like material carry minute little crystals and stellate accretions scattered about in the plant tissue. These grow into small pellets that uniting together produce firm layers. Thin laminæ of carbonate of lime also form between the membranes, and a compact deposit of travertine results from a combination of the two.

Siliceous sinter—The hot springs and geysers of the Yellowstone are surrounded by large areas of siliceous sinter that often entirely cover the floor of the geyser basins. About the spouting vents this material has been built up into mounds and cones of unique forms and great beauty. The more quiet pools have built up more or less regular mounds of white sinter which are in places as much as 20 feet in height above the surrounding level. Besides these deposits, the alkaline waters of the geyser regions have left deposits of silica wherever they have flowed, and many square miles within the park are covered by white and glistening deposit of this material.

Until the Yellowstone deposits were studied it was the generally accepted theory that the geyser waters reached the surface heavily charged with silica, which by relief of pressure, by cooling, and by evaporation was precipitated out and deposited by the waters. Observation of the natural conditions under which the Yellowstone deposits are forming, together with experiments and a study of the chemical analyses of the geyser waters, showed that the silica brought to the surface by the geyser waters was not separated out and deposited by the first two causes, but that deposits are formed about the geysers and the margins of springs by evaporation, producing a true *geyserite*. A new mode of deposition was then recognized, namely, the separation of silica by plant life, by the algæ that are abundant in the hot waters of the region. It is by this agency that by far the largest part of the sinter deposits of the region have been formed.

This algous vegetation is sure to be observed by every visitor to the region. Its varied tints of pink, yellow, orange, red, brown, and green adorn the slopes of geyser cones, flush the white silica of the little basins with their tints, and mark the waterways with their brilliant colors. It is ever present where the temperature does not exceed 185° F., often lining the great bowls of the cooler springs and *laugs* with leathery sheets of brown or green. Where a constant overflow prevails, the channel is often filled by a vigorous growth in which an algæ mat is formed having the consistency of a firm jelly, and most beautifully colored. In whatever form it is found, and no matter how brilliantly tinted, this algous material if removed from the water and dried in the hot sun of the region rapidly loses its color, shrinks in size and, becomes an opaque white mass of silica, whose weight is not one per cent of its former state. Chemical analysis shows this dried material to be silica and water, viz:

SiO$_2$.. 93.37
H$_2$O.. 4.17
Organic matter.. 1.50

Experiments showed the writer that the growing algæ form a jelly of hydrous silica; it is of this material that the algæ filaments are formed, and the algæ slime of other waters is here a hydrous silica binding the threads together. The nature of this separation may be seen under the microscope, though the fresh hydrous silica is difficult to study, and the dried material becomes opaque. In most cases the glassy rods can be readily distinguished and the inclosing paste usually shows globules and pellets of the dehydrated silica.

The process of sinter formation is best illustrated, and its importance most apt to be appreciated, by an examination of an area covered by a large and constant overflow, such as that from the Black Sand whose connected overflow pools are known as Specimen lake. Here the algæ growing in the waters rapidly choke up the channel and cause the main supply to be diverted. Basins are formed by the algous growths, and in them pillars grow up from the bottom that are often a foot in height. These increasing in number finally fill the pool, their tops reaching the surface coalesce and roof over the basin until the waters, becoming choked, seek other outlets. The gradual lessening of this supply of water causes the final death of the algæ. In the cool waters that fill the space between the pillars the hydrous silica begins to harden. Aided by the acids of the decomposing vegetable matter this process is quite rapid, and more silica is separated from the cold water to form a coral-like coating, and finally the former soft algous jelly becomes a hard and firm rock. Eventually diversion of the hot waters builds up another growth upon the old one and thus the channel, swinging around from side to side, successively forms new basins, new growths, and new deposits of silica.

Every step of this process has been patiently studied for many years. In the majority of cases it is less easy to recognize than at Specimen lake. In the channels that carry off the water ejected from Old Faithful geyser, for instance, a different species of algæ from that building up the deposits of Specimen lake forms a velvety furze upon the channel floor. Its color is a brilliant orange to a cedar-red or dark seal-brown, depending upon the temperature. This species, identified as *Calothrix* by Wolle, forms a delicately fibrous but close-grained sinter, quite unlike the coralline masses of Specimen lake. The tangled silky skeins seen floating in the overflow of the Giantess and other geysers, and abundant as a thin orange-colored mat about the great Prismatic Spring, is a *Leptothrix* and produces a thatch-like or straw-like form of sinter.

A collection of these hot water algæ made by the writer is being studied by Prof. W. G. Farlow, of Harvard University.

The importance of these plant growths in building up sinter deposits may be realized when it is stated that in the walls of the great Excelsior geyser a section of 15 feet in thickness is exposed, of which over 12 feet is recognized as clearly and undoubtedly of algous formation, and the remainder of cemented fragments of weathered sinter. Even where the deposit does not reveal its origin by its structure, as is the case in the glassy compact sinter whose thin layers compose the platform about the Giant geyser, it is probable, judging from present conditions, that it is but an algous sinter altered by the long continued action of steam and percolating waters.

When the varied conditions of life and of evaporation are observed, it becomes at once apparent that any attempt to estimate the age of a geyser by the thickness of its deposits is a most difficult problem and a wholly unreliable foundation for other than comparative statements. Sinter formed by evaporation is produced very slowly under the most favorable conditions at the Firehole Geyser basins; one-twentieth of an inch a year is the maximum. Sinter formed by plant life may attain a thickness of eight inches a year in limited areas.

In general the sinters produced in these two modes may be readily distinguished by their physical structure. In chemical composition they are so alike that they cannot be separated.

Of less importance, because of its greater rarity, is the production of a siliceous deposit by a moss, *Hypnum aduncum*, observed at the Upper Geyser basin and at the Madison or Terrace springs. The moss grows only in the cooled waters that have already had considerable silica extracted while hot by the algous growths mentioned, but the moss growing in the cold waters is rapidly incrusted, in fact appears to build its structure largely of silica, and the resulting deposits cover several acres at each locality.

Diatom beds are common throughout the Park, but the deposits now forming are all in cool marshes supplied by hot spring waters. The resulting diatom earth, beds of which are sometimes six feet thick, generally contains more or less glassy silica separated from the waters by decomposing vegetable matter.

LIVINGSTON TO THE SNAKE PLAINS.

ITINERARY.

Station.	Distance.		Elevation.		Station.	Distance.		Elevation.	
	Miles.	Kilo-meters.	Feet.	Meters.		Miles.	Kilo-meters.	Feet.	Meters.
Livingston	0	....	4,487	1,368	Willow Creek ..	......	......	4,132	1,259
Coal Spur.......	5	8	4,735	1,443	Sappington	68	109	4,178	1,273
Hoppers........	9	14	5,175	1,577	Whitehall	87	140	4,343	1,324
Muir	12	19	5,500	1,676	Pipestone.......	94	151	4,690	1,429
Tunnel			5,565	1,696	Homestake	110	177	6,310	1,923
Sum. above tunnel............			5,835	1,778	Homestake Tunnel............			6,380	1,945
West End	14	23	5,540	1,689	Sum. above tunnel............			6,435	1,961
Timberline			5,500	1,676					
Mountainside ..	16	26	5,275	1,608	Butte[1]..........	120	193	5,570	1,698
Fort Ellis.......	22	35	4,860	1,481	Silver Bow	127	204	5,344	1,629
Bozeman	25	41	4,752	1,448	Melrose.........	159	256	......	......
Storey	30	48	......	......	Dillon	189	304	......	......
Belgrade........	35	56	4,435	1,352	Beaver Canyon .	265	426	6,025	1,836
Central Park ...	41	66	......	......	Eagle Rock.....	332	534	......	......
Moreland.......	45	72	4,240	1,292	Ross Fork	371	597	......	......
Logan	49	79	4,094	1,248	Pocatello	383	616	4,468	1,362
Three Forks....	55	89	4,053	1,236					

[1] Population, 10,723.

[By WALTER H. WEED.]

At **Livingston** the railroad leaves the valley of the Yellowstone and passes due west up the valley of Billman Creek over Livingston beds, whose somber sandstones form the ridges on either side. At **Coal Spur** a branch line runs up Coke creek to the mines and ovens of Cokedale, where the Laramie coal seams are extensively worked, " a place where the relations of the Laramie and underlying Cretaceous to the Livingston beds are splendidly exposed. A dike of analcite basalt may be seen cutting the Livingston beds in a hill north of the railroad, a short distance beyond **Coal Spur**. From this point westward the ascent is very steep, 1,000 feet (305 m.) in 12 miles (18 km.), to the **Muir** tunnel, by which the railroad crosses the divide between the waters of the Yellowstone and the Missouri. Emerging from the tunnel, cut through the Livingston rocks, the railroad enters a mountain valley, passing the station of **Timberline**, where a narrow-gauge road runs to coal mines in Laramie rocks, two miles to the southward, that supply the railroad with fuel. The coal-bearing rocks are seen at **Mountain-side**, where the coals are also mined. West of here the road enters a narrow gorge known as Rocky canyon, cut across a sharply folded anticline pitching steeply to the north. The various formations of the Cretaceous are seen passing up the hillside to the north and, together with the Jurassic, curving around the massive limestones of the Carboniferous. The latter rocks form the picturesque pinnacles and towers of the central portion of the canyon. A little beyond, the Mesozoic strata are again seen, their sandstone beds curving about the Carboniferous in sharp flexures. The ridge cut through by Rocky canyon connects the Gallatin range to the south with the uplift of the Bridger mountains to the north, and is really a low part of the Front range of the Rocky Mountains, which is here broken down into a number of low uplifts arranged en échelon. The valleys are usually eroded in the soft Cretaceous shales; the ridges show the resistant limestones of the Carboniferous.

Leaving Rocky canyon the road enters the broad intermountain basin known as the Gallatin valley. Immediately north of the railroad the bluffs of the East Gallatin river show fine exposures of Neocene lake beds, the deposits here being conglomerates and coarse sandstones dipping at an angle of 3° to the northward. **Fort Ellis**, an abandoned military post, is built upon these lake beds that form the gently sloping table-land southward, beyond which the peak of Mount Ellis, formed of Carboniferous limestones, is seen.

[By Dr. A. C. PEALE.]

Between **Bozeman** and **Central Park** the road passes over the alluvial valley of the Gallatin river. On the east side of this valley is the Bridger range, in which the nearest foothills are composed of gneisses. The portion of the Bridger range in sight from the railroad is a monocline, mainly of Paleozoic rocks, Carboniferous limestones forming the crest line, with Archean gneisses on the west. At the south end of the range the beds are overturned, the Cambrian, Devonian, and Carboniferous beds inclining to the westward, with the older beds on top dipping under the gneisses.

Bridger peak is the prominent point at this end of the range, while farther north Ross peak, although not the highest, is the most rugged prominent mountain seen from the railroad. South of the valley the Gallatin mountains are seen. They are composed mainly of gneisses and eruptive rocks. The most prominent peak, almost due south of Bozeman, is Mount Blackmore, which is composed of andesite, while farther to the west is the gneissic Gallatin peak.

The Gallatin valley is one of the old lake basins, of which a large number are found in Montana. Its deposits were largely derived from showers of volcanic dust, which, falling into the quiet waters of the lake, were arranged in beds of very pure dust some 20 feet (6 m.) in thickness. Above these is a very considerable thickness of beds, evidently made up from the same material, which was washed into the lake from the shores and surrounding country and rearranged in beds of a rusty color.

At **Central Park** the railroad crosses the Gallatin river, whose head-branches have their source in the northwestern corner of the Yellowstone National Park.

From **Central Park** to **Logan** alluvium and Quaternary gravels, resting on lake beds, are passed over. As **Logan** is approached very good exposures of the Paleozoic and Algonkian beds are seen across the Gallatin river on the right, and the road passes through a cut in the former just before reaching **Logan**. Immediately opposite **Logan**, on the north side of the Gallatin river, there is a fine exposure of Cambrian, Devonian, and Carboniferous limestones in the bluffs that rise from the water's edge.

A few outcrops of the Paleozoic are noted on the left side, but the road soon comes out on the lake beds after leaving the Gallatin river below **Logan**. These beds are also well shown bordering the east side of Madison River valley just before **Three Forks** station is reached. They also show, in the distance to the south, between **Three Forks** and **Willow creek**. The Madison river is crossed just before the **Three Forks** station is reached. The Three Forks valley, at the lower end of which the Gallatin, Madison, and Jefferson rivers unite to form the Missouri river, is one of the most interesting geographical points in Montana. These streams were named by Lewis and Clarke, who first saw this valley in July, 1805, on their way to the Columbia river and the Pacific coast.

On the northwest or right side of the Jefferson river above **Three Forks** the hills are mainly of Cambrian, Devonian, and Carboniferous, the Algonkian beds forming the lower foothills farther up the river.

A short distance above **Willow creek** the road passes through a canyon, almost two miles in length, which the Jefferson river has cut through Carboniferous limestones. At the entrance to this canyon, on the north side of the river, stands a butte of basalt.

From the canyon the road comes out into the alluvial valley of the Jefferson, in which lies **Sappington** station. On the left (south) the hills or mountains are mainly of Carboniferous rocks. On the right (north) the nearest small ridges are of Jura, Trias, and Cretaceous. Above **Sappington** the Jefferson river is crossed and the road enters a second canyon cut in upper Carboniferous and Jura-Trias rocks, from which it emerges into a narrow valley in which the rocks are of Cretaceous age. These Cretaceous beds on the right (north) are faulted down against very somber-colored beds of Algonkian age. The hills on the left (south) are of Carboniferous, Jura, Trias, and Cretaceous. At the upper end of the valley the Cretaceous is faulted down against the Carboniferous, and the fault line is crossed by the road as it enters a third canyon, which is mainly in Carboniferous.

On the south side of this canyon placer-mining operations are carried on; the flume which conducts water for washing the gravels can be traced on the side of the hills for many miles, cutting through the rocks by tunnels in several places. A second fault-line is crossed a short time before the canyon is left, by which Carboniferous rocks are brought down against the Algonkian. The somber greenish beds of the latter are well shown on the right just before the Boulder river is crossed. Beyond, the road is located on a broad island in the Jefferson river, after leaving which the valley of Big Pipestone creek is followed. Here Pliocene lake beds rest upon granitic and eruptive rocks. As the road is followed a short distance above Pipestone springs it leaves the lake beds and comes out on a body of eruptive rock (porphyrite?) in which there are several cuts. This rock is in contact with the granite which occupies all the country in the vicinity of Homestake tunnel. From the tunnel to Butte everything is granitic.

[By S. F. EMMONS.]

The granite hills form the divide or watershed between streams flowing into the Atlantic Ocean on the one side and the Pacific Ocean on the other. The age of these granites is not definitely known, but they are of eruptive origin and probably pre-Paleozoic. Having left behind the various valleys tributary to the Missouri river, the railroad now descends into the valley of Silverbow creek, which is one of the southeastern sources of Clarkes Fork of the Columbia river.

The interior portion of Montana was early discovered to contain the precious metals; rich placer gravels, from which great quantities of gold have been washed out, abounding in its valleys. They are still worked here and there, but their greatest production was in the decade 1860–1870. In the last two decades deep-mining has been steadily increasing, and has developed many famous mines, such as the Granite Mountain, Drumlummond, and others, which have paid millions in dividends to their fortunate owners.

Butte City,[45] the most important mining center of Montana, had a population in 1890 of 10,723, an increase of 218 per cent over that of 1880. Actual mining upon the silver and copper lodes which constitute its present wealth may be said to have commenced in 1875, though gold had been extracted from the placer sands of the neighboring valleys since 1864, when the town was first founded. The earliest large mining operations were in the silver mines; the Alice, Moulton, and Lexington mines being the most prominent. The latter was purchased in 1881 by the "Société Anonyme des mines de Lexington," of Paris, France. During the last decade the copper mines have gradually increased in importance, and within the past few years the value of the copper produced has far exceeded that of silver. The production for the year 1890 was:

	Kilograms.	Value.
Gold, 25,704 ounces	800	$513,316
Silver, 7,500,000 ounces	233,264	9,696,750
Copper, 112,700,000 pounds	51,117,204	16,623,250

The city is situated on the southern slope of rounded granite hills, included within the angle of Silverbow creek as it flows first south and then west. It lies just east of a round conical hill, known as the "Butte," from which it takes its name. The greater part of the silver and copper mines lie in the hills back of and north of the city and to the east of the Butte, but a few important silver mines, notably the Bluebird and the Nettie, are situated a few miles to the westward, beyond the Butte.

Of reduction works for extracting the metals from their ores, the stamp mills are generally located on the hills near the mines, while the several smelting works lie in the valley of Silverbow creek. The smelting works of the largest copper mine, the Anaconda, are situated 27 miles (43 km.) to the northwest, where a town of the same name, of several thousand inhabitants, has been built around them, which is connected with the mines by a railroad.

The geological formation of the district is almost entirely of granite and rhyolite. There are two varieties of granite. The most widespread variety, which forms the country rock of the mines around the city and to the east of "the Butte," is an unusually basic rock, carrying a relatively large proportion of plagioclase feldspar ; its basic minerals are mica, hornblende, and augite, and much of the hornblende appears to be only a paramorphic alteration of augite. The other granite occurs to the west of "the Butte" and forms the principal country rock of the Bluebird, Nettie, and other mines. It is a light-colored rock and consists almost exclusively of quartz and orthoclase feldspar, with a few minute grains of biotite. At its contact with the Butte granite it is found to send veins into it and include fragments of it, and is hence assumed to be of later age. The Butte itself consists of rhyolite, a fine-grained, granular rock consisting of quartz, sanidine,

and mica in a subordinate ground mass. This is evidently a later eruption, and sends out tongues or dike-like masses into the surrounding
granite, which are not always visible on the surface but are disclosed
in the deep drifts of the mines. Rhyolite of the typical banded variety
also forms a series of low ridges about 5 miles (8 km.) west of Butte.

The ore deposits of the district occur in a series of veins in the granites, standing nearly vertical and having a general east and west trend.
None so far as known have been found in the rhyolite. These veins
are rock fractures produced by dynamic force and show evidence of a
faulting movement in striated surfaces and zones of broken or crushed
country rock, but have comparatively little selvage matter or clay
walls. Each great mine consists as a rule of several parallel fractures
or fissures, with minor cross-courses or fractures nearly at right angles
with the main fissures; these sometimes appear to fault them, but it is
by no means certain that they are later fractures. The ore is largely a
replacement of the country rock from the original fissures outwards, it
being often possible to trace a gradual transition from solid ore, through
partially altered, into entirely unaltered granite, with no defined plane
of demarcation or wall. In the silver mines the ore body or vein matter is often six to eight feet wide. In the copper mines much greater
thicknesses of solid copper glance (chalcocite), sometimes 20 to 30 feet
(6 to 9 m.) wide, are found.

The most common minerals in the silver mines are pyrite, sphalerite
or zincblende, galena, and sulphides of silver, with manganese in the
form of rhodocrosite or rhodonite, and little or no copper. Gold forms
an important value in some ores, but it is rarely visible.

In the copper mines chalcocite, chalcopyrite, bornite, and enargite
are the prevailing valuable minerals, but zincblende and the manganese
minerals are almost entirely absent. Quartz is the general gangue
material, and has apparently been derived by lateral secretion from the
adjoining granites. Barite and fluorspar are also found. The oxidation of the manganese minerals, which form a remarkably regular constituent in the silver mines, produces very prominent outcrops or
gossans, stained by the black oxide of manganese.

The larger mines have been opened to a depth of over 1,000 feet
(300 m.) and are still productive, though the expense of mining and of
treatment has in some cases led to a suspension of exploitation. The
increasing proportion of zincblende is very disadvantageous in the
amalgamation processes by which these ores are reduced.

Distinct methods of treatment are pursued for the two different
classes of ore. The silver ores are pulverized in dry stamp-mills, subjected to desulphurizing-roasting, chloridized by mixture with salt,
and amalgamated in revolving pans, after the Nevada system. For ores
that are extremely refractory, owing to a large proportion of zinc-

blende, the Russell process of lixiviation, a modification of the hyposulphite process, has recently been introduced. The amalgamation mills, in which this process is carried on, are the most complicated and expensive of their kind. The finest here is that of the Bluebird mine, about 3 miles (5 km.) west of Butte, which cost over $300,000 and has 70 stamps, two Stetefeldt shaft furnaces, each 60 feet (18 m.) high, for the desulphurizing and chloridizing of the pulverized ore, and a lixiviation plant. The copper ores are subjected to preliminary concentration in ore-dressing works, roasted, and smelted to a rich matte. This matte is sometimes shipped East, sometimes reduced to metallic copper here. One smelter, the Colorado, smelts both copper and silver ores, producing a matte which has absorbed all the silver. This silver-bearing matte is sent to the great smelting words at Argo near Denver, to be treated by the Augustine-Ziervogel process. The salt used for chloridizing is brought by railroad from Great Salt lake.

From Butte the road runs westward 7 miles along the valley of Silverbow Creek, passing the Bluebird mill about midway in this distance, and crossing low ridges of recent rhyolite just before reaching Silverbow station, which lies in a broad valley of Quaternary gravels. The route now follows the line of the Utah and Northern division of the Union Pacific Railroad southward, traversing a series of low mountain ridges mainly made up of Paleozoic rocks, which form the geological connection between the north and south trending northern extension of the Wasatch mountains, and the northwest trending Bitter Root mountains. These ranges not only form the continental divide between east and west flowing waters, but represent in a general way the western limits of the mediterranean ocean of Mesozoic times. No geological examination has yet been made of the region, so that it is not possible to give exact indications of the age or character of all rock exposures along the route. It may, however, be safely inferred that they are mostly of Paleozoic strata and recent eruptive rocks, with a covering of Pleistocene gravels in the larger valleys.

Before reaching Melrose the road crosses a divide into the valley of a tributary of Jefferson river.

From Dillon it ascends the valley of Beaver Head creek, one of the main tributaries of this river, for a distance of over 50 miles; it then crosses the continental divide, which is the boundary between Montana and Idaho, on to the waters of the Snake river.

Beaver canyon, which is a gorge in basaltic lavas, is the point from which a stage line about 100 miles long runs to the Lower Geyser basin in Yellowstone National Park.[46] The road now traverses the northeastern portion of the Snake Plains and crosses the Snake river at Eagle Rock, from which it follows southward the eastern edge of the Snake River desert.

At **Ross Fork** is the agency of the Fort Hall Indian reservation, a tract of land set aside by Government for portions of the Bannock and Shoshone tribes. These Indians, who number 3,000 to 4,000, are peaceful and largely devoted to agricultural pursuits, raising hay which they sell to cattle owners whose herds graze on the neighboring hills.

Pocatello, near the southern edge of the Indian reservation, is at the junction with the Oregon Short Line which runs westward through Idaho to Portland, Oregon. It is on the Portneuf river, where it debouches from a narrow valley between rounded hills of Paleozoic rocks, partly filled by very recent flows of basalt, which are readily visible from the train as it ascends to the more open valley of Marsh creek.

EXCURSION TO SHOSHONE FALLS.

THE SNAKE PLAINS.

By S. F. EMMONS.

The valley of the Snake river presents a most interesting study to the vulcanologist. From its source in the mountains around the Yellowstone Park to its junction with Clarkes fork of the Columbia, a distance of about 800 miles (1,287 km.), the rocks which environ it are mostly recent eruptive rocks or actual lavas. The Columbia river, formed by the junction of these two great streams, flows westward across the lava plains of eastern Oregon for about 120 miles (193 km.), and then debouches on to the lowlands of the Pacific coast, through the stately portal of the Cascade mountains, where, on either side, walls of basalt rise to 3,000 feet (914 m.) almost vertically, and at a little distance north or south the great extinct volcanoes of the Cascade range raise their snow-capped summits to altitudes of 10,000 to 14,500 feet (3,000 to 4,420 m.) above sea level. This interesting region has not yet been systematically surveyed and the little that is known of its geology has been obtained by reconnoissances of individuals and small exploring parties.

The Snake plains, properly speaking, occupy an area, whose dimensions are not accurately known, extending from the sources of the river some 250 miles (400 km.) slightly south of west, in its broader portion nearly 100 miles (160 km.) wide. This area, for the most part practically a plain of basaltic lava, is incidented here and there by low hills, either some slight preexisting elevation of underlying older rocks, or inequalities in the lava itself, but contains no high mountains. A general depression, or area without high mountains, extends westward beyond the valley of the Snake, and on the same general line, to 121°

of west longitude, or 10° west of the Yellowstone Park. This area is occupied by recent deposits of Pliocene or Pleistocene age and large sheets of recent eruptive rocks. The present bed of the Snake river leaves this depression at the eastern boundary of Oregon and flows northward 200 miles (322 km.), cutting a deep gorge in the mountains which lie across its path. There is some reason for assuming that this sudden change of direction is of geologically recent date and that an earlier valley existed further westward. The solution of this question, as to earlier draining and as to the age of the recent deposits in this valley, is of ethnological[47] as well as geological interest.

It will be seen later that the interior basin now occupied by Great Salt lake was once filled by a much larger lake called Lake Bonneville. At one period in its history Lake Bonneville overflowed and its superfluous water drained out through the Snake River valley. Whether the lavas, which now form the surface of the Snake Plains, were poured out before or after the dessication of this lake has not yet been definitely ascertained.

ITINERARY.

By S. F. EMMONS. *

Station.	Distance.		Elevation.	
	Miles.	Kilometres.	Feet.	Metres.
Pocatello	0	0	4,466	1,360
American Falls	25	41	4,341	1,323
Shoshone	103	166	3,973	1,211
(By stage) to Shoshone Falls	26	42		

Those who take this excursion leave the Utah Northern train at **Pocatello** and take the regular train on the Oregon Short Line to **Shoshone** station, from which they reach the falls by stage.

At **American falls** the Snake river is reached. The falls can be seen on the left or below the bridge. A short distance above the bridge, wide, alluvial bottoms border the Snake River, which support a luxuriant growth of grass and large cottonwood trees, where cattle, which pasture on the hills further east, are taken during the winter.

From **American falls** downward, the river runs in an ever-deepening gorge in the basalt, whose walls at the bridge are about 70 feet (21 m.) high, and at Shoshone falls, about 150 miles (241 km.) lower down the river, are 400 feet (122 m.) and 620 feet (190 m.) above and below the falls, respectively. After crossing the river, the railroad keeps a westerly course, slowly diverging from the course of the stream

---

* From notes furnished by J. P. Iddings and J. S. Diller.

which flows somewhat to the south of west. The country now assumes the desert aspect characteristic of the lava plains, treeless and waterless, the rocks bare or covered in the hollows with the common desert shrub, the sage-brush (*Artemisia*). No rivers flow into the Snake river from the north, although many bold mountain streams have flowed out upon the borders of the lava plains from the mountains which form its northern boundary. These are known as Lost rivers; and, while much of their water has undoubtedly been dissipated by evaporation, no inconsiderable portion must have found its way downwards through the cracks and rifts in the lava to the rocky bed over which it was originally spread out. This water, gathered into subterranean streams of considerable volume, pours out along the walls of the Snake river canyon to the southwest, below Shoshone falls, where the corrasion of the stream has cut down below the base of the basalt into the underlying andesite or dacite.

From Shoshone station the stage route leads south over the same monotonous barren plain, which has a scarcely perceptible inclination toward the river. Long before the canyon in which it runs can be distinguished in the level monotony of the plain, the presence of the stream is indicated by the thundering roar of waters as they fall. Suddenly one comes upon the brink of the gorge. A steep descent of 400 feet (122 m.) leads down to the banks of the river above the Shoshone falls.[48] The travelers are ferried across the river to a comfortable little hotel near the brink of the falls, having a view down the canyon below, which is 620 feet (190 m.) deep. These falls are more broken and varied than Niagara; their height is greater, being 212 feet (65 m.) instead of 156 feet (48 m.), but the volume of water is less, though still very great, especially in the early summer, and the country around has the desolate grandeur of the desert instead of the brilliant verdure of the thickly-populated region in the vicinity of Niagara.

The canyon presents a striking contrast to that of the Yellowstone river both in coloring and form. The Yellowstone canyon is brilliant and light-colored, with innumerable vertical pinnacles and spires; the Snake River canyon is somber and black, with predominantly horizontal lines marking the successive sheets of basalt, which are, however, vertically columnar. The first is cut in a rhyolite plateau, the second in a basalt plain.

The basalt consists of three sheets, which form the upper 250 feet of the canyon. The lower 250 feet, just below the falls, is massive andesite of an abnormal type, approaching dacite in chemical composition. Its upper portion near the ferry is glassy and resembles certain modifications of the rhyolite of the Yellowstone National Park. Where the plateau of the Yellowstone Park descends into the plain of the valley of the Falls river, a tributary of the Snake, the same geological structure exists. An acid lava (rhyolite) forms the bed of the stream and is overlain by a sheet of basalt.

A short distance above Shoshone falls are the Upper or Twin falls, about 180 feet high, which drop from a cliff formed by the two lower sheets of basalt. Red tuff and volcanic agglomerates are found in places between the successive sheets of lava.

A hundred miles or more further down the stream the basalts have been observed to rest directly on sedimentary beds of recent age, probably deposited in some inclosed fresh-water lake. Other cascades and falls are known to exist along the course of the Snake River, but its canyon has never, so far as known, been followed continuously, and their number and its extent is unknown. The stream is unusually rapid for its size, in spite of these many leaps in its course, and is cutting down its bed very fast, though yet far from reaching a baselevel of erosion. From longitude 112° W. to longitude 117° W. its total descent has been about 2,500 feet (762 m.).

GREAT SALT LAKE AND LAKE BONNEVILLE. [49, 50]

By G. K. GILBERT.

A large district of interior drainage, lying to the west of the Wasatch mountains and the Plateau region, is known as the Great basin. It includes nearly the whole of the state of Nevada, the western half of Utah, and smaller portions of Idaho, Oregon, and California. It is naturally subdivided into a number of smaller basins, from each of which the entire product of precipitation is evaporated, so that there is no discharge to the ocean. In most of the basins there are no permanent lakes, but temporary lakes are produced by the waters of each great storm. In a few basins there are permanent lakes with saline waters. The largest of these is Great Salt lake, which receives the waters of the Bear, Weber, and Jordan rivers, and has an area of about 1,800 square miles (4,500 sq. km). The extent and depth of the lake are determined by the balance between inflow and evaporation. In years of great rainfall the surface of the lake rises, and in dry years the waters recede. During the past thirty-five years the water height has several times oscillated through a range of 11 feet (3·3 m.), and it is now (1891) near its lowest observed stage. The salinity undergoes corresponding changes, being greatest when the lake is low. The solid contents now amount to about 20 per cent, of which four-fifths is sodium chloride. Sodium sulphate is naturally precipitated by the cold of each winter, and afterwards redissolved. The lake is very shallow, having a mean depth of 13 feet (4 m.) and a maximum depth of less than 40 feet (12 m.). It is inhabited by a brine shrimp and the larva of a fly.

In Pleistocene time the lakes of the Great basin were larger, and many perennial lakes were formed in valleys whose floors are now saline and desert. Great Salt lake was expanded so as to coalesce with the lakes of contiguous basins, producing a body of water 19,750 square miles (51,000 sq. km.) in extent, which has been named Lake Bonneville. This lake was twice formed and twice dried away, each time depositing over the plain a sheet of calcareous clay with fresh-water fossils. The two sheets of clay are separated by an unconformity, the first having been eroded before the second was laid down. In some places, moreover, a wedge of alluvial gravel intervenes between the two clays, showing that the lacustrine epochs were separated by an arid interval, during which alluvial deposition took place, as at the present time.

The highest water stage was attained during the second lacustrine epoch, and is recorded in a conspicuous series of sea cliffs, terraces, and beaches known as the Bonneville shore line. This shore line has a general altitude of 1,000 feet (300 m.) above Great Salt lake, or 5,200 feet (1,580 m.) above the ocean, but its height varies from place to place, ranging in the vicinity of Great Salt lake from 960 feet (290 m.) to 1,050 feet (320 m.) above the modern water surface. As all parts of the shore line were produced at the same time, their present differences in altitude indicate a warping of the earth's crust since the period of their formation.

At this stage the lake overflowed the northern rim of its basin, and the channel of outflow was eroded to a depth of nearly 400 feet (120 m.), when the cutting was arrested by a ledge of limestone and the water was held for a long period at one level, giving the waves time to sculpture a second series of terraces, etc., constituting the Provo shore line. At many heights between the Bonneville shore line and the Great Salt Lake shore line, the waves of the oscillating water have left their traces, so that the number of fossil shore lines is very great; but the Provo shore line is distinguished from all of these by the magnitude of its features. Its terraces are broader, its cliffs are higher, its spits are greater, and with it are associated great delta terraces built by tributary creeks and rivers during the Provo epoch.

During the long period for which the lake maintained an outlet its water must have been completely freshened, so that the salt contained in the modern lake has all been accumulated in recent times. A comparison of the high salinity of the modern lake with the approximate purity of its tributaries enables one to realize the antiquity of the Provo epoch, and yet in this arid climate the vestiges of wave action have been preserved almost unscathed.

FAULT SCARPS. [49]

By G. K. Gilbert.

The mountains of the Great basin are, in large part, carved from orogenic blocks uplifted along fault planes. The displacements, which were probably initiated in Mesozoic time, were continued during various Cenozoic epochs, and are now in progress. The steeper faces of most of the mountain ranges are rugged escarpments primarily due to faulting, and at their bases are frequently to be found smaller escarpments of so recent date that the traces of subsequent erosion are scarcely perceptible. In 1872 the production of such a fault scarp along the base of the Sierra Nevada was accompanied by an earthquake. In connection with the earthquake in Sonora, Mexico, in 1885, other scarps were produced. In the Salt Lake basin there is no historical record of their formation, but many of them intersect the beaches and deltas of the Bonneville shores, and some are so fresh that vegetation does not yet clothe them, and it is hard to believe their antiquity is measured by centuries rather than decades. They have been found along the bases of a dozen ranges of the Salt Lake basin, but they are most persistent and have greatest magnitude at the western base of the main ridge of the Wasatch chain, where they have been traced almost continuously for a hundred miles.

As a great orogenic block, separated from another by a fault plane, rises, the débris resulting from its sculpture is thrown upon the block beyond the fault, and rests as an alluvial bank against the cliff produced by the faulting. When subsequent movements occur on the same fault plane they are superficially manifested either in the alluvium, or at its plane of junction with the rock. The forms of the alluvium, being determined by the laws of fluvial deposition, are regular, and the cliffs produced by the faulting are thus rendered conspicuous and unmistakable. Sometimes a single scarp is seen to cross an alluvial slope, rising and falling as the slope rises and falls; sometimes two or more scarps are seen to run parallel to each other, and in such case the intervening surfaces of alluvium have new attitudes, their tendency being to incline toward the mountain face; sometimes a wedge of alluvium has fallen into the fissure due to faulting, so as to produce a trench on the alluvial surface.

All these special phenomena are illustrated in the localities to be visited by the Excursion party, and the fault scarps of the Wasatch can also be observed, at a distance, from the windows of the train.

OUTLET OF LAKE BONNEVILLE AT RED ROCK PASS.

POCATELLO, IDAHO, TO SALT LAKE CITY, UTAH.

ITINERARY.

By G. K. GILBERT.

Station.	Distance.		Elevation.		Station.	Distance.		Elevation.	
	Miles.	Kilometers.	Feet.	Meters.		Miles.	Kilometers.	Feet.	Meters.
Pocatello	0	0	4,460	1,360	Dewey	97	156	4,318	1,316
Portneuf	6	10	4,495	1,370	Brigham	113	182	4,313	1,315
McCammon	23	37	4,753	1,449	Utah Hot Springs	125	202	4,275	1,303
Thatcher	34	55	4,818	1,468	Ogden *	134	216	4,301	1,311
Oxford	53	85	4,774	1,455	Kaysville	151	243	4,298	1,310
Battle Creek	64	102	4,490	1,368	Lake Shore	158	254	......	......
Cache Junction	85	137	......	......	Salt Lake City †	170	274	4,228	1,289
Collinston	92	148	4,689	1,429					

* Population, 14,899. † Population, 44,843.

From **Pocatello** the railroad line runs a few miles eastward up the narrow defile of the lower Portneuf valley, then bends southward, following for 35 miles (56 km.) the broad Pleistocene valley of Marsh creek. In the middle of these valleys, and sometimes filling their bottoms so as to block up the mouths of the tributary ravines, are recent flows of basaltic lava.

Marsh valley lies between parallel mountain ranges trending with the meridian and consisting, so far as known, of Paleozoic rocks. They may be considered as northern members of the Wasatch chain. They belong to an ancient topography whose drainage system has been considerably modified in geologically recent time. The Portneuf river breaks into the valley from the east at a point about midway, follows it to its northern end, and escapes by the defile just mentioned. Late in Tertiary or early in Pleistocene time a flow of basaltic lava followed the course of the river into the valley. Subsequent erosion, chiefly by the outlet of Lake Bonneville, lowered the drainage system of the valley so that the basaltic coulée stands at the top of a steep-sided mesa. The Bonneville channel follows the western margin of this mesa, and the Portneuf river follows the eastern, breaking across it at the northern end of the valley. The train follows the river for several miles, then

377

rises to the eastern alluvial slope of the valley, and finally descends to the Bonneville channel at a point beyond the lava tables.

Where the Portneuf skirts the lava bed its channel is obstructed by a series of low dams of travertine, which seems to be rapidly deposited by the water of the river.

At its southern end Marsh valley joins the northern end of Cache valley, being separated by a low divide known as Red Rock pass. Through this pass Lake Bonneville discharged its surplus water, and it is here that the train passes from the basin of the Columbia river to the basin of Great Salt lake. A halt will be made for the purpose of examining the channel of outflow (Plate v).

At this point a low ridge of Carboniferous limestone lies athwart the valley trough, its crest projecting above the alluvium in two buttes, whose iron-stained cliffs give name to the pass. Near them are Pliocene lake beds, upturned at a high angle, but these are seen only where the alluvium has been washed away. The alluvium derived from the mountain ranges rests in great conical heaps against them, the cones joining along the middle of the valley. At the pass is an exceptionally large alluvial cone, built by Marsh creek, which issues from a canyon at the east (left). Before Lake Bonneville existed this formed the summit of the pass, and when the lake broke over its edge the alluvium was washed away with speed, letting the imprisoned waters escape to Marsh valley in a debacle of tremendous power. Marsh creek then cut a channel through the cone it had previously built, and in this channel it still flows. It has built a small cone in the abandoned river channel, which it is feebly laboring to fill. A little farther south other creeks have built alluvial cones in the Bonneville channel, partitioning it into little basins occupied by swamps and ponds. In time of flood Marsh creek turns northward and follows the old channel to the Portneuf, but it ordinarily sinks in its own alluvium near the pass.

The Mormon town of **Oxford** lies just beyond Red Rock pass in the northern end of Cache valley. Cache valley also lies between mountain ranges of the Wasatch system, trending north and south. It is traversed by the Bear river, the largest tributary of Great Salt lake, which enters the valley through a canyon at the northeast and, after traversing half its length, escapes to the west through a narrow gorge in Paleozoic limestone known as "The Gates." It is joined by many smaller streams issuing from the eastern range. The mountains on either side are constituted chiefly of Paleozoic strata, but fresh-water beds referred to the Pliocene (Humboldt beds) rest against their flanks at several points and have shared in the later displacements. Bonneville shore lines encircle the valley, and marl deposits of the same date occupy its lower levels. Associated with these are great delta deposits accumulated chiefly at the date of the Provo shore line.

5ᵐᵉ CONGRÈS GÉOLOGIQUE INTERNATIONAL.

THE GATES OF BEAR RIVER.

The largest ancient delta is that of the Bear river, but those formed by some of the smaller streams are more symmetrical. That built by Logan river has the form of a semicircular terrace projecting into the valley, its upper surface constituting an obtuse cone whose apex is at the mouth of the mountain gorge whence issues the river. By subsequent action the river has divided the terrace into two parts, and there are other delta terraces at lower levels. Upon these terraces stands the city of Logan on the east side of the valley Near the base of the mountain the Logan delta is traversed by a fault scarp, 6 feet (2 m.) high. Cache valley lies chiefly in Utah and contains a large number of thriving settlements founded by the Mormons. Their chief industry is agriculture, and this is carried on by the aid of artificial irrigation, the waters of the streams being diverted from their channels and carried by ditches to the farming land. The principal town, Logan, has a population of 4,565.

Northward from Red Rock pass the railroad traverses for several miles a plain little below the level of the pass and then descends to the immediate valley of Bear river, which is followed to the "Gates" and beyond (Plate VI). The passage opened by the river through the western mountain ridge exhibits a nucleus of Paleozoic limestone, against which on each side rest Humboldt beds. These last are upturned to 45° at the east and to 15° at the west, and are in turn overlain by marls, sands and tufas of Bonneville date. The erosion of the river gorge was pre-Bonneville, and its sides are sheeted by calcareous tufa to a depth of several feet. In sections opened by the railway the tufa is seen to have been deposited in two or more sheets separated by bodies of talus believed to represent one or more inter-lacustrine epochs. Here a great engineering work is in progress. By a dam near the head of the gorge the water of the river is diverted into two canals, which have been carried, partly through tunnels, along the walls of the gorge and then led out along the upper benches of the plain to the west. By their aid a large tract of desert land will be reclaimed to agriculture.

After passing through "The Gates" the road runs southward, between the steeper mountain slope on the left and the valley of Bear river on the right. A fine view is obtained, across the valley, of the Promontory range, which forms its western boundary, 40 miles (65 km.) away. This range is so called because it extends into Great Salt lake, dividing its northern portion into two great bays.

The road passes the Mormon towns of Willard, Box Elder, and **Brigham**, resting on the gentler slopes of the valley and surrounded by fields irrigated by waters from the mountain streams issuing from ravines behind them. Beyond **Brigham** the road passes round a projecting point of the Wasatch mountains, opposite the north end of Great Salt lake, to **Utah** (Bonneville) **Hot springs**, which issue from the great Wasatch fault. They have a maximum temperature of 130° (58° C.) and leave a ferruginous deposit.

At **Ogden**, an important Mormon town of 15,000 inhabitants, the road crosses the Union Pacific railroad, the pioneer transcontinental line, which reached here in 1869, crossing the Wasatch range through the deep gorge of Weber river. In Bonneville time this river built an immense delta of sand and gravel on the margin of the plain.

From **Ogden** the road makes a detour to the westward around this delta and follows the lower level of the valley, near the lake, past the Mormon towns of **Kaysville, Farmington,** and **Centerville.** Fault scarps are continuously visible along the base of the mountains, being especially conspicuous near Farmington.

FIG. 17.—Shore lines and fault scarps near Farmington.

The road now passes around another low promontory of the Wasatch, opposite the southern end of the lake, to **Salt Lake** city, which lies to the south of the promontory. Near **Salt Lake** city are both warm and hot springs, also issuing from the great fault, with a maximum temperature of 123° F. (53° C.).

THE WASATCH MOUNTAINS.

By S. F. Emmons.

The Wasatch Range, whose imposing western front the party will pass in review, is one of the most important single chains in the whole Cordilleran system. Within its mass may be found representatives of all the great geological formations recognized in the western United States, and as a rule developed in greater thickness than the same horizons show elsewhere in the system. Its topographical relations are also somewhat interesting. All the waters drained from its slopes, whether on the east or on the west, flow finally into Great Salt Lake, which lies at its western base, and these waters are practically the only feeders of the lake. Thus the lake constitutes a great hydrometer, which measures the relation between the amount of water precipitated upon the mountain range and that taken up by the dry desert winds which sweep over its glassy surface. The general level of the lake surface is about 4,200 feet, but the actual rocky bottom of the great valley in which it lies must be far deeper, for borings through the lake deposits, which fill up the inequalities of its surface, have gone down 1,500 feet in places without reaching the underlying rocks. The western or higher crest of the range rises abruptly from 5,000 to nearly 8,000 feet (2439 m.) above the present surface of the valley plains.

Its present form, geologically considered, may be said to represent only the eastern portion of the original range, the western half having been sheared off and sunk below the valley level by a great fault, which followed closely the line of its present western base.

The internal geological structure of the Wasatch range is most complicated, and the various dynamical movements by which it was produced, if thoroughly and accurately worked out, would present an epitome of the geological history of the greater part of the Cordilleran system. The range was first systematically studied in 1869 by the geologists of the exploration of the Fortieth parallel, and the preliminary determination of horizons thus made served as a basis for later determinations throughout the whole breadth of the Cordilleran system. The following table shows the general geological column of the Cordilleran system as a whole, as represented here.

Group.	System.	Series.	Where found.
Cenozoic	Pleistocene	Bonneville Lake beds	Great basin.
	Pliocene	Wyoming conglomerate	Uinta mountains.
		Humboldt	Cache valley.
	Miocene	Truckee	In Nevada.
		White River	East of Rocky mountains.
	Eocene	Bridger	Wasatch and eastward.
		Green River	Wasatch and eastward.
		Wasatch (Vermillion Creek)	Wasatch and eastward.
Mesozoic	Cretaceous	Laramie	Wasatch and eastward.
		Montana (Fox Hills)	Wasatch and eastward.
		Colorado	Wasatch and eastward.
		Dakota	Wasatch and eastward.
	Jurassic		Wasatch and eastward.
	Triassic	Red beds	Wasatch and eastward.
Paleozoic	Carboniferous	Permo-Carboniferous	Wasatch and eastward.
		Upper Carboniferous	Wasatch east and west.
		Weber (grits and quartzites)	Wasatch east and west.
		Lower Carboniferous	Wasatch east and west.
		Sub-Carboniferous	Wasatch east and west.
	Devonian	Nevada (limestone)	Wasatch and westward.
		Ogden (quartzite)	Wasatch and westward.
	Silurian	Ute-Pogonip (limestone)	Wasatch east and west.
	Cambrian	Middle Cambrian	Wasatch and westward.
		Lower Cambrian	Wasatch and westward.
Pre-Paleozoic	Huronian (Algonkian)		Wasatch east and west.
	Archean		Wasatch east and west.

While in the more recent phase of its geological history the range is a great faulted block, the internal structure of that block proves the existence of a succession of more ancient mountain ranges of complicated structure produced by a succession of orographic movements at different periods in the earth's history, of which only the most brief and scanty outline can here be given.

The following great transgressions or unconformities have been observed, which mark critical epochs in its geological history:

First. At the close of the Archean.

Second. During Algonkian or Pre-Cambrian times.

Third. At the close of the Paleozoic.

Fourth. Toward the end of the Jurassic.

Fifth. At the close of the Mesozoic.

Sixth. The final uplift in Tertiary times, which is continuing to the present day.

Of these movements the First, Third, and Fifth were the most widespread, and have left the most definite evidence of their existence, not only here, but in other parts of the Cordilleran system.

The post-Archean transgression is here, as elsewhere, most distinct and easily recognizable, and the positions which the succeeding sedimentary beds bear to the massives of ancient crystalline rocks, show that they must have been deposited on the flanks of lofty and precipitous mountain masses. Of the rocks which were first deposited against these shores the outcrops are very limited and have been but slightly studied. The great thickness of Paleozoic beds, which form the principal mass or main crest of the range, rest upon the projecting massives of Archean rocks; not in regular folds with parallel axes, as in the Appalachian system, but wrapping around them, with curving strike and ever-changing dip, in anticlines and synclines with axes of varying trends, which, on the eastern flanks of the range, are still partly buried beneath the beds of the Tertiary transgression.

The movement at the close of the Paleozoic is proved by no discernible angular discrepancy in the position of beds deposited before and after it, but by the change in the character of sedimentation and by the striking fact that the crest of the range marks the western limit of deposition of Mesozoic beds in this latitude. From the meridian of this range (112° W. long.) to long 117° 30′ W. (from Greenwich), no trace of Mesozoic beds has been found, and those that exist west of the latter are of entirely different character, both lithological and paleontological, from those found east of the Wasatch range.

The transgression about the close of the Jurassic is shown by a discrepancy of strike rather than of dip, and even this, in the Wasatch mountains, is not very marked but is well developed further east, especially in the Rocky mountains of Colorado.[59] A considerable thickness of Lower Cretaceous beds is found to the north, in British Columbia (Kootanie beds), and in Texas, east of the mountains and south of the Great Plains (Comanche beds), which are wanting in the Wasatch and in Colorado.

The transgression at the close of the Laramie (coal-bearing) Cretaceous is the most distinctly marked and most readily observed, next to that at the close of the Archean. It is generally (though not always) shown by a marked unconformity of angle between the Laramie and succeeding deposits. The earlier deposits with unimportant exceptions are marine, the Laramie brackish water, and all succeeding deposits in the interior of the Cordilleran system of fresh-water origin.

In the Wasatch region the succeeding Eocene Tertiary conglomerates overlap the eroded edges of all the earlier series, and even rest upon denuded Archean high up on the western slopes of the range. They generally constitute the greater portion of the high table-lands or mesas which form the eastern and southern continuation of the Wasatch uplift.

The Tertiary transgression, or transgressions (for there have been several), are less readily defined and are apparently rather local in their nature. In places, especially on the immediate flanks of the mountain massives, the elsewhere horizontal Tertiary beds are found to be upturned at considerable angles. In the great Tertiary basin of Green river, east of the Wasatch and north of the Uinta mountains, there is evidence of dynamic movement, and of a limited amount of erosion, at the close of each of the three Eocene epochs there represented.

The principal plication of the Wasatch range must have taken place at the close of the Cretaceous, and at this time also was formed the great east and west anticline of the Uinta Mountains, which stretches 150 miles eastward from the eastern flank of the Wasatch, opposite the great granite mass of Little Cottonwood canyon. That the movement of uplift of this range has been, in a measure, continued in Tertiary times, is proved by the fact that the Tertiary beds resting upon its flanks are also bent upwards; but the angle at which they are upturned, even in the lowest and most disturbed of these Eocene series, is much lower than that of the Cretaceous beds, while those of the two upper series often pass in a nearly horizontal position completely over the eroded edges of the upturned Cretaceous strata.

Erosion, since the deposition of the Eocene Tertiary beds, has carved out many interior valleys to the east of, and generally parallel with, the main crest of the Wasatch range, and this erosion has exposed portions of the underlying upturned beds. Over the greater part of the plateau region immediately west of the crest, however, the Tertiary covering still remains and masks their structure.

The broad general features of structure of the older part of the range, as far as it has been made out, are as follows:

At the north is a great syncline trending to the west of north, in whose eroded axis lies the great depression of Cache valley. The western flank of this syncline forms the front of the range from "The Gates" south to Brigham city. From here south to beyond Farmington stretches the Farmington Archean body, one covered by an arch of sedimentary beds. The western portion has been cut off by the great Wasatch fault, and the remainder in great measure denuded of its covering of sedimentary beds. Southeast of Salt Lake city is the great granite body of Little Cottonwood canyon, assumed to be of post-Archean age, though the crystalline rocks through which it was intruded are now considered Algonkian. The position of the overlying sedimentary beds, and the fact that their lowest members contain fragments of granite in the immediate vicinity of that body, prove that it is certainly pre-Cambrian. Included between the Farmington and the Cottonwood masses is a syncline of beds ranging from Cambrian to Cretaceous, whose axis runs nearly east and west. These upturned beds wrap round either of the older masses on the east; that is, the strikes of beds forming either member of the synclinal fold diverge fan-like to the eastward and curve respectively to the north and south around either body. The western extension of the syncline, were it not for the fault, would pass above Salt Lake city. . It probably lies at some depth beneath it.

South of Lone Peak the granite sinks beneath the surface, and the range opposite the valley of Utah Lake is formed of a flat arch of Paleozoic beds, half of which has been cut off by the fault and depressed beneath the valley level.

The accompanying map (Fig. 18) is a reduced copy of the geological map of the range in the atlas of the Fortieth-parallel reports, of which Dana [52] says: "It is the grandest exhibition of facts pertaining to an individual case of mountain building in geological literature." *

GEOLOGICAL PANORAMA OF THE WASATCH RANGE

AS SEEN FROM THE RAILROAD.

By S. F. EMMONS.

From the northern point of the range to **Brigham** its east front shows a series of northwesterly dipping beds, ranging from Lower Carboniferous limestones, through Devonian and Silurian, to Cambrian at **Brigham.**

From a distance one can distinguish the different series of beds by their colors, the limestones being very dark, the quartzites light in color. Thus in this part of the range the white Cambrian quartzites at the base of the sedimentary series are sharply contrasted with the dark band of Silurian limestone, and this again from the overlying light Ogden quartzite at the base of the Devonian.

The depression in the range south of **Brigham** marks a line of strike-fault, running southeast with downthrow to the southwest, by which the Cambrian and Silurian beds are repeated on the mountain-mass projecting westward from the main line of the range, round which the railroad bends in going to **Hot Springs.**

At **Hot Springs** the waters issue from outcrops of Cambrian quartzite, forming part of a mass of quartzite and limestone, broken down by the great Wasatch fault from the beds which now cap the high peak to the east. On this peak the white line of Cambrian beds, with a thin cap of

---

* The small scale of this map necessitates the omission of many topographical and some geological details. It serves, however, to illustrate the general outlines of structure as above described, and shows the western end of the broad anticlinal arch of the Uinta mountains.

On the western base of the range B. C. indicates the mouth of Big Cottonwood canyon; L. C., that of Little Cottonwood canyon. Figures denote elevation in feet above sea level.

451 GE——25

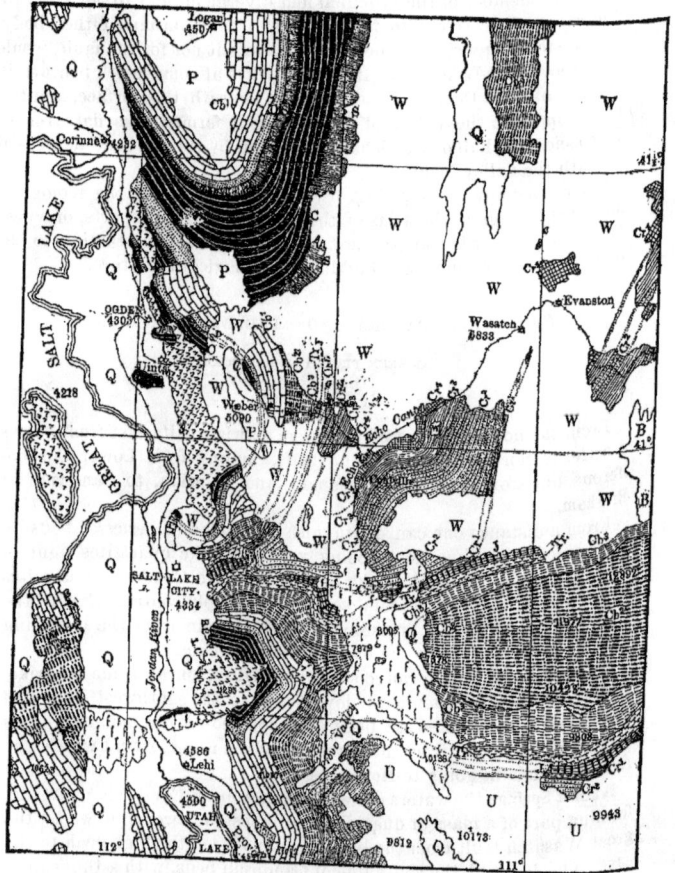

FIG. 18—Outline geological map of the Wasatch mountains.

darker limestone above, can be readily distinguished at the very crest, while the lower part of its steep western face is formed of darker structureless Archean rocks.

At a re-entering angle of the mountain front, just south of the **Hot Springs**, known as Ogden's Hole, is a transverse fault at right angles to the trend of the range, whose movement has been a downthrow on the south, by which movement the Cambrian and Silurian are brought down to the foot of the steep western face of the mountains, while Lower Carboniferous limestones form the crest of the ridge.

Directly east of **Ogden city** is Ogden canyon, a narrow gorge cutting entirely across the range, and connecting with a small interior valley, once a bay of Lake Bonneville, which affords an admirable exposure of the whole series of rocks from the Archean up to and through the Lower Carboniferous.

Just south of Ogden canyon a smaller ravine marks the line of another transverse fault, with upthrow to the south, by whose movement the Archean rocks on the south are brought nearly to the crest of the range, with a thin covering of Cambrian quartzite which disappears to the south.

A few miles south of **Ogden** is the still deeper gorge cut by Weber river, which drains a large portion of the eastern slopes of the Wasatch. The Rio Grande Western railroad passes round a delta of the Bonneville Lake beds that in great measure hides this fine gorge. The gorge affords an admirable section from Archean up to Cretaceous and unconformable Tertiary beds, which can be seen along the line of the Union Pacific railway as it crosses the mountains.

From Weber canyon southward for nearly 20 miles (32 km.) the crest and west front of the range is of Archean rocks; while coarse Tertiary conglomerates rest high up on the eastern flanks and conceal all the lower beds. At the southern end of this Archean mass the older sedimentary beds appear again, through denudation of the conglomerates, striking nearly east and west, and dipping southeast and south away from the Archean. These upturned beds, broken through by eruptive rocks and partly covered by Tertiary beds, jut out to the westward of the main front of the range, forming a projecting promontory between the Mormon town of Centerville (east of **Lake Shore** station) and **Salt Lake city**.

This promontory is formed of steeply upturned Paleozoic beds, partly covered by a coarse conglomerate of supposed Tertiary age and an eruptive body of trachyte or andesite. The **Warm Springs** at the point of the promontory issue from the much broken Paleozoic limestones. The canyon of City creek, directly north of **Salt Lake city**, cuts through the conglomerate and eruptive body, and near its head, beyond the conglomerate, primordial trilobites have been found in the shales at the base of the limestones.

To the north of Camp Douglas, the United States military post three miles (4.8 km.) east of the city, are upper Carboniferous and Permian beds, mostly limestones, carrying characteristic fossils. Red Butte canyon, northeast of Camp Douglas, marks the dividing line between the shaly limestones of the Permian and the pink red sandstones of the Trias. The latter furnish much of the building stone for **Salt Lake city**. Above these, dipping 40° to the southeast, are the drab limestones and argillaceous shales of the Jurassic. These beds strike about N. 40° E., and thus cross the ravines whose direction is more nearly east and west. Emigration canyon, directly back of Camp Douglas, lies nearly in the axis of the synclinal fold included between the two great Archean masses to the north and south (3 and 4 on the map).

Going south from Emigration canyon, along the foothills, one crosses the ends of a series of beds dipping north and northwest, ranging in geological horizon from the Jurassic down to the base of the Cambrian, a thickness in round numbers of 35,000 feet (10,750 m.).

Between Emigration and Parleys canyon a secondary anticline brings up the Permian beds from under the Trias. At Parleys canyon, up which runs the narrow-gauge railway to Park city, the Permian beds are again exposed with the regular northerly dip. They carry abundant Permian forms, *Aviculopecten, Eumicrotis, Myalina*, etc.

Mill Creek canyon, the next to the south, is in the Weber quartzites of the Middle Carboniferous. South of this is a re-entering angle of the foothills in which the outcrops are more covered by débris and the succession less easy to trace. The high projecting spur beyond is formed of the great mass of Cambrian [53] quartzites and slates striking northwest and dipping 45° to the northeast.

An excellent section across these beds and up into the Carboniferous is obtained by following up the next canyon gorge, called Big Cottonwood, as the beds curve in strike more to the southward at the upper part of the canyon, wrapping around the granite body of Little Cottonwood canyon.

The section (Fig. 19), on a nearly east and west line through Twin Peak, a little south of Big Cottonwood canyon, shows the position these beds occupy relatively to the granite. The lowest Cambrian beds, near the contact with the granite, contain fragments of the latter, showing its eruption to have been at least Pre-Cambrian, and that all these beds were deposited around it, and have been subsequently uplifted into their present position.

The next canyon south, known as Little Cottonwood, is a deep glacier-carved gorge cut almost its entire length in granite, but showing easterly dipping Paleozoic beds at its very head. The steep face of Twin

Peak (11,560 feet or 3,523 m.
elevation) forms its northern
wall, which is mostly granite,
but with Cambrian quartzites
forming the summit ridge as
shown in the section. The
foothills between the mouths
of Little and Big Cottonwood
canyons are formed of a series
of crystalline schists dipping
westward away from the granite body, which are represented by the blank space at
left of section. At the contact the granite sends veins
into these schists, and also
includes fragments of them
within its mass, which proves
its later age. This series of
beds was considered by the
geologists of the Fortieth parallel survey to probably represent the Huronian division
of the Archean; further
study may lead to their inclusion in the Algonkian.

South of this canyon the
second highest point of this
portion of the range, Lone
peak [51] (11,205 feet (3415 m.)
elevation) is about the center
of the great granite mass
whose exposed area measures
8 by 12 miles (13 by 19 km.).
An encircling series of lower
Paleozoic beds wraps around
its northern, eastern, and
southern sides, and a thin
shell of crystalline schists
rests upon its western flanks
forming the extreme foothills
toward the Jordan Valley.

On the ridge forming the
southern wall of the entrance
to Little Cottonwood canyon

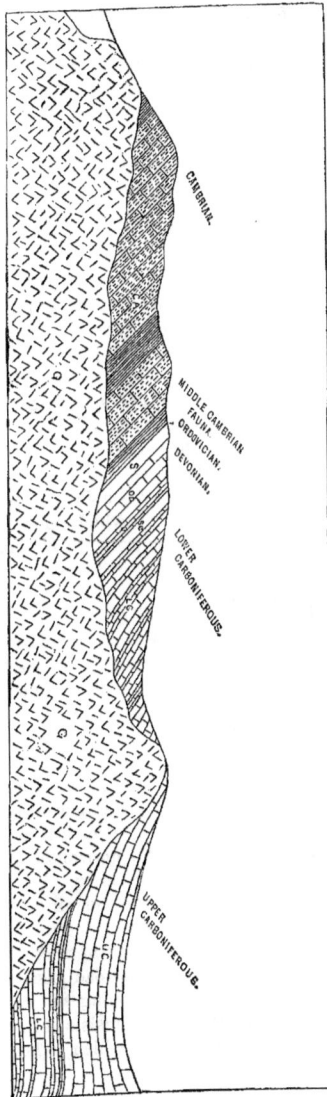

Fig. 19.—Section between Big and Little Cottonwood canyons.

is a fragment of quartzites and limestones to the west of the outcrop of crystalline schists, which represents a portion of the overlying Paleozoic beds brought down by the great Wasatch fault. These lie somewhat to the eastward of the line of faulting in the Pleistocene deposits of the valley.

At the head of the two Cottonwood canyons is Claytons peak, a boss of granite projecting through the Paleozoic beds, and to the east of this, on the east slopes of the Wasatch range, are extensive areas of eruptive rocks. It is in this portion of the range alone that important mines have thus far been developed. The most prominent have been the Emma mine, in Paleozoic limestones at the head of Little Cotton-wood canyon, and the Ontario mine, a strong fissure vein worked to a depth of over 1,000 feet (305 m.) in Middle Carboniferous quartzites, with associated porphyries, which has produced over twenty millions of silver and is still paying dividends.

From the southwestern extremity of the Lone peak granite mass a low ridge of eruptive rocks, resting on altered sedimentaries of unknown age, stretches across the valley from the Wasatch to the Oquirrh range. This separates the valley of Salt Lake from that of the fresh water Utah Lake. This lake is fed by four principal streams, which drain interior valleys to the east of the main crest of the Wasatch, and cut deep trans-verse gorges across it, which afford admirable sections of the range. These are American Fork, Provo river, Hobble's creek, and Spanish Fork. The former shows a broad anticlinal arch, and the great mass of Timpanogos peak south of it, nearly 12,000 feet (3,667 m.) in height, is formed of horizontal Paleozoic beds forming the crest of this arch. The axis of this fold runs somewhat east of south and to the west of the crest of this peak, following the line of a projecting shoulder; it is somewhat broken, and in Provo canyon, next south, is only visible as a fault line with the beds somewhat curved upward near to it.

South of Provo canyon a second line of folding and faulting, en échelon and set off a little to the west of this, is developed back of the town of Provo at the base of Provo peak. The section seen in Rock canyon, back of this town, shows lower Paleozoic beds sharply upturned in an S-fold, which sometimes pass the vertical in dip, but a short distance eastward shallow almost to the horizontal position. Quartzites and schists occur at the base of the series. While the lines of strike continue in a direction east of south the range itself assumes topographically a direction more and more to the west of south.

In Hobble's canyon, and in Spanish Fork canyon through which the Rio Grande Western Railroad passes, only upper members of the Pale-ozoic series are seen, dipping somewhat south of east at 25°. To the south of this the range consists of mountain masses set off successively more to the westward, which culminate in Mount Nebo, 12,000 feet

(3,657 m.), 20 miles (32 km.) to the south and about 12 miles (19 km.) to the west of the Timpanogos line of elevation. This peak is formed by a sharp anticline of Paleozoic beds, the axis of which is near the summit. The eastern member of the anticline stands almost perpendicularly and soon disappears under unconformable Mesozoic and Tertiary beds, while the western member dipping 40° to 45° west forms a long and more gentle slope. This is the geological termination of the Wasatch range, as a complex mass of folded and faulted Paleozoic strata. Its topographical continuation to the south is found in a series of high tablelands or mesas, formed of Mesozoic and Tertiary beds, known as the High Plateaux. [54]

GREAT SALT LAKE VALLEY.

By G. K. GILBERT.

Salt Lake city.—A body of Mormons, or members of the Church of Jesus Christ of Latter-Day Saints, founded this city in the midst of a wilderness in 1848, more than twenty years before the construction of the first railway. From that time it has received a continuous growth by Mormon immigration, and from it have been colonized nearly the whole of Utah and portions of adjacent Territories. Agriculture and grazing are the chief industries, and the development of mines has been discouraged by the church. But mines were, nevertheless, developed, and the mining industry drew into the Territory a considerable body of Gentiles, by which title non-Mormons are known. Municipal control in Salt Lake city has recently passed for the first time from the hands of the Mormons into the hands of the Gentiles.

The head of the Mormon church is called president, and the office is now filled by Wilford Woodruff. The members of a high council of twelve are called apostles, and other officers are known as elders, hundreds, and bishops. The bishops have important secular functions, one being placed in charge of each settlement and of each ward of the city. The church is sustained by tithes, a tax of 10 per cent on the incomes of its members. Payment is made chiefly in produce, which is gathered in tithing houses and placed on sale at fixed rates. The principal meeting house of Salt Lake city is called the Tabernacle, and its auditorium has a seating capacity of about 8,000. It is architecturally unique, having the form of half an egg shell. A more imposing structure, known as the Temple, is used for secret ordinances of the church, and visitors are not permitted to examine the interior. Other objects of interest connected with the church are: the Lion house, where a large part of the late Brigham Young's polygamous family resided; the Emma palace, built by him as a residence for his favorite wife, but never used for that purpose; and Zion's Cooperative Mercantile Institution, the headquarters of a commercial system organized and conducted by the church.

Polygamy, which was formerly enjoined by the church upon its members as a duty, has, after long and bitter conflict with the United States Government, been abandoned.

The system of irrigation can be conveniently examined in the suburbs of the city.

Fault scarps near the Thermal springs.—Along the base of a great range like the Wasatch a vast amount of alluvium is deposited. On the steep slopes of the mountain gorges erosion is rapid, and the resulting detritus is rolled forward by the torrents and deposited on the plain outside the line of mountain front. Between the canyon mouths the waste from the mountain face produces talus, and this combines with the alluvium in the formation of a sloping apron of débris. In the case of the Wasatch range this tendency is partially counteracted by progressive movement along the great fault line which everywhere bounds the range. From time to time the mountain block rises, and the adjacent part of the valley block, with its load of detritus, sinks, and the height of the alluvial slope is thus diminished.

Near the Warm springs there is a point where the vertical rock face of the mountain is laid bare and the valley alluvium meets it as a horizontal plain instead of a foot slope. A few feet up the rock cliff a line of adhering cement marks former contact with the alluvium, the separation being presumably due to a recent fault. Near by, the alluvium is locally disposed in a series of steps marking details of displacement. A little farther northward between the Warm springs and the Hot springs, a symmetrical cone of alluvium is built upon the valley plain, and recent faulting has dropped one portion of this cone below another so that their relations can be clearly seen. The flood plain of the wet-weather stream descending the alluvial cone shows a series of terraces of different heights, illustrating the fact that the fault scarp was produced by a number of successive movements.

Garfield.—Garfield is a bathing resort on the shore of Great Salt lake, and will be visited in order to give members of the party an opportunity to bathe in the lake, and to examine the terraces and cliffs of Lake Bonneville.

The brine of the lake has a density at the present time of about 1·15, and its consequent buoyancy gives a new sensation to the bather. He readily floats, with head, feet, and hands out of the water. In swimming he is annoyed by the tendency of his feet to rise to the surface, and if he brings them beneath him so as to assume an erect position in the water, he finds not only his head but his shoulders above the surface.

The Oquirrh range is a mountain ridge rising several thousand feet above the plain and composed of Paleozoic rocks. At the northern end these are of Carboniferous limestones and quartzites. The range ends abruptly at the southern margin of the lake and its strata dip steeply toward the water. At low stages of Lake Bonneville it projected as a bold promontory, and at its northern face received the impact of waves generated in a broad sheet of water 1,000 feet (300 m.) in depth. As a result it is sculptured in the most elaborate and beautiful manner. Above the Bonneville level tower sea cliffs hundreds of feet in height. At the Provo level terraces hundreds of feet in width are carved from the limestone. At intermediate levels the face presents a succession of minor cliffs and terraces, and at lower levels near the shore of the modern lake are fossil beaches and spits of shingle. A climb of 600 feet (180 m.) to the Provo terrace will be repaid by a view of the phenomena of fossil shores such as can have few equals. [49]

Tooele valley.—Twenty miles west of the Oquirrh range the plain is interrupted by a similar ridge, the Aqui range, and this, too, ends abruptly at the shore of the lake. The lowland between the two mountain ridges is called Tooele valley at the north and Rush valley farther south. The two valleys are separated by a small cross range except at one point where a low pass connects them. Before the formation of Lake Bonneville Rush valley was drained northward to Tooele valley and traces of the drainage channel are still to be seen at the pass, but the waves and currents of Lake Bonneville transported an immense quantity of detritus to the pass and built across it a spit so massive that the waters of Rush valley have never been able to reopen the way. The Utah Western Railway will carry the party nearly to the northern base of this spit [45] and an opportunity will be afforded to walk or drive to and along its crest.

Little Cottonwood canyon.—The loftiest portion of the Wasatch range is about 20 miles (32 km.) south of Salt Lake city, and includes peaks rising 7,000 feet (2,100 m.) above the valley. Little Cottonwood is one of a group of creeks which head in this lofty region, course through deep gorges, and issue upon the plain below, where they become tributary to the Jordan river. In Pleistocene time the high mountain region gathered snows to form a dozen glaciers, and one of these, following Little Cottonwood canyon, protruded a short distance beyond its mouth, depositing lateral and terminal moraines of Alpine type. The ice foot was washed by the water of Lake Bonneville, and the terminal moraines were partly buried by delta deposits. The southern lateral moraine stands as a narrow embankment of typical form; the northern appears to have been afterward overridden by the ice, so that its material is spread into a low flat hill. The walls of the canyon at its mouth are of dark quartzite, but the moraines consist chiefly of white granite derived from the upper portion of the canyon. Their boulders have furnished the material for the Temple at Salt Lake City.

Just south of Little Cottonwood canyon a shallower valley called Dry Cottonwood descends the mountain face, and this, too, carried a Pleistocene glacier descending to the mountain base. The lateral moraines in this case coalesce with a massive terminal which the modern creek has notched but not yet divided.

Across the two pairs of lateral moraines and across the flood plains of the creeks runs a line of faulting characterized by the settling of great belts of earth. A portion of the north moraine of Little Cottonwood creek has dropped between parallel fault planes as much as 50 feet (15 m.), and this so recently that the escarpments are not yet covered by vegetation (Pl. VII). The longitudinal profiles of the other moraines, originally simple curves, have been made serrate, and the inner

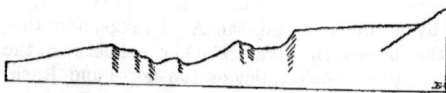

FIG. 20.—Profile of South Moraine at mouth of Little Cottonwood canyon.

faces of the moraines, originally smooth, have been furrowed and ridged in sympathy with the serrations. One of the fault-caused terraces of the northern flood plain has been utilized in the construction of an ore-smelting establishment. It is an impressive fact that the stream, a roaring torrent every spring, has not yet been able to reconstruct its dislocated flood plain.

We are here brought face to face with a process of mountain-making in actual progress. The Wasatch has grown perceptibly within a few years, and the gathering orogenic strains may culminate any day in another convulsive leap. It is a dictum of dynamic geology that great mountains are young mountains; [55] and the greatest of the mountains of the Salt Lake basin has not yet ceased to grow.

TROUGH PRODUCED BY FAULTING, NORTH OF MOUTH OF LITTLE COTTONWOOD CANYON.

SALT LAKE CITY, UTAH, TO GRAND JUNCTION, COLORADO.

ITINERARY.

Station.	Distance.		Elevation.		Station.	Distance.		Elevation.	
	Miles.	Kilometers.	Feet.	Meters.		Miles.	Kilometers.	Feet.	Meters.
Salt Lake City [1]	0	.....	4,228	1,289	Soldiers Summit	90	145	7,465	2,275
Francklyn	7	11	4,291	1,308	Pleasant Valley	98	158	7,182	2,189
Germania	10	16	4,296	1,309	Kyune	103	164	7,052	2,149
Bingham Junction	13	21	4,366	1,351	Castle Gate	112	180	6,151	1,926
Draper	17	27	4,394	1,339	Price	122	196	5,547	1,691
Jordan Narrows	23	37	....	....	Farnham	133	214	5,534	1,687
Lehi	29	46	4,544	1,385	Sunnyside	142	228	5,270	1,606
American Fork	32	51	4,554	1,388	Lower Crossing	161	259	4,630	1,411
Battle Creek	36	58	4,497	1,371	Desert	174	280	....	....
Provo	46	74	4,517	1,377	Sphinx	181	291	....	....
Springvile	51	82	4,566	1,392	Green River	187	304	4,086	1,226
Spanish Fork	55	88	4,865	1,483	Cisco	237	383	4,447	1,355
Castilla Springs	61	98	....	....	Agate	243	391	4,425	1,349
Thistle	66	106	5,043	1,537	Cottonwood	249	401	4,602	1,403
Red Narrows	72	116	5,342	1,689	Utah line	258	415	4,661	1,422
Mill Fork	77	124	5,791	1,765	Ruby	....	....	....	....
Clear Creek	84	135	6,228	1,898	Grand Junction	292	470	4,500	1,390

[1] Population, 44,843.

[By G. K. GILBERT.]

Leaving **Salt Lake** city in the morning the party resumes its journey, following the line of the Rio Grande Western Railroad through the valley of the Jordan, Utah valley, the canyons of Spanish fork and Soldiers fork, the valley of Price river, and the great monoclinal valley at the base of Book cliffs. For 75 miles southward from Salt Lake City the bold façade of the Wasatch range faces westward.

Twenty miles west of it lie smaller ranges, the Oquirrh, the Cedar, and the Tintic, and at its foot lie Jordan and Utah valleys. The Jordan valley is largely devoted to agriculture, but it contains also some of the principal establishments for the smelting of the silver ores of Utah.

The Jordan and Utah valleys are separated from one another by spurs of volcanic rock, stretching from the Wasatch and Oquirrh ranges until they nearly meet. The narrow pass between their extremities is occupied by the Jordan river, whose bank the railroad follows. In

Bonneville times the eastern spur was exposed to an energetic attack of waves from the north. A deep notch was cut in its side, producing a cliff hundreds of feet in height, and the material excavated was piled as a spit of shingle in the pass, partially closing it. As the train moves up the Jordan valley the shore terrace and the sea cliff can be seen from a distance, and, as it threads the pass, the gravels of the spit appear near at hand. In the pass are the headworks for the diversion of the Jordan water into canals for the irrigation of the valley.

The mountains towering above Utah valley are of wonderful boldness and beauty, and their precipitous faces are such as result from no orogenic process save that of faulting. Back of the visible mountain crests are extensive uplands, serving as a gathering ground for the streams which here and there break through the main ridge in defiles and fertilize the valley. Their waters are finally gathered in Utah lake, from which the Jordan river issues, flowing, like that other Jordan after which it is named, from a lake teeming with life to a sea of death.

The Bonneville and Provo shore lines are to be seen all about the valley, and the Provo shore line is characterized by numerous delta terraces. The delta of the Provo river has a radius of more than 4 miles (7 km.), and crowds the railroad close to the lake. Along the mountain base and across the delta terraces run fault scarps, occasionally giving one of the shore lines a sudden change of altitude.

The two sides of the valley are in striking contrast. The eastern, receiving the drainage of the lofty Wasatch, enjoys the condition essential to fertility in an arid region, and is dotted with thriving villages: Lehi, American Fork, Battle Creek, Provo, Springville, and Spanish Fork are passed by the train. On the west, the low ridge called Cedar mountain induces little precipitation and possesses no permanent stream. Even springs are lacking, and its slopes, clothed only by a scant growth of low bushes, sustain no human home.

Due east of Provo the face of the Wasatch is unusually steep, rising in a slope of nearly 35° to Provo peaks, and presenting a remarkable section of Paleozoic rocks over 10,000 feet in thickness, from Cambrian up to Middle Carboniferous, resting on crystalline schists, of which a small exposure is found at the very base of the slopes. These beds all dip eastward, at first very steeply, but with decreasing angle to the east.

Hobble creek, which waters the town of Springville, built a broad delta at the level of the Provo shore line. The Spanish fork, as the next stream is called, built a still broader delta at the same level, and the two are confluent. The Spanish fork built one also at the Bonneville level with a radius of 4,000 feet (1.2 km.). This was widely trenched during the building of the lower delta, so that the head of the lower lies within a gorge excavated from the upper. Both deltas are profoundly dislocated, the zone of faulting being a mile wide, but many of the faults traverse the upper delta only, showing that the lake epoch, like the recent epoch, witnessed mountain growth. The train, on its way from **Springville** to Spanish Fork canyon, climbs and crosses the confluent lower terrace and traverses the district of faults.

[By S. F. EMMONS.]

In Spanish Fork canyon the main Wasatch uplift is crossed in a southeasterly direction at a point where it is much narrower than in either of the transverse gorges further north. No systematic geological examination of this portion of the range has yet been made, and only the broader outlines of its structure can be given.

About 6,000 feet (1,829 m.) of Upper Paleozoic beds, mainly siliceous members of the Upper Carboniferous, are first crossed, which dip 25° to 30° to the southeast. These form the main front range through which the Spanish Fork gorge is cut. Overlying these and generally with somewhat steeper dip, are conglomerates, shales, and red sandstones, presumably of Triassic age. Two streams, running in monoclinal valleys eroded out of these rocks, unite to form that of Spanish Fork. Beyond the junction the road bends to the south up the valley of one of the streams, known as Thistle Creek. On the east side of this valley are fine cliff exposures of the characteristic red sandstones of the Trias, overlain by massive yellowish white sandstones, remarkable for their cross-bedding and the curious forms into which they weather in the overhanging cliffs. The estimated thickness of these rocks above the Carboniferous is 4,000 feet (1,219 m.). [56] A branch line follows this valley southward to the Mormon towns in the San Pete and Sevier valleys.

The main line at **Thistle** bends eastward, across the strike of the beds, up the valley of Soldiers' Fork. A short distance beyond this station are seen the drab shales and thin-bedded limestones of the Jurassic, resting in apparent conformity on the cross-bedded sandstones, in which, along the railroad cut, may be found Pectens and *Pentacrinus Asteriscus*. Ascending the valley of Soldiers' Fork the upturned edges of the Jurassic strata are covered by nearly horizontal beds of reddish conglomerates and coarse sandstones which, in geological position, composition and manner of weathering, resemble those of the Wasatch Eocene, as exposed in Echo canyon along the line of the Union Pacific Railroad. After ascending for some miles past the castellated cliffs formed by those conglomerates, an overlying series of light-colored calcareous marls and shales is reached, which are less massive, and probably represent the Green River Eocene of the Wyoming Basin, since they are said to rest unconformably on the conglomerates and to overlap on to the upturned Jurassic and Triassic strata nearer the mountains.

The road ascends rapidly through these Tertiary beds, the gradient, in the last 7 miles (11 km.) before reaching **Soldiers Summit**, being $3\frac{3}{4}$ per cent, or 200 feet (61 m.) to the mile (1.61 km.).

Soldiers Summit is at the crest of the Tertiary watershed, which lies from 20 to 40 miles (32 to 64 km.) to the east of the crest of the Wasatch range. The road now descends the valley of Price river nearly to the Colorado river, through a region of almost horizontal Mesozoic and Tertiary beds which are uninfluenced by the Wasatch upheaval. This region is not only of modern geological formation, in that it is traversed by no known lines of long continued dynamic movement, and hence owes its topographical form entirely to erosion, but its erosion is of the type peculiar to an arid or practically rainless region whose streams are fed only by the precipitation on high areas outside of the region. This erosion is produced primarily by the streams in their narrow beds, whose winding course was first determined on a comparatively level surface of unconsolidated and easily abraded beds. The first characteristic of such a drainage system, as distinguished from one that is built up on the drainage systems of earlier geological periods, is its comparative independence of the obstacles opposed to its course by the relative hardness, or the position of older strata, which it has reached after cutting its bed down through the uniform and softer overlying strata. This produces the effect designated "inconsequent drainage."

From **Pleasant Valley** a branch runs southward up the main fork of Price river to the coal mines in the Laramie Cretaceous at Scofield. In the reddish beds about a mile (1.61 km.) north of **Pleasant Valley** station are found veins of ozocerite or mineral wax. This and allied hydrocarbons are found abundantly in the beds of the Green River Eocene at various points in the basin region between the Wasatch and the Rocky mountains. A gray limestone near the station contains fresh water mollusks, which may prove to belong to this period.

The road descends rapidly in geological horizon, as well as topographically, and in a few miles enters the massive gray sandstones of the Laramie Cretaceous, in which Price river has cut a narrow, winding, and ever-deepening gorge, whose walls show a great variety of castellated forms due to erosion. The finest of these is just above **Castlegate**, where a narrow column of sandstone, over 500 feet (152 m.) high, stands at the end of a sharp, narrow ridge like the watch-tower of a castle.

At **Castlegate** is a coal mine with coke ovens, which obtains coal from a seam near the base of these massive sandstones.

As the road descends below the coal horizon of the Laramie, through the more thinly bedded sandstones of the Fox Hills formation which contain an ever-increasing proportion of clay shales, into the readily eroded clay beds below, the sandstones are left behind forming in curving lines of precipitous cliffs on either side.

At **Price** the road emerges into the great open monoclinal valley in the clays of the Middle Cretaceous, which extends eastward to and beyond **Grand Junction**, a distance of nearly 200 miles (322 km.). The Laramie sandstones, dipping gently 7° to 10° north, retreat to the northward, and gradually merge into the great mural escarpment of the Roan or Book cliffs, which, with an average height of over 2,000 feet (610 m.), stretch entirely across the Colorado basin, in a curve convex to the south, connecting the plateau system on the east flank of the Wasatch uplift with that on the west of the Rocky Mountain system of Colorado. Through this whole distance the Laramie Cretaceous forms a continuous outcrop. Above it, sometimes forming second lines of cliffs at a little distance back, and sometimes capping the main escarpment, rest successively the coarse reddish or roan-colored sandstones of the Wasatch Eocene, and the drab calcareous shales of the Green River Eocene. The latter are characterized by the thinness of their strata and the great definition of their bedding lines, so that their cliffs resemble the leaves of a book, whence the name "Book Cliffs." Throughout the interior of the basin these Eocene beds rest in parallel transgression upon the Laramie Cretaceous, but along its periphery, especially on the flanks of the Uinta mountains, which form its northern border, they overlap its upturned edges, high up on the flanks of the range, to a contact with Jurassic, Triassic, and even Carboniferous rocks.

The erosion which has formed these cliffs is peculiar to a practically rainless region surrounded, as is the Colorado basin, by high mountain masses. It is produced by the undermining of rock faces in the softer strata beneath by sudden floods, resulting from violent showers, popularly known as cloud-bursts, starting at long intervals and under favorable meteorological conditions from the surrounding elevated regions and spreading out locally over different portions of the arid basin region. These in a few moments change dry stream beds into boiling, muddy torrents, which carry away an enormous amount of material that dry disintegration, aided by great diurnal variations of temperature, had already loosened. Thus the Book cliffs, which, owing to the northerly dip of their component strata, are eroded almost entirely on their southern face, have retreated most rapidly at either extremity, owing to its proximity to the mountains, and now stand in a curve convex to the south, which in the center of the basin is 40 miles (64 km.) farther south than its western end.

The Cretaceous in this region consists of a series of peculiarly resisting quartzitic sandstones (Dakota) at the base, succeeded by several thousand feet of clay shales, with a few thin sandstones and limestones, all of marine formation, and capped by the massive sandstones and shales of the coal-bearing Laramie. The monoclinal valley followed by the road shows well the peculiar topography which everywhere characterizes wide exposures of these Middle Cretaceous shales.

At **Lower Crossing** may be observed an interesting example of inconsequent drainage, where the Price river leaves the wide, open valley in the soft clays and boldly cuts its way in a deep gorge through the Book cliffs, in an easterly direction, to the Green river, which it reaches 16 miles (26 km.) above the railroad crossing.

The road now rises slightly before descending into the valley of Green river, and from favorable points one can see, far to the south, remarkable castellated forms of erosion in harder beds of the lower portion of the Mesozoic, brought up by the monoclinal uplift of the San Rafael swell (Plate VIII). In clear weather a glimpse may also be had, on the distant southern horizon, of the sharp laccolitic peaks of the Henry mountains.

Back of **Green River** station is a mound of dark shales, probably belonging to the horizon known as Fort Benton Cretaceous, which abounds in fish remains and inocerami.

Green river, which is crossed at a point where but little can be seen of its characteristic canyon scenery, is the main tributary of the Colorado river of the West. It takes its rise nearly 300 miles (483 km.) due north of the crossing, in the Wind River mountains, and flowing south across the interior Tertiary basin of Wyoming, famous for its vertebrate remains, it cuts through the heart of the Uinta mountains in a series of deep, winding gorges carved out of the very hardest siliceous rocks. It then flows south through the Colorado basin, and about 60 miles (97 km.) below the crossing is joined by the Grand river, which drains the western slopes of the Rocky mountains. It is only after the confluence of the latter stream that it is called the Colorado river.

In the vicinity of Green river, and especially on the east side, the characteristic scenery of the Cretaceous shales is most marked, the country being absolutely bare of vegetation of any kind for long distances. The stations in this region are appropriately named **Desert, Sphinx, Solitude,** etc. The accompanying sketch (Fig. 21) of the latter was made from the train by Mr. Cadell.

The Valley of Desolation in these shales is followed for 50 miles (80 km.) east of Green river. The road gradually rises over low divides, being at times more than a thousand feet (305 m.) above Green river, and then descends into the valley of Grand River. In clear weather fine views are had, from elevated points, of the laccolitic group of peaks lying 30 miles (48 km.) to the south and east, known as the Sierra la Sal, [57] whose summits have elevations of 12,000 to 13,000 feet (3,658 to 3,962 m.).

At **Cisco** an artesian well has a considerable flow of water, the surplus

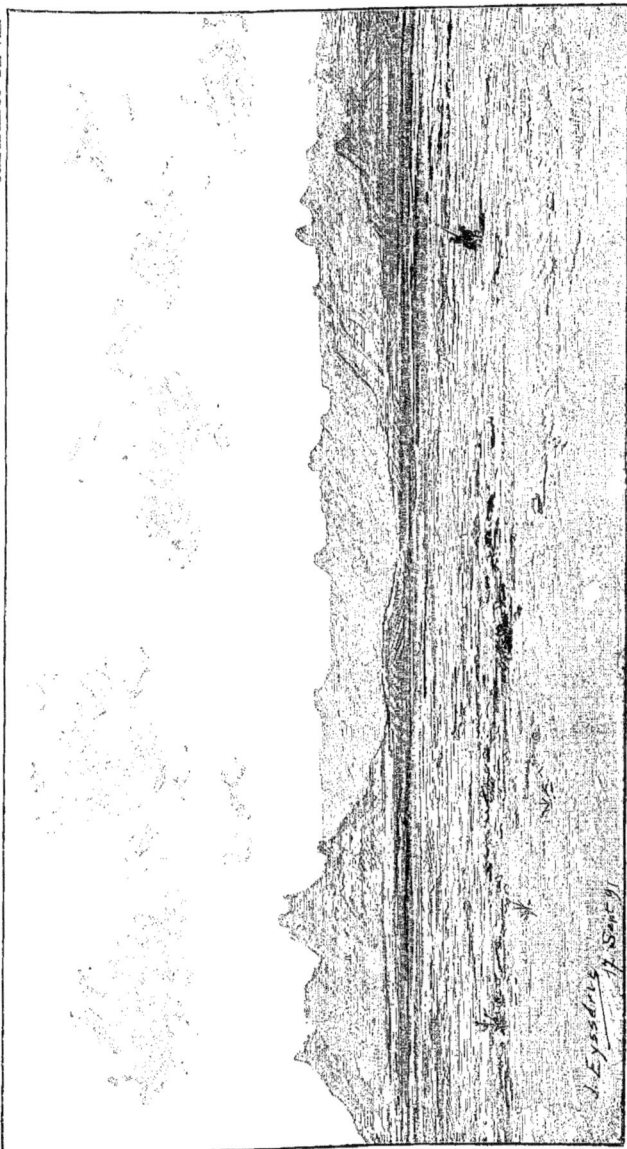

SAN RAFAEL SWELL.

of which, over and above that used by the railroad, has has been utilized for irrigation, but as yet without much beneficial effect upon the clayey soil.

Between this station and **Agate** the northwesterly dipping sandstones of the Dakota Cretaceous are crossed, and the railroad then descends through variegated clays and sandstones of the Jurassic into the thin-bedded sandstones of the Trias, which it reaches at the Grand river.

The road now runs for over 15 miles (24 km.) along the banks of the river, following its winding canyon gorge, cut in the massive sandstones of the Trias, which form the northern edge of the great Uncompahgre plateau. This plateau is an area extending 80 to 100 miles (129 to 161

FIG. 21.—Solitude Station.

km.) northwest from the base of the San Juan mountains, of nearly horizontal or gently folded strata of Lower Mesozoic and Upper Paleozoic age resting unconformably upon Archean granites and gneisses. The latter rocks are only exposed in the bottom of some of the many canyon gorges which intersect the plateau. The northern edge of the plateau is formed by an abrupt monoclinal fold, along which the strata bend down so abruptly that they seem to be faulted. The canyon is cut mainly in the massive and nearly horizontal sandstones on the north side of this fold, but at times in its meanderings discloses views of the fold itself.

Going east, after reaching the valley of the Grand, the more massive beds of the Trias slowly rise from beneath the thinner beds at the top. They present on the north of the road magnificent walls formed by vertical cleavage planes and which show fine examples of cross-bedding.

451 GE——26

At **Utah** line the boundary between Utah and Colorado is marked by a white line running up the face of the cliff.

A few miles further east, after passing round a northward projecting point of the cliffs, one can distinguish along the banks of the river and projecting above its surface in midstream, rounded knobs of Archean rocks, blackened, polished, and singularly channeled by the action of the water. These rocks form apparently an east and west ridge in the line of the monoclinal fold, which the road now follows for some distance on a tangent.

Just before reaching **Ruby** an excellent section of the monoclinal fold may be seen on the north side of the river, where a small tributary ravine is cut along its axis.

The road now passes beyond the fold into the gray sandstones and intercalated shales of the Jurassic and Dakota Cretaceous, in which the effects of undermining by the action of the stream are well displayed. It soon leaves the valley of the Grand river, passing up a tributary ravine, through a tunnel in Dakota sandstone, and out on to the broad valley of Cretaceous shales again. Twelve miles (19 km.) distant, forming the southern boundary of this valley, are the Little Book cliffs,[58] the eastern continuation of the line of cliffs which has been followed from Utah. It is formed here entirely of Cretaceous beds, the overlying Tertiary beds forming a second line of cliffs further back, trending to the northeast, for all the beds are now commencing to rise to the east on the west flanks of the Rocky mountains.

The clay valley is followed to **Grand Junction**, a growing town that is now becoming prominent for the excellent quality of its fruits which have been raised by its citizens on the neighboring mesas under the beneficent influence of irrigation.

THE ROCKY MOUNTAINS OF COLORADO.

By S. F. Emmons.

Upon leaving Grand Junction the route passes from the basin of the Colorado river into the Rocky mountain region, or, according to Powell's and Gilbert's nomenclature, from the Plateau region into the Park province. The name "Rocky Mountains" has been retained for this most eastern and most elevated group of mountain ranges of the Cordilleran system, because it was the one which was first seen by early explorers after their weary journeys across the Great Plains, and is therefore the one to which the name was first applied. It was called the Park province from its great interior valleys, entirely closed in by mountain ridges, which constituted the natural game preserves of the various Indian tribes that formerly inhabited the surrounding lowlands.

The route chosen for the party does not traverse any of the greater of these parks or valleys, but follows the deeper and narrower drainage channels of more modern date, from whose fresher excavations and cuts the geologists may more readily obtain by a passing glance an insight into the internal structure of the region than he could from the older, gentler, and more covered slopes of the parks. A description of the region proceeds naturally from the east westward, rather than in the direction by which it will be approached by the present party, since on the eastern side its characteristic features are more strongly and distinctly marked. While the general physiographical features of the group as a whole arrange themselves along north and south lines, the prevailing northwest and southeast trend of the Cordilleran system is manifested in many details of the topography, as it is in the internal geological structure.

The eastern front of the group, facing the Great Plains, is formed by the Colorado or Front range from the northern boundary of Colorado south to Pueblo, a distance of about 200 miles. Further south it is formed by the Wet Mountain and the Sangre de Cristo ranges, set off successively a little to the west of each other *en echelon*, and separated by northwest-trending indentations or bays.

Back or to the west of these come the Parks—the North, Middle, and South Parks, lying opposite the Colorado range; Wet Mountain valley, on the flanks of the mountains of the same name, and San Luis park, stretching along the western foot of the Sangre de Cristo range, whose culminating point, Blanca Peak, is the highest mountain in Colorado.

The western wall of the North and Middle parks is formed by a single chain known as the Park range; opposite the South Park this is represented by the double elevation of the Mosquito and Sawatch ranges, which in earlier geological time formed a single massive, but are now separated by the great meridional valley of the Upper Arkansas. To the south these two ranges coalesce in the single ridge of the Sangre de Cristo range, which trends first to the southeast, dividing Wet Mountain valley from San Luis park, then southward forming the eastern front of the whole group, and finally disappears beneath the desert plains of New Mexico.

The westernmost tier of mountain uplifts, which is still more broken and irregular, is constituted by the White Mountain plateau, on the north, opposite the Park range; the complicated and lofty group of the Elk mountains, which adjoin the western flanks of the Sawatch uplift; and to the south, separated from the latter by the broad mesa slopes of the Gunnison valley, are the San Juan mountains, stretching westward from the San Luis park to the Uncompaghre plateau.

With the exception of Middle and San Luis parks, all the interior valleys send their waters across or around the component parts of the Front range into the Mississippi valley; the outlet of San Luis park is southward through New Mexico into the Gulf of Mexico; Middle park is drained by the Grand river, which threads its way in deep narrow gorges, between the White River plateau on the north and the Sawatch and Elk mountains on the south, to the Colorado river.

The geological structure of these different mountain masses is, as a rule, very complicated, but as the study of the region progresses it becomes more and more evident that its broad general features must have been outlined at very early dates in its geological history. In each of the great mountain masses are found representatives of the Archean, Paleozoic, and Mesozoic groups. In but few have Algonkian beds yet been recognized. Eruptive rocks of different ages are found in all, but the greater manifestations of eruptive activity, in Tertiary times, have occurred mainly in the southern and western portions of the region.

Five great transgressions, indicating as many orographic[59,60] movements, have thus far been recognized, and future study may disclose others. These are: The post-Archean, the post-Algonkian, the late-Paleozoic, the late-Jurassic, the post-Cretaceous. Minor movements undoubtedly occurred in Tertiary times, as in the Wasatch region, whose general effect has been to raise the whole group, and to compensate for the degradation produced by erosion.

In the middle belt, which is traversed by the party and may be taken as a type of the structure of the whole group, the central uplift is formed by the Sawatch massive, a great oval area of Archean rocks, surrounded by an almost completely encircling fringe of Paleozoic and

Mesozoic beds dipping away from it at various angles. These beds give internal evidence of having been deposited around a land mass, and from their present position it is evident that the central Archean area was never completly covered by them. They also show little or no discrepancy in their angle of dip.

The Elk mountain massive, which lies immediately to the west of the Sawatch, is a region of intense disturbance. The sedimentary beds are profoundly plicated and faulted; and, though the peaks are as high as those of the Sawatch, and the intermediate valleys even more profoundly eroded, only an extremely limited area of Archean is exposed. Many of the higher peaks are immense masses of diorite, up to 15 miles in diameter, which have been protruded through the sedimentary beds during the post-Cretaceous movement. The evidence of the various transgressions, which is remarkably distinct in these mountains, is found in discrepancies of strike rather than of dip. Thus, where the upturned sedimentary series is crossed by the train, on the Grand river above Newcastle, the strata seem perfectly conformable from Laramie Cretaceous down to Cambrian. Yet the sandstone series of the Dakota Cretaceous (which includes also some beds of probable late Jurassic age, hence sometimes called Jura-Dakota sandstones), if followed southward along their strike into the Elk mountains, would be found to rest on successively lower beds from Paleozoic to Archean. Besides the diorite masses in the more disturbed central region, numerous picturesque mountain masses are formed by laccolites of quartz-porphyry in the nearly horizontal Cretaceous strata of the southern and western portion of this remarkable group of mountains. On the east, between the Elk mountains and the Sawatch. is a zone of profound faulting, following north and south lines.

To the east of the Sawatch mountains, and separated from them by the great longitudinal valley of the Upper Arkansas, lies the Mosquito or Park range, which once formed an integral part of the Sawatch and has been uplifted into its present position by plication and faulting, which was initiated during the late Jurassic movement. The fault follows a north and south line just west of the main crest, and has been traced for over 50 miles. The Upper Arkansas valley is thus a comparatively modern topographical feature.

The wide valley of South park, on the other hand, which lies between the Mosquito range and the broad Archean mass of the Front or Colorado range, is underlaid by the beds of the entire Mesozoic and Paleozoic groups, which descend with decreasing angle of dip from the crest of the former and disappear beneath the recent beds which form its present floor. They are probably cut off by a fault on its eastern edge, but this does not appear farther south, on the line traversed by the train along the valley of the Lower Arkansas.

On the eastern flanks of the Colorado or Front range the Paleozoic and earlier Mesozoic beds appear only in disconnected patches, being covered in the intermediate portions by beds deposited during the late Jurassic, post-Cretaceous, and Tertiary transgressions. A narrow zone of sharply-upturned beds, mostly of Mesozoic age, is generally found at the immediate base of the mountains, the harder beds forming narrow monoclinal ridges called "Hogbacks," which are separated from the main mountain slopes by longitudinal valleys eroded out of the softer beds of the series. These sharply-upturned beds, whose angles of dip are as high as 60° or 70°, change sharply to an approximately horizontal position in very short distances east of the mountain foot. In some cases the outer or upper beds of the upturned series stand at a steeper angle than those below, or nearer the Archean base, producing thus a partial fan structure. The overlying Tertiary beds sometimes partake to a limited extent in the upward curve of the underlying Mesozoic strata, thus evidencing a comparatively recent movement of uplift of the mountain mass, or a sinking of the plain area.

The coal-bearing beds of the Laramie Cretaceous, which were deposited as the ocean waters were finally leaving the western United States, spread not only over all the Great Plain areas around the mountain groups but also over many of its interior valleys. These beds being the first of the series to be acted on by the forces of erosion, have been removed from considerable areas that they once covered, especially along the eastern portion of the Great plains and the southern portion of the Colorado basin. In other areas they have been deeply buried beneath succeeding deposits of Tertiary age. In spite of these facts, the available coal-bearing areas that still remain are of enormous extent, and already play an important part in the industry of the region. The Census reports for 1890 show that in that year the coal mines of Colorado produced 2,360,000 tons of coal and gave employment to 4,645 persons. The coal varies in character from a rather light, dry, porous coal, with high percentage of water and of volatile matter, to fairly dense caking or coking coal. The latter coals are generally found to the south and west, in regions where eruptive rocks abound. In the southwestern Elk mountains, moreover, are beds of excellent anthracite.

The aggregate thickness of the various geological series has, as a rule, very materially decreased from that shown in the Wasatch region. A greater mass of Archean is shown because it has been more deeply eroded. The Ouray beds (assumed to be Algonkian) of the San Juan mountains are over 10,000 feet thick. On the other hand, but a few hundred feet of Paleozoic beds below the Carboniferous are found, the lower members of the Cambrian and the entire Devonian being apparently unrepresented. The Carboniferous series is the most generously developed of any in the Paleozoic group, but this averages only about

5,000 feet, as against 15,000 feet in the Wasatch. The various members of the Mesozoic have a wide range of local variation in thickness, but it is always less than that of corresponding horizons in the Wasatch. The marine Jurassic of the Wasatch is apparently wanting, as is the Lower Cretaceous of Texas and British Columbia. The age of many of the disconnected Tertiary deposits has not yet been determined, though Eocene and Pliocene and probably Miocene are represented.

GRAND JUNCTION TO GLENWOOD SPRINGS.

By S. F. EMMONS.

Station.	Distance.		Elevation.		Station.	Distance.		Elevation.	
	Miles.	Kilometers.	Feet.	Meters		Miles.	Kilometers.	Feet.	Meters.
Grand Junction ..	0	0	4,560	1,390	Rifle	63	101	....	....
Palisades	12	19	....	....	Newcastle	77	124	5,555	1,693
De Beque	33	53	....	....	Chacra	81	130	....	....
Parachute	46	74	....	....	Glenwood.........	89	143	5,767	1,758

From **Grand Junction** the road goes northward for 15 miles (24 km.) across the clay plains to the mouth of Hogback canyon. On the east may be seen the line of the Book cliffs continued in those of Grand mesa, a high plateau formed of basalt, capping Wasatch and Green River Eocene beds.

Hogback canyon, above **Palisade** station, is cut in the heavy gray sandstones of the Laramie Cretaceous, which dip gently north and west, so that higher horizons are constantly coming in view. The canyon shows fine cliff sculpture and excellent instances of the erosion due to undermining, which is peculiar to the Colorado basin or Plateau province. In one case the ruins of a spur of Laramie sandstone, which had been undermined on either side, are piled up along the valley in huge broken masses over a quarter of a mile in width and nearly a mile in length.

Beyond the canyon a broad alluvial valley opens out, from which, just before reaching **Debeque** station, can be seen, to the northwest, the main Book cliffs, which rise 3,000 feet (914 m.) above the river, formed of the massive roan-colored beds of the Wasatch Eocene, capped by the thinly stratified, shaly beds of the Green River Eocene. The latter contain abundant fish remains, and, in consequence, much bituminous matter, some of the shales being so rich in hydrocarbons that they may be ignited by a match. Deposits of an asphaltum mineral, Uintaite or Grahamite, are found in considerable quantities at various points. On the east may be distinguished the Mam mountains (11,000 feet, 3,353 m.), twin peaks of basalt capping Green River beds.

Of the streams entering the Grand river from the north, Roan and Parachute creeks drain the interior of the Book plateau, but Rifle creek drains the monoclinal valley within the Hogback ridge of Laramie sandstones which forms a semicircle around the west flanks of the great White River plateau, breaking through this ridge in a narrow gorge six

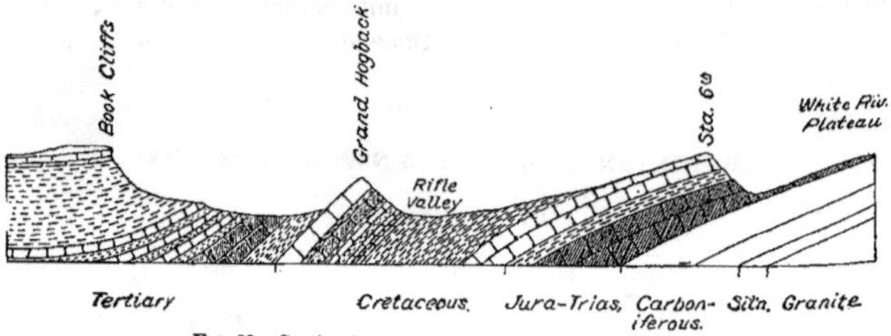

FIG. 22.—Section from Book cliffs to White River plateau. [59]

miles (10 km.) due north of **Rifle** station. This great Hogback ridge, which here rises 2,000 feet (610 m.) above the valley level, can be traced, almost continuously, northward along the west flanks of the Elk mountains, then bending in a curve around the White River plateau to White river, and then westward along the southern flanks of the Uinta mountain uplift to the Wasatch, a distance of about 250 miles (402 km.). According to Mr. R. C. Hills, who has made a special study of the coal-bearing beds, the Wasatch Eocene is also upturned on the flanks of this ridge, whence he reasons that the Eocene movement (post-Bridger) played an important part in the Rocky mountain uplift. [60]

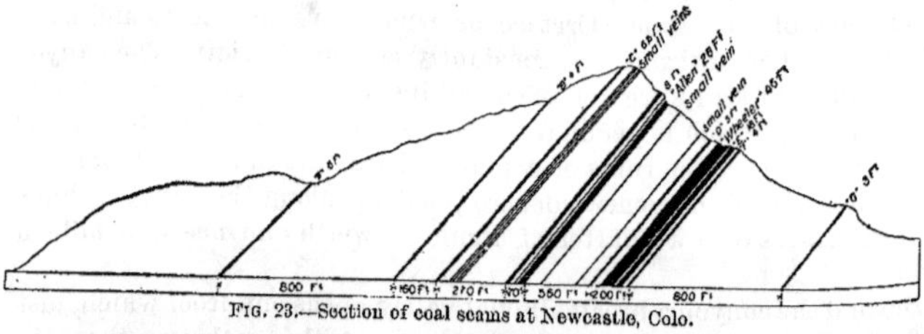

FIG. 23.—Section of coal seams at Newcastle, Colo.

At **Newcastle** [61] the Hogback ridge is seen in section to the west of the town, where coal mines are opened on some of the numerous coal veins, which contain an aggregate thickness of over 125 feet of coal. These are shown in the cross section (Fig. 23), furnished by the manager of the mines, Mr. W. B. Devereux.

About 2 miles (3 km.) south of the town, the great 45-foot vein has been burning for ages deep down beneath the surface. At many other points, the baked and reddened rocks along the outcrops show that coal seams have been similarly consumed near the surface. From Coal Bridge station, a short distance beyond **Newcastle**, an inclined tramway can be seen on the east side of the river, which brings coal down from seams in the Laramie sandstones high up on the steep northern face of the ridge.* These mines belong to the Elk Mountain Fuel Company.

The monoclinal valley of Elk creek, which enters the Grand river from the northwest at **Newcastle**, is cut in the clayey beds between the sandstones of the Laramie above, and the Dakota-Jura below. Some of these are shown in section in a railroad cut on the west side of the river above **Newcastle**. The valley of the Grand, which is here crossing the beds, bends to the southeast when it reaches the Triassic sandstones, following them irregularly with and diagonally across the strike for about five miles (8 km.), and then, at **South Canyon** station, bending again across the strike in a narrow gorge, comes out into a more open valley eroded out of the shaly limestones and gypsiferous beds of the Upper Carboniferous, which it follows eastward to **Glenwood Springs**.

The sedimentary beds are here all upturned against the southwestern flanks of the White River plateau, which is an elevated area of Paleozoic rocks resting on Archean and capped by flows of basalt. It is nearly circular in shape, 30 to 40 miles (48 to 64 km.) in extent and 10,000 to 13,000 feet (3,048 to 3,962 m.) above sea level. It is almost completely surrounded by an encircling fringe of Mesozoic beds (for the most part steeply upturned in monoclinal folds), some and probably all of which once covered the plateau, but were eroded away before the outpouring of the basalt.

The pretty town of **Glenwood Springs** is situated at the mouth of the Roaring Fork, an important stream flowing southwest from its sources on the west slopes of the Sawatch range, and on the south bank of the Grand river, which has a general southwest course from its head in the Middle park. It is also at the junction of two important lines of railway, the Colorado Midland, which after crossing the summit of the Sawatch range descends the valley of the Roaring Fork, and the Denver and Rio Grande, which comes down the valley of Eagle and Grand rivers. Its owes its importance in no less degree to its thermal springs and fine bathing establishment.

The springs issue from the Lower Carboniferous limestones which here dip southward away from the flanks of the White River plateau, while the open valley of the Roaring Fork above the town, and that

---

* On the maps of the Hayden Geological Atlas of Colorado, the color of the Fox Hill Cretaceous is spread over many areas on the west flanks of the mountains which have since been proved to be true coal-bearing Laramie.

of Grand river for a few miles below, are eroded out of the overlying softer, gypsiferous beds of the Carboniferous. The latter can be seen in section by following up a steep narrow ravine, directly east of the town, which is cut in the flanks of a basalt-capped plateau of Triassic and Carboniferous rocks. The harder limestones and quartzites of the Lower Paleozoic may be well seen by following the Denver and Rio Grande Railroad a mile or two above the town through the first tunnel, beyond which the Cambrian quartzites rest on the Archèan.

The bathing establishment is on an island in the river opposite the town, and consists of handsomely appointed bath buildings, back of which is an open-air bathing pool of masonry, nearly 700 feet (213 m.) long, at one end of which a stream of hot water from the springs* is constantly pouring in, while at the other is a fountain of cold mountain water, so that the bather can find any desired temperature. There is also a natural steam bath, arranged in a cave in the limestone on the south bank of the river just above the town.

From the hill slopes on the north side of the river, opposite the town, a beautiful distant view may be had, up the Roaring Fork valley, of the snow-capped Sopris Peak, 12,823 feet (3,600 m), one of the great diorite masses breaking up through much disturbed Paleozoic strata, which are a characteristic feature of the Elk mountains.

* *Analyses of Glenwood Spring water in grains per United States gallon.*

[By C. F. Chandler, PH. D.]

	Yampa spring.	Unnamed spring.		Yampa spring.	Unnamed spring.
Chloride of sodium.......	1,089·8307	1,086·9449	Bicarbonate of iron......	Trace.	Trace.
Chloride of magnesium ..	13·0994	13·4011	Phosphate of soda	Trace.	Trace.
Bromide of sodium.......	0·5635	0·8203	Biborate of soda	Trace.	Trace.
Iodide of sodium........	Trace.	Trace.	Alumina.................	Trace.	Trace.
Fluoride of calcium......	Trace.	Trace.	Silica	1·9712	2·0119
Sulphate of potash.......	24·0434	24·5971	Organic matter..........	Trace.	Trace.
Sulphate of lime.........	82·3861	80·2499	Total	1,250·0411	1,243·4303
Bicarbonate of lithia.....	0·2209	0·2872			
Bicarbonate of magnesia.	13·5532	13·7634	Temperature	124·2° F.	124·2° F.
Bicarbonate of lime......	24·3727	21·3545			

GLENWOOD SPRINGS TO ASPEN.

ITINERARY.

By S. F. EMMONS.

Station.	Distance.		Elevation.		Station.	Distance.		Elevation.	
	Miles.	Kilometers.	Feet.	Meters.		Miles.	Kilometers.	Feet.	Meters.
Glenwood			5,743	1,750	Aspen Junction	24	39	6,585	2,007
Cardiff	4	6	5,925	1,806	Watson	32	51		
Sands			6,081	1,853	Rathbone	36	58	7,653	2,333
Wheeler	18	29	6,357	1,937	Maroon	40	64		
Sherman					Aspen	42	68	7,935	2,419

The valley of Roaring Fork, for some distance south of Glenwood, is an anticlinal valley. That is, the bounding ridges on the west have the western dip of the upturned fringe of beds along the western flank of the Rocky Mountain uplift, while in the hills on the east the beds have a slight dip east toward the triangular synclinal area included between the Elk mountains, the Sawatch range, and the White River plateau. For the first part of the way the valley bottom is in the softer beds of probable Carboniferous age, immediately underlying the red sandstones of the Trias, which are seen in the hills on either side. The valley contains a great amount of glacial débris, often arranged in flood-plain terraces, and increasing toward the upper part of the valley.

At **Cardiff** station are extensive coke ovens. Here a short branch line comes in from the west which brings coal from large mines at Sunshine and Jerome Park, in a monoclinal valley in Cretaceous strata beyond and parallel to the ridge bounding Roaring Fork valley on the west.

About 12 miles (19 km.) from **Glenwood** the valley forks, Rock creek coming in from the north to join Roaring Fork. From here one has an uninterrupted view of Sopris Peak, which lies between these two streams. The general structure of the peak,[62] with the sedimentary beds sharply upturned around the dioritic core, can readily be distinguished in clear weather, and in the distance, up Rock Creek valley, one can see the high summits of the Ragged Mountain group, composed of laccolitic bodies of porphyry intruded into Cretaceous strata.

All the way up the valley of Roaring Fork, above Rock Creek, occasional glimpses are obtained over the intervening ridges, of the graceful summit of Sopris, which is most beautiful in the spring and autumn when it is freshly covered with snow.

For about 12 miles (19 km.) above Rock Creek the valley bottom is still in soft Upper Carboniferous beds, somewhat faulted and contorted, but in general occupying a gentle anticline between the synclinal fold on the west, at the east base of Sopris, and the triangular syncline to the eastward already mentioned. The latter has been considerably eroded and subsequently covered with basalt flows, parts of which still cover the higher flat-topped ridges, and their débris may be distinguished on the east side of the valley, mingled with that of the red sandstones which form the bounding cliffs.

On the west side of the valley the lighter-colored Jurassic beds may be distinguished above the red sandstones, which now come down to the valley bottom with a westerly dip.

At **Aspen Junction** the main line of the Colorado Midland leaves the Roaring Fork valley and follows up that of Frying Pan creek in an easterly direction. For many miles above the junction this line follows a winding gorge cut in red sandstones of Triassic and Upper Carboniferous age, which present grand cliff exposures. Beyond these it passes along more open valleys cut in Lower Paleozoic limestones and quartzites, and finally reaches the Archean granite and gneiss which forms the core of the Sawatch range. A long tunnel in these rocks, under the crest of the range, is mainly in granite.

In the Roaring Fork valley, above **Aspen Junction**, the railroad follows a flood-plain bench through a widening of the valley, and then passes through narrows formed by red sandstones, massive below, thin-bedded and lighter-colored above. The structure is here somewhat complicated, showing sharp folds and some faulting. In general the route now lies along the eastern side of a synclinal in Cretaceous and Jurassic beds, the northern continuation of the synclinal fold, already mentioned as formed by the uplift around Sopris Peak. The beds are mostly argillaceous shales with some impure limestones. On the east side of a second widening of the valley, which shows considerable flood-plain terraces, is the mouth of Woody creek, another tributary draining the west slopes of the Sawatch.

The road now passes along the strike of the westerly dipping Cretaceous shales and limestones, the hills on the east being formed of underlying red sandstones, while on the west the sharp synclinal fold can be distinguished in a harder limestone bed (probably Niobrara Cretaceous).

Just before reaching **Aspen** a flood-plain terrace is crossed between Maroon and Castle creeks, up whose valleys, to the west and north respectively, distant glimpses may be had of the castellated forms around Maroon and Pyramid peaks in the Elk Mountains.

The valley gorge of Maroon creek is cut, in an east and west direction, across the strike of easterly dipping Carboniferous, Triassic, Jurassic, and Cretaceous beds. That of Castle creek has a north and south course, and follows closely the line of a great fault and reversed fold, which brings Cambrian, Silurian, and Carboniferous beds up into juxtaposition with the red sandstones of the Trias. The reversed fold structure is well seen in a little hill on the east side of the Roaring Fork valley at the mouth of Castle creek, where the Jura-Trias beds rest upon the Dakota Cretaceous, as shown in Fig. 24, copied from Holmes. [62]

Immediately beyond Castle creek, in the angle between it and the Roaring Fork, lies Aspen mountain, a steep sharp ridge rising 2,000 feet (610 m.) above the valley, made up of faulted Archean, Cambrian, Silurian, and Carboniferous rocks.

The pretty mining town of **Aspen**, the second in importance in the State, lies in the valley of Roaring Fork at the east base of Aspen mountain. It is built upon a flood plain of the valley where it emerges from the Archean rocks of the Sawatch into the upturned Paleozoic beds which rest upon their flanks. These beds strike diagonally across

FIG. 24.—Reversed fold and fault on Roaring Fork near Aspen, Colo.

the valley in a northeast direction, dipping 45° to 50° to the northwest. The narrowest portion of the valley is at the upper end of the town, where the harder Cambrian quartzites project on either side, forming a sort of gateway into the Archean area beyond. Below this gateway the valley widens rapidly, and moraine material covers the lower slopes of the hills to an elevation of 1,000 feet, forming a broad bench on the east side of the valley.

The ores are sulphides, arsenides, and antimonides of silver, and sulphides of lead and iron, with their decomposition products. Barite is a common gangue material. They occur as a replacement of the Paleozoic limestones, the greater part being found in those of the Lower Carboniferous horizon, which are in great measure dolomitized in the vicinity of the ore bodies. Immediately above this limestone occurs a considerable thickness of black shales, in which is an intrusive sheet of diorite, generally decomposed.

The complicated geological structure of the region is the result of extreme compression of sedimentary beds against an unyielding Archean mass, which has thrust the beds into monoclinal and reversed folds, and broken them by thrust faults. Along the upper ridge of Aspen mountain, to the southwest of the town, the rocks strike north and south and are broken into blocks by a series of north and south faults. Along Spar gulch, in which rich ore was first discovered, the strata bend in strike to the northeast, and continue in this direction across the valley into and beyond Smuggler mountain, which is the spur between the valley of Roaring Fork and that of Hunters creek. This portion is also broken by thrust faults, which run so nearly parallel with the stratification of the steeply dipping beds that they are difficult to trace.

The principal mines are in Aspen and Smuggler mountains, overlooking the town on either side, but rich developments are found for many miles along the horizon of the Lower Carboniferous, both north and south, always within a few hundred feet of the underlying Archean. The production of the region has been from $6,000,000 to $8,000,000 per annum for some years, but is now (1891) increasing. The Mollie Gibson mine on Smuggler Mountain has a most remarkable body of nearly pure polybasite, from which dividends are being paid at the rate of $100,000 per month.

GLENWOOD SPRINGS TO LEADVILLE.

ITINERARY.

By S. F. EMMONS.

Station.	Distance.		Elevation.		Station.	Distance.		Elevation.	
	Miles.	Kilo-meters.	Feet.	Meters.		Miles.	Kilo-meters.	Feet.	Meters.
Glenwood			5,767	1,758	Avon	54	87		
Shoshone	10	16	6,104	1,860	Minturn	60	97	7,809	2,380
Dotsero	18	29	6,139	1,871	Rock Creek	65	105	8,289	2,527
Gypsum	25	41	6,310	1,923	Red Cliff	67	108	8,656	2,638
Rio Aquila	32	52			Pando	74	119	9,227	2,812
Sherwood	40	64	6,886	2,099	Tennessee Pass	83	134	10,418	3,175
Wolcott	43	69			Keeldar	88	142	9,955	3,034
Allenton	47	76	7,129	2,173	Leadville*	90	145	10,185	3,104

* Population 11,212.

From **Glenwood** for 12 miles the road follows a narrow canyon cut by the Grand river in the Lower Paleozoic and Archean rocks at the southern base of the White River plateau. This canyon, which is bounded by vertical cliffs weathered into picturesque castellated forms, reaches a depth of 3,500 feet, the White River plateau immediately north of it being 5,000 feet above the river bed.

After leaving **Glenwood** the strata gradually rise, and beyond the tunnels the Archean rocks are suddenly brought up, apparently by a fault. They consist of red and gray granites, inclosing masses of gray gneiss and dark amphibolite, and traversed by irregular veins of pegmatite. These rise, midway in the canyon, 500 to 1,000 feet above the river bed, and gradually descend to the eastward. Their unconformable contact with the overlying Paleozoic rocks can be distinctly traced for a long distance on either side of the river. At the east end of the canyon the harder Lower Paleozoic rocks (Cambrian, Silurian, and Lower Carboniferous) gradually sink below the river level, and the valley widens out in the overlying gypsiferous beds.

At **Dotsero** the road leaves the valley of the Grand, which bends to the north, to follow that of its main tributary, Eagle river,[63] which comes in from the east. Above the junction the hills on either side are capped by Triassic beds, forming the southeast fringe of the White River plateau. On the summit of those to the north is a basalt vent of extremely recent age, whose lava has run down the southwest face of the cliff and flowed for a considerable distance along the valley bottom, crowding the stream to its southern edge. The track passes for a half mile close to this remarkable coulée, and the singularly rough, ropy, and scoriaceous nature of its surface can be readily seen from the train. The absence of any evidence of erosion, and its relation to the river gravels unite to prove it the most recent known lava flow in Colorado.[64]

Before reaching **Gypsum** station the road enters the considerable east and west valley of the same name, so called from the abundant beds of gypsum in the soft shales and limestones of Carboniferous age, out of which it has been eroded. It is about 15 miles (24 km.) in length and occupies a gentle anticline in the contorted and folded gypsiferous beds, which form part of a general synclinorium included between the Park and Sawatch ranges on the east, and the White River plateau and Elk mountains on the west. These beds are peculiar to this region, the same horizon in other parts of Colorado being represented mainly by coarse grits with a very subordinate development of shales and limestones.

The valley has at one time been occupied by a fresh-water lake, presumably of Pliocene age, in which were deposited a series of soft pink marls, remnants of which can be seen on the north side of the river. Still more recent deposits of waterworn gravels, probably brought down by the floods which occurred at the close of the Glacial period, are exposed in the railroad cuts.

At the eastern end of the valley the gypsiferous beds sink down below the valley level and give place to sharply upturned Red Beds (Trias). The valley now bends to the northeast, and then takes a sharp turn to the southeast in Elbow canyon, crossing the upturned Red Beds and entering a narrow syncline in Cretaceous rocks. In this syncline the Dakota sandstones are strongly developed and contain some coal seams. They are overlaid by the black shales of the Middle Cretaceous; the uppermost hard bed seen in the axis of the syncline is probably the Niobrara limestone, which is always a well marked horizon.[63]

The southeast course is maintained until the Red Beds appear again from under the syncline, when the road takes a more easterly course for about 10 miles (16 km.) across an area of rounded hills formed of Upper and Middle Carboniferous beds, already showing a larger proportion of grits in the vicinity of the ancient shore line of the Sawatch upheaval.

The road then enters Eagle River canyon, which it follows in a southeast course for over 15 miles (24 km.). This picturesque gorge is cut along the north and northeast flanks of the Archean uplift of the Sawatch. Its bottom, which is from 1,000 to 1,500 feet (305 to 457 m.) below the level of the surrounding country, is generally in the underlying granite and gneiss, the summits of the cliffs being formed of Cambrian quartzites, and Silurian and Lower Carboniferous limestones, which dip $10°$ to $15°$ to the northeast, succeeded in the higher hills to the north by several thousand feet of grits and sandstones of Middle and Upper Carboniferous age. The aggregate thickness of the Lower Paleozoic series, up to and including the Lower Carboniferous, is only about 600 feet (183 m.), of which about 300 feet (91 m.) are quartzites.

Shortly after entering the canyon, shaft houses of various mines are seen perched high up on the cliffs to the north of the road, from which long aerial wire tramways for the transportation of ore reach down to the level of the railroad. These are a portion of the mines of Battle mountain in the Red Cliff mining district, whose ores are shipped from **Rock Creek** and **Red Cliff** stations to various smelting works throughout the country.

Red Cliff Mining District.—The ores of this district are found at two distinct horizons, without any visible connection between them.

The first are bodies of iron pyrite with argentiferous galena, in the Lower Carboniferous or Blue dolomitic limestone, beneath an intrusive sheet of quartz-porphyry. They occur in immense bodies as a replacement of the limestone, but are generally of rather low grade. The manner of successive alteration of such ores by surface or oxidizing waters through sulphates to oxides can be remarkably well seen in some of the mines.[65]

The second horizon is 200 to 300 feet (61 to 91 m.) lower geologically, at the top of the white Cambrian quartzite. The ores are in smaller volume and more irregular in distribution, but very much richer. They are fine-grained ochreous material, consisting largely of basic sulphate of iron, containing silver and gold, which is in general not visible to the naked eye, though many fine nuggets of gold have been found. There is good ground for assuming that these metals have, in part at least, been leached from the ore bodies of the higher horizons in solutions of persulphate of iron.[66]

The mines are mostly situated near the mining town of Clinton or Battle mountain, on a shoulder of the cliffs about two miles north of Red Cliff station.

Near Red Cliff the Cambrian quartzites, which have come down to the valley level, are stained by iron oxides to a pinkish color. The narrow gorge continues for some miles beyond Red Cliff, and the valley then widens. In the more open valley two curved ridges can be distinguished, which are terminal moraines marking halts in the retreat of a glacier which once stretched down from the base of the Mosquito range, 15 miles (24 km.) to the eastward.

The railroad ascends gradually along the southeast wall of this valley, from which good exposures of the Paleozoic series up to the Upper Carboniferous, with sheets of intrusive porphyry, can be seen on the opposite side.

It then bends southward up a side valley in Archean gneiss to Tennessee pass (10,400 feet—3,170 m.) which forms part of the Continental divide separating the waters of the Pacific Ocean from those flowing into the Gulf of Mexico.

It passes the summit of the pass through a long tunnel, beneath a thin sheet of Cambrian quartzite which forms the summit of the pass, and descends on the other side, past exposures of Cambrian quartzite and Silurian limestone on the east, into the broad Pleistocene valley, known as Tennessee Park, at the head of the valley of the Arkansas. At the lower end of this park it bends to the eastward across the East fork of the Arkansas, up whose glacial-carved gorge may be distinguished some of the high peaks of the Mosquito range in which it takes its rise. It then rises over a gently sloping mesa formed of glacial lake beds to the city of Leadville, which is situated on the edge of the mesa, at the base of the western spurs of the Mosquito range.

451 GE——27

LEADVILLE.[67]

By S. F. EMMONS.

This is the most important mining district in Colorado, and since the falling-off in the production of the Comstock, has been the greatest silver-producer in the whole West. Like so many other silver districts, it owes its existence to the restless wanderings of the early pioneers in search of gold. Gold was discovered in the neighboring gulches in 1860; the richest portions of the placers were exhausted in a few years, and the settlements that had sprung up around them were practically abandoned. When, fifteen years later, it became generally known that the heavy iron-stained stones, which had so much annoyed the gold washers, were rich silver ores, a new excitement, or "boom," among miners resulted; and, where in 1878 a few scattered log houses were all that were to be seen of human occupancy, in 1880 already a city of nearly 15,000 inhabitants had sprung up, with smelting works, banks, theaters, and all the adjuncts of a prosperous mining center. The town of to-day, though more substantially built, and with a population which, though somewhat less numerous, has also a smaller proportion of restless prospectors and adventurers and more substantial business men, still shows many characteristics of that which first sprung into existence with such marvellous rapidity.

The value of Leadville's product, in the thirteen years since it became a silver camp, has been between one hundred and fifty and one hundred and sixty millions of dollars. The annual product now varies from ten to fifteen million dollars, the principal value of the ores being in silver and lead, with a small proportion of gold.

The city is situated at the western base of the Mosquito range, on the upper edge of a gently-sloping mesa, about four miles east of and 600 feet above the bottom of the Arkansas valley. This mesa is formed of rudely-stratified beds of coarse gravel and sand, washed down from the adjoining mountains at the close of the first glacial period and deposited in a lake which, at that time, filled the Upper Arkansas valley above the present canyon at Granite. The moraines left by the glaciers of the second glacial period extended out over these beds, and their rearranged material, locally known as "wash," an unstratified drift, covers the lower portions of the hills and the intermediate valley. The depth of these successive gravel deposits immediately under Leadville is 300 to 400 feet and upwards.

From the city a fine view is had of the Arkansas valley and of the Sawatch mountains beyond it. That of the Mosquito range is much less satisfactory, owing to foreshortening. To the east of the city the ground slopes upward in a succession of benches or gently-rounded hills, known successively as Fryer Hill and Carbonate Hill, Yankee Hill and Iron Hill, etc., in which, as will be readily perceived from the unsightly dumps of waste that disfigure their surfaces, mining has been most extensively carried on. Although the greatest amount of underground work has been done within a mile or two of the town, within which distance there is an almost continuous network of subterranean passages reaching down several hundred feet from the surface, rich mines are by no means confined to this area, but are found up to the very crest of the range, at altitudes of 12,000 to 13,000 feet, and also beyond the crest on its eastern slopes.

The various smelting works, in which a portion of the ores produced here are smelted, are situated along the streams issuing from the mountains at either extremity of the city.

The greater part of the ores come from ore bodies in the Lower Carboniferous limestone, generally known as the Blue or ore-bearing limestone. During the first few years the productive minerals were carbonates of lead, and chlorides and bromides of silver, in a gangue of iron and manganese oxides mixed with clay and silica. These ochreous-looking ores are generally known among miners as "Carbonates," irrespective of whether the lead occurs, as here, mainly in the form of carbonate, or, as at Red Cliff in similar deposits, largely as sulphate. All these ores, being the result of secondary alteration by surface waters —a natural process of concentration that has removed a larger proportion of worthless materials, such as sulphur and zinc, than of the precious metals—were not only richer but better adapted for smelting than those which are obtained at the present day from the unaltered ore bodies below the zone of oxidation. While, therefore, the aggregate annual product of the district in tons of ore has regularly increased, its intrinsic value has fallen somewhat below that of early days.

As at present mined, the ore bodies, which are of enormous size, consist mainly of galena, iron pyrite, and zinc blende. Copper has, as a rule, been conspicuously absent, but of late years a deep-seated ore body, at a lower geological horizon than those generally worked, has been found to carry a considerable proportion of its value in this metal in an admixture of copper pyrite. In some of the large pyrite bodies small but appreciable quantities of tellurium occur. In others, bismuth has been found with the lead carbonates, and these also carry an unusually large proportion of free gold. Gold is found generally in deposits occurring exclusively in the porphyry bodies and quartzites; in them it is not infrequently associated with galena. The limestone ores, except in certain limited areas, do not contain an appreciable amount of this metal.

The geological structure of this district is extremely complicated. Its principal mines are situated on one of the western spurs of the Mosquito range, which range, as has already been explained, is a portion of the Sawatch massive, uplifted into its present position by faulting. This faulting was not, however, a simple uplifting of the strata without plication or compression, as has been described as the characteristic structure of the Plateau region. It was, on the contrary, primarily a plication produced by powerful compression against the unyielding Archean massive (Horst, Butoir) of the Sawatch. Faulting has been the final result, where the limit of plasticity of the involved rock masses had been reached. The relative plasticity of different portions of these rock masses appears to have borne an inverse relation to the amount of intrusive beds and laccolitic masses, which already formed an integral part of the sedimentary beds involved before the inception of the plication. Where these were only in thin sheets, as on the eastern flanks of this range, sharp anticlinal folds, generally much steeper on the western side, were formed before actual fracturing and displacement took place. The faulting in such cases occurred on the steeper side of the fold, where the tension was greatest. With a relatively greater proportion of eruptive rocks to sedimentary strata, plication was less marked and faulting more frequent. In the immediate vicinity of Leadville, where the eruptive sheets greatly exceed in volume the sedimentary beds above the Archean, with which they are practically interstratified, faulting has been a much larger factor in the uplift than plication, and the spur in which the principal ore deposits occur consists of a number of blocks, each faulted up to the eastward above the other, and traceable in the present topography as shoulders or hills. These faulted blocks are not, however, as might appear at a first glance, simple uplifted monoclinals, but form part of a system of synclines and anticlines, whose axes bear a definite, though sometimes rather obscure, relation to the planes of the various faults.

The prominent phases in the geological history of the region, as far as they have yet been deciphered, are—

1. Successive intrusions of porphyry, the earliest of which (White porphyry) was generally parallel to the stratification, spreading out in places into laccolitic bodies. The later intrusions (Gray porphyry) were more frequently transverse to the bedding, and formed thinner and more dike-like bodies, probably following to some extent planes of fracture produced by the shattering attendant upon the earlier intrusions.

2. During the second phase occurred the original ore deposition, which was a concentration of the metals disseminated through the rock

masses along channels formed during the preceding phase—contact planes between eruptive and sedimentary rock masses, and planes of fracture traversing either. The deposition of ore took place mainly in the Blue or Carboniferous limestone, and was a process of replacement of this rock by metallic minerals in the neighborhood of these channels, the original rock structure being in many cases preserved in the ore body.

3. The third phase was produced by the dynamic movements which resulted in folding, faulting, and displacement both of the sedimentary and eruptive rocks, and of their included ore bodies.

4. The final phase has been the gradual wearing away and degradation of the uplifted masses and the carving out of the present mountain forms, so that the ore bodies, formerly deeply buried, have been brought near the surface and exposed to the action of atmospheric waters. These reached them readily along the great fault lines and down the gently dipping stratification planes, and changed not only their mineralogical and chemical composition, but to a certain extent their position. The great accumulations of gravels, which cover all the lower portions of the hills and valleys, have probably rendered especially active the decomposing action of surface waters, as they contain a sufficient admixture of clay to render them capable of holding within their mass a great deal of water, after the manner of a sponge, and thus giving it time to act more thoroughly and deeply than if it were rapidly and freely drained away.

LEADVILLE TO MANITOU.

ITINERARY.

Station.	Distance.		Elevation.		Station.	Distance.		Elevation.	
	Miles.	Kilo-meters.	Feet.	Meters.		Miles.	Kilo-meters.	Feet.	Meters.
Leadville*........	0	0	10,185	3,104	Vallie............	78	126	6,534	1,992
Malta............	4	6	9,565	2,915	Cotopaxi........	84	134	6,371	1,942
Crystal Lake.....	7	11	9,317	2,840	Texas Creek.....	91	146	6,203	1,891
Hayden	12	19	9,143	2,787	Parkdale........	106	171	5,722	1,744
Twin Lakes	16	26	9,012	2,747	Canyon City.....	116	187	5,329	1,624
Granite..........	18	29	8,930	2,722	Florence........	124	200	5,184	1,580
Pine Creek	22	35	8,640	2,633	Beaver..........	134	216	4,983	1,519
Riverside........	27	43	8,357	2,547	Carlile..........	137	220	4,936	1,504
Americus........	31	50	8,125	2,476	Swallows........	142	228	4,863	1,481
Buena Vista......	35	56	7,955	2,425	Meadows........	147	237	4,797	1,461
Midway	38	61	7,837	2,388	Goodnight.......	153	246	4,713	1,436
Nathrop.........	43	69	7,680	2,341	Pueblo..........	157	253	4,653	1,418
Hecla Junction...	51	82	7,356	2,242	Eden	165	266		
Browns Canyon ..	53	85	7,307	2,227	Buttes..........	183	294	5,355	1,632
Salida...........	60	97	7,034	2,144	Fountain........	188	303	5,552	1,692
Cleora	62	100	7,000	2,134	Colorado Springs	202	325	5,978	1,822
Swissvale(Badger)	68	109	6,852	2,088	Manitou........	207	333	6,309	1,923
Howard	72	116	6,699	2,042					

* Population, 11.212.

[By S. F. EMMONS.]

From **Leadville** the road descends over the gently sloping mesa of Pleistocene lake beds to the Arkansas valley bottom at **Malta**. It then follows the level river bottom southward for 12 miles and enters a narrow, winding gorge in Archean granite. The old outlet of the lake that once filled the upper portion of the Arkansas valley lies to the west of the present course of the stream, nearer the base of the Sawatch mountains. It is now filled with gravels and crossed by magnificent moraine ridges of the later glacial epoch. These gravels carry gold, and it was this that attracted the first miners to this region in 1860. A flume carrying water for hydraulic washing crosses the railroad on the right. As one descends the valley, many abandoned placer washings in the coarse glacial gravels can be seen.

Twin Lakes is the station from which the beautiful glacial lakes a few miles to the westward are reached. These lakes lie at the opening of a magnificent gorge in the Sawatch range and are formed by terminal moraines, a moraine ridge separating the one from the other.[68]

Granite was the first town founded in this region, in 1860, by the gold-placer miners. Above it on the east are the abrupt granite slopes of the Mosquito range, cut through by frequent dikes of porphyry. The road follows a narrow gorge cut by the stream on the east side of the valley and then enters a broad, open valley of gently sloping Pleistocene gravels, which it follows for over 30 miles.

From **Buena Vista** fine views are had of the high peaks of the Sawatch range to the west. Mount Harvard (14,375 feet—4,381 m.), is an enormous mass; next to it, Mount Yale (14,187 feet—4,324 m.); to the south of west rise Mount Princeton (14,246 feet—4,342 m.), Mount Antero (14,246 feet) and Mount Shavano (14,239 feet—4,340 m.). On the slopes of the Mosquito range, east of the town, can be distinguished the line of the Colorado Midland road, which passes up the valley of Trout creek and crosses the Mosquito range into the South Park. Between **Midway** and **Nathrop** the road passes for a short distance through a narrow gorge cut in the granite, on the east side of the valley.

[By WHITMAN CROSS.]

At **Nathrop** are several rhyolite masses forming oblong hills with trend north-northwest to south-southeast. That these masses are huge dikes is shown by steeply inclined contacts with gneiss, parallel to which the eruptive rock is banded, the outer zone being glassy. The rhyolite of Ruby hill, on the eastern bank of the Arkansas directly opposite the station, is a lilac-colored banded rock with lithophysal cavities arranged on certain planes, containing many beautiful crystals of manganese garnet (spessartite) and topaz.[66] Good specimens can easily be obtained, and many have been collected by mineral dealers. On the west side of the river, near the railroad and north of the station, a small ridge of rhyolite rises out of the valley bottom, which is made of a rock containing phenocrysts of smoky quartz and feldspar and small lithophysæ containing minute topaz and garnet crystals.

At **Browns canyon** the road traverses another gorge on the east of the valley cut in Archean rocks, mainly coarse granite and amphibolite.

At **Salida** a considerable stream, known as the South Arkansas, joins the main river from the west. The original western line of the Denver and Rio Grande railroad follows up the valley of the South Arkansas, crossing the southern end of the Sawatch range at Marshall pass (10,481 feet—3,304 m.), and descending along the valley of the Gunnison to **Grand Junction**.

At **Poncho**, four miles west, are thermal springs. On the western side of the valley, from here northward to **Buena Vista**, are gently sloping mesa-topped ridges, covered by Pleistocene beds and underlaid by upturned beds of earlier Tertiary, whose age has not yet been determined. The view of the Sawatch Mountain peaks to the northwest and Mount Ouray on the west is here very fine.

The hills on the eastern side of the Arkansas, for several miles above and below **Salida**, are mainly composed of a series of crystalline schists and some more massive bands, which have been referred to the Algonkian period.* The known exposures of this series of rocks extend from a point about four miles (6 km.) north of **Salida** southward along the eastern bank of the river to where it bends east into the sedimentaries, which cover them, and thence westward into the north end of the Sangre de Cristo range. Immediately opposite **Salida** andesitic breccia conceals the schists for nearly one mile (1.61 km.) back from the river, and at six miles (10 km.) or more the sedimentary rocks in remnants are found on the edges of the schists. Their outcrops to the north are limited by andesitic breccia.

As now known, these schists are derived from eruptive rocks of at least two kinds, one acidic and one basic, which originally succeeded each other in a long series of alternating flows, probably with tuffs or fragmental deposits of the same materials in certain places. The most massive rocks now preserved are those occurring in the Arkansas valley below **Salida**, and the most altered schistose forms are at the northern extremity of the known section. Here are fine schistose rocks variously characterized by mica, chlorite, garnet, staurolite, and various amphiboles. One loose-fibered actinolite schist is locally so impregnated by copper ores that the material has been mined at a profit. The principal mine, the "Sedalia," is situated on the hillside facing the Arkansas valley, about three miles (5 km.) north of **Salida**. In a chlorite schist adjoining the copper-bearing layer occur the huge dodecahedral garnets which have been so widely distributed over the world through mineral dealers.

[By S. F. EMMONS.]

Below **Salida** the Arkansas valley changes from a southern to an eastern course, across the southern extension of the Mosquito or Park range to the Colorado or Front range.

At **Cleora** the valley narrows and the road passes over the more massive rocks of the schistose series into Lower Paleozoic beds, dipping to

---

*In the month following the excursion of the Congress the writer was able to examine, in some detail, the schistose series in question, and the statements as to its character here given are based on this recent study. The strata visited by some of the geologists in the ravine east of Salida belong to the most thoroughly metamorphosed part of the series, and their origin is clear only through the intermediate stages found in other parts of the section.

the northeast. The Cambrian quartzite is not distinctly recognized. The Silurian is represented by grayish limestones, the Lower Carboniferous by dark blue or gray limestones. Above these come a series of black shales and impure limestones, belonging to the Middle Carboniferous, followed by a great thickness of purplish-red beds, sandstones, conglomerates, and shales, which constitute the Middle and Upper Carboniferous, a horizon whose lithological constitution varies much in the Rocky Mountain region. As yet only plant remains have been obtained from these beds. Their apparent thickness is magnified by a secondary roll, and probably by some faults. Still above them are the finer-grained, bright-red sandstones which here constitute the Red Beds of probable Triassic age.

At **Badger** station the valley widens, and rocks in the immediate vicinity of the road are obscured by Pleistocene gravels.

From **Howard** station a fine view is obtained, through the hills to the southwest, of the peaks of the Sangre de Cristo range, which forms the southern continuation of the Sawatch and Mosquito uplifts, though not in a direct line with either.

Beyond **Vallie** station the road leaves the open valley and passes into a gorge of Archean, on the west flanks of which rest thin-bedded limestones dipping westward. The Archean rocks consist mainly of dark gneisses and amphibolites, with intrusive red granite, showing considerable regularity of structure.

From **Cotopaxi** station a stage line runs south into the Wet Mountain valley, a great interior depression to the southward between the Sangre de Cristo range on the west and the Wet mountains (which face the Great Plains) on the east, in which is the Silver Cliff mining district.[70]

The road now passes for several miles through a peculiarly crumbling, massive red granite, which is found in most of the deep cuts across the Archean mass of the Rocky Mountains, and is apparently a lower portion of the Archean series.

Texas Creek, the next station, is at the mouth of a stream of that name which drains the northwestern portion of the Wet Mountain valley. The Arkansas valley here opens out in the crumbling red granite, but narrows again a few miles to the eastward, passing into granite-gneiss, with an apparent bedded structure dipping to the eastward. It contains large pegmatite veins and a considerable development of amphibolite schists. The road then passes into another small open valley above **Echo** station, from which, looking up the northern tributary valleys, can be distinguished hills capped by dark lavas.

Beyond **Echo** the road passes again into a winding gorge of granite-gneiss, whose structure lines dip 50° to the east. Through this are intrusive masses of coarse grey granite and pegmatite veins.

At **Parkdale** is a considerable open valley, in which are nearly horizontal Cretaceous beds lying unconformably upon Silurian limestones. These were deposited in an ancient bay, which connected with the ocean to the eastward, both to the north and south of the Archean mass through which the railroad now passes.

For nearly 10 miles the road now follows a deep canyon, known as the Royal Gorge, whose walls rise almost perpendicularly above the track two or three thousand feet. This is probably the most imposing canyon gorge that is traversed by railroads in Colorado. The rocks are largely dark gneiss and amphibolite, cut by eruptive granite, pegmatite veins, and narrow diabase dikes, and show distinct structure lines which stand almost perpendicular. Just as the river emerges from the canyon it is joined by Grape creek, a rapid stream which comes down another narrow, winding gorge from the northwest portion of Wet Mountain valley. Just east of the mouth of Grape creek, and on the south side of the river, is a low ridge of Silurian sandstones and limestones resting directly upon the Archean. At the northern end of this ridge are hot springs, where are bath houses and a hotel adjoining. The thermal waters issue from the Silurian limestones.

From the Archean gateway of the Royal Gorge the road passes for a mile across an interior monoclinal valley between the foothills and a hogback ridge, formed by the hard sandstones of the Dakota Cretaceous. This monoclinal valley stretches ten miles due north, parallel to the eastern slope of the mountains. It is bounded on the east by a very steep wall, formed by the edges of the Dakota sandstones, and slopes gently up to the west against the flanks of the Archean. These slopes are made up of Silurian and Lower Carboniferous sandstones and limestones resting uncomformably upon the Archean, and in turn uncomformably overlaid by limestone breccia and soft arkose sandstones, whose age has not yet been definitely determined, but is either Carboniferous or Trias. Out of the softer rocks the bottom of the valley has been eroded. Above them rest the Jura-Dakota sandstones, again unconformable, though where crossed by the railroad there is no marked discrepancy in the angle of dip. East of the Hogback ridge are quarries in Niobrara limestone, worked by convicts from the State penitentiary at **Canyon City.**

SILURIAN VERTEBRATE LIFE AT CANYON CITY.

[By C. D. Walcott.]

Canyon City is situated near the southwestern shore of a great bay of early Silurian (Ordovician) time, an arm of which reached inland as far as Parkdale. Along the western shore of this bay sediments were deposited that now form the sandstones and limestones of the monoclinal valley above mentioned.

The lower division of the Silurian consists of sandstones (86 feet). In these, 11 genera and 19 species of invertebrate fossils have been found and an immense number of fragmentary remains of ganoid fishes. The invertebrate fauna is of the type of the basal Trenton of the New York section. The icthyic fauna includes fragments of a Placoderm closely allied to *Asterolepis*, numerous scales of the character of those of *Holoptychius*, and what is considered to be the calcified chordal sheath of a form allied to the recent *Chimara monstrosa*. This fauna was the subject of a paper read before the Geological Society of America at its recent session.[71]

The typical section is seen at Harding's quarry, which is about one mile northwest of the State penitentiary at Canyon City.

The invertebrate fauna, occurring three feet above the uppermost fish-bearing stratum, includes 34 genera and 55 species, of which 27 have been identified. At an horizon 180 feet higher in the section, 33 genera and 57 species occur, of which 33 species have been identified. These faunas are respectively of the types of those of the Lower and Upper Trenton faunas of the New York section, or the Lower and Upper Bala of Wales.

The character of the fauna at the lower horizon is shown by *Receptaculites Oweni, Halysites catenulatus* (a Lower Bala and Llandeilo species); *Columnaria alveolata; Strophomena alternata; Streptorhynchus filitextum; S. sulcatum; Orthis biforata; O. flabellum; O. subquadrata; O. tricenaria; Rhynchonella capax* var. *increbescens; R. dentata* Hall; *Ambonychia bellastriata*, Hall; *Modiolopsis plana*, Hall; *Murchisonia tricarinata*, Hall; *Cyclonema bilex; Bellerophon bilobatus,* Sow; *Endoceras proteiforme,* Hall; *Ormoceras tenuifilum; O. crebriseptum; Orthoceras vertebrale,* Hall; *O. multicameratum,* Hall; *Gomphoceras powersi,* James; *Asaphus,* like *A. platycephalus; Illænus crassicauda; I. milleri.* Of these, 11 species pass up into the fauna 180 feet above.

Six miles north of Canyon City limestones are exposed beneath the

Harding sandstone that in the section near **Canyon City** are concealed by the overlap of the Harding sandstone. The section is as follows:

	Feet.
5. Fremont limestone (estimate)	80
4. Halysites limestone (estimate)	100
3. Harding sandstone	145
2. Calciferous reddish limestone	4
1. Cherty, reddish Cambrian limestone	28

Characteristic Cambrian fossils were found in No. 1, and in No. 2 the genera *Ophileta* and *Bathyurus* occur.

The finest exposure of this series is found in Garden Park, north of **Canyon City**, where a plateau formed by the Halysites limestone is traversed by numerous canyons that cut down into the crystalline rocks beneath the Cambrian.

[By S. F. EMMONS.]

In the beds immediately underlying the Dakota sandstones about 10 miles north of Canyon City abundant Saurian remains of upper Jurassic types (*Atlantosaurus*, *Stegosaurus*, etc.)[72] have been found.

From **Canyon City** the road passes out upon the Great Plains, whose surface from here to Pueblo is formed of the nearly horizontal clays and limestones of the Montana and Colorado Cretaceous. A short distance southeast of **Canyon City**, in comparatively close proximity to the foothills, a small synclinal basin of coal-bearing Laramie rocks has escaped erosion, and coal is actively mined. Just before reaching **Florence** the northern end of this basin is crossed, and the Laramie sandstones can be distinguished capping some high bluffs to the north of the river.

About **Florence** is an oil field whose product reaches 2,000 barrels per day. The oil comes mainly from the Pierre shales of the Montana Cretaceous, and is obtained from depths of 900 to 2,000 feet (275–610 m.). The accompanying map (Pl. IX), from a paper on the Florence oil field, by George H. Eldridge,[73] shows the geological structure of this region. It is on a scale of 3 miles to the inch, or about $\frac{1}{190000}$. The contours are at vertical intervals of 25 feet. The letters indicating geological subdivisions are: A = Archean; S = Silurian; C = Carboniferous; T = Trias; J = Jura; Kd = Dakota (Cretaceous); Kb = Benton; Kn = Niobrara; M = Montana; Kl = Laramie Cretaceous; q = Quaternary.

Beyond **Florence** the rocks rise slowly, and the road follows the river bottom a hundred feet or more below the level of the surrounding plains. In the bluffs which border this bottom can be distinguished the harder beds of Niobrara limestone, which rise and fall in gentle undulations along the route.

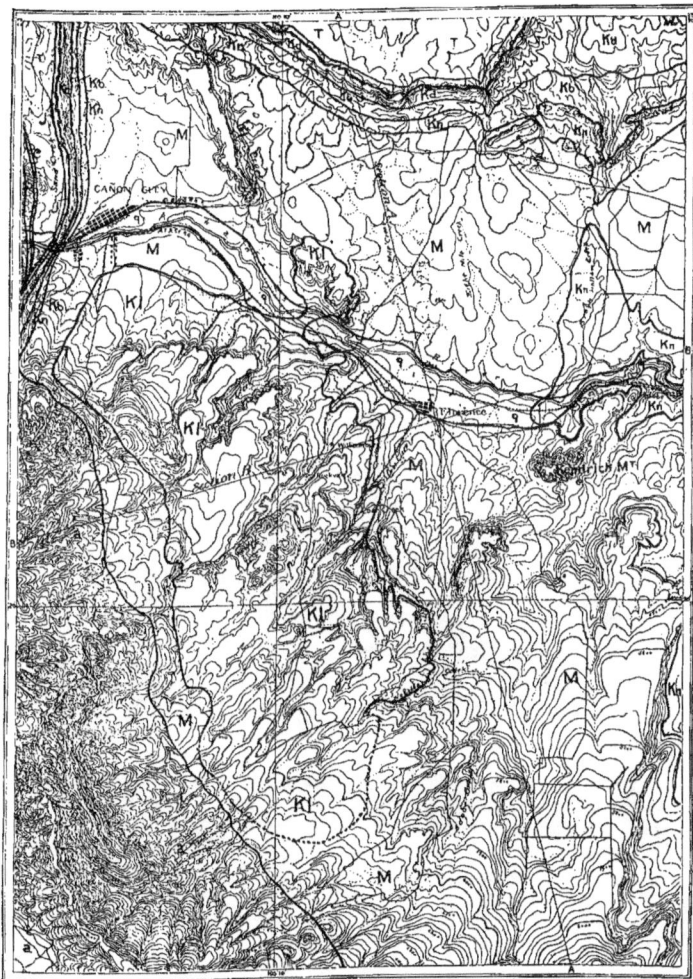

MAP OF THE FLORENCE OIL FIELD, COLORADO.

The city of **Pueblo** is important as a railroad center, and contains metallurgical works for reducing the various ores which are brought down grade from the mining districts of the mountains. On the mesa to the south, at Bessemer, are the iron and steel works of the Colorado Coal and Iron Company, where there is an extensive Bessemer plant for making rails. Nearer the town, on the mesa, are the lead-smelting works of the Colorado Smelting Company, managed by Mr. Anton Eilers, and to the north of the town and adjoining the railroad are the lead-smelting and refining works of the Pueblo Smelting and Refining Company, Herman Geist, manager. A permanent exposition building, known as the Colorado Mineral Palace, has recently been opened here for the purpose of exhibiting collections illustrative of the mineral resources of the State.

The road now bends northward up the valley of Fountain creek, about 12 miles (19 km.) to the east of the mountains, still remaining upon the clays of the Middle Cretaceous. As it gradually approaches the mountains, good views of the Pikes Peak group are obtained. The stream was originally named by Canadian trappers *Fontaine qui bouille*, from the effervescent springs near its source.

Colorado Springs is a flourishing city of 7,000 inhabitants situated upon the mesa, on the east bank of Monument creek, three miles from the mountains, whose mild, yet invigorating climate has made it famous as a health resort, especially for consumptives. The traveler has here a magnificent panorama spread out before him, the central feature of which is Pikes Peak, 12 miles (19 km.) away, towering above the lesser mountains and foothills. To the south is Cheyenne mountain (9,407 feet—2,867 m.), a precipitous peak rising abruptly from the plains, while to the west of it is St. Peters dome, a celebrated mineral locality. Between Cheyenne mountain and Pikes Peak is the picturesque gorge of Cheyenne canyon. A beautiful casino near by is connected by tramway with the city of **Colorado Springs**.

From **Colorado Springs** the train follows a branch road up the valley of the Fountain to **Manitou**, five miles to the westward, in a sheltered nook at the very base of the mountains.

Just beyond **Colorado City**, the first capital of Colorado, the upturned Niobrara Cretaceous limestone is seen, and from here the route crosses in succession the Lower Cretaceous, Jura, Trias, and Carboniferous, all steeply upturned. To the south of the track, the jagged edges of red Triassic sandstones are seen striking for the Archean foothills. A northwest-southeast fault cuts them off, as indicated on the Hayden map. To the north, the towering masses of the "Garden of the Gods" are seen from certain points.

MANITOU.

By Whitman Cross.

Manitou is a watering place and summer resort, celebrated for its springs of mineral water and the many objects of interest in its immediate vicinity. It has five large hotels and many smaller ones. The town lies in the valley of Fountain creek, in a little bay indenting the foothills of the Archean mountains. The sedimentary rocks occupying this bay are of Silurian and Lower Carboniferous ages, and do not appear elsewhere for many miles north or south. Although the details of the local geology have not been accurately worked out, the general structure is shown by the Hayden map. [74] To the north the strata are seen resting on Archean gneiss, with a southerly dip of 10° to 35° and running well up on the mountain slopes, but only small patches appear to the north of the stream which issues from the foothills at Glen Eyrie, three miles from Manitou. The disappearance of the formations in this direction seems to be caused by erosion preceding the Trias, for the shore line of the latter formation shows a transgression from the Archean obliquely across the Silurian and Carboniferous.

To the south of Fountain creek the Paleozoic, Triassic, Jurassic, and Dakota Cretaceous beds are successively cut off by a fault line with northwest-southeast trend. This line crosses Ruxton creek west of town near the Midland Railroad bridge. In the line of this fault to the northwest is Manitou park, a long, isolated basin in the mountains, occupied by the Paleozoic formations seen at Manitou, but with a somewhat different development.

The best opportunity to see the Silurian beds and the contact with the underlying Archean is in Williams canyon, a narrow, picturesque gorge, whose mouth is close by the Cliff House. The canyon extends almost due north for about two miles, its nearly vertical walls, 300 to 500 feet in height, being mainly made up of Silurian limestone. Ascending the canyon from Manitou one descends geologically, owing to the southeasterly dip of the strata, and at about one mile from the town the contact of the thin quartzite under the limestone with the gneiss is very plainly seen. In the cliffs on the western side are extensive caves in the most massive stratum of limestone. The "Cave of the Winds" is open to visitors. There are several other caves at the same horizon in the vicinity of Manitou.

By passing up Ute creek to Rainbow falls, one mile from the bath house, another good view of the lower part of the Silurian section and the Archean contact is obtainable. The line of the fault mentioned above is seen in going up Ruxton Creek to the Ute Iron Springs and to the depot of the Pikes Peak railroad.

The mineral springs, for which Manitou is famous, issue at several horizons in the sedimentary rocks and from the Archean. They are all carbonated and saline, while one of them is strongly chalybeate. The temperature of the water varies from 43° to 60° F. (6° to 15° C.). There is a bath house in connection with the larger soda springs. The Ute Iron Springs are in Ruxton valley.*

The "Garden of the Gods" is situated about one mile north of Fountain creek and three miles northeast of Manitou. In it rise towers and pinnacles consisting of the vertical strata of the white Dakota sandstone, or of the red Triassic sandstones below. Some of these huge masses rise vertically for 200 or 300 feet, and serve as fine examples of the erosion of steeply-upturned strata of varying consistency. Lower white ridges of Jurassic gypsum or of Cretaceous limestone afford strong contrasts in color when compared with the red sandstones.

Pikes Peak.—Pikes Peak rises to an elevation of 14,147 feet (4,312 m.), or 7,838 feet (2,389 m.) above Manitou. It is seven miles distant in air-line; is reached by cog-railroad, by horseback on path, or by carriage from Cascade station on the Colorado Midland railroad.

Pikes Peak is the center of a group of mountains which are mainly made up of a coarse reddish biotite granite. In the granite are many coarse pegmatite veins, and in some localities pockets lined by crystals of smoky quartz, amazon stone, and a large number of less abundant minerals. While these mineral occurrences have made the name of

----

* The following analyses of the waters of the principal springs are taken from Dr. A. C. Peale's Mineral Springs of the United States: [75]

MANITOU SPRINGS.

[Parts in 100,000.]

Constituents.	Oscar Loew, analyst.					
	Iron Ute.	Little Chief.	Manitou.	Navajo.	Ute Soda.	Shoshone.
Sodium carbonate..........	59·34	15·16	52·26	124·69	23·82	88·80
Calcium carbonate..........	59·04	75·20	111·00	129·40	40·00	108·50
Magnesium carbonate	14·50	13·01	20·51	31·66	6·10	
Lithium carbonate..........	Trace.	Trace.	0·21	0·24	Trace.	Trace.
Iron carbonate.............	5·78	1·80	Trace.	....	1·40	....
Sodium sulphate...........	30·86	51·88	19·71	18·42	12·24	37·08
Potassium sulphate.........	7·01	6·24	13·35	16·21	Trace.	5·12
Sodium chloride	31·59	47·97	40·95	39·78	13·93	42·12
Silica	2·69	2·22	2·01	1·47	Trace.	Trace.
Total	210·87	213·48	260·00	361·87	97·49	281·62

Pikes Peak known the world over, the mountain proper has yielded but few specimens of note. The chief interest in ascending Pikes Peak centers in the magnificent view of mountain and of plain to be obtained from its summit and from the cars in making the ascent. A beautiful alpine flora may be found below the débris of the summit, and especially on the southern slope.

FIG. 25.—Pikes Peak railway.

The above sketch by M. Eysséric, of the visiting geologists, was taken on the southern slope above timberline.

The list of minerals found in the granite of the Pike's Peak region embraces orthoclase, microcline, albite, biotite, muscovite, quartz (smoky and clear), topaz, phenacite, kaolinite, arfvedsonite, astrophyllite, hematite, limonite (pseudomorph after siderite), goethite, turgite, cassiterite, rutile, zircon, fluorite, cryolite, pachnolite, thomsenolite, gearksutite, ralstonite, prosopite, elpasolite, tysonite, bastnäsite, allanite, xenotime, gadolinite, samarskite.[76]

A large number of minerals, including all the fluorides, zircon, astrophyllite, and arfvedsonite, occur near St. Peters dome, 10 miles (16 km.) southeast of Pikes Peak. This locality is reached by carriage from Colorado Springs, a full day being necessary for the trip. The same road leads to the Seven lakes at the south base of Pikes Peak, and along its course fine mountain views are obtained at many places.

Other points at which fine specimens have been found are scattered all over the foothills of Pikes Peak, but only at St. Peters dome will the casual visitor be likely to find valuable specimens himself. A fine collection of minerals for sale can be found at the Iron Spring in Ruxton creek, below the depot of the Pikes Peak railroad.

The locality which has furnishd the finest amazonstone crystals, and also phenacite and topaz, is near Florissant, a station on the Colorado Midland railroad, 30 miles (48 km.) from Manitou and 15 miles (24 km.) northwest of Pikes Peak. Fine topaz crystals have also been obtained at Devils Head (Platte mountain), 30 miles to the northwest.

In the vicinity of the village of Florissant is the small isolated lake basin of Upper Eocene (Oligocene) beds, which have become celebrated for their wonderful insect fauna and fossil flora. Fishes and, more rarely, birds have also been found here. The strata of this deposit are composed of volcanic ashes, and they can not as yet be accurately correlated with any beds known elsewhere. The geology of the basin has been described by S. H. Scudder.[77]

451 GE——28

COLORADO SPRINGS TO DENVER.

ITINERARY.

By WHITMAN CROSS.

Station.	Distance.		Elevation.		Station.	Distance.		Elevation.	
	Miles.	Kilo-metres.	Feet.	Meters.		Miles.	Kilo-meters.	Feet.	Meters.
Colorado Springs ..	0	0	5,978	1,822	Larkspur	31	50	6,656	2,029
Edgerton	8	13	6,402	1,951	Castle Rock	40	64	6,205	1,891
Husted	13	21	6,582	2,006	Sedalia	49	79	5,822	1,775
Monument.........	19	31	6,960	2,121	Acequia	56	90	5,515	1,681
Palmer Lake.......	22	35	7,222	2,201	Littleton	63	101	5,357	1,632
Greenland	28	45	6,906	2,105	Denver	73	117	5,182	1,579

From **Colorado Springs** northward the route follows up the valley of Monument creek. For several miles its course is in Montana shales, but below **Edgerton** it enters the Laramie formation, which presents a line of bluffs facing to the southwest. Along the banks one sees from the train a tendency to the production of curious forms by the unequal erosion of the sandstone containing hard and usually ferruginous layers. In the district lying to the west of the railroad is the celebrated Monument park, where the erosional forms are most abundant and noteworthy in shape and size.

The road gradually approaches the Archean foothills, and at the head of Monument creek reaches the divide between the waters of the Arkansas and Platte rivers. This divide extends for many miles eastward with a general elevation of about 7,500 feet. It is timbered in certain areas and is composed of Tertiary strata, most of which are assigned to the Monument Creek formation of Hayden, though detailed studies may very probably show that several distinct formations have been grouped as one. Owing to a discovery of Miocene vertebrate remains somewhere in this series the whole has been supposed to be of that age.

The railroad crosses the divide at **Palmer Lake**, within a hundred yards of the Archean line, and just west of the station may be seen strata of the Monument Creek formation resting on granite, with a gentle easterly dip. East of the station is a small lake on the summit of the divide. It has been artificially enlarged. East of the lake rises a hill in which are good exposures of the coarse-grained grits and sandstones of the Monument Creek. On either side of the divide the usual zone of upturned sedimentary rocks, following the Archean contact, is concealed by the Tertiary strata which abut against the foothills. On the northern side of the divide the road again diverges from the foothill line and passes rapidly down into the valley of Plum creek, which it follows to the Platte river. The most striking feature of the landscape, for a considerable distance, is the number of isolated flat-topped buttes, which are made up of Monument Creek strata with, in many cases, a capping of pinkish rhyolitic tuff, a rock much used as a building stone in Denver. These buttes are specially well shown in the vicinity of Larkspur, Douglas, and Castle Rock. The quarries from which the tuff is obtained are near the last two stations. In Castle Rock, a small hill with abrupt cliffs at the top, is evidence of an erosion which followed the tuff deposition, and indicates a division line in the Tertiary series, the importance of which has not yet been demonstrated. Tertiary beds extend somewhat farther north than is represented by the Hayden map, and it is not until within a short distance of the Platte river that the underlying beds are exposed.

Below Sedalia the line of upturned strata following the mountain base becomes again visible, and the Dakota hogback in particular can be traced for miles. Shortly before reaching Acequia a large irrigation ditch is crossed, by which water from the Platte is conducted to the dry plains immediately east of Denver. Opposite Acequia is seen the gap through which the Platte river issues from the mountains. From the mouth of Plum creek to Denver the route passes over the two post-Laramie formations—the Arapahoe and the Denver—which will be referred to in detail below.

DENVER.

By WHITMAN CROSS.

In 1859 the first log house was built where Denver now stands. In 1870 the city had 4,730 inhabitants; in 1880, 35,628; in 1890, about 140,000. In its business blocks, theaters, school buildings, and private residences Denver stands on a par with many older Eastern cities of a larger population.

The city is built on slightly terraced plains, along either bank of the South Platte river, at about 10 miles from the mountains. From a favorable point the observer commands a magnificent view of the Front range of the Rocky Mountains, extending from the northern line of the State southward to Pikes Peak, a distance of 150 miles (240 km.). The most prominent peaks visible from Denver are, commencing at the north, Longs Peak (14,271 feet—4,350 m.); Grays Peak (14,341 feet—4,371 m.); Mount Evans (14,330 feet—4,368 m.), and Pikes Peak

(14,147 feet—4,312 m.). A notable feature of the landscape is Table mountain, a gently-sloping mesa at the base of the mountains, directly west of Denver, where a basalt sheet has protected the soft underlying strata from erosion. Behind the mesa is the city of Golden.

Denver is the commercial center of a very large territory. It is of special importance through its relations to the mining industry of the mountainous region to the westward, for it is the greatest ore market of the Rocky Mountains. From mining camps, not only in Colorado, but also in the surrounding States and Territories, ore is brought to Denver, sampled, and sold in the public market.

There are in the city three very large smelting establishments. Of these the Boston and Colorado Smelting Company, whose works are in the suburbs at Argo, deals principally with the more refractory ores, and pays especial attention to ores containing copper, silver, and gold. The process is a modification of the Ziervogel process. The gold is separated from the copper by an operation which is kept secret. The Omaha and Grant Smelting and Refining Company, whose refining works are at Omaha, is said to be the largest establishment of its kind in the world. Argentiferous lead ores, such as those of Leadville and Aspen, are the ones most sought for here, as also at the Globe Smelting Works, near Argo, where some of the largest lead furnaces in the world are in operation.

Among the public buildings worthy of note is the State capitol, now being erected on a beautiful site on Capitol hill; the Arapahoe County court-house on Sixteenth street, built of a grayish sandstone from the Laramie Cretaceous, and the High School building, one of the largest school edifices in the United States. In the upper corridors of the last-named building are temporarily displayed the collections of the Colorado Scientific Society, which are well worthy of a visit, especially for the mineralogists.

A mining-stock exchange, started a few years ago, is now erecting a handsome building for its use.

The visitor will, undoubtedly, note the great variety of beautiful building stones used in Denver. These are nearly all the product of the State, and many of them come from quarries near the city. From the Archean comes a coarse-grained red granite and a fine-grained gray granite. The red sandstones of the Trias and Jura are much used. White sandstones come from the Dakota Cretaceous, or much more commonly from the Fox Hills or Laramie Cretaceous. The latter rocks are very easily worked and have various soft tints of gray, brown, or yellow. Beautiful marbles of white and variegated colors occur in Colorado, but have not yet been used to any great extent. One of the most popular building stones, especially adapted for residences, but also used in the Union Depot and in other large buildings is the light pinkish rhyolitic tuff intercalated in the Tertiary strata south of Denver. The fine flagstones used in the sidewalks of Denver come from the red beds of the Trias or from the Dakota Cretaceous.

The geological formation immediately underlying and surrounding the city is the Denver beds, [79] a fresh-water lake deposit whose sandstones and conglomerates are characteristically, and in the lower portion almost exclusively, made up of volcanic rocks, representing many varieties of andesite. The Monument Creek beds rest unconformably upon the Denver strata in the highlands to the southeast.

Below the Denver beds occurs another fresh-water lake deposit, the Arapahoe beds, whose most prominent member is a conglomerate free from volcanic materials, but containing pebbles of sedimentary rocks recognized as belonging to various horizons from the Laramie down to the red sandstones of the Trias. Below the Arapahoe beds come the normal Laramie Cretaceous clays and coal-bearing sandstones. Unconformities of deposition occur between the Laramie and Arapahoe, and between the latter and the Denver beds. [81]

Both the Arapahoe and the Denver beds contain numerous fragments of *Dinosaurian* remains, the majority belonging to the recently discovered family of the *Ceratopsidæ*. The fossil flora of the Denver beds is very rich, but paleobotanists have not as yet differentiated it from that of the coal-bearing Laramie of this region. The molluscan fauna is small, but also shows a close relationship between the Denver and the Laramie. For these reasons paleontologists have held that the two lake deposits in question belong to the Laramie Cretaceous, but the stratigrapher finds, in the evidence of enormous erosion preceding the Arapahoe epoch and of a long period of eruptive activity before the Denver epoch, grounds for holding that these formations should be distinguished from the Laramie proper, and that they are either of earliest Eocene age or of a Cretaceous period succeeding the Laramie.

The Denver strata are exposed in the banks of the Platte river, and in many of its smaller tributaries; but a good idea of the characteristics of the series can only be obtained by examining the sections shown on the line of folding near the mountains. The rusty brown sandstones appearing on the plains are often indurated by a large amount of zeolitic cement, and the soils resulting from the decomposition of the eruptive materials are quite fertile.

A few years ago artesian water of remarkable purity was struck in Denver while boring for coal. The water was found to come from several horizons between 175 feet and 1,200 feet in depth, mainly in the Arapahoe beds. As the supply basin was a small one the large number of wells which were sunk soon decreased the pressure, and to-day nearly all the wells in use are pumping wells.

The formations constituting the plains about Denver are chiefly a

Loess-like Pleistocene deposit, which forms a very rich soil, though from lack of moisture the plains appear to be arid, and until irrigated produce little beside the prickly-pear cactus (*Opuntia*), the soap weed (*Yucca*), the sagebrush (*Artemisia*), greasewood (*Sarcobatus*), and scanty grasses. This Loess deposit is widely distributed over eastern Colorado, and seems to connect directly with that of the Missouri valley. [20]

From Denver a number of very interesting excursions can easily be made to points illustrating clearly various features of the foothill geology. At Platte canyon, Morrison, Golden, Ralston creek, Boulder, and St. Vrain's creek, all reached by railroad, one can study beautiful sections of the upturned sedimentary series from the Archean to the Colorado Cretaceous, or, in some cases, to the post-Laramie of the Denver formation. The excursions to Morrison, Golden, and Ralston creek are easily made from Denver in one day, returning to the city at night. There are good hotels both at Morrison and Golden.

EXCURSION TO MORRISON.

By Whitman Cross.

Morrison is a little village 12 miles (19 km.) southwest of Denver, at the end of a branch of the Denver, Leadville and Gunnison railroad. It is picturesquely situated on the banks of Bear creek, which here issues from the mountains, crosses the monoclinal valley between the Archean and the Dakota hogback, and cuts its way through the latter to the plains. Just within this gap is the town, with fine exposures of the Mesozoic section below the Colorado Cretaceous close at hand.*

Dark red Triassic sandstones at the base of the sedimentary series form the most prominent outcrops seen in the section on the north bank of Bear creek. Above them comes a creamy white sandstone, and above that, forming a low ridge at the base of the Dakota hogback, is a limestone which is burnt for quicklime. This limestone occurs in the midst of red clays and shales. Mr. Eldridge has included these beds in the Triassic part of the section, differing in this from the division adopted by the Hayden survey.

The smooth western slope of the great Dakota hogback exhibits very beautifully the variegated marls, clays, and shales of the Jura. At a

---

*Fig. 26, on page 439, is a reproduction on half scale of the drawing by W. H. Holmes, published in the Hayden report for 1874, p. 32. The section given is on the north bank of Bear creek and shows the relationships of the various formations which can be easily visited, from the Archean gneisses of Bear canyon or Mount Morrison on the west, through the sedimentary section beginning with the red beds of the Trias, and extending to the highest known Denver beds in the summit of Green mountain on the east.

horizon not far below the Dakota, and on the slope facing the town of Morrison, Prof. Arthur Lakes found the first gigantic Dinosaur bones upon which Prof. Marsh established the genus Atlantosaurus. Other remains of the same animal have been found at this horizon, both to the north and south of Morrison, and also at a corresponding position near Canyon City.

In the white Dakota sandstones and in a fire-clay bed in the middle of the sandstone section there has been found an extensive fossil flora which has been made the subject-matter of a monograph by Prof. Lesquereux, published by the U. S. Geological Survey. The section of Fort Benton shales immediately overlying the Dakota is most clearly seen on the south bank of Bear creek, along the line of an irrigating ditch.

The small ridge at the base of the Dakota hogback is caused by the Niobrara limestones and the entire section of the Niobrara is very well shown in the part of the section illustrated.

Following the Niobrara comes a great thickness of Fort Pierre shales, which, along Bear creek, have a thickness of over 5,000 feet, while in the section running from the summit of Green mountain, about three miles to the northward of Bear creek, their thickness is reduced to a few hundred feet. From the observed variation in the thickness of these beds, together with the visible divergence of strike in this interval, Mr. Eldridge has established a considerable unconformity between the Niobrara and Fort Pierre.[81]

Above the Fort Pierre come the shales of the Fox Hills in their normal development. At the western base of Green mountain is a small but distinct ridge of the Laramie sandstones (d, Fig. 26) in vertical position, containing coal beds, here of less than the normal thickness. This line of the Laramie beds is clearly traceable along the southern slopes of Green mountain, running parallel to the band representing the Fox Hills, and crosses Bear creek at a hill called Mount Carbon

FIG. 26.—Section at Bear Creek.
a, Mt. Morrison. b, Morrison. c, Green Mt. (Denver beds). d, Line of Coal outcrop-vertical.

three miles east of Morrison. Here the coal banks have become thicker and have been opened in several mines.

Green mountain will always remain a classic locality for the post-Laramie lake beds which have been named the Arapahoe and Denver formations. In a small ravine on the southwest slope of Green mountain, whose course is almost toward the point of view in the section given, one may see an almost continuous exposure of the Laramie and Arapahoe beds in vertical position. It was in this ravine that the first pebbles of Dakota conglomerate, and other Mesozoic sandstones were detected in the conglomerate, which is the most conspicuous feature of the Arapahoe series. From this ravine the Arapahoe conglomerate can be traced southward in practical continuity to a point several miles south of the Platte canyon, and northward to Golden, although in the latter direction it is less well marked, owing to the smaller size of the pebbles.

The main mass of Green mountain is made up of Denver beds. Contact with the Arapahoe beds is most clearly indicated in the ravine first mentioned and in others on the western slope of the mountain; the lower 200 or 300 feet of the formation are, however, not very continuously exposed in any section. But a characteristic conglomerate composed of dark andesitic pebbles is clearly shown at the western base of the more abrupt slopes of the mountain, and from this horizon to the summit there is an almost continuous section illustrating very clearly the characteristics of the formation. The coarse conglomerates of the upper part of the Denver formation are here so loosely consolidated that the mountain summit and upper slopes are covered with loose boulders, which were not originally identified as belonging to a definite formation in place, but considered drift boulders. For details concerning the exposures of the Denver and Arapahoe formations in Green mountain the reader is referred to the original article [79] on the Denver Tertiary formation and the accompanying map, in which the location of the section ravine and the best exposures are clearly indicated. The outcrops of the Denver formation are shown at intervals along the banks of Bear creek and its junction with Platte river. Many Dinosaur bones, referred by Prof. Marsh to the Ceratopsidæ, have been found in the Denver beds on the western slopes of Green mountain and in the plains area between that and the Platte river.

EXCURSION TO GOLDEN.

By Whitman Cross.

Golden lies 10 miles (16 km.) west of Denver, on Clear creek, at the point where it issues from the mountains. The position of Golden corresponds to that of Morrison, except that the elevations separating it from the plains, North and South Table mountains, have an entirely different character from the Dakota ridge of the other locality. The section of the sedimentary rocks exposed at Golden is comparatively very narrow through the thinning of several formations between the red beds of the Trias and the Laramie. Some of the formations, such as the Dakota, entirely disappear from the section at the point where it is crossed by Clear creek.[81]

The first formation to preserve its normal thickness and position in the section at Golden is the Laramie Cretaceous, which here has a thickness of about 800 feet (244 m.), standing in vertical position. The coal measures are well developed, and are explored by several coal mines. An abundant fossil flora has been found in the sandstones adjoining the coal.

To the east of Golden, not more than one mile (1·61 km.) from the Archean foothills, are the two Table mountains, separated by a gorge made by Clear creek in cutting its way to the plains. The Table mountains have a basaltic capping of varying thickness, up to a maximum of 300 feet (91 m.). Basalt protects the soft friable sandstones of the Denver beds, which contain a very well preserved and abundant fossil flora in many horizons. The Denver strata in Table mountain represent the characteristic development of that formation in its lower portion. There are exposures in South Table mountain in the neighborhood of Castle Rock and at the northeastern corner of the mountain.[79]

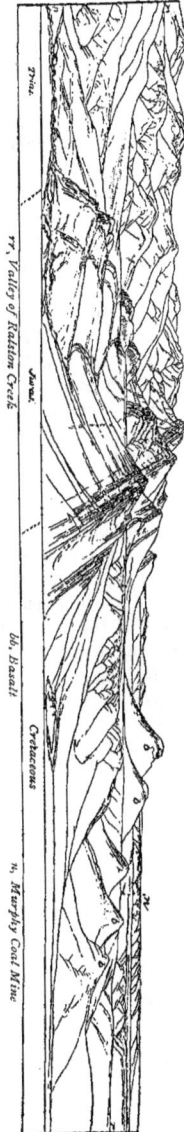

Fig. 27.—Section at Van Bibber creek.

rr, Valley of Ralston Creek

tt, Basalt

n, Murphy Coal Mine

Trias   Jura   Cretaceous

Between the Laramie and the Denver beds is a narrow outcrop of the Arapahoe formation, which is, however, exposed in very obscure outcrops in this region.   It occurs as a narrow band of steeply upturned strata apparently conformable with the Laramie.

In a wide amygdaloidal zone in North Table mountain there is a succession of zeolites, which have been described in detail.[76]   The occurrence is interesting for the distinct succession of species and for the unusual development of certain forms.

In the Dakota hogbacks north and south of Golden are also fire-clay beds.   These are used in manufacturing a very fine quality of fire brick, furnace muffles, crucibles, etc.   There is also a large manufactory of ornamental brick in Golden, the material used being largely the plastic clays of the Laramie.   The Colorado State School of Mines is located at Golden.   It has a fair museum, and is rapidly growing.

In Clear Creek canyon, above Golden, are found exposures of the Archean gneisses and schists.   At about 6 miles (10 km.) north of Golden, Ralston creek issues from the mountains, cutting across the sedimentary beds and affording another excellent section which very nearly resembles that at Morrison.   Fig. 27, p. 441 (after Holmes,) represents the section displayed on the north bank of Van Bibber creek, a small stream north of North Table mountain.   This illustrates the section exposed in better detail along the line of Ralston creek, which crosses the field of view along the middle-ground.   There is here a great development of the Triassic red beds, the Jura, and the Dakota. There are extensive quarries of the lower sandstones.   On the right hand of the figure are seen a number of conical hills of basalt, representing necks or dikes, through which the lavas of Table mountain ascended.   The terrace formations of the plains are very clearly represented.

THE GREAT PLAINS OF COLORADO AND KANSAS.

By S. F. Emmons.

From Denver to Kansas City the entire width of the Great Plains is again crossed, which present the same, or even greater, monotony of scenery than where they were crossed on the outward journey in Dakota. Their surface descends in a gentle but imperceptible slope about 10 feet in the mile, or 5,000 feet (1,250 m.) in the 500 miles (800 km.) that lie between these points. There is but little variety of erosional forms. The modern rivers, which in the spring and early summer are rapid, muddy torrents, constantly changing their meandering courses in their wide bottoms, gradually decrease in volume during the summer and autumn, and through the greater part of the year are shallow, inconsiderable streams. There is evidence that the larger of these streams have followed the same general course in early Tertiary times that they do now, for Tertiary deposits have filled them up in some places, and later erosion has not entirely removed these deposits from the sides of the later channels. There is little to be seen of the characteristically glacial topography shown on the more northern route, nor are any typical *mauvaises terres* to be seen along the route of travel. The streams whose valleys are traversed during the night and succeeding day do not have their sources in the Rocky Mountains, as do the Platte and Arkansas rivers, but head in springs between these two rivers and at some distance from the mountains.

The substructure of the Plains in this latitude is formed by strata of Paleozoic, Mesozoic, and Tertiary ages, which lie nearly in position of original deposition, and show no perceptible discordance of stratification between the beds of successive series. The surface is masked to some extent, especially along the more important valleys, by a loess-like deposit which gives great fertility when rainfall is sufficient for cultivation, or when, as near the mountains, a sufficient supply of water may be obtained for irrigation. In general, lower beds in the geological series are disclosed in going east; but as the region has not yet been systematically surveyed, except in limited areas, only the broader general features of its geology are known. The transgression which is so marked in the Appalachians between Paleozoic and Mesozic is not distinguished here. On the contrary, there seems to be a gradual passage, not only in the sedimentation but in the succession of life, from the Carboniferous into the Red Beds of the Trias. Between the latter and

443

the Dakota Cretaceous, however, the transgression is more distinct, and that between Laramie Cretaceous and Tertiary is very marked. In the area passed over during the night, after leaving the Laramie exposures which form the eastern boundary of the Denver basin, Tertiary beds extend eastward to Phillipsburg which is reached in the early morning. From there to Clay Center, Cretaceous beds furnish the only visible outcrops, the Red Beds of the Trias which come to the surface in southern Kansas not appearing along the line of travel. The coal measures of the Missouri-Iowa coal basin, which also extend into and are of economic importance in Kansas, pass upward into a series of shales and limestones, with a few unimportant sheets of sandstones, the whole containing a fauna of newer facies than that of the uppermost members of the Carboniferous remaining in the Appalachians. These have been commonly referred by local geologists to the Permian, or Permo-Carboniferous. The Tertiary beds which once covered the eastern portion of the plains area have been almost entirely removed by erosion. They are partly replaced by the drift deposits of the ice sheet, whose western margin extended a short distance to the west of the present valley of the Missouri river, where it forms the boundary between the States of Iowa and Missouri on the east and Nebraska and Kansas on the west.

Mr. Robert Hay contributes the following on the surface geology of the Plains in Kansas and Nebraska where traversed by the various railroads:

The geology of the country is substantially the same along these various routes. From the longitude of about 103° 30' eastward there are two formations that are conspicuous. The highest I call the Plains marl. Though it has variations it is remarkable for its lithologic similarity over vast areas, and samples not to be distinguished from each other could be obtained from the northern plains of Nebraska, the midplains (Platte-Arkansas) region and the Panhandle of Texas. It is argillaceous, arenaceous, and calcareous everywhere, and its varieties are due simply to the predominance of one or the other of its principal ingredients. Where the lime and clay have been weathered out, sand dunes are left. It has few fossils, but horse teeth have been found in it. In at least a part of the plains it is the Equus beds of Cope. Its origin probably began in the Pliocene era and stretched all through Pleistocene time. It forms the smooth floor of the unbroken high prairie of the West. The short gramma and other grasses have given its surface a compact sod that turns the water off it and in time of storm causes rapid flooding of the sandy arroyos and river valleys. On the one hundred and second meridian and from the thirty-fifth to the fortieth parallel it is from 50 to 200 feet deep, increasing in thickness north of the Republican and decreasing eastward. Broken by the plow it makes a fertile soil, and taken from excavations it is of the same quality to the bottom. Vegetation has not been sufficient to more than slightly discolor its surface. There is no black soil on the prairie. Owing to erosion immediately preceding its deposition, it is found in many valleys where, vegetation having since been ranker, the marl is more humus-like under the grass roots. Erosion of the modern age having made more impression as we proceed east, the narrowing area of the uplands has less of the Plains marl, and east of the ninety-eighth meridian it is only recognized in small isolated areas. Coming east it becomes more and more loess-like and in Nebraska and northern Kansas it merges into that formation, which is typically developed in the bluffs of the Missouri River. In eastern Colorado and western Nebraska and Kansas the Plains marl is usually underlaid by well-defined Tertiary formations. About the one hundred and fourth meridian on both sides of the Platte River these are the White River strata principally, which may be named, from their lithologic character, the Mortar beds. Coming east and south these thin out and give place to the Loup Fork, into which they merge. This last formation has within it beds of conglomerate, but whether as fresh-water lime beds, mortar beds, or conglomerates, the Loup Fork is everywhere a water holder; and this is true also of the thicker White River beds stretching toward Wyoming and north-west Nebraska. The more impervious Plains marl lets the meteoric water percolate slowly through it, and the looser arenaceous or gravelly texture of the White River or the Loup Fork beds holds the water till it is reached by wells from above, or escapes in springs in ravines where erosion has cut sufficiently deep. The ever-present arenaceous character of these Tertiary beds, whether as mortar beds or conglomerates, and their consequent water-bearing capacity have suggested the term Tertiary grit as a designation showing their relation to economic geology. These two formations, the Plains marl and Tertiary grit, make the essential features of the mid-plains geology. The mammalian and reptilian (turtle) fossils of the grit have been described by Marsh and Cope. Some floral remains and fresh-water univalves have also been found. Underneath the marl and grit in all the mid-plain region lie Cretaceous formations. These are all more impervious than the Tertiary grit and so help to make the water-bearing character of that formation more decided.

DENVER TO KANSAS CITY

ITINERARY.

By S. F. EMMONS.[*]

Station.	Distance.		Elevation.		Station.	Distance.		Elevation.	
	Miles.	Kilometers.	Feet.	Meters.		Miles.	Kilometers.	Feet.	Meters.
Denver a	0	0	5,182	1,579	Morganville	469	755	1,233	376
Limon	90	145	5,354	1,632	Clay Center	476	766	1,198	365
Goodland	198	319	3,687	1,124	Broughton	482	776	1,185	361
Norton	304	489	2,270	692	Bala	488	785	1,266	386
Phillipsburg	338	544	1,939	591	Riley	495	797	1,274	389
Gretna	344	554			Keats	504	811	1,124	343
Agra	349	562	1,756	535	Manhattan	513	826	1,014	309
Kensington	353	568	1,773	540	Zeandale	520	837	1,007	307
Athol	359	578	1,786	544	Wabaunsee	525	845	1,044	318
Smith Center	367	591	1,804	550	McFarland	534	859	1,020	311
Bellaire	373	600	1,866	569	Paxico	538	867	991	302
Lebanon	379	610	1,816	554	Maple Hill	546	879	957	292
Esbon	386	621	1,829	557	Williard	551	887	912	278
Otego	391	629	1,792	546	Valencia	555	893	904	276
Mankato	399	642	1,787	544	Topeka b	566	911	877	267
Montrose	406	653	1,658	505	Grantville	572	921	812	247
Formosa	411	661	1,515	462	Newman	578	930	796	243
Courtland	416	669	1,499	457	Medina	580	933	789	240
Scandia	422	679	1,431	436	Williamstown	585	941	789	240
Belleville	436	702	1,514	461	Lawrence c	596	959	763	233
Cuba	441	710	1,588	476	Linwood	607	977	733	223
Agenda	447	719	1,409	429	Bonner Springs	618	995	730	225
Clyde	455	732	1,295	395	Armstrong	633	1,019	690	210
Clifton	461	742	1,265	386	Kansas City d	635	1,022	750	229

a Population, 106,713.   b Population, 31,007.   c Population, 9,997.   d Population, 132,716.

From **Denver** to **Limon** the train runs on the tracks of the Kansas Pacific Railroad, first eastward, then bending to the south-southeast. The main line of the Chicago and Rock Island Railroad starts from **Colorado Springs,** and in its course to **Limon** runs a little north of east along the southern base of the mesa region which forms the divide between the waters of the Platte and Arkansas rivers. These mesas have not yet been systematically studied, but as far as known consist

[*]From notes furnished by R. Hay and R. T. Hill.

mainly of beds of the Tertiary system already described as the Monument Creek series, which may include several formations.

A few miles east of **Denver** the Denver beds are lost to view, and the surface is formed by those of the Laramie Cretaceous, also occupying a practically horizontal position. The higher elevations to the southward, especially as **Limon** is approached, are occupied by remnants of the Tertiary formations, while the broader valleys often contain local developments of more recent formations.

From **Limon**, which is on the Big Sandy creek, a tributary of the Arkansas River, the road takes an easterly course passing on to the head waters of the Republican river, one of the important streams which takes its rise in springs on the plains at a considerable distance from the mountains. Eighteen miles west of **Goodland** it passes into the State of Kansas, and follows the northern edge of the divide between the drainage of the Republican and Solomon rivers nearly to **Phillipsburg**, which is within the drainage system of the latter stream.

The surface of the country between **Limon** and **Phillipsburg** is, presumably, mostly covered by Tertiary formations of as yet undetermined age, the earlier being a grit or series of sands and conglomerates (supposed to be Miocene), with a recent marl or loess deposited on its eroded surface. The latter is well seen near **Norton, Phillipsburg**, and **Smith Center**. Wherever older Mesozoic outcrops are exposed by erosion of these overlying beds, they are found as a rule to be successively older as one goes farther east.

Phillipsburg is situated near the eastern edge of the great Tertiary plains. The Tertiary beds which once stretched east of it have been in great part removed by the drainage system of the Republican, Solomon, and Smoky Hill rivers. Just before reaching it a prominent mound, known as the (inverted) Bread Bowl, is formed by a protecting top of the hard, conglomerate grit of the Miocene. Fifteen miles to the north, at Long Island, in what are called the mortar beds of the Miocene, is the bone bed from which Prof. O. C. Marsh obtained many mammalian remains.

Farther east, in Smith county (**Smith Center**, county seat), the beds of Colorado Cretaceous are seen; first, the soft magnesian Niobrara limestone, then occasionally a blue shale. At the crossing of the Republican River, near **Scandia**, the bluffs are of Fort Benton age.

Near **Belleville**, and south from there, the sandstones and colored shales of the Dakota are passed over, but the outcrops are few and inconspicuous.

Although no outcrops of Mesozoic beds lower than the Dakota have been recognized, red sandstones, which are presumably Triassic, have been pierced in boring for salt along the line of the Kansas Pacific Railway in Ellsworth County, still farther southward, in about the middle of the State. The salt measures are barren gray beds above the highest light-colored Permian, and contain beds of rock salt from 100 to 200 feet in thickness.

The route from Clay Center to Kansas City, and also that running northeastward across Nebraska to Omaha, lies mainly upon limestones and shales of Permian and Carboniferous age.

Between Clyde and Clay Center the road follows a valley which had been enormously widened in Quaternary times and cut below the level of the plains, as preserved in the dividing ridges between the streams. Leaving this valley at Clay Center the road ascends the divide between the Republican and Blue rivers, and enters upon the Paleozoic area. The underlying rock is now the horizontal yellow limestone of the Permian, to whose existence the persistence of the divide is due.

The road enters the valley of the Kansas or Kaw River at Manhattan. Twenty miles up this river, in a southwesterly direction, on a point overlooking the junction of the Republican and Smoky Hill rivers and within the Fort Riley military reservation, there is a stone monument marking the geographic center of the United States as determined in 1880.

At Manhattan a good cross section of the Kansas Valley is obtained. On the opposite side of the river are heavy beds of orange-colored loess. Near this station, and again near McFarland, are outcrops of the flint beds, which are a conspicuous feature of the Permian throughout several counties. The road now passes into the Coal Measures, which constitute the prevailing formation of the uplands as far as Kansas City. Drift boulders of red quartzite from Minnesota, supposed to come from the drift in northern Kansas, are seen close to the road before entering McFarland.

Between Topeka and Kansas City drift is abundant, and its presence is shown by the form of the low hills to the north of the road.

At Lawrence a considerable morainal deposit is seen on both sides of the river, across which it once próbably formed a dam.

THE PRAIRIES.

By W J McGee.

When first explored by white men the eastern United States was wooded; much of the interior was woodless—the "prairie" of the French explorers, whose designation promises to outlive the condition described by the term; still farther westward lay the Great Plains, with which the prairies merged, and beyond lay the mountains. During this day's journey the route traverses a representative prairie land—the identical tract to which the name was originally applied. Now it is diversified, sometimes by natural groves or belts of woodland along the rivers, sprung up since the day of aboriginal prairie fires ended; elsewhere by artificial groves, hedges, and wind-breaks, such as abound over most of the interior region.

The prairie soil is commonly the long-weathered surface of one of the two great glacial drift sheets, antedating the terminal moraine. Generally the drift grades upward into a loam, sometimes loess-like, but more commonly displaying the characters of clay. Near the Mississippi, however, this loam deposit has developed into a fairly well-defined loess, which is not, however, so distinctive or so abundantly developed as on the Missouri. In the interior the loam is commonly so thin that drift boulders may be seen approaching the soil in the railway cuts, and not infrequently they lie scattered over the surface in considerable numbers. All, or nearly all, of these boulders are far-traveled erratics, carried down from near or beyond the northern boundary of the United States.

The relief is low, the surface undulating gently; the configuration is the product of faint hydrodynamic sculpture, for the most part directed by the antecedent glacial configuration and by the temporary waterways born of the melting ice. It is noteworthy that this surface, primarily ice fashioned, is now so far modified by waterwork that practically the entire surface is well drained. Marshes occur only rarely and lakes never. In this respect the extra-morainal surface is strongly distinguished from the intra-morainal area, in which lakes abound and marshes are innumerable.

FROM KANSAS CITY, MISSOURI, TO CHICAGO, ILLINOIS.

ITINERARY.

By W J McGee.

Station.	Distance.		Elevation.		Popula-tion.	Station.	Distance.		Elevation.		Popula-tion.
	Miles.	Kilometers.	Feet.	Meters.			Miles.	Kilometers.	Feet.	Meters.	
Kansas City	0	0	750	229	132,716	Bureau	404	650	670	204	....
Cameron..........	53	84	1,022	301	....	La Salle	419	674	665	203	9,855
Gallatin	75	121	.....	.....	....	Ottawa	434	698	688	210	9,985
Jamesport.......	84	135	.....	.....	....	Morris..........	456	734	722	220	....
Belknap.........	210	338	887	267	....	Joliet...........	478	769	540	165	....
Muscatine.......	306	492	562	165	11,454	Blue Island Junc	501	806	807	246	....
Davenport.......	335	539	595	181	26,872	Auburn Junc...	510	821	795	212	....
Rock Island	337	542	470	143	13,634	Englewood	511	822	603	183	....
Moline..........	339	546	773	236	12,000	Chicago	518	833	594	181	1,099,850

At **Kansas City** are drift and loess, both resting on Carboniferous beds. The loess is here, as elsewhere, a clayey loam, distinguished by its tendency to cleave vertically and to stand in perpendicular walls as a result of erosion, thus producing characteristic topographical forms.

The road now takes a northwesterly direction through the State of Missouri. After leaving **Kansas City** the loess continues as a conspicuous feature for miles, grading into loam and then into drift. The substructure is formed by upper Carboniferous beds, which contain coal seams of variable thickness and considerable extent, but which have not been exploited on account of the proximity of the Iowa-Missouri coal basin, in which the seams of the middle Coal Measures are found at less depth. Midway between the Missouri and the Mississippi the route crosses this basin, and gradually descends geologically over the lower Coal Measures and the sub-Carboniferous formations, the latter being well developed about the Des Moines river.

At **Muscatine** the road reaches Mississippi river, whose valley it follows to the crossing at **Davenport**.

The subterrane from the Mississippi to Lake Michigan is Paleozoic, including Carboniferous, Devonian, and Silurian rocks. About **Davenport** and **Rock Island** the Carboniferous rocks are brown sandstones, commonly considered to have been deposited in outlying basins near the shore line of the middle Carboniferous sea; but within 25 miles east of Mississippi river the rocks come to resemble more closely those characteristic of the Coal Measures of the Illinois-Indiana basin. The sandstones near the river, and the shales and limestones (with a few coal seams) of the interior, rest unconformably on the older strata; for the calcareous series of deposits, commonly referred to the sub-Carboniferous, is absent, and the Devonian, broadly developed in Iowa, soon fails, leaving the Carboniferous and Silurian strata nearly or quite contiguous.

On approaching **Davenport** the railroad passes through deep cuts in Pleistocene deposits, which afford interesting sections. The uppermost deposit is loess, somewhat more clayey than that of Missouri river, yet abounding in similar fossils; its thickness ranging from 10 to 30 feet. At its base it commonly becomes gravelly or grades into an attenuated drift sheet; and loess and drift alike rest on an ancient soil, or "forest bed," which has yielded not only abundant remains of coniferous woods, but also bones and tusks of the elephant. Below lies a dense, tenacious drift sheet, representing the earliest well-defined ice invasion of the Pleistocene.

Between **Davenport** and **Rock Island** the Mississippi flows in what may be styled, for that stream, a contracted gorge, half a mile to a mile in width. This gorge opens immediately below the cities; upstream it extends for some 25 miles to the town of Le Claire. Throughout this stretch of relatively narrow channel the current is exceptionally strong, particularly at the upper and lower extremities. Thus the terms "Rock Islands Rapids" and "Le Claire Rapids" have long been familiar to the pilots and captains of the packets plying between St. Louis and St. Paul, and, indeed, to other rivermen. Formerly the rocky islet, from which the Illinois city takes its name, was separated from the mainland of Illinois by a navigable channel, and indeed represented nothing more than the largest of a series of "reefs" rising from the river bottom. Subsequently the Iowa channel was deepened and the Illinois channel was finally dammed to afford waterpower for the United States Arsenal located upon the island.

The only rocks exposed in the immediate vicinity of **Davenport** and **Rock Island** are Devonian; they are fossiliferous, and, for the most part, have been correlated with the Hamilton of New York; but their exact position in the geological scale can not be said to be finally determined. Exposures of brown Carboniferous sandstone occur within a few miles of both cities.

A few miles east of **Rock Island** the railroad approaches and finally crosses Rock River, which, it may be observed, flows in a disproportionately broad gorge. Moreover, it may be noted that this valley bifurcates, sending its principal arm northward to be occupied only by a trifling stream, while the narrower is marked by the course of Rock river.

This broad valley represents one of a plexus of channels formed by the floods derived from the melting ice about the close of the Pliocene; it is the *Marais d'Ogee* of the French missionaries, which, traced against the general direction of river flow, extends northeastward and then northward for 40 miles, where it divides once more, this time into three branches. The smallest of the branches coincides with the present course of the Mississippi about Clinton; the next in size coincides with the Wapsipinnicon River for 50 miles; and the principal branch diverges from the Wapsipinnicon a few miles above its mouth and then cuts directly across the interior of eastern Iowa to reunite with the Mississippi at the mouth of the Maquoketa, 100 miles (by river) above **Rock Island.**

East of the plexus of ancient channels the road traverses one of the low divides characteristic of northern-central Illinois—a typical prairie land. On approaching Illinois river the surface becomes more rugose, and rock outcrops appear. About **La Salle** and **Ottawa** the river bluffs are 100 to 200 feet high, and expose a variety of formations within a limited area, ranging from the coal-bearing Carboniferous through the Niagara (the characteristic Upper Silurian formation of the interior) and down to the Oneota (or "Lower Magnesian"), or the St. Peter. In this vicinity the structure is more complex than elsewhere in Illinois save in two localities, i. e., at the mouth of Illinois river, and near the confluence of the Ohio; it is a region of decided deformation of unconformable strata, afterward planed down to base level. Noteworthy fossil localities occur in the vicinity. The rocks are mantled with glacial drift so deeply that exposures are rare, except in the river bluffs.

Nearly all the way from **Ottawa** to **Chicago** the route traverses monotonous prairie land, faintly relieved here and there by moraines, inconspicuous in comparison with those of the northern route. There are few rock exposures, the most notable being at **Joliet**, where there are extensive quarries. The subterrane from **Ottawa** to **Chicago** is deeply buried beneath the drift, but is probably almost wholly Upper Silurian, and chiefly the interior representative of the New York Niagara.

CHICAGO TO NIAGARA FALLS.

ITINERARY.

By G. K. GILBERT.

Station.	Distance.		Elevation.		Popula- tion.	Station.	Distance.		Elevation.		Popula- tion.
	Miles.	Kilometers.	Feet.	Meters.			Miles.	Kilometers.	Feet.	Meters.	
Chicago..........	0	.....	594	181	1,099,850	Schoolcraft	146	235	878	267	....
Elsdon..........	8	13	.....	.....		Battle Creek....	175	282	823	251	....
Blue Island Junc	20	32	.....	.....		Lansing	220	354	836	256	13,197
Harvey..........	23	38	.....	.....		Durand.........	252	406	794	247	13,102
Thornton Junc ..	25	42	.....	.....		Port Huron.....	335	539	584	178	....
Maynard	31	51	.....	.....		Sarnia..........	336	541	587	179	13,543
Griffith	36	59	.....	.....		London..........	397	639	806	246	....
Redesdale	39	64	.....	.....		Woodstock	425	685	951	291	....
Ainsworth	45	74	.....	.....		Harrisburg......	453	731	734	224	....
Sedley..........	50	82	694	212	....	Dundas.........	.....	.....	.....	.....	
Valparaiso	55	90	806	245	....	Hamilton.......	472	760	255	78	....
South Bend......	100	162	713	217	21,819	Niagara Falls :					
Mishawaka......	104	167	721	220	....	Canadian side ..	516	830	573	174	....
Cassapolis	122	196	880	268	....	American side ..	518	841	574	175	....

Leaving Chicago at 3 o'clock p. m., we cross before night the corners of Illinois and Indiana and a part of Michigan. In passing about the head of Lake Michigan the underlying Silurian and Devonian rocks are not seen, but only the Champlain sands accumulated by winds and waves at the end of the lake when its water stood at higher levels. The winds are still busy with them, piling them in traveling dunes which block the drainage, converting much of the land into marsh.

Soon after entering the State of Michigan we pass from Devonian to Carboniferous terranes, entering the Michigan coal basin. That portion of the State lying between lakes Michigan and Huron, called the Lower Peninsula, is characterized by a synclinal basin of gentle dips, carrying the Silurian rocks below the level of the sea and bringing the Carboniferous below the plane of denudation. In this structural basin brines are preserved, on which an important salt industry is based.

It is of interest to note the relation of the Michigan syncline to the general structure and to the basins of the Laurentian lakes. Throughout a large region, including the lake district, the general strike of outcrops is east and west, the older rocks lying at the north and the newer at the south. In latitudes 85° to 87° a double interruption to this arrangement occurs. The line of Devonian outcrop swings 250 miles (400 km.) southward in a great loop about the Cincinnati upward arch; the Silurian outcrop swings an equal distance northward about the Michigan downward arch; and between the two arches there is a belt in which the dips are northward. The most resistant members of the Paleozoic are the Trenton limestone, at the base of the Silurian, and the Niagara limestone, near the top of the same system. Between these are shales and soft sands of the Utica, Hudson River, and Medina series, and above the Niagara are equally soft shales of the Salina, Hamilton, and Chemung series. The basins of lakes Michigan and Huron are carved from the monocline of soft rocks above the Niagara, where it sweeps around the Michigan syncline, and are thus made to embrace the coal basin. Lake Erie lies in a trough carved from the same monocline where its trend is nearly normal. Lake Ontario lies in the monocline below the Niagara limestone, where its course is normal, and the same monocline, where it curves about the Michigan coal basin, holds Georgian and Manitoulin bays, dependencies of Lake Huron, and Green bay, a dependency of Lake Michigan. It is believed that the excavation of the basins was accomplished partly by rains and rivers during pre-Pleistocene and interglacial times when the district stood at a higher level than now, and partly by Pleistocene ice currents. As to the relative importance of the two agencies geologists are not agreed, and it may safely be said that the determination of the question belongs to the future.[82, 83]

The surface features in Michigan depend largely on the drift, consisting of ground moraine traversed by numerous marginal moraines. Lakelets, ponds, and swamps are numerous. On both sides of the St. Clair river, connecting Lake Huron with Lake St. Clair, the till is overlain by laminated Champlain clays, here called Erie clay, and the surface of the country is smooth.

From **Sarnia** to **Hamilton**, in the Province of Ontario, Dominion of Canada,[84] the route continues on the eastern limb of the Michigan syncline, gradually descending in the geologic scale through the Devonian and Upper Silurian. The Hamilton shales are succeeded by the Corniferous limestone near **London**; the Onondaga follows, and at **St. George** the Guelph, an upper member of the Niagara limestone. The main mass of the Niagara is met at **Capetown**, the underlying Clinton at **Dundas**, and the Medina at **Hamilton**. Nearly the whole country is heavily sheeted by drift deposits. The Erie clay, which occupies the entire surface for the first hundred miles and appears at intervals beyond, is a laminated calcareous clay, with erratic pebbles and boulders, but no fossils, apparently the deposit of a great lake at the margin of the ice. Smaller bodies are traversed of the Algoma sand

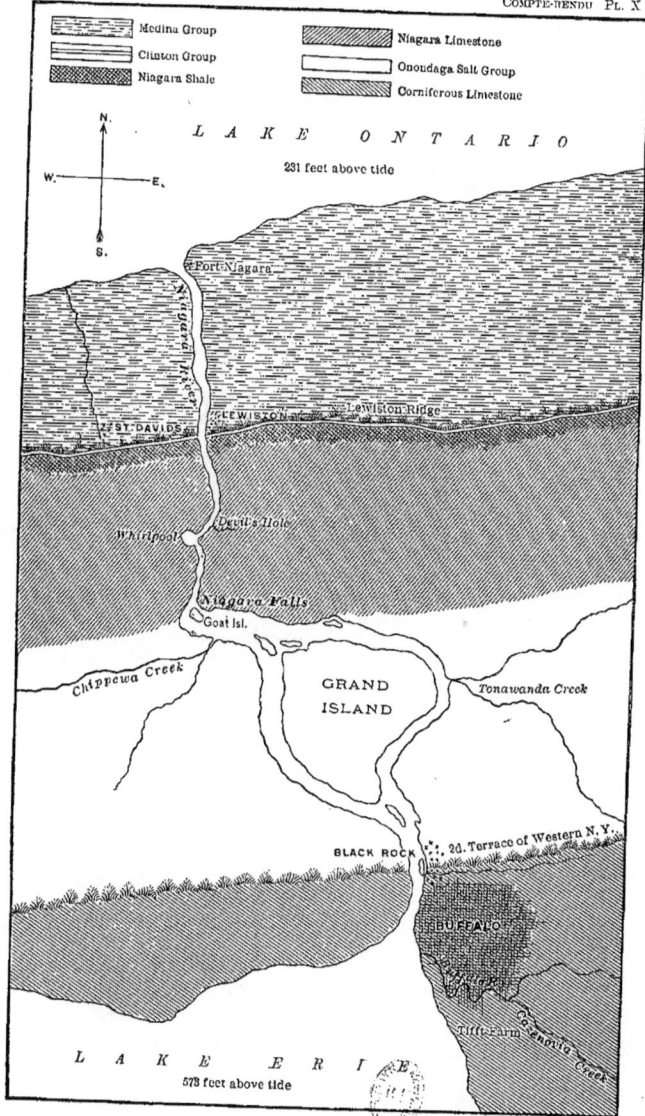

MAP OF NIAGARA RIVER FROM LAKE TO LAKE.

and the Artemisia gravel, deposits likewise of great extent and containing material of distant origin, but not yet satisfactorily interpreted.

From **Hamilton**, at the head of Lake Ontario, the train runs eastward over a broad escarpment of red Medina shales (Upper Silurian) superficially sheeted with till and clay similar to the Champlain. At the left lies Lake Ontario; at the right the plain rises to the foot of an escarpment several hundred feet in height, which is capped by Niagara limestone. The plain is contoured by an old shore-line of Lake Ontario, known as the Iroquois beach. In the city of **Hamilton** at the head of the lake this beach takes the form of an immense free spit or embankment more than 100 feet in height. At other points it appears as a low barrier of sand and shingle, and yet at other points as a low bluff undermined by the waves. Since the date of the Iroquois beach the lake water has also stood at a level lower than the present, and during the period of low water the small streams which traverse the plain opened valleys in the lacustrine clay. These valleys are now partly occupied by lake water, being marked by small bays, to each of which a small stream is tributary.

Leaving the littoral plain, the train climbs the escarpment, so as to approach **Niagara falls** on the plain constituted by the upper surface of Niagara limestone.

NIAGARA FALLS. [55, 56]

By G. K. GILBERT.

The Niagara river flows northward from Lake Erie to Lake Ontario. The region is floored by Paleozoic strata, which dip at a low angle toward the south or upstream. Two limestones are of physiographic importance, the Corniferous and the Niagara. Beneath the Corniferous and above the Niagara are several hundred feet of shaly beds (Onondaga salt group), yielding readily to erosive agencies. Beneath the Niagara limestone are feebly resistant beds, known as the Niagara shale, the Clinton beds, and the Medina shale. As a result of this alternation of hard and soft strata the district consists topographically of two sloping plateaus, each limited toward the north (downstream) by an escarpment. The Corniferous escarpment, lying near Lake Erie, is relatively low; the Niagara escarpment is about 200 feet high and faces toward Lake Ontario. Lake Erie rests on the Corniferous plateau, and the Corniferous limestone determines the height of its water surface. Across this limestone the river flows with a rapid current. In the region of shales beyond it travels more slowly and spreads out broadly. It traverses the plateau of Niagara limestone in a narrow canyon, at the head of which is a cataract. The passage through the canyon is by a series of violent rapids, the water tumbling over a rough bottom composed of limestone blocks fallen from the walls. Beyond the Niagara escarpment the river traverses a low plain with deep and quiet current.

In the walls of the canyon the strata are finely displayed. At the top the Niagara limestone has a thickness of about 80 feet (24 m.) near the cataract, and this gradually diminishes to the edge of the plateau, the difference being due to the general degradation of the surface. The full thickness of the limestone previous to erosion was about 140 feet (42 m.). Beneath the limestone is the gray Niagara shale, about 80 feet (24 m.) in thickness; then come the Clinton beds, gray limestones and shales, with a sandstone at base, and a total thickness of 35 feet (10 m.), and finally the Medina shales and sandstones, here consisting chiefly of red arenaceous shale rarely interrupted by ledges of sandstone. At the foot of the cataract, the Clinton is near the water's edge; northward it rises at the rate of 25 feet (7 m.) to the mile, and the river falls at a much more rapid rate, so that a basal exposure of the red Medina increases rapidly from the cataract to the escarpment. All these beds, except the Niagara limestone, are in places more or less obscured by talus, but the complete section can be seen on the American side, just below the lower suspension bridge.

The basins occupied by Lake Erie and Lake Ontario had a different system of drainage previous to Pleistocene time, and were remodeled by the work of the ice sheet, which modified the geography of the Great Lake region in important ways. Some regions of soft strata suffered notable erosion, and the old drainage lines were in many cases completely obliterated by deposits of the glacial drift. When the ice melted the waters were compelled to find new ways, and the drainage of the glaciated region was imperfect or immature, in that it included an immense number of lakes, large and small. Lake Erie and Lake Ontario came into existence at that time, and so did the Niagara river.

The erosion of the Niagara gorge from the escarpment back to the cataract is therefore a post-glacial work, and as a measure of post-glacial time it has attracted great attention. The length of the gorge eroded, about six miles, is readily measured. The present rate of erosion by the cataract is susceptible of measurement, and observation has already given it a value with valid claims for consideration. From a survey made in 1842, and subsequent surveys made in 1875, 1886, and 1890, it appears that the central portion of the main cataract, the Horseshoe fall, is receding 4 or 5 feet per annum, and that the American fall, which carries much less water, is receding much less rapidly. For those who are willing to postulate a uniform rate of recession through the whole extent of the gorge, it is easy to estimate the age of the river from these data, but there are important reasons for questioning the validity of the postulate.

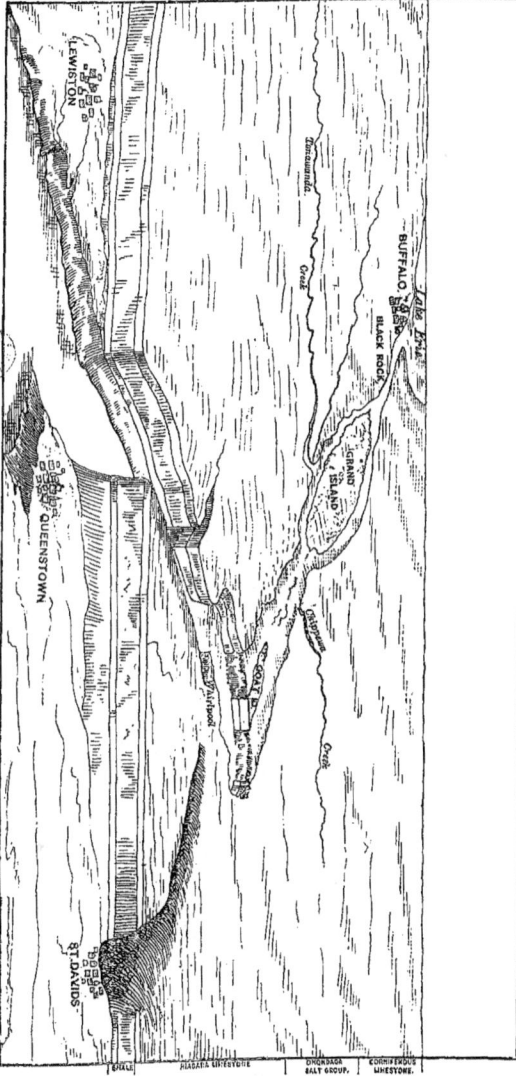

BIRD'S-EYE VIEW OF THE NIAGARA RIVER.

LEWISTON

BUFFALO

BLACK ROCK

Lake Erie

Tonawanda Creek

GRAND ISLAND

QUEENSTOWN

Bloody Whirlpool

FORT

ST. DAVIDS

SHALE | NIAGARA LIMESTONE. | ONONDAGA SALT GROUP. | CORNIFEROUS LIMESTONE.

The continuity of the geologic section in the walls of the canyon is interrupted at one point on the Canadian side. At the locality known as the Whirlpool there is an embayment of the wall, and at the head of that embayment a body of drift is exposed from top to bottom of the bluff, replacing the Palæozoic strata. How much deeper it extends is not known, but the river has here gouged out a deep pool, in which the current is temporarily slackened, and it appears probable that this excavation was in the soft drift. Associated with this feature is an embayment of the Niagara escarpment, near the town of St. Davids, not far away. For the space of a mile the limestone cliff disappears, and is partially replaced by glacial drift. It is believed that this embayment and the preglacial cavity at the Whirlpool constitute parts of the same preglacial valley, a valley opening to the northward and terminating southward within the present river canyon, between the Whirlpool and the cataract. So far as this valley extended the river had an easy task, for its canyon was already dug.

On the other hand, it is doubted that the river has at all points been able to work as rapidly as now. The height of the Falls is about 160 feet (48 m.); the pool below the cataract has a depth, at the nearest point where sounding has been successful, of 185 feet (56 m.). This great depth of pool appears to be essential to rapidity of recession, for under the American fall there is no pool and there the recession is slow. But the river has a depth comparable with that of the pool for only a mile or two below the cataract, and at most other points the present cross section renders it improbable that a deep pool was ever formed. It is surmised that at such points the rate of recession may have been slow.

Accordant with this view is the hypothesis, not yet fully tested, that all of the upper lakes, except Lake Erie, once found discharge to the St. Lawrence river by other routes, so that for an unknown fraction of postglacial time the Niagara river drained a district only one-eighth as large as that which it now drains.

The train approaches the Niagara falls from the Canadian side across the suspension bridge, and follows the American bank to the village of Niagara Falls, where it halts for the day. One can conveniently cross again to the Canadian side by the steamer *Maid of the Mist*, or by the upper suspension bridge, and the geologist will wish to see the cataract from both sides and from the river brink below. Goat island, which divides the American fall from Horseshoe fall is reached by a bridge from the American shore, and a spiral stairway leads one thence to the water's edge between the two falls. At the head of this stairway guides and waterproof suits can be secured for a visit to the Cave of the Winds, in which one passes beneath a sheet of falling water constituting a portion of the American fall. The path lies partly on talus, partly on limestone ledges of the Clinton, and partly on fallen blocks of Niagara limestone. From the Canadian side one can enter a tunnel dug in the Niagara shale, and follow it to a point beneath the Horse-shoe fall, whence he can look out at the descending water. At numerous points farther down stream stairways, elevators, and inclined railways are constructed to enable the visitor to reach the water's edge. Those of most value to the geologist are on the American side at the Whirlpool Rapids, and on the Canadian side at the Whirlpool. From the latter a few minutes' walk takes one to the mass of drift filling the preglacial channel, and its contact with the older rocks can be seen.

The accompanying sketch (Fig. 28) was made by M. Golliez, of the visiting geologists.

FIG. 28.—The Whirlpool of the Niagara River.

NIAGARA FALLS TO NEW YORK CITY.

ITINERARY.

By Chas. D. Walcott.

Station.	Distance. Miles.	Kilometers.	Eleva-tion. Feet.	Meters.	Popula-tion.	Station.	Distance. Miles.	Kilometers.	Eleva-tion. Feet.	Meters.	Popula-tion.
Niagara Falls.....	0	0	574	175	....	Kingston........	362	583	185	56	21,261
Buffalo	22	35	624	190	255,664	Milton...........	383	616	10	3	....
Bergen	74	119	575	174	....	Newburg........	394	634	31	9	23,087
Genesee Junction.	87	140	524	160	....	Cornwall	398	640			....
Rochester........	91	140	494	151	133,896	West Point......	403	649	8	2	....
Newark...........	121	195	436	133	....	Ft. Montgomery.	408	657	8	2	....
Port Byron	147	236	403	123	....	Stony Point	416	669			....
Syracuse	172	277	400	122	88,143	West Nyack.....	426	686			....
Canastota	193	311	434	132	....	Tappan..........	432	695	75	23	....
Utica	219	352	523	159	44,007	Bergenfields	439	706	66	20	....
Frankfort	229	369	399	121	....	Hackensack	443	713	47	14	....
Little Falls.......	241	389			....	Weehawken	451	726	7	2	....
South Schenectady	298	480	349	106	....	New York City, (foot of Jay street)........	....	.....	....	....	1,515,301
Coeymans Junction.............	322	518	180	55	....						
Catskill...........	340	547	97	30	....						

This day's journey is over a region classical in the annals of the development of North American geology. It was along this line that Hall, Emmons, Vanuxen, Mather, and Conrad conducted their investigations, the results of which are published in the first four volumes of the geology of New York and in the annual reports which preceded them. The nomenclature established by them, on the basis of stratigraphy and paleontology, became a part of American geologic science that was extended from State to State by subsequent surveys.

From **Niagara Falls** to the valley of the Hudson at **Schenectady** there are no marked topographic features; the country is undulating, and the route crosses the slightly southward dipping rocks of Lower Devonian, Silurian, and Lower Silurian (Ordovician) age. They are for the most part subjacent to heavy deposits of drift, although numerous fine sections are shown in the various streams that flow in the north and south drainage lines across which the train passes.

459

From **Schenectady** to **Newburg** the valley of the Hudson River is open and broad, and its undulating surface constitutes part of an ancient base-level, originating in Tertiary time. Afterward continental elevation led the river to corrade its channel deeply, so that the immediate valley of the river lies several hundred feet below the plain of the general valley. The rocks in which this plain is carved are Lower Silurian in age, including the Hudson River, Utica, and Trenton series. They are greatly disturbed, and beautiful sections are shown on the cliffs and in the railroad cuts. Shaly portions have received a cleavage structure, and metamorphism has extended so far that through considerable areas the several series have not yet been discriminated. Uplands visible at the right are due to the superior resistance of the horizontal Lower Helderberg limestone, and loftier uplands seen beyond them are due to the endurance of the Catskill sandstones. The mountain from which the name Catskill (Kaaterskill) is derived is a conspicuous feature at the right (west). At **Newburg** the Trenton limestone occurs with but little alteration, and just beyond it the train enters the area of crystalline schists, to which the mountain range known as the Highlands belongs.

The train then passes through the gorge of the Highlands, keeping close to the water's edge. Above it the old base level plain holds place as a terrace within the gorge.

Beyond the gorge glimpses are obtained of disturbed and altered Paleozoic rocks, and then the train approaches the Palisades, a ridge of trap, originally a sheet or dike injected in the great New Jersey series of red shales and sandstones commonly referred to the Trias. This ridge borders the lower portion of the Hudson on the west for many miles, and at its northern end swings westward. The railroad, passing through the curved northern extremity by a tunnel, follows the western base of the ridge for more than half its length, and then by another tunnel reaches its eastern base and the bank of the river opposite New York.

In passing from **Niagara Falls** to **Buffalo** the train crosses the southern outcrop of the Niagara limestone, the entire width of the Onondaga Salt group, and enters upon the Corniferous limestone of the Lower Devonian. Turning eastward it passes over the latter formation for about 50 miles, when it recrosses the Onondaga Salt group and continues on the Niagara limestone to **Rochester**. From a little distance east of **Rochester** it follows the Onondaga Salt group for over 100 miles, when it crosses the thin eastern extension of the Niagara limestone and the broad development of the Clinton formation before entering upon the Utica and Lorraine shales, upon which it follows for 105 miles except at **Little Falls**, between **Utica** and **Schenectady**, where a spur of bedded Algonkian gneiss crosses the line of the railroad; a fine geologic section is exposed along the cuts and the Mohawk River.

The Onondaga, Niagara, and Clinton terranes have been fully described by the veteran geologist and paleontologist, James Hall, and the two latter terranes are mentioned in this Guide as they are seen in the canyon of the Niagara. The immediate subjacent formation, as it occurs in the vicinity of Utica, includes 800 feet of argillaceous shales in which numerous sandy layers occur near the summit. The lower 710 feet is the Utica formation, and the upper 90 feet the equivalent of the Lorraine shales and sandstones. The Trenton limestone has a thickness of about 150 feet at Little Falls, and the immediately subjacent Calciferous sandrock, 190 feet. A thin bed of sandstone and shale just above the Algonkian gneiss has been referred to the Upper Cambrian (Potsdam sandstone) zone, but on evidence that is not conclusive. A fine section of the bedded Algonkian gneisses is shown in the cliffs below the sandstones, and in the river, at the upper end of the narrows, a massive gneiss or granite is to be seen. The almost horizontally bedded Algonkian strata, although crystalline, have frequently been taken to be the downward extension of the superjacent Calciferous formation. Other exposures of the Algonkian, Calciferous, and Trenton terranes occur in the cliffs of the northern side of the valley, between Little Falls and South Schenectady, while with slight exception the hills on the south side are formed of the Utica shale, with the Trenton limestones at the base.

Turning southeast from South Schenectady the road enters the valley of the Hudson and an area where the geologic structure is entirely unlike that passed over from Niagara Falls to South Schenectady. The Silurian (Ordovician) rocks are upturned, compressed, and more or less broken by the westward thrust of the masses of rock disturbed by the crumpling and folding of the strata of New England. Within the valley there remains to be solved one of the most complicated local geologic problems in North America geology. The higher, outer western sides of the valley are formed of the horizontal Lower Helderberg limestones, with a thin band of the Niagara corralline limestone, beneath which a great thickness of alternating sandstones and shales extend down to a limestone, found in deep wells at 3,475 feet beneath the upper limestone. It is this series of shales and sandstones that are so plicated and altered in the valley to the eastward. As far as known there are no exposures of the undisturbed strata below a point 600 feet beneath the Helderberg limestone. The upper 600 feet exposed has been correlated with the Lorraine, the limestones at the bottom of the deep wells with the Trenton, and the strata between with the Utica and Lorraine, or all above the supposed Utica, just above the limestone, with the "Hudson River group," of the New York Survey. As the latter occurs out in the valley it contains a strongly marked graptolitic fauna, usually called Normans Kill fauna. This fauna is considered by Prof. Lapworth to be of about Trenton age and pre-Utican—a view sustained by Mr. Ami in his studies at Quebec and by Dr. Gurley in his review of the graptolitic faunas. Certain it is that the graptolitic fauna is not the same as that of the Utica shale of the Mohawk Valley, as has been advanced by Hall, Whitfield, and Walcott. The term Hudson has been applied to these beds between the Lower Helderberg limestone and the supposed Trenton limestone beneath, and the Lorraine and Cincinnati formations correlated as equivalent. This can not be done logically to-day, for the series of shales and sandstones, with occasional interbedded lentiles of limestones, includes all the formations from the Calciferous to Lorraine inclusive. It is, as stated by Sir William Logan, practically the equivalent of the Quebec group, although it includes more at its upper limit in taking in the Lorraine strata beneath the Lower Helderberg limestone.

The sedimentation on the outer limits of the eastern side of the valley includes the Berlin grits and great thicknesses of interbedded purple shales, and again red and green slates with dark argillaceous shales, carrying the Normans Kill graptolitic fauna. Lentiles of limestone occur bedded in the shales, in which the Calciferous, Chazy, and Trenton faunas are found—sometimes one or two in the same lentile, as in Washington County, New York. On the east side of the Hudson Valley 5,000 feet of shales, slates, sandstones and limestones, of very irregular succession, may be referred to the Hudson terrane.

The evidence now at hand leads to the conclusion that in the valley of the Hudson, or the northern portion of the Appalachian trough, the sedimentation, from the Upper Cambrian to Lower Helderberg time, was unlike that of the region to the westward. It was greater in quantity and variety, and during Trenton time a graptolitic fauna was buried in it, such as is unknown elsewhere in New York and the southern Appalachians, although present in Arkansas and Nevada. All attempts to correlate the Hudson series with sections elsewhere must be more or less defective, except that the great mass is of Lower Silurian (Ordovician) age. The best region to study the Hudson series is from Fort Edward, above Albany, to the vicinity of Catskill.

In Dutchess County, opposite Kingston, the Lower Cambrian quartzite rests on the Algonkian gneisses, and the stratigraphic section is represented from this horizon to the Trenton and Lorraine. South of Poughkeepsie, on the east side of the river, the sedimentary beds are crowded out to the river by the pre-Cambrian rocks, and below Newburg the crystalline schists of the Highlands occupy the western shore until the trap ridge of the Palisades is encountered. At **Weehawken** the characters of the trap are finely shown in the quarries.

The drift features, from **Niagara Falls to Schenectady,** include at first ground moraine with undulating surface, and afterward, between **Rochester** and **Syracuse,** a tract of drumlins of the more elongated type. At **Little Falls** the Mohawk River appears to have cut a postglacial channel, and fine river terraces occur above the gorge through which the river passes. Below the falls the river silt and gravels nearly fill the valley from side to side.

The most important Pleistocene deposit of the Hudson Valley is a great bed of laminated clay referred to the Champlain epoch and known locally as the Albany clay. It is the sedimentary record of a large estuary occupying the Hudson Valley after its abandonment by the ice, and it appears originally to have stretched from side to side of the valley, filling the river channel and masking the rugosity of the baselevel plain. Near **Schenectady** its upper surface bears a heavy layer of sand. After its deposition the land rose temporarily to a height greater than the present, permitting the river to carve its channel to such depth that it has not since been refilled with alluvium. The so-called river is still an estuary for 150 miles (240 km.) from its mouth, transmitting ocean tides as far as Albany.

From the terminus of the West Shore Railway at **Weehawken** travelers are transferred to **New York City** by ferry.

EXCURSION TO THE CANYON OF THE COLORADO.

DENVER, COLORADO, TO FLAGSTAFF, ARIZONA.

ITINERARY.

Station.	Distance Miles.	Kilometers.	Elevation Feet.	Meters.	Station.	Distance Miles.	Kilometers.	Elevation Feet.	Meters.
Denver *	0	.....	5,182	1,579	Azul	389	626	6,672	2,034
Colorado Springs ‡	73	117	5,978	1,822	Las Vegas	395	636	6,383	1,945
Fountain	86	138	5,552	1,692	Romero	400	644	6,288	1,917
Buttes	93	150	5,353	1,632	Sulzbacher	409	658	5,975	1,821
Piñon	104	167	5,022	1,531	Tecolote	411	661	5,846	1,782
Pueblo †	117	188	4,653	1,418	Bernal	414	666	6,068	1,849
Baxter	123	198	4,602	1,403	San Miguel	425	685	6,021	1,835
Chico	129	208	4,532	1,381	Sands	430	693	6,388	1,947
Boone	136	219	4,460	1,359	Fulton	432	696	6,527	1,989
Neposta	144	232	4,356	1,328	Rowe	441	710	.....	.....
Rocky Ford	169	272	4,102	1,269	Pecos	446	718	6,366	1,940
La Junta	180	290	4,046	1,233	Glorieta	451	726	7,417	2,261
Benton	189	304	4,263	1,299	Canoncito	455	732	6,855	2,089
Timpas	198	319	.....	.....	Lamy	460	740	6,460	1,969
Iron Springs	208	335	4,676	1,425	Ortiz	472	760	5,821	1,774
Delhi	217	349	5,042	1,536	Los Cerrillos	478	769	.....	.....
Thatcher	225	363	5,401	1,646	Wallace	491	790	5,248	1,600
Tyrone	235	378	5,520	1,682	Elota	498	801	5,125	1,563
Earl	244	393	5,673	1,729	Algodones	502	808	5,099	1,554
Hoehnes	253	407	5,706	1,739	Bernalillo	511	822	5,099	1,554
Trinidad	262	422	5,967	1,819	Alameda	520	837	4,981	1,518
Starkville	267	430	6,333	1,930	Albuquerque	528	850	4,935	1,504
Morley	272	438	6,748	2,057	Laguna	594	956	5,869	1,789
Wooton	277	446	.....	.....	McCartys	611	983	.....	.....
New Mexico line	277	446	.....	.....	Bluewater	635	1,022	.....	.....
Raton	285	459	6,622	2,018	Continental Divide	.....	.....	.....	.....
Dillon	288	463	6,456	1,968	Coolidge	664	1,068	6,977	2,127
Maxwell City	311	500	.....	.....	Wingate	674	1,084	6,714	2,046
Dover	319	513	5,819	1,774	Defiance	694	1,117	.....	.....
Springer	325	524	5,768	1,758	Manuelito	702	1,130	.....	.....
Wagon Mound	351	565	6,178	1,883	Carrizo	766	1,233	5,199	1,584
Tipton	361	581	6,365	1,940	Holbrook	781	1,257	5,069	1,545
Shoemaker	368	592	6,256	1,907	Winslow	813	1,308	4,824	1,470
Watrous	376	605	6,398	1,950	Canyon Diablo	839	1,350	5,399	1,645
Onava	385	620	6,730	2,051	Flagstaff	872	1,403	6,864	2,092

* Population, 106,713.   † Population, 11,140.   ‡ Population, 24,558.

[By S. F. EMMONS.]

The route between **Denver** and **Colorado Springs** has already been described (p. 434).

From **Colorado Springs** nearly to **Trinidad** the road crosses open, unincidented plains of Middle Cretaceous shales, in which the only geological landmarks are occasional outcrops of the harder beds of the Niobrara limestones carrying abundant casts of *Inocerami*.

The road first runs south along the alluvial bottom of Fountain Creek to Pueblo, then bends eastward and follows the bottom lands of the Arkansas river to **La Junta**, from which point it takes a southwest course, leaving the river bottom and following the gently rolling plains and the beds of various streams which rise in the Sangre de Cristo range to the southwest. Over all these barren-looking plains large herds of cattle and sheep are grazed, and, wherever there is sufficient water for irrigation, the various cereals and many varieties of fruits are successfully cultivated.

Before reaching **Trinidad** the beautiful eruptive mountain group of the Spanish Peaks can be seen about 35 miles to the westward.[87]

They consist of two distinct peaks—an eastern (12,720 feet, 3,877 m.) and western (13,620 feet, 4,151 m.)—which rise out of a platform of Laramie Cretaceous and recently-discovered Eocene Tertiary beds (known as the Huerfano beds), about 10 miles east of the Sangre de Cristo mountains. They are of the laccolitic type, but not so regular or symmetrical as the Henry mountains. The laccolite, which spreads out in the softer shaly beds of the Colorado Cretaceous, is about 2,000 feet thick in its central portion, and sends out an intricate system of dikes through the overlying beds, which are so thoroughly metamorphosed that they were assumed by the first observers in this region to be of Carboniferous age.

The San Juan branch of the Denver and Rio Grande road crosses the Sangre de Cristo range into the San Luis park at Veta Pass just to the right of the Spanish Peaks.

At El Moro, to the right before reaching **Trininad**, are the coke ovens of the Denver and Rio Grande Railroad.

Trinidad owes its importance to the vicinity of most valuable beds of excellent coking coal, admirably situated for economical exploitation.

The thickness of Laramie measures, reckoning from the top of the sandstones to the Fort Pierre shales which outcrop at their base, is estimated to be about 1,800 feet. They contain 32 coal seams, which have an aggregate thickness of 105 feet, though the seams are by no means continuous throughout the field. The areal extent of the coal field is about a million acres.[88] The coal is either a slightly caking or else a coking coal, differing thus from the coals of the Denver basin at the same horizon, which are non-coking and quite porous and hygroscopic. The Laramie sandstones lie in an approximately horizontal position, and are capped, to the east of the road, by overflows of basalt, the greater mass of which forms Fishers Peak (9,460 feet, 2,843 m.), which is about 3,300 feet above the town to the southwest. This is the culminating point of the Raton hills, a broad, flat-topped ridge which extends eastward from the base of the mountains and forms the divide between waters flowing into the Arkansas on the north, and those flowing southward through New Mexico and Texas directly into the Gulf of Mexico.

The difference between slightly caking and coking coal in this field bears an evident relation to the magnitude of the neighboring eruptive masses, the coking coal occurring in the portion underlying the Fishers Peak overflow. In many parts of the field the injection of lava along a coal seam has produced either a dense natural coke or an impure powdery graphite. The outcrop of natural coke near Trinidad is probably two miles long. In other parts of the field outcrops of coke have been traced 4 and 5 miles. In a few places limited quantities of semi-anthracite have been produced. The neighboring sandstones are altered to quartzites.

From Trinidad the road rises, in a valley bordered by bluffs of Laramie sandstones and shales, to Starkville, at the west base of Fishers Peak.

Between Morley and Lansing the boundary line between Colorado and New Mexico is crossed. The edges of the coulées of basalt, capping the mesa, can be distinguished on the east.

The road now descends rapidly to Raton, and passes out into the broad, open valley of the Canadian river, eroded out of Middle Cretaceous shales.

At Maxwell the hills to the east are capped by basalt. This is on the well-known Maxwell Grant, one of the grants of land made by the Spanish authorities before New Mexico was ceded to the United States. At the time these grants were made land had little value, and the boundaries of the grants were very loosely defined by natural features, such as streams and watersheds, whose names have since been changed. The treaty of cession provided that the U. S. Government should confirm titles to lands thus granted. This particular grant, as surveyed for the persons who purchased it from the original grantees, covered in the neighborhood of a million acres. It was sold by them to Dutch capitalists. Since the sale there has been long litigation, based upon an asserted fraud in making the surveys of the boundaries. As these surveys had been accepted by the U. S. Land Office, the title of the present holders was finally confirmed.

Cimarron Creek, a tributary of the Canadian, which drains an interior monoclinal valley of the southern portion of the Sangre de Cristo range, is crossed at **Springer** station.

At **Wagon Mound** the road traverses a gap in basalt between Ocate mesa on the west and the Canadian hills on the east, both of which are capped by basalt. South of the Ocate mesa and west of the road are the Turkey mountains, which are formed of Carboniferous strata surrounded by Dakota sandstones, which dip gently away in all directions.

At **Tipton** the road has passed on to the Dakota sandstones, which underlie the shales. Directly west of this station, on the southern point of the Turkey mountains, is an extinct basaltic volcano with extremely perfect crater, whose rim is broken only by a narrow gap on the south side. According to Prof J. J. Stevenson [89] a coulée from this crater flowed south down Cherry Creek to the canyon of the Mora river, and then east along the bottom of this canyon to its junction with Canadian river, 30 miles to the eastward. In the upper part of the canyon only fragments of the coulée remain, but below it is continuous for nearly 20 miles. This flow occurred at a time when the Mora canyon, at its lower end, had been eroded to a depth of 860 feet (262 m.) below the top of its present walls. The basalt coulée then filled the bottom of this chasm to a depth of 400 feet; since which time the stream has eroded a new channel, partly in the basalt, and partly in the sandstone on one side of it. This latter channel, at the mouth of the canyon, is 230 (70 m.) feet below the base of the lava and 1,090 feet (322 m.) below the plain.

From **Shoemaker** to **Watrous** the road follows the valley of Mora river in Dakota sandstones. It then bends southwestward across a plain of Middle Cretaceous shales to **Las Vegas**, whose fine thermal springs lie a few miles west of the main line, at the foot of the steeper slope of the mountains. Here the railroad company has built a bathing establishment and a handsome hotel, which has several times been burned down.

From **Las Vegas** the road runs southward into Dakota sandstones, resting against the upturned Carboniferous beds which form the southeastern extremity of the Sangre de Cristo range.

At **Bernal** it turns westward, along the northern base of a mesa of Dakota sandstones, and passes into the valley of the Pecos river. It then bends northwestward and follows along the south side of the valley of the Pecos to near its source at **Glorieta**. It then bends southwestward, cutting through a projecting tongue of the Dakota mesa at **Canoncito**, and passes into the valley of Gallisteo creek.

At **Manzanares** it touches the southern point of the southwest extremity of the Sangre de Cristo range, which is formed of Archean with an encircling fringe of Carboniferous beds.

From **Lamy** a branch runs north across Tertiary beds (Santa Fe marls) 40 miles (65 km.), to Santa Fe, one of the oldest settlements in the United States. Like many other Spanish towns of the Southwest, it occupies the site of an Indian pueblo.

The road now passes into the Laramie coal-bearing rocks, in which some mines have been opened not far from **Ortiz.** The beds are, however, much broken by eruptive rocks, and the coal in some cases has been changed to anthracite. In the valley to the south, around the Placer mountains, is a considerable accumulation of gold-bearing gravels which might be profitably worked if it were not for the absence of water.

To the north of **Los Cerrillos,** in the hills of the same name, turquoise is found in rhyolite. The mines from which this mineral is obtained are supposed to have been worked by the Aztecs before the advent of the Spaniards.

Beyond **Wallace** the road enters the valley of the Rio Grande del Norte. This stream takes its rise in the various mountains which surround the great interior valley of San Luis Park. After leaving this well-watered and fertile valley it passes through narrows formed by coulées of basalt into the arid regions of New Mexico. To one coming from the east the portion of the valley followed by the railroad has a general aspect suggestive of that of the Nile. The river flows in a broad alluvial bottom, bounded by low bluffs at considerable distances back from the river. In the early summer, when the snows melt in the mountains, its waters spread out over the bottoms and leave a thin deposit of fine alluvial soil, which soon becomes brilliantly green with growing crops and fruits. As the river falls, the heat of summer gradually turns this verdure to a somber yellow or drab, except in a few favored spots. The old-world aspect of the valley is heightened by the quaint old Spanish towns, largely built of adobe or sun-dried bricks, and still more by the villages of the Pueblo Indians, built of stone, but plastered over the surface with mud.

[By G. K. GILBERT.]

East of **Albuquerque** stand the Zandia mountains[90] overlooking the Rio Grande with a bold mural front, even and straight, and little gashed by canyons. From the water to the crest the rise is 7,000 feet (2,100 m.). Except the crest the whole front is Archean, but from end to end there is a cornice of Carboniferous limestone a few hundred feet thick, that by its continuity shows the whole was raised in a single unshattered mass. The eastern face is of easier slope, but is less regular. The limestone band, that forms the persistent and almost level line of crest, is the edge of an eastward-dipping bed that is succeeded in that direction by superior Carboniferous and Mesozoic strata, all dipping from the mountain. But going westward from the Archean belt the unaltered rocks are not found in the same order. The tough Carboniferous limestone that holds its own so valiantly on the summit does not appear at the west, as it should if the structure of the mountain were anticlinal; but the first strata seen, after passing the valley gravels, which bury the base of the Archean wall, are of Cretaceous age, and they dip toward, rather than from, the ridge. The mountain is a great but simple monoclinal mass, bounded on the west by a profound fault, along the line of which is the river valley. The difference of level between the Carboniferous strata on the crest of the mountain and the dissevered fraction of the same strata, buried far below the Cretaceous rocks in the valley, is not less than 11,000 feet (3,300 m.), and something greater than this must have been the throw of the fault that separated them.

Thence, westward to **Flagstaff** and northward to the Grand canyon of the Colorado, the route lies exclusively within the Plateau region. The rocks are Cretaceous, Jura-Trias, Carboniferous, and volcanic. The Cretaceous system includes alternations of yellow sandstones with gray argillaceous shales, and there are occasional beds of coal. The maximum thickness is about 4,000 feet (1,200 m). The Jura-Trias is composed of sandstones with sandy shales and marls, and is everywhere characterized by brilliant colors. In the upper part of the system lenses of gypsum occur, and further west beds of salt are associated with them. At the west, calcareous beds have been found in the upper part, with marine shells, called Jurassic. Farther east, beds near the summit of the series have yielded plants referred to the Trias, and bones referred to the Jura. About the Zuñi uplift[91] the most conspicuous member of the system is the Wingate sandstone, a massive bed nearly 500 feet (150 m.) thick occurring near the middle of the system, which has here a total depth of about 3,500 feet (1,000 m.). Farther west another massive bed appears near the upper limit, and acquires topographic prominence. The Carboniferous system is characterized by two great beds of limestone, which weather but slowly and are thus rendered prominent in the topography. The lower appears in the Grand canyon of the Colorado; the upper, known as the Aubrey limestone, is seen in the Zuñi mountains, and constitutes a large portion of the plateau traversed between **Flagstaff** and the Grand canyon. Above this are a few hundred feet of bright-colored shales and sandstones, resembling the Jura-Trias rocks, but classed with the Paleozoic by reason of fossils of Permian type discovered in southern Utah. The same Permian facies characterizes fossils of the upper layer of the Aubrey limestone.

Associated with these are volcanic rocks with many modes of occurrence. Basic lavas, chiefly andesitic, rest upon the plateau in great cones built in Tertiary time, and left by the progressive degradation of the region upon pedestals of less resistant rock. Next in importance are lava flows, partly andesitic, but largely basaltic, and of such antiquity that the country has been eroded about them, leaving them as caps of small plateaus or mesas. In some cases, where the outpoured lava has been removed by erosion, the congealed lava of the conduit remains as a volcanic neck. Of more modern date, and often of extreme recency, although not historical, are basaltic cinder cones and basaltic coulées, which diversify the plateaus and course through the valleys.

Normally the strata are approximately level and interest centers in the variations from this attitude. The Zuni mountains were produced by a moderate uparching, the axis of which trends northwest and southeast. On the northeast side the dips are gentle; on the southwest steep. From the northwestern end of the uplift, a monoclinal fold— the Nutria fold [81]—continues for several miles, and then gradually fades out. Where a monoclinal fold is normally developed it is the equivalent of a fault, except that strata are flexed instead of fractured; in this case flexure and fracture are combined (Fig. 29). Farther westward another monocline, the Defiance fold, is seen, as well as minor flexures, and near the Grand canyon yet others.

The San Mateo Plateau is occupied by an immense composite sheet of lava 700 square miles (1,800 sq. km.) in area. By the degradation of the surrounding Cretaceous rocks it has received a relative altitude of more than 1,000 feet. Upon it stand numerous cinder cones and the great andesitic mass of Mount Taylor, and around about it a multitude of volcanic necks testify to its greater original extent. Mount San Francisco, likewise a great cone of andesite, was built upon Jura-Trias rocks, but these have been worn from the surrounding plain, together with all other strata down to the Aubrey limestone, and the talus of andesite almost completely conceals the sedimentary pedestal, so that the peak seems to spring 5,000 feet (1,500 m.) into the air from a plain of Aubrey limestone. All about it are more recent basaltic cinder cones and lavas.

Where the train enters the Plateau region, soon after crossing the Rio Grande, it follows a valley cut through the Cretaceous and into the Jura-Trias. Near Laguna it rises to the Cretaceous, but before that point is reached a fresh black lava stream appears in the valley, and this is kept in sight for many miles. Thence to McCartys the visible sedimentary rock is all Cretaceous. Northward at a short distance appears the San Mateo plateau, with Mount Taylor on its back, and in the face of the plateau are to be seen folds of Cretaceous strata formed previous to the volcanic eruption. Nearer by are several buttes, constituted by volcanic necks, and just beyond Cubero station is a dome-like lava cone. At McCartys another fresh lava stream is encountered in the valley, and this remains in sight until Bluewater is passed. A few miles beyond McCartys the train passes from Cretaceous to Jura-Trias rocks, but the characteristic features of the system do not appear until the lava is left behind. From near **Bluewater** to Mineral spring the train follows a monoclinal valley belonging to the northeastern flank of the Zuñi uplift. Beneath the track is the lower division of the Jura-Trias. At the right are a series of picturesque vermilion and orange bluffs and towers marking the outcrop of the Wingate sandstone. Farther back a second cliff line marks the outcrop of the basal sandstone of the Cretaceous. On the left rises the Zuñi range, exhibiting over large areas the upper surface of the Aubrey limestone, denuded of all superior strata and embodying on its surface the details of mountain structure. With favorable light, a system of minor faults and monoclinal flexures can be confidently traced from a distance.

The slopes of the valley are gentle, but its drainage is curiously divided, a part going to the Atlantic and another part to the Pacific. Quite unconscious that his train is surmounting a summit, the traveler here crosses the **Continental divide**.

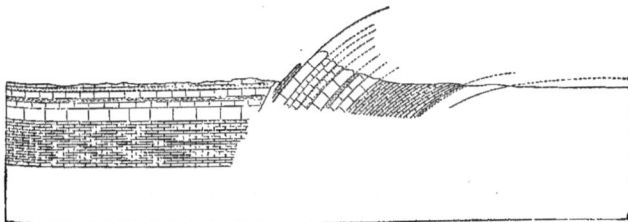

FIG. 20.—The Nutria fold.

At Mineral spring the Nutria fold is crossed. To the northward the strata can be seen to arch over and then suddenly descend. To the southward the fold is marked by a line of rocky crags. Thence to **Manuelito** the way lies among mesas of Cretaceous rock. A little west of **Manuelito** the Defiance fold is encountered—a monoclinal similar to that of Mineral spring, but of opposite throw. The Jura-Trias is again brought to the surface, and upon it the train continues to the crossing of the Little Colorado at **Winslow**. The beds here seen belong chiefly to the lower portion of the system, including perhaps also the upper of the strata referred to the Permian. Their varied hues in this desolate region have given name to the Painted desert, but near the Little Colorado they are partly concealed by alluvium.

From **Winslow** to **Flagstaff** the prevailing rock is Aubrey limestone, which rises westward from beneath the Jura-Trias. Its surface is disturbed by minor faults and folds, and diversified by basaltic mesa and cinder cones; and about San Francisco Mountain it is clothed by a noble forest of yellow pine (*Pinus ponderosa*).

Where the drainage lines cross low anticlines of the limestone they are sharply incised, and two such trenches are crossed by the railway.

The greater of these, Canyon Diablo, has a depth of 250 feet (75 m.) and has given name to the railway station just east of it.

At **Flagstaff** the mode of travel changes; the party is conveyed by wagons and saddle horses, takes its meals out of doors, and sleeps in tents.*

FLAGSTAFF TO THE GRAND CANYON.[92]

By G. K. GILBERT.

Flagstaff stands at the southern base of San Francisco Mountain. The road to the brink of the Grand canyon curves eastward about the mountain and then takes a northerly course. In the vicinity of the mountain are a great number of basaltic cinder cones from 500 to 1,500 feet (150 to 450 m.) in height, and most of these are so newly formed that their craters are well preserved. A few are not yet clothed with vegetation, and one, Sunset peak, is associated with a black lava field equally barren. The sides of this cone are of black lapilli, but its crest is tipped with red in a way to suggest that it catches the last rays of the setting sun. In the crest of another cone are artificial caves dug by Indians to serve as dwellings, but long abandoned.

The general altitude of the plateau is 7,000 feet (2,100 m.), and it is beautified by forests of pine, which give peculiar delight to eyes wearied with treeless plains and mesas, but water is nevertheless scanty. There are no streams, and springs are rare. Hull spring, the first one seen by the party, is a day's journey from Flagstaff and determines a point of encampment. The degradation of the country has here progressed several hundred feet since the spreading of a great field of basaltic lava, and the beds of resistant basalt cap a mesa facing toward the north. Beneath are soft shales of Permian age, and the water stored in the crevices of the basalt escapes slowly at the plane of contact.

The sloping Permian outcrop is sheathed by fragments of the basalt, which breaks away in huge blocks as it is sapped. One of these blocks, separated from the main cliff by a chasm a hundred yards across, was chosen as the site of an Indian village and covered with stone houses. The ruined walls remain, with fragments of pottery, and chips of flint and obsidian.

From Hull spring northward the road descends below the zone of trees and for 20 miles (30 km.) traverses a prairie floored by Aubrey limestone. Continuing on the same terrane, it then rises again into the zone of pine forest, and there remains till the brink of the canyon is reached. This timbered upland is the Cosnino plateau, the companion and counterpart of the Kaibab plateau north of the river. Indeed the two are parts of one uplift divided by the corrading river.

---

* The prophecy of "tents" was not verified; the party bivouacked, and was so unfortunate as to encounter storms of rain, snow, and wind.

The Aubrey limestone withstands erosion so much better than the Shinarump (Permian) shales and sands above it, that its surface has been denuded over a vast area and constitutes the floor of the country. Each great orogenic block stands as a plateau, each fault is marked by a cliff, each rock flexure is revealed in a topographic profile. Through this grand tectonic model runs the river's trench, revealing its anatomy in either wall.

The brink of the chasm is reached at a point nearly opposite Point Sublime, and the view does not differ in character from that sketched by Holmes.[93,94] The canyon is here broad and its walls are elaborately sculptured in sinuate terraces and cliffs, with buttresses, alcoves, pyramids, and spires innumerable. The Aubrey limestone and a firm sandstone beneath it, both pale in tint, constitute the first cliff, and a broad sloping terrace below it reveals a series of bright red shales and sandstones likewise of the Aubrey group. The foundation of this terrace and the material of the next cliff is a massive gray limestone, named the Red Wall because generally stained by pigment washed from above. The cliff is 1,000 feet (300 m.) high, and can be scaled only here and there in a deep recess. The next terrace is due to sandy shales of dingy hues, green, gray, and brown—the Tonto shales; and the Tonto sandstone forms a chocolate-colored cliff beneath.

The Aubrey beds and the Red Wall limestone carry Carboniferous fossils; there are Cambrian fossils in the Tonto shales and Tonto sandstones. Devonian and Silurian claim a narrow zone at the base of the Red Wall. Beneath the Tonto sandstone is a profound unconformity; it rests partly on Archean schists and granite, partly on basset edges of two great systems of Algonkian strata, comprising all ordinary types of clastic rocks, but yielding only tantalizing traces of contemporary organisms. These systems are themselves separated by an unconformity, and a still greater break divides the lower from the Archean.

PUEBLOS.

To the ethnologist this whole region is of special interest by reason of the opportunities afforded to study the institutions, arts, and architecture of Pueblo Indians. From the train may be seen the villages of **Isleta** and **Laguna**, besides a number of outlying farms and hamlets belonging to **Laguna**, and the whole region abounds with ruins and other vestiges of more extended occupation. Through an immense area, comprising the half of Colorado and Utah and the greater part of Arizona and New Mexico, there is scarcely an acre on which shards of Pueblo pottery may not be found. Though the houses are of stone, the mortar employed has no lithifying principle and yields to the storms, so that the walls of abandoned houses are apt to fall, but a multitude of structures built in shallow caves on the faces of cliffs have been preserved, enabling the student to assure himself of the identity of the culture represented by the ruins with that of the modern villages. A group of cliff dwellings is readily accessible from **Flagstaff**.

NOTES AND SKETCHES BY VISITING GEOLOGISTS.

NOTE ON WALNUT CANYON AND ITS CLIFF DWELLINGS.

By Prof. T. McK. Hughes.

FIG. 30.—Cliff dwellings in Walnut canyon.

Walnut canyon, 8 miles (12 km.) southeast from Flagstaff, Ariz., is a dry canyon in summer, but after the rains the channel is a roaring torrent. It is cut to a depth of 250 feet (85 m.) in the Aubrey or Upper Carboniferous rocks, which yielded to our party two species of *Productus* and some mytiloid lamellibranchs. The lower half (Fig. 31-2) consists of false-bedded sandstones of very uniform character, the breadth of the bands picked out by the principal divisional planes, whether of bedding or cross-bedding, being approximately the same from top to bottom, and the flat clean-cut wall showing no evidence of alternations of harder and softer beds. Not so the upper half of the side of the gorge, which consists of irregular beds of limestone, some of which protrude in ledges of varying thickness, while others have flaked off under the influence of the weather and receded into continuous rock-shelters, like those in the mountain limestone along the vale of Clwyd in North

Wales or the *abris* along the Vézère in Mesozoic rocks. As in the cliff of Dordogne, so in the Walnut canyon, these shelters have been occupied by people whom the necessities of primeval life or the exigencies of war drove to easily defended fastnesses in the rocks. They do not imply that the races who availed themselves of these strongholds were in a more rude stage than those who lived in huts upon the plateau. In the Middle Ages the Rock of Tayac was long held by the English, and so the remains left by the Indian tribe which occupied these cliff dwellings in the Walnut canyon did not lead us to infer that it was in an early stage of civilization. The fragments of pottery showed much artistic taste and skill. The stone arrowheads were highly finished and of the same type as those used by the Indians of historic time. The walls of the dwellings were of stone cemented by mortar, in which were pieces of pottery, showing that the building was still going on after the tribes had lived there for some time and scattered household rubbish about. Cobs of Indian corn told of cultivation, while their state of preservation confirmed the impression, derived from the mortar and other remains, that there was no ground for assigning the occupation of the cliff to any remote antiquity.

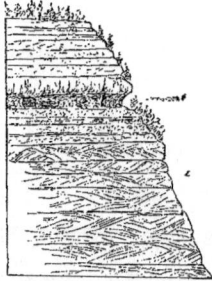

Fig. 30.—Section in Walnut canyon.

SECTION IN CONGRESS CANYON OPPOSITE POINT SUBLIME.

By Dr. FRITZ FRECH.

The succession of strata exposed in the Grand Canyon of the Colorado has already been several times described ([95,96,97]).

But all the sections hitherto published differ somewhat in regard to the petrographical character, the relative thickness and the disturbances of the strata, and as the section shown in Congress canyon has not been investigated before by any geologist, it may well be described at some length. The section is interesting not alone for the opportunity it affords to make a diagnosis of the petrographical character of the strata. There are probably few places on the earth where the geological phenomena of folding, faulting, uplifting, as well as numerous transgressions, may be so easily taken in at a single glance.

As the geologist passes upward from the bottom of the canyon to Hance's cabin he crosses the terranes of the following principal geological divisions: Archean, Algonkian, Cambrian, and Carboniferous. The more important divisions of these groups are marked by the Roman

numerals 1 to VII, in ascending order, in Fig. 32.

I. *A r c h e a n.* Gneiss with intrusive dikes of granite, pegmatite, and later diabase.

II. *A l g o n k i a n.* Grand Canyon series,* lying unconformably on the gneiss. Coarse red sandstones, shales, and conglomerates, with a sheet of diabase in the lower part (intrusive or surface flow?). Its total thickness, as observed at other points, is over 13,000 feet (4,000 m.); in this section only 300 to 400 metres are exposed.

III and IV. *Cambrian,* (about 1,000 feet (305 m.) thick) lying unconformably on the upturned edges of the Algonkian beds. The faults which traverse the latter beds terminate at the base of the Cambrian.

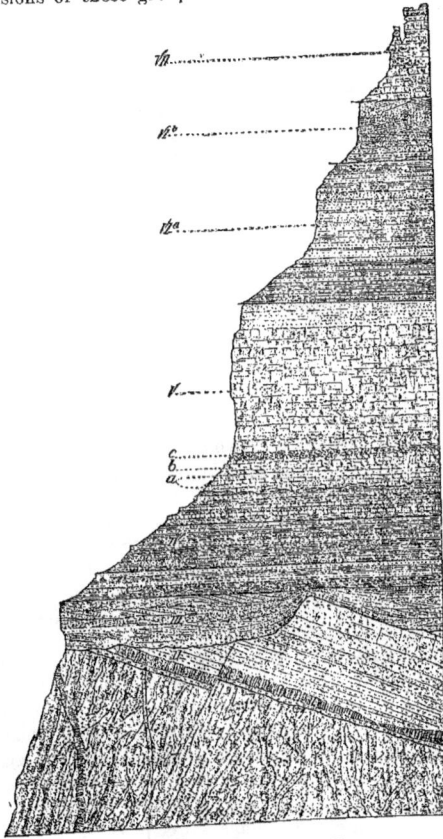

FIG. 32.—Section in Congress canyon.

---

*The Grand Canyon series is fully developed only in the main canyon and thins out in the smaller valley through which the trail from Hance's cabin descends. The Cambrian sandstone rests directly on the gneiss in the picturesque niche or panel called St. Gabriel's Cathedral (25 m. high), where the geologists camped the first night.

The Cambrian is subdivided into—

III. *Lower Tonto,* (about 300 feet) massive red sandstones with Scolithus at the top.

IV. *Upper Tonto,* (about 475 feet) greenish shales and shaly sandstones with impure limestones in the upper part.

The Devonian, which has been observed by Walcott[95] on the other side of the Grand canyon, between Marble canyon and Kaibab plateau, is altogether wanting at this point. The unconformity of erosion at the base of the Red Wall limestone, observed by the same geologist, is also very obscure here. There seems to be a gradual passage from the arenaceous sediments of the Tonto to the Carboniferous limestones above, but the fact that a sandstone bed is intercalated at the base of the Carboniferous limestones is consistent with Walcott's observations.

V. *Lower Carboniferous,* ca. 1,000 feet. Red Wall limestone.

(*a*) At the base: alternating sandstone and limestone.

(*b*) Red and white stratified (thin-bedded ?) limestone, in part crinoidal limestone with casts of *Spirifer striatus.*

(*c*) Bed of bluish brecciated limestone.

Above these comes the cliff of the Red Wall; it consists of a massive or obscurely bedded pure white limestone, which is superficially colored by waters seeping through the red beds of the Aubrey group.

VI and VII. *Upper Carboniferous* or Aubrey group.

VI (*a*). Lower Aubrey sandstone and shale, ca. 1,000 feet. The upper and lower parts are formed by thin-bedded sandstones and shales; in the middle there is a well defined cliff of massive sandstones.

VI (*b*). White Aubrey sandstone, ca. 400 feet, cross-bedded and forming a steep cliff distinctly visible from a long distance as a white band in the midst of the red rocks.

VII. Aubrey limestone and dolomite, ca. 500 feet (800 feet on the other side of the canyon). In the lower part is found a pure limestone which rests directly on the sandstone without any transition. The upper part consists of cherty limestones which contain a large *Allorisma* and some ill-preserved gasteropoda (*Euomphalus* and *Pleurotomaria*). In Coconino or Walnut canyon, near Flagstaff, the following Brachiopoda were found in the faint pink-colored dolomite in which are the famous cliff dwellings. (See p. 475.)

Productus Ivesii, Newberry, very common.

Productus aff., *scabriculus,* rare.

Spirifer (*Martinia*) *lineata,* Marsh, rare.

The specimens of *P. Ivesii* are very clearly related to the Upper Carboniferous form of *P. semireticulatus*, which Dr. E. Schellwien[*] has named var. *bathycolpos;* but in *P. Ivesii* the ribs of the shell and the spines are more strongly developed and the medial septum in the smaller valve is divided. *P. Ivesii* should be considered a variety of *P. semireticulatus* and not a distinct species.

The principal terraces are developed at the horizons of the Upper Tonto (IV), and at the base and top of the Lower Aubrey sandstone (VI *a*). They are indicated by the reduced slope of the cliff in the accompanying section (Fig. 32). The actual slope of these terraces could not be given for want of space.

The principal cliffs are formed by the Archean, the Grand Canyon and Lower Tonto series and the Red Wall. The picturesque carved (or incised) forms of the latter have a remarkable resemblance to the Dolomite cliffs of the Upper Trias in southern Tyrol.

The following table gives the results of a more detailed examination of the Tonto series (III and IV):

		Feet.
III.	1. Coarse, red, cross-bedded sandstones, with pebbles of quartz at the base and *Scolithus* at the top..ca..	260
	2. White sandstone, spotted black	30
IV.	3. Thin-bedded, brown, quartzitic sandstones and shales................	14
	4. Yellow and chocolate-colored, sandy shales, alternating with cross-bedded sandstones and conglomerates	45
	5. Well-defined bed of brown sandstone with glauconite(?), containing *Obollela*..	4
	6. Greenish or snuff-colored, shaly sandstone with worm tracks (*Cruziana*), ripple-marks and glauconite (?) (*Obollela polita* Hall ? *Lingula monticula* Walcott ?)...	65
	7. Same rocks as in 5 and 6. Large ripple-marks in the lower part, and in the upper part a glauconite layer 5 to 15 cm. thick. *Obollela* sp. ?..	32
	8. Snuff-colored sandstone, forming a well-defined cliff 8 feet high. In the upper part some calcareous shale..............................	85
	9. Greenish or snuff-colored shales with small ripple-marks, forming a gentle slope...	85
	10. Same beds as in 8. In the upper, a bed of limestone forming a small cliff..	44
	11. Greenish shales, with small cliff in the midst formed by a greenish sandstone..	85

The section shown in Congress canyon is in many respects incomplete. The following table, made by C. D. Walcott[96] at the Kaibab Plateau on the north side of the Grand canyon, gives the complete section (Roman numerals denote corresponding beds observed in Congress canyon):

	Feet.
Tertiary	815
Cretaceous...	3,095

---

[*] To whom I am indebted for the determination of the two Productus. Compare the figures of *P. semireticulatus* White, G. and G. Surveys W. of 100th meridian, Vol. IV. p. 111, Pl. VIII, fig. 1, and *Productus Sp.* ? Meek. U. S. G. Explor. 40th Par. Vol. IV p. 67, Pl. VII, fig. 6.

Jurassic (identified)		960
Jura-Trias		3,430

Carboniferous	Permian	854	
	VII. Upper Aubrey limestone	805	
	VI. Lower Aubrey limestone	1,485	4,106
	V. Red Wall limestone	962	
Devonian	Temple Butte limestone		94
Cambrian	IV. Tonto (calcareous and arenaceous shale)		1,050
	III. Tonto (sandstone)		
Algonkian	Chuar (shales and limestones)	5,120	
	II. Grand Canyon (sandstones with lava flows in upper part)	6,830	12,950
	Vishnu (bedded quartzites and schists)	1,000	

26,500

GENERAL CONCLUSIONS.

The interest afforded by the Grand canyon section is not restricted to the mere petrographical and stratigraphical diagnosis of its beds. It would be difficult to find another locality where the geological changes induced by faulting, folding, and volcanic eruptions can be so easily observed at a single glance. As Capt. Dutton remarks: [94] "Probably there is no instance to be found in the world where an unconformity is revealed upon such a magnificent scale, and certainly not amid such impressive surroundings." If we attempt to read the pages of the gigantic manual of geology, which is revealed to us from the brink of the great chasm, we may decipher the following episodes in its ancient history:

1. Energetic folding of the gneiss, and simultaneous or subsequent intrusion of pegmatite dikes, which have also been folded.

2. Complete erosion and planing down of the pre-Algonkian mountains; deposition of 13,000 feet (4,000 m.) of Algonkian sandstones and shales.

3. Eruption of diabases [the diabase dikes which cross the pegmatites lie conformably between the Algonkian sandstones (whether surface flows or intrusive sheets?) but do not penetrate the younger rocks].

4. Upheaval and faulting of the Algonkian sediments and inclosed eruptive beds.

5. Transgression of the Upper Cambrian (Tonto) sandstones; incomplete erosion and planing off of the Algonkian land surface. [The Algonkian beds are partially or totally wanting between the gneiss and the Tonto series; in other words, the sediments thin out over the ancient reefs of the Cambrian sea.]

6. The Silurian is wanting, and at the top of the irregularly distributed Devonian there is an unconformity by erosion (without discordance of stratification). These facts may be explained in either of three ways: (1) The Canyon area became dry land before and after Devonian time; (2) or, after deposition of the Silurian strata, they

CONGRESS CANYON BELOW HANCE'S CABIN.

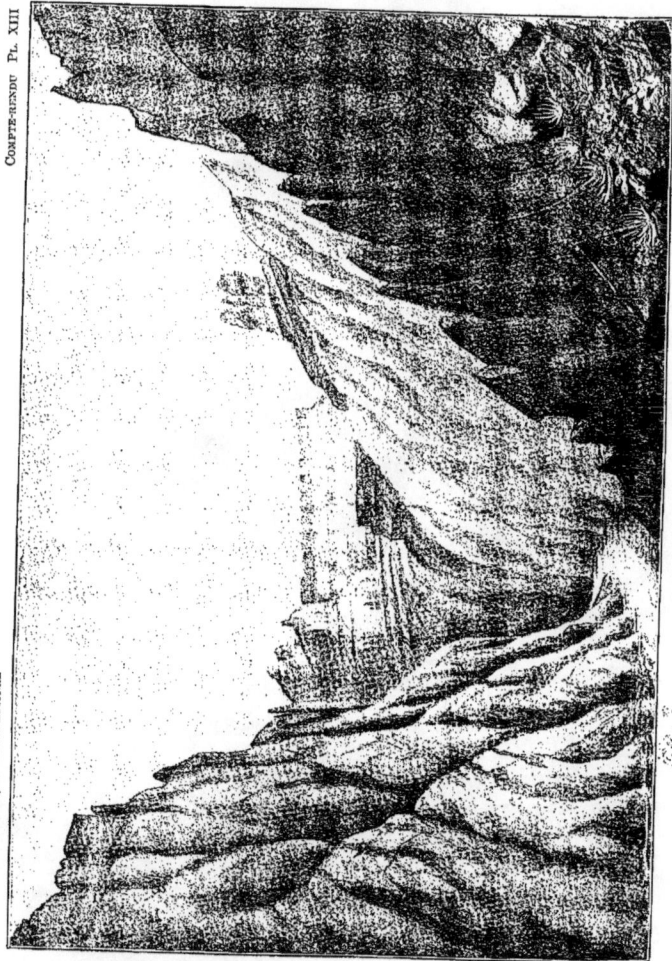

GRAND CANYON ABOVE THE MOUTH OF CONGRESS CANYON.

were subsequently removed by erosion; (3) or, no sedimentation whatever took place in the Silurian sea. In either case the changes observed were effected without any dynamic movement in the earth's crust. On the other hand, the unconformity between the Devonian and the Carboniferous is evidently due to some change in the sea level.

7. With the earlier Carboniferous begins a period of regular marine sedimentation, which went on uninterruptedly until the close of Permian time. Between Permian and Trias (in the Triassic, Shinarump, conglomerate), and again in Trias and Jura "we find instances of these peculiar unconformities by erosion without any unconformity of dip in the beds." [94]

In early Tertiary time a period of disturbance (folding and faulting) again set in in the Grand Canyon region, where no such changes had taken place since the Algonkian epoch. A typical instance of a flexure in the massive Red Wall limestone was observed in descending through the upper part of Congress canyon. The same phenomenon has been observed by Walcott, who pointed out the interesting fact that the pre-Cambrian and Tertiary movements took place on the same line of displacements. The upturning of the strata on the western side of the fault was effected during pre-Cambrian time; that on the eastern side during Tertiary time. [97]

It was at this time, probably, that the period of volcanic activity commenced, during which the San Francisco mountains were formed. After the formation of these andesitic mountains, basaltic eruptions took place, in which Dutton distinguishes an earlier and a later period. During the earlier, the plateau surface, now formed by the Aubrey limestone, was covered with Permian shales and clays. The basaltic flows of this time protected these Permian beds from subaerial erosion. Red Butte, which was passed during the second day's journey to the canyon, and consists of Permian capped by basalt, is a characteristic instance of such protection.

The cinder cones, lava flows, and ash beds near Flagstaff, passed during the first day's journey, belong to the very latest eruption and may be of very recent origin.

451 GE——31

BIBLIOGRAPHY.

[References in the text are made by means of numbers corresponding to the numbers appended to titles in this text.]

1. "On the Physical Structure of the Appalachian Chain, etc.," by W. B. and H. D. Rogers. Am. Jour. Sci. (1st series), Vol. XLIII, p. 177; Vol. XLIV, p. 359. 1842.
2. "On the surface geology of the Basin of the Great Lakes and the Valley of the Mississippi," by J. S. Newberry. Annals Lyceum Nat. Hist., Vol. IX. New York, 1869.
3. "Changes of level of the Great Lakes," by G. K. Gilbert. The Forum, Vol. V, June, 1888.
4. Report on the Geology of portions of Nevada, Utah, California, and Arizona, examined in the years 1871 and 1872, by G. K. Gilbert. U. S. Geographical Surveys west of the 100th Meridian, Vol. III, Part 1, 1875. pp. 21–156.
5. Results of a Biological Survey of the San Francisco Mountain Region and Desert of the Little Colorado, Arizona, by Dr. C. Hart Merriam. U. S. Department of Agriculture, North American Fauna, No. 3. Washington, 1890.

APPALACHIANS.

6. "Three formations of the Middle Atlantic slope," by W J McGee. Am. Jour. Sci. (3d series), Vol. XXXV, 1888.
7. "The Geology of the head of Chesapeake Bay," by W J McGee. U. S. Geol. Survey, Seventh Ann. Rept. 1888.
8. "Paleontology of New York," Vol. III. Introduction.
9. "Observations on the origin of some of the earth's features," by J. D. Dana. Am. Jour. Sci. (2d series), Vol. XLII, 1866, pp. 205, 252.
10. Report of the Brit. Assn. for 1842; Trans. Roy. Soc. of Edinburgh, Vol. XXII, p. 463, 1856, and Geology of Pennsylvania, Vol. II, p. 911, 1858.
11. "Coal and its Topography," by J. P. Lesley. Philadelphia, 1856.
12. "Rivers and valleys of Pennsylvania," by Wm. M. Davis. National Geographic Magazine, Vol, I, 1889.
13. "The Petrography and Structure of the Piedmont Plateau in Maryland," by Geo. H. Williams. Bull. Geol. Soc. Am., Vol. II, 1890, p. 301.
14. "Celestite from West Virginia," by G. H. Williams. Amer. Jour. Sci. (3d series), Vol. XXXIX, 1890, p. 183.; Zeitsch. für Krystallographie, Vol. XVIII, 1890, p. 1.
15. "The Mannington Oil Field," by I. C. White. Bull. Geol. Soc. Amer., Vol. III, pp. 187–216.
16. Stratigraphy of the Bituminous Coal field in Pennsylvania, Ohio, and West Virginia, by I. C. White. U. S. Geol. Survey, Bull. No. 65. Washington, 1891.

OHIO.

17. Ohio Geological Survey, Vol. I, 1873, p. 314 et seq. (Zanesville), p. 611 (Tiffin).
18. Ohio Geological Survey, Vol. II, 1874, p. 452 (Defiance).
19. Ohio Geological Survey, Vol. III, p. 543 (Bellaire).

20. Ohio Geological Survey, Vol, v, 1884, p. 618 (Bellaire), 96, 718 (Zanesville), 750 et seq. (Newark).
21. Ohio Geological Survey, Vol. vi, 1888, p. 60 (gas and oil), 104, 192 (Fostoria), 197, 784 (Tiffin), 251 (Defiance), 253 (Deshler), 288 (N. Baltimore), 314 (Zanesville), 350 (Chicago Junction), 365 (Mansfield), 366 (Mount Vernon), 623 (Quaker City).
22. Geological Survey of Ohio, First Annual Report, 1890, 60 et seq. (gas and oil), 94–104 (rock pressure), 254 (Barnesville), 231 (Newark gas field), 244 (Mount Vernon), 245 (Mansfield), 140, 305, et seq. (N. Baltimore). U. S. Geological Survey, Eighth Ann. Rept., Part II, pp. 453 et seq., Plates LIV–LVIII.
23. Indiana Geological Survey, Thirteenth and Fourteenth Ann. Repts., 1884–'85.
24. Second Geological Survey of Pennsylvania, PP. "Permian flora."

MISSISSIPPI VALLEY.

25. "Terminal moraine of the second glacial epoch," by T. C. Chamberlin. U. S. Geol. Survey, Third Ann. Rept., pp. 291–402.
26. "Driftless area of the Upper Mississippi Valley," by T. C. Chamberlin and R. D. Salisbury. U. S. Geol. Survey, Sixth Ann. Rept., pp. 199–322.

MINNESOTA.

27. "Catalogue of the flora of Minnesota," by Warren Upham. Pt. IV of the Twelfth (1883) Ann. Rep. and Bull. No. 3 of the Geol. and Nat. Hist. Survey of Minnesota.
28. "The iron ores of Minnesota," by N. H. and H. V. Winchell. Bull. No. 6, Geol. and Nat. Hist. Survey of Minnesota, 1891. Amer. Geol., Nov., 1889.
29. "On some peculiarly spotted rocks from Pigeon Point, Minn.," by W. S. Bayley. Amer. Jour. Sci., 1888 (3d ser.), Vol. XXXV, pp. 388–393. "A quartz keratophyre from Pigeon Point and Irving's augite syenite," by W. S. Bayley. Amer. Jour. Sci., 1889 (3d ser.), Vol. XXXVIII, pp. 54–63.
30. "Paleontology," by Leo Lesquereux. Geol. Survey of Minnesota, Vol. III of the final report.
31. Fifth (1876) Ann. Rept. of the Geol. and Nat. Hist. Survey of Minnesota, by N. H. Winchell. pp. 35–37.
32. "The upper beaches and deltas of the glacial lake Agassiz," by Warren Upham. U. S. Geol. Survey, Bull. No. 39, 1887.
33. "The recession of the Falls of St. Anthony," by N. H. Winchell. Q. J. S. of London, vol. 34, 1878, pp. 886–901; 5th (1876) Ann. Rept. Geol. Survey of Minnesota, pp. 156–189; Final Rept. Geol. Survey of Minnesota, vol. 2, 1888, pp. 313–341.
34. "Account of a deserted gorge of the Mississippi near Minnehaha Falls," by Uly S. Grant. Amer. Geol., Vol. v, 1890, p. 1.

Minnesota (General reference).

Geological and Natural History Survey of Minnesota.
(a) Annual Reports, I (1872) to XVIII (1889).
(b) Bulletins I to VI.
(c) Final Reports I and III, edited and largely written by N. H. Winchell, assisted by Warren Upham.
Minnesota Academy of Natural Sciences.
Bulletin, Vols. I to III.
Hall, C. W.; "The distribution of the granites of the Northwestern States and their lithological characters;" A. A. A. S., Cleveland Meeting, 1889, p. 189.

Irving, R. D.; "The copper-bearing rocks of Lake Superior;" Mon. v., U. S. G. S., 1893. "Preliminary paper on an investigation of the Archæan formations of the Northwestern States;" 5th An. Rept. U. S. G. S., 1885, pp. 175–242.

Kloos, J. H.; "Geologische Notizen aus Minnesota;" Zeitsch. d. d. geol. Gesell., 1871. pp. 417–448. Translated in 10th (1881) Ann. Rept. Geol. Sur. Minn., pp. 175–200.

Lawson, A. C.; "Report on the Geology of the Lake of the Woods region, with special reference to the Kewatin (Huronian?) belt of Archæan rocks;" Geol. and Nat. Hist. Sur. Canada, Vol. I, n. s. C. C., 1886, pp. 1–151. "Report on the Geology of the Rainy Lake region;" Geol. and Nat. Hist. Sur. Canada, Part F, Ann. Rept., III, 1887.

Owen, R. D.; Report of a geological survey of Wisconsin, Iowa, and Minnesota. 4to. 1852.

Streng, A. and J. H. Kloos; "Ueber die krystallinischen Gesteine von Minnesota in Nord-Amerika;" Neues Jahrb. f. Min. Geol. u. Pal., 1877, pp. 31–56, 113–138, 225–242. Translated in the 11th (1882) Ann. Rept. Geol. Sur. Minn., pp. 30–85,

Winchell, A.; "Conglomerates inclosed in gneissic terranes;" Amer. Geol., Vol. III, 1889, pp. 153–165, 256–262. "Some results of Archæan studies;" Bul. Geol. Soc. Amer., I, pp. 357–394. "A last word with the Huronian;" Bul. Geol. Soc. Amer., Vol. II, pp. 85–124.

Winchell, H. V.; "The diabasic schists containing the jaspilyte beds of northeastern Minnesota;" Amer. Geol., Vol. III, 1889, pp. 18–22.

Winchell, N. H.; "Some thoughts on eruptive rocks, with special reference to those of Minnesota;" A. A. A. S., Cleveland Meeting, 1889, pp. 212–221.

MONTANA AND THE YELLOWSTONE PARK.

35. "Geology of the Crazy Mountains, Montana," by J. E. Wolff. Bull. Geol. Soc. Amer., Vol. III, pp. 445–452.

36. "Laramie and overlying Livingston formation," by Walter H. Weed. U. S. Geol. Survey, Bull. No. 105.

37. "Cinnabar and Bozeman coal fields of Montana," by Walter H. Weed. Bull. Geol. Soc. Amer., Vol. II, pp. 349–364. "Two Montana Coal Fields," by Walter H. Weed. Bull. Geol. Soc. Am., Vol. III, pp. 301–330.

38. "Glaciation of the Yellowstone Valley north of the National Park," by Walter H. Weed. U. S. Geol. Survey, Bull. No. 104.

39. Fifth Ann. Rept. U. S. Geological and Geographical Survey, 1872, p. 162.

40. "Geology of the Yellowstone Park," by W. H. Holmes; "Thermal Springs of the Yellowstone Park," by Dr. A. C. Peale. Twelfth Ann. Rept. U. S. G. and G. Survey, Part II. Washington, 1883.

41. "Geological history of the Yellowstone National Park," by Arnold Hague. Trans. Am. Inst. Mg. Engrs., Vol. XVI, 1888, p. 783.

42. "Obsidian Cliff," by J. P. Iddings. U. S. Geol. Survey, Seventh Ann. Rept., Washington, 1888, pp. 249–295. "Spherulitic crystallization," by J. P. Iddings. Bull. Phil. Soc. of Wash., Vol. II, Washington, 1891, pp. 445–464.

43. "Formation of travertine and siliceous sinter," by Walter H. Weed. U. S. Geol. Survey, Ninth Ann. Rept., 1890, p. 613. "The Diatom beds of the Yellowstone Park," by Walter H. Weed, Bot. Gazette, Vol. XIV, 1889, pp. 117.

44. "Montana coal fields," by G. H. Eldridge. Tenth Census, Vol. XV. Mining Industry, pp. 739–757.

45. "Butte, Mont.," vide: "Grangrevier von Butte," by G. vom Rath. N. Jahrb. f. Min., Geol., u. Pal., Jahrg. 1885, Bd. 1. "Western Montana," by C. O. Ziegenfuss. Butte, Mont., 1886. "Silver mining and milling at Butte," by W. P.

Blake.. Trans. Am. Inst. Mg. Engrs., Vol. XVI, 1887, p. 38. "Notes on the geology of Butte," by S. F. Emmons. Ibid., p, 49; also p. 830. "Association of minerals in the Gagnon vein, Butte," by Richard Pearce. Ibid., p. 62. "The Rainbow lode, Butte City," by W. P. Blake. Ibid., p. 65.

IDAHO.

46. "Geological sketches," by A. Geikie, 1882, p. 237.
47. "The Nampa Image." Proc. Boston Soc. Nat. Hist., Vol. XXIV, June 1890, p. 424.
48. "Shoshone Falls," U. S. Geol. Exploration of the 40th Parallel, Vol. I, 1878, p. 592, Pls. XVI to XIX.
49. "Lake Bonneville," by G. K. Gilbert. U. S. Geol. Survey, Mon. I, Washington, 1880, pp. 97, 137, 184, 306–311, 340–349, Pls. I, IX, XX, XXIX, XLIII, XLIV.
50. "Utah," by Marcus E. Jones. Rep. of Bur. of Stat. (Treas. Dept.) on Internal Commerce, Washington, 1890, pp. 933–936.
51. "Wasatch Mountains." Reports of the U. S. Geol. Exploration of the 40th Parallel. "Systematic Geology," by C. King, Vol. I, 1878, pp. 44, 154, 264, 293, 296, 304, 316, 434, 458, 490, 745, 758, Pls. I, X, XII, XXII. "Descriptive Geology," by A. Hague and S. F. Emmons, vol. II (1877), pp. 340–488, Pls. XIV, XV, XVI.
52. "Rocky Mountain Protaxis and the Post-Cretaceous Mountain making along its course," by J. D. Dana. Am. Jour. Sci., 3d ser., Vol. XL, Sept., 1890, p. 181.
53. "Cambrian faunas of N. America," by C. D. Walcott. U. S. Geol. Survey, Bull. 30, 1886, p. 38.
54. "Geology of the high plateaus of Utah," by C. E. Dutton. U. S. G. and G. Survey, Washington, 1880.
55. "Geology of the eastern portion of the Uinta Mountains," by J. W Powell. U. S. G. and G. Survey, Washington, 1876, p. 196.
56. E. E. Howell, U. S. Geog. Surveys west of the 100th Meridian. Vol. III, Washington, 1875, p. 235.

ROCKY MOUNTAINS OF COLORADO.

57. "Report on the Grand River district," by A. C. Peale. Ninth Ann. Rept. of the Hayden Survey for 1875, pp. 46–49, 59–62, 92–97, Pls. VII, VIII.
58. "Report on the Grand River district," by A. C. Peale. "View of Little Rock Cliffs," by W. H. Holmes. Tenth Ann. Rept. of the Hayden Survey for 1876, pp. 74, 161, 195, Pl. IV, sec. 3, X, fig. 3.
59. "Orographic movements in the Rocky Mountains," by S. F. Emmons. Bull. Geol. Soc. Amer., Vol. I, 1889, pp. 245–286.
60. "Orographic and structural features of Rocky Mountain geology," by R. C. Hills. Proc. Colo. Sci. Soc., Vol. III, 1890, pp. 362–458.
61. "Geology of Colorado coal deposits," by Arthur Lakes. Ann. Rept. State School of Mines, Golden, Colo., 1889, pp. 137, 154, 158.
62. "Geology of the Elk Mountains," by W. H. Holmes. Eighth Ann. Rept. of the Hayden Survey for 1874, pp. 59–72.
63. "Geology of Eagle River," by A. C. Peale. Ibid, pp. 79–84, Pl. II.
64. "Extinct volcanos in Colorado," by Arthur Lakes. Amer. Geol., Vol. V, 1890, p. 38.
65. "Some Colorado ore deposits," by S. F. Emmons. Proc. Col. Sci. Soc., Vol. II, 1886, p. 100.
66. "Battle Mountain gold deposits in quartzite," by Franklin Guiterman. Proc. Col. Sci. Soc., Vol. III, 1890, supplement.
67. "Geology and mining industry of Leadville," by S. F. Emmons. U. S. Geol. Survey, Second Ann. Rept., 1882, pp. 203–287. U. S. Geol. Survey, Mon. XII, 1886, with atlas. See also; Trans. Am. Inst. Mg. Engrs., Vol. XIII. p. 384; Vol. XIV, pp. 181–273; Vol. XV, p. 125; Vol. XVI, p. 805; Vol. XVIII, p. 145.

68. "Moraines of Upper Arkansas valley," by F. V. Hayden and W. H. Holmes. Eighth Ann. Rept. of the Hayden Survey for 1874, pp. 47–53.
69. "Topaz and garnet in lithophyses of rhyolite," by Whitman Cross. Proc. Col. Sci. Soc., Vol. II, p. 61; and Amer. Jour. Sci., Vol. XXXI, p. 432, 1886.
70. "Wahrnehmungen in der Umgebung von Silver Cliff, Salida, Leadville, und Gunnison, Colo.," by G. vom Rath. Verhandlungen des naturhistorischen Vereins, Bonn, 1885, Correspondenz Blatt No. 2.
71. "Preliminary notes on the discovery of a vertebrate fauna in Silurian (Ordovician) strata," by Chas. D. Walcott. Bull. Geol. Soc. Am., Vol. III, pp. 153–172.
72. "American Jurassic Dinosaurs," by O. C. Marsh. Amer. Jour. Sci., Vol. XIV, 1877, p. 513; Vol. XV, p. 241; Vol. XVI, p. 411; Vol. XVII, p. 86; Vol. XVIII, p. 501; Vol. XIX, p. 253.
73. "The Florence oil field, Colorado," by G. H. Eldridge. Trans. Am. Inst. Mg. Engrs., Vol. XX, pp. 442–462, 1891. With geological map including the Canyon City coal field and profile sections.
74. "Geological map of Colorado Springs and vicinity." Eighth Ann. Rept. of the Hayden Survey for 1874, p. 40. Also, Hayden Atlas of Colorado, sheet XIII.
75. "Mineral springs of the United States," by Dr. A. C. Peale," U. S. Geol. Survey, Bull. 32, 1886, p. 192.
76. "Contributions to the mineralogy of the Rocky Mountains," by Whitman Cross and W. F. Hillebrand. U. S. Geol. Survey, Bull. 20. Describes particularly zeolites of Table Mountain at Golden and the fluorides of St. Peter's dome, with some other species of the Pike's Peak region.
77. "The Tertiary Lake basin at Florissant, Colo.," by S. H. Scudder. U. S. G. and G. S., Bulletin Vol. 2, No. VI, 1881, pp. 279–300.
78. "Geologische Briefe aus America," by G. vom Rath. Sitzungsberichte der niederrheinischen Gesellschaft für Natur. und Heilkunde zu Bonn. January 7, 1884.
79. "The Denver Tertiary formation," by Whitman Cross. Proc. Col. Sci. Soc., Vol. III, 1888, pp. 118–133. Also: Amer. Jour. Sci. Vol. XXXVII, 1889, pp. 261–282, with a geological map of the Denver region.
80. "The Quaternary of the Denver basin," by Geo. L. Cannon, jr. Proc. Colo., Sci. Soc., Vol. III, pp. 48–70, 1888.
81. "On some stratigraphical and structural features of the country about Denver, Colo.," by G. H. Eldridge. Proc. Colo. Sci. Soc., Vol. III, p. 86. Bull. Philos. Soc. Wash. Vol. XL, p. 247.

KANSAS AND ADJOINING STATES (GENERAL REFERENCE).

Ninth Ann. Rept. Hayden Survey for 1875, p. 277.
Biennial reports of Kansas State Board of Agriculture. First, by B. F. Mudge. Third and fourth, by O. St. John. Fifth, sixth, and seventh, by R. Hay.
Trans. Acad. of Sciences, vols. 9, 10, 11, and 12.
"Artesian wells on the Great Plains," by Profs. White and Aughey, U. S. Dept. of Agriculture, 1882.
Artesian well investigation, Senate Ex. Doc. 222, Fifty-first Congress, 1890.
Artesian well investigation, final report, Senate. Ex. Doc. No. 1, first session Fifty-second Congress.
"Geological reconnaissance in southwestern Kansas," by Robert Hay, U. S. Geol. Survey, Bull. No. 57, Wash., 1890.
"An old lake bottom," by L. E. Hicks, Bull. Geol. Soc. Amer., Vol. II, p. 25.
"Carboniferous Cephalopods," by A. Hyatt, Geol. Survey of Texas, Vol. II, 1890, p. 329.

IOWA (GENERAL REFERENCE).

Reports of Davenport Academy of Sciences.
Reports of Second Geological Survey of Iowa.
U. S. Geol. Survey, Eleventh Ann. Rept. 1891, pp. 189–577.

THE REGION OF THE GREAT LAKES.

82. "The surface geology of the basin of the Great Lakes and the valley of the Mississippi," by J. S. Newberry. Annals of the Lyceum of Natural History, New York, 1869. Republished in Am. Nat., Vol. IV, June, 1870, pp. 210–213, Salem, Mass., 1871.
83. "Origin of the basin of the Great Lakes," by J. W. Spencer. Quar. Jour. Geol. Soc. (London), Vol. XLVI, 1890. Reprinted in Amer. Geol., Vol. IV, 1891.
84. Geological Survey of Canada, report of progress to 1863. Sir W. E. Logan, director, Montreal, 1863.
85. "The history of the Niagara river," by G. K. Gilbert. Sixth Ann. Rept. of the Commissioners of the State Reservation at Niagara, Albany, N. Y., 1890.
86. "The Niagara gorge," by Dr. Julius Pohlman. Proc. A. A. A. S., 35th meeting, (Buffalo), pp. 221–222. (Abstract) Trans. Am. Inst. Mg. Engrs., Vol. XVII, p. 322.

GRAND CANYON EXCURSION.

87. "Spanish Peaks region", R. C. Hills. Proc. Col. Sci. Soc., Vol. III, pp. 148, 224.
88. "Coal measures of Southeastern Colorado and Northeastern New Mexico," by P. H. Van Diest. Proc. Col. Sci. Soc., Vol. III, p. 185.
89. U. S. Geog. Surveys west of 100th Meridian, Vol. III (supplement), Washington, 1881, p. 171.
90. "Mount Taylor to Zandia Mountains," U. S. Geog. Surveys west of 100th Meridian, Vol. III, p. 288, Fig. 122.
91. "Mount Taylor and the Zuñi Plateau," by Clarence E. Dutton. U. S. Geol. Survey, Sixth Ann. Rept., pp. 111–183, Washington, 1885.
92. Explorations of the Colorado River of the West, 1869–1872, by J. W. Powell, Washington, 1875.
93. U. S. Geol. Survey. Second Ann. Rept. Plates XXXI, XXXII.
94. Tertiary History of the Grand Canyon district, with atlas, by C. E. Dutton. U. S. Geol. Survey, Monograph II, Washington, 1882, pp. 207, 211, 256, Pl. XLI.
95. Amer. Jour. Sci. III, Vol. XX, 1880, p. 221. Ibid., Vol. XXVI, 1883, p. 438.
96. Cambrian Correlation Papers, by C. D. Walcott. U. S. Geol. Survey, Bull. 81, pp. 220–221.
97. "Study of a Line of Displacement in the Grand Canyon," by C. D. Walcott. Bull. Geol. Soc. Amer., Vol. I, p. 49.

PUEBLOS (GENERAL REFERENCE).

"Houses and House-life of the American Aborigines," by Lewis H. Morgan. Contributions to North American Ethnology, Vol. IV, Washington, 1881.
Papers (American series) and Reports of the Archæological Institute of America, containing articles of Ad. F. Bandelier. Boston and Cambridge, 1880–1890.
History of Arizona and New Mexico, by Hubert H. Bancroft. San Francisco, 1889.
"Study of Pueblo Architecture, Tusayan and Cibola," by Victor Mindeleff. Bureau of Ethnology, Eighth Ann. Rept., Washington, 1892.

C.

EXCURSION TO LAKE SUPERIOR.

PRE-CAMBRIAN GEOLOGY OF THE LAKE SUPERIOR REGION.

By C. R. Van Hise.

489

TABLE OF CONTENTS.

PLATES.

FIGURES.

N. B.—Figures above the line in the text refer to titles in the Bibliographic list page 511.

PLATE XIV.

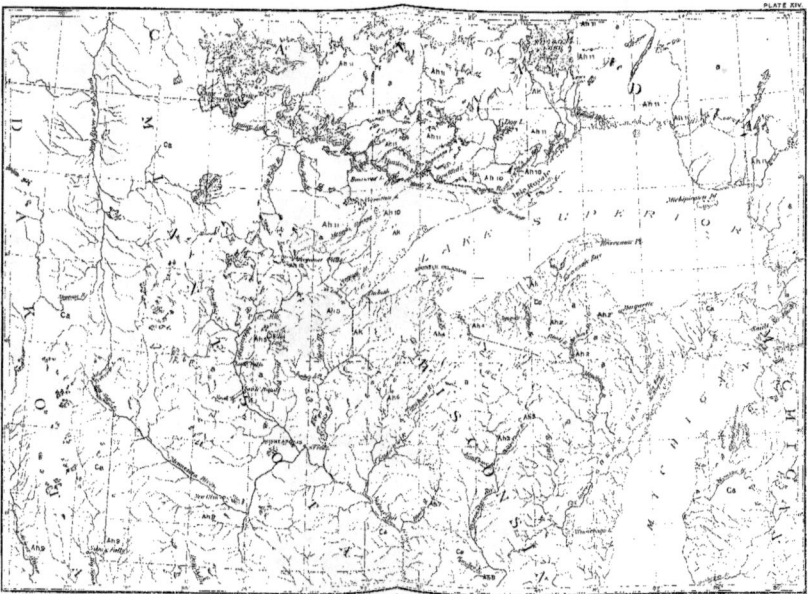

GEOLOGICAL MAP OF THE LAKE SUPERIOR REGION
SHOWING THE CAMBRIAN AND CRYSTALLINE ROCKS.
Compiled from Official Methed U. S. and Canadian Surveys

SKETCH OF THE PRE-CAMBRIAN GEOLOGY SOUTH OF LAKE SUPERIOR, WITH REFERENCES TO ILLUSTRATIVE LOCALITIES.[*]

By C. R. Van Hise.

The ancient formations south of Lake Superior may be grouped into five great divisions: The Basement Complex, the Lower Huronian, the Upper Huronian, the Keweenawan, and the Lake Superior Sandstone. These five divisions are separated by unconformities of great magnitude, two of the unconformities at least being of the first order. According to the classification adopted by the United States Geological Survey, the Basement Complex is Archean; the Lower Huronian, Upper Huronian, and Keweenawan constitute the Algonkian for this region, and the Lake Superior sandstone is Cambrian.

Basement Complex.—The characteristic rocks of the Basement Complex are (1) light-colored granites and gneissoid granites; and (2) dark-colored, finely foliated or banded gneisses or schists. These are cut by various basic and acid intrusives, many of which are not different from eruptives found in the later series, with which they are doubtless in part continuous.

The granites and gneissoid granites are placed together, because between the two there are constant gradations. If one speaks accurately and includes among granites only those rocks which are completely massive, the gneissoid granites include the major portion of the granitic rocks; for in large exposures it is usually possible to find some evidence of foliation. The granitoid areas are of greatly varying sizes, running from small patches to those many miles in diameter. When everywhere surrounded by the schistose divison of the Basement Complex they frequently have oval or ovoid forms. In nearing the outer border of the granitoid areas the foliation often becomes more and more prominent, and near the edge of an area the rock frequently passes into a well-laminated gneiss.

The schistose rocks include fine-grained hornblende-gneisses, mica-gneisses, chlorite-gneisses, and various green schists formerly supposed to be sedimentary but now known to be greatly modified basic and acid igneous rocks. The schists have usually a dark-green or black color, are strongly foliated, and the variations in strike and dip of this foliation within small areas is very great. Not infrequently these schistose rocks are traced with all gradations into massive igneous rocks. The contacts between the schistose division and the gneissoid division of the Basement Complex are usually those of intrusion, the granitoid

---

[*] The illustrations are from the publications of Brooks, Irving, Williams (G. H.), Pumpelly, and Van Hise.

rocks being the later. In passing from a schistose to a granitoid area, small pegmatitic looking veins of the granite are usually first found. In going onward these veins become more numerous, and after a time appear unmistakable dikes of granite, which multiply in number and size in approaching the granite area, until the granite is found in great bosses. Here we have, perhaps, a nearly equal quantity of schistose and granitoid rocks, and in this intermediate zone the schists may be found as a mass of blocks within the granite, sometimes at but small distances from their original positions, the whole having frequently a somewhat conglomeratic appearance. However, these pseudo-conglomerates, so well described by Lawson,[14] grade more or less rapidly on the one hand into the schists, and on the other into the solid gneissoid granite. The complete change may occur within a short distance or it may take a mile or more.

The Basement Complex is then composed of intricately interlocking areas of granitoid rocks and schistose rocks. Moreover all of these rocks are completely crystalline. None of them show any unmistakable evidence of having been derived from sedimentaries, but many can be traced with gradations into massive rocks, and therefore the greater proportion of them are igneous, if a completely massive granular structure be proof of such an origin.

The Basement Complex is the most widespread of any of the Lake Superior systems, and in many areas good sections may be seen which well illustrate the characters of the rocks and their relations. A few of the more accessible may perhaps be mentioned. Both divisions of the Basement Complex may be readily studied near the Menominee, Marquette, and Penokee iron districts.

In the Menominee district (Pl. xv) the schists and granites of the Basement Complex may be seen in typical exposures both south and north of the Huronian rocks. The northern schist area is well exposed at Twin and Four Foot Falls (Pl. xvi), 3 to 5 miles northwest of Iron mountain, while the northern granite appears some miles northeast of Iron mountain.[1,27] The southern schists are finely exposed southeast of Iron mountain at the Upper Quinnesec falls (Pl. xvii) and Lower Quinnesec falls (Pl. xviii), at Sturgeon falls (Pl. xix), and also at and near the crossing of the Menominee by the Milwaukee and Northern Railway.[1,27] At the Horse Race above Upper Quinnesec falls and about one-half mile to the south, before the solid granite is reached, numerous dikes of this rock may be seen cutting the schists.[27] On the Milwaukee and Northern Railway the granite rocks appear about four miles south of Iron mountain. At this point and east and west of the railroad the intrusive relations between the granitic and schistose rocks are finely shown.

In the Marquette district, on the railways north of Republic, Mich.,

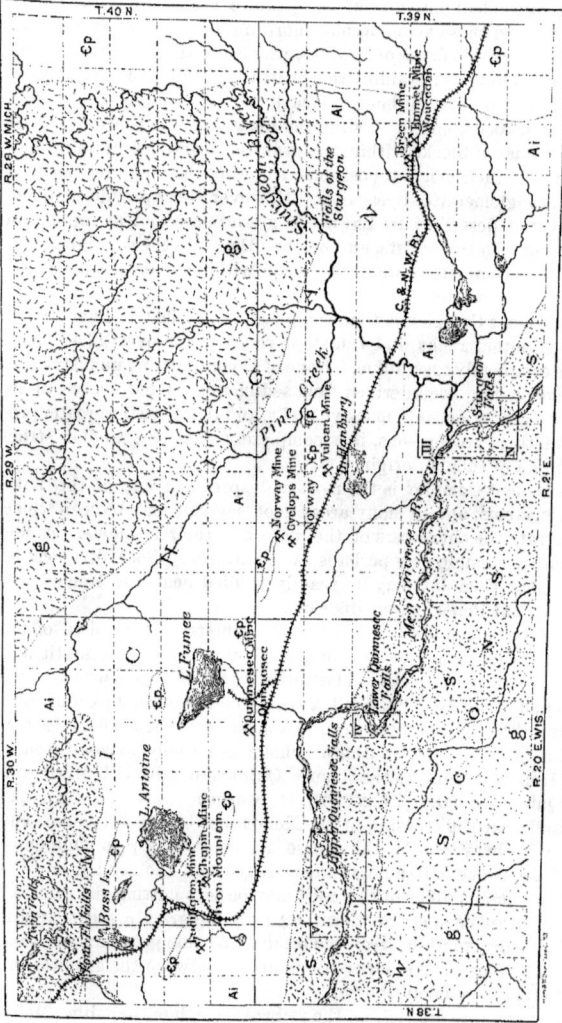

OUTLINE GEOLOGICAL MAP OF THE MENOMINEE IRON REGION.

Compiled by R. D. Irving from maps by T. B. Brooks, C. E. Wright and C. Rominger and from original observations.

ARCHEAN.

Granite and Gneiss.   Greenstone Schists.

g   S

ALGONKIAN.

Iron Bearing Series.
(Schists, Limestones & Ferruginous Schists.)

Ai

CAMBRIAN.

Potsdam Sandstone.

€p

Scale of miles.

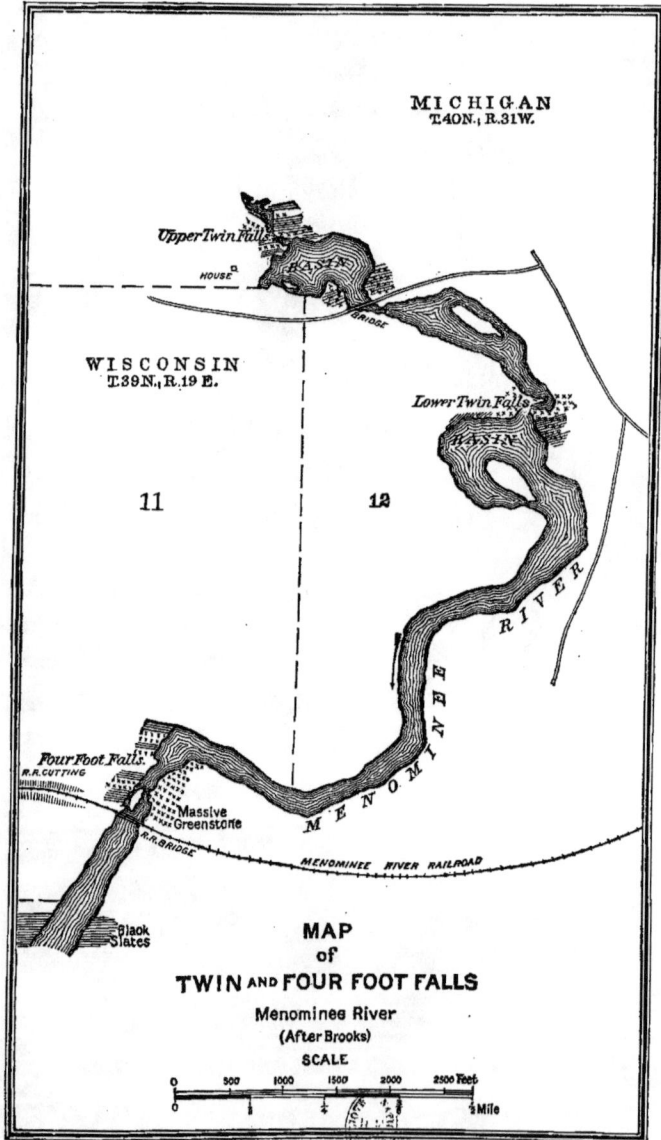

MICHIGAN
T.40N.; R.31W.

Upper Twin Falls

HOUSE

BRIDGE

WISCONSIN
T.39N.; R.19 E.

Lower Twin Falls

11   12

MENOMINEE RIVER

Four Foot Falls
R.R.CUTTING

Massive
Greenstone

R.R.BRIDGE

MENOMINEE RIVER RAILROAD

Black
Slates

MAP
of
TWIN AND FOUR FOOT FALLS
Menominee River
(After Brooks)
SCALE

0   500   1000   1500   2000   2500 Feet
0   ½ Mile

MAP
of
UPPER QUINNESEC FALLS.
Menominee River
(After Brooks)
SCALE

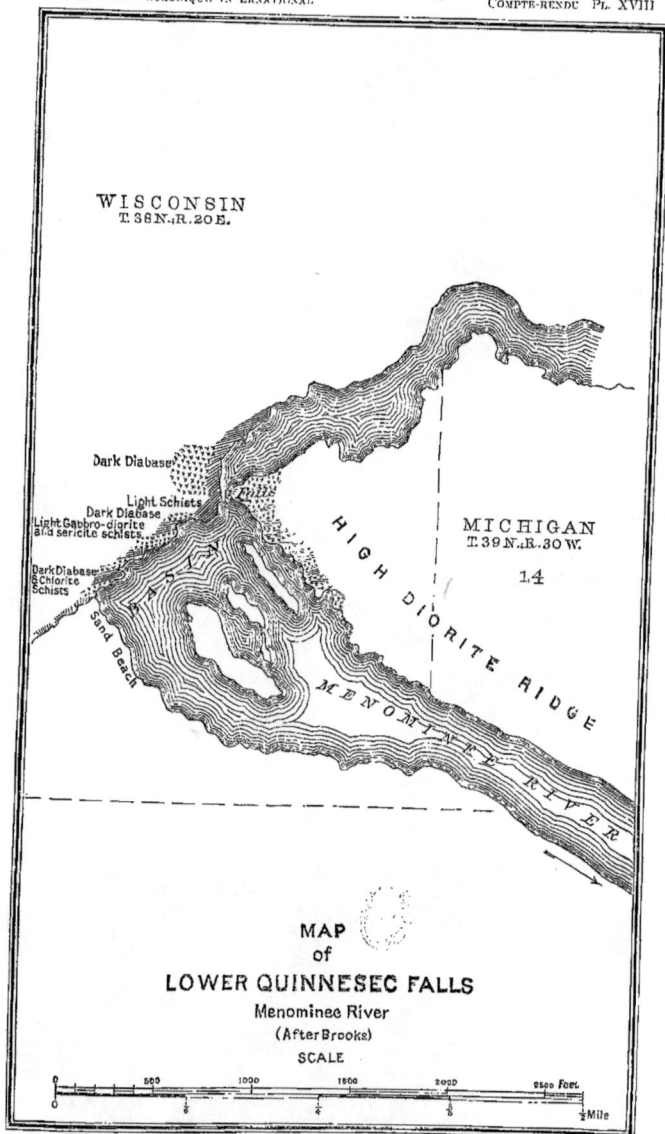

MAP
of
LOWER QUINNESEC FALLS
Menominee River
(After Brooks)
SCALE

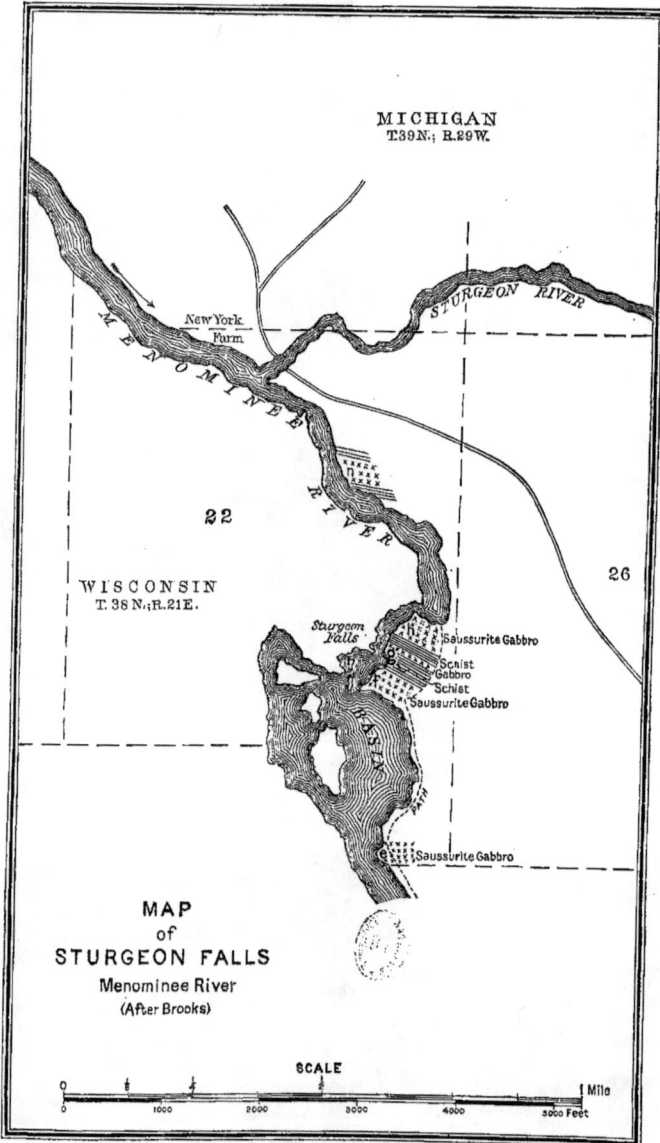

MICHIGAN
T.39 N.; R.29 W.

STURGEON RIVER

New York
Farm.

MENOMINEE RIVER

22

WISCONSIN
T. 38 N.; R. 21 E.

26

Sturgeon
Falls

Saussurite Gabbro
Schist
Gabbro
Schist
Saussurite Gabbro

Saussurite Gabbro

MAP
of
STURGEON FALLS
Menominee River
(After Brooks)

SCALE

0   1000   2000   3000   4000   5000 Feet   1 Mile

and along the railroad between Michigamme and Summit, the granite and gneiss are beautifully exposed. East and west of Summit, for instance, the grey gneissoid granite and the black hornblendic gneisses are cut through and through by red and pink granite dikes and veins of various sizes. In some places the granite is greatly predominant. In others, considerable exposures of the gneisses are wholly free from the granite veins, and between these two there are all gradations. Near Marquette (Pl. xx) the fine-grained dark-colored green schists, the granites, and the intrusive relations of the latter to the former may be finely seen. At Light-House point (Fig. 33) the schists cut by both basic and acid dikes occur in typical development. On Picnic islands near by, granites and ancient diorites are both found, their intricately intrusive relations being such as to lead Williams [27] to the conclusion that both must have been plastic at the same time. Along and near the carriage road to Presque Isle the granite-gneisses are well exposed. Northeast of Negaunee (Pl. xx) the schists extend for 2 or 3 miles, after passing which the granite-gneiss is found. About one-half mile north of Baldwin's Kilns, Sec. 21, T. 48 N., R. 26 W., dikes of granite may be seen cutting the schists.

In the Penokee district (Pl. xxi) the schists of the Basement Complex may be well seen almost anywhere in the eastern schist area, or in the western schist area at Potato river.[9] (Fig. 34). The granite of the Basement Complex is well exposed south of the row of mines running east from Ironwood to the Palms mine, and on the Wisconsin Central Railway south of Penokee Gap. The intrusions, of the schists by the granite are perhaps best seen along the contact zone near the west branch of Black river.

All of the above areas have not been mapped in detail, and a closer study may show that some of the localities, here described as Archean, are really Lower Huronian. This is particularly true of the surface volcanics which are cut by granite veins and on that account have been placed with the Basement Complex.

Lower Huronian.—The well-known, characteristic rocks of the Lower Huronian are, (1) conglomerates, quartzites, quartz-schists, and mica-schists; (2) limestones; (3) various ferruginous schists; (4) basic and acid eruptives, which occur both as deep-seated and as volcanic rocks. The order given, with the exception of the eruptives, is, from the base upward, that of the formations of this series.

The inferior formation is usually a quartzite or feldspathic quartzite. Where metamorphism has been severe it passes into a quartz-schist or a mica-schist. The lowest horizon of the formation is in places a coarse conglomerate, and this when metamorphosed may become a conglomerate-schist. This conglomerate is of two types, depending upon the character of the underlying formation, whether granitic or schistose.

The limestone formation, when at its maximum, is of very considerable thickness. The limestone is usually very crystalline, and may often be properly called marble. It frequently contains a considerable amount of chert. In places it may be divided into two horizons, one of which is a nearly pure chert and the other nearly pure marble. At other times the limestone becomes very siliceous by a mingling of fragmental quartz, while zones of wholly fragmental material may occur. These impure phases are often at the lower part of the limestone, where they may be considered as a transition from the underlying formation. The formation overlying the limestone is usually known as the iron-bearing member, since it contains all the iron ore of the Lower Huronian. It has varied aspects but the different varieties grade into one another both vertically and laterally, so that when one becomes familiar with them the rocks of the formation may invariably be recognized. Here are included hematitic and magnetitic schists, cherts, jaspers, ferruginous carbonates, and other forms. The formation always differs from the limestone in carrying a very considerable content of iron, and it differs from the quartzite in being largely, and sometimes wholly, a chemical or organic sediment rather than a mechanical one.

In the Lower Huronian basic eruptive rocks are abundant and locally cover considerable areas. Frequently acid eruptives also occur. These eruptives include both contemporaneous volcanics and subsequent intrusives.

The three sedimentary members of the Lower Huronian are not often seen in a single section. This may be due to lack of exposures, but in some cases is undoubtedly due to the absence of one or more of the formations themselves. In the Lower Huronian series are to be placed the following iron-bearing districts: Lower Marquette, Felch Mountain, Lower Menominee, and the Cherty Limestone member of the Penokee.

In the Menominee district (Pl. XV) the Lower Huronian can be best studied at Iron mountain and vicinity, and at Norway and vicinity. At these places the iron-bearing member and the limestone are well exposed.

In the Felch Mountain district the series is best studied in the neighborhood of Metropolitan.

In the Marquette district (Pl. XX) the iron-bearing member can best be studied at Republic mountain and in the vicinity of the two towns of Ishpeming and Negaunee. The lower quartzite and conglomerate may be well seen below the iron-bearing formation at Republic and near Palmer, between that village and the Platt mine.

The cherty limestone of the Penokee district (Pl. XXI) occurs only at a single readily accessible locality, near and under a bridge of the Wisconsin Central Railroad, just south of Penokee Gap.

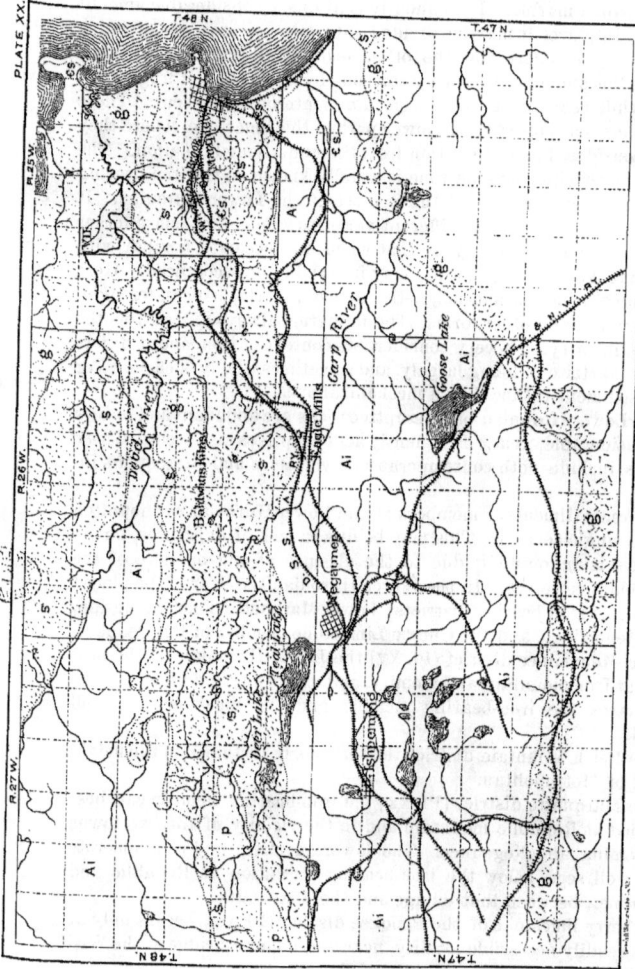

PLATE XX.

OUTLINE GEOLOGICAL MAP OF THE MARQUETTE REGION.

Compiled by R.D. Irving from maps by T.B. Brooks and C. Rominger.

Scale of miles.

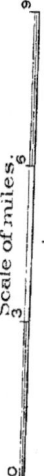

190080

ARCHEAN. Granite and Gneiss. Greenstone Schists.

AGE UNKNOWN. Altered Peridotite.

ALGONKIAN. Iron Bearing Series. (Detritals, Limestones and Ferruginous Schists, with interbedded Greenstones.)

CAMBRIAN. Lake Superior S.S.

Relations of Basement Complex and Lower Huronian.—At the open-ing of this paper it was said that the Lower Huronian is separated from the Basement Complex by an unconformity. The evidence of this uncon-formity consists in the intricately folded character of the Basement Complex as compared with the much more regularly folded Lower Huronian; in the completely crystalline character of the first as compared with the evident fragmental character of the second; and, finally, at a number of places at the base of the lowest member of the Huronian, the occurrence of basal conglomerates, the fragments of which consist mainly of the subjacent schists and granites. The first two points may be appreciated by the contrasting lithological characters of the rocks and their field relations at almost any place along the contact of the Basement Complex and the Lower Huronian. As the facts at actual contacts will, however, carry to many minds more weight than such general relations, the particular localities will be pointed out at which the junction of the Lower Huronian and the Basement Complex may be seen.

In the Marquette district (Pl. xx) basal conglomerates are found both north and south of the Lower Huronian. At various places along the contact line of the Lower Huronian and the basement complex in T. 47 N., R. 25 W., the basal horizon of the quartzite becomes conglom-eratic, and contains a large amount of detrital feldspar. Near the southwest corner of Sec. 22, T. 47 N., R. 26 W., the Huronian rocks contain numerous granitic fragments, and near the center of this sec-tion, for some distance east and west along and near the quarter line, is exposed a magnificent basal conglomerate, great boulders of granite being closely packed together, and the matrix being mainly composed of finer material of the same kind. At one point appears the solid ledge of subjacent granite, from which the material is derived.[19] On the north side of the Huronian belt, on the State road, a short distance north of a small lake (Fig. 33), in the SW. $\frac{1}{4}$ of the SE. $\frac{1}{4}$ of Sec. 29, T. 48 N., R. 25 W., is a beautiful basal conglomerate, the abundant fragments of which are chiefly of granite and of green schist, some of the former being two feet in diameter. The actual contact between the conglomerate and green schist is seen. The granite itself was not found in this immediate vicinity, but occurs to the northward.

Of the localities described in the literature as showing basal con-glomerates or contacts between the Basement Complex and the Huron-ian, it can not be positively asserted that there are any at the base of the Lower Huronian rather than at the base of the Upper Huronian. There are, however, two localities where there is no doubt upon this point, the first near Cascade and the second near Republic.

According to Dr. Wadsworth,[25] "on sec. 32, T. 47 N., R. 26 W., about 1,850 paces north and 1,250 paces west of the southeast corner,

451 GE——32

one can see the coarse conglomerate resting upon the gneisses to the
south, and overlain to the north by quartzite, fragmental jaspilite, and
quartz-schist. The dip is about 40°, with a strike north 70° west. The
conglomerate here contains numerous pebbles of gneiss, as well as some
of granite, diorite, schist, and of quartz veinstones."

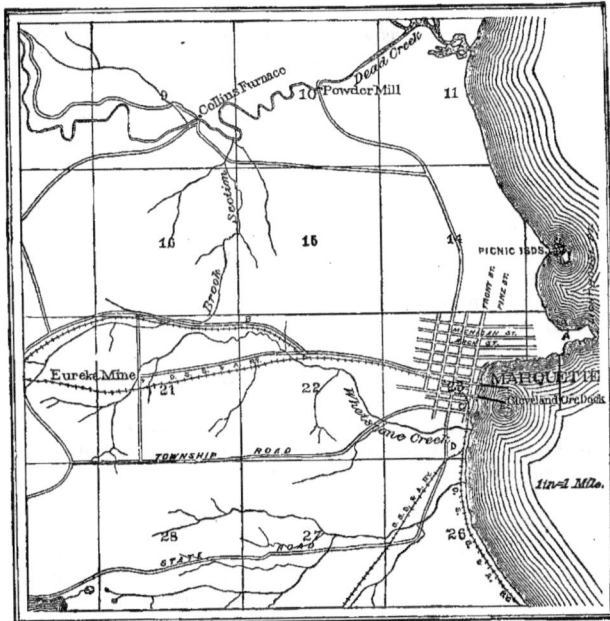

Fig. 33.—Map of the environs of Marquette.

At Republic, contacts between the Basement Complex and Lower
Huronian have been found at two points. The first is near the sharp
bend of the southeast corner of the Republic "horse-shoe," where the
granite-gneiss occurs in actual contact with the quartzite and quartz-
schist of the Lower Huronian. Here, as a result of powerful dynamic
action, the latter has become more than usually crystalline, and the
contact is of such a character as to make the relations somewhat
obscure. Taking this point alone, the phenomena could be interpreted
in either of two ways. It may be considered that, as a result of shearing and metamorphism, a common secondary foliation has developed in

both the gneissoid-granite and the quartzite derived from it; or that the gneissoid-granite is a subsequent intrusive, which has metamorphosed the adjacent quartzites. The first explanation is regarded as far more probable. At another place farther west[2] the quartzite and gneissoid-granite occur very close to each other, and here the strike of the foliation in the granite is discordant with that of the bedding of the quartzite.

The second locality, discovered by Mr. Smyth, is at the southwest corner of the Republic "horse-shoe." Facing the north side of the granite is a layer of conglomerate several feet in thickness, containing numerous large boulders and small fragments of granite, schist, and vein-quartz. One of the granite boulders is 3 or 4 feet in greatest diameter. Many are as much as a foot in diameter, and these are usually very well rounded. The granite fragments are identical with the solid mass of granite adjacent. The schist or gneiss fragments resemble the gneiss of the Basement Complex. The conglomerate has been considerably sheared. As viewed in one direction, the boulders and pebbles are elongated, but where they can be seen in a section transverse to the foliation, they present a rounded appearance, having reasonably sharp contacts with the matrix, although the metamorphism has been so severe as to cause the outer zones of the boulders to merge to some extent into the matrix. Near the conglomerate, the solid granite contains narrow layers or stringers of conglomerate, which run into the granite for some feet. It appears that these stringers represent detritus sifted into the granite ledge, broken, and creviced at the time of the deposition of the conglomerate. A little way from this conglomerate is an exposure of the ordinary typical lower quartzite, and immediately above this follows the actinolite-magnetite-schist of the Lower Huronian. The Upper Huronian contact and basal conglomerate is some distance to the north. As the dips are high, it does not appear possible that in any way this conglomerate can belong at the base of the Upper Huronian rather than at the base of the Lower Huronian.

In the Menominee district, the only locality at which the supposed Lower Huronian is known to be in direct contact with the Basement Complex is at the falls of Sturgeon River. (Pl. xv.) Here the Basement Complex has its typical character. It consists of coarse black hornblendic gneiss, which is cut through and through by red granite. Both occur in large masses, and along the numerous contacts the granite and gneiss are often minutely interlaminated, evidently by the intrusion of the latter. At many places dikes of granite may be seen passing from large granite masses and penetrating the schists, and then gradually dying out. In places the schists are so cut by stringers of granite as to have a genuine pegmatized appearance. Upon the irregular eroded surface of this Basement Complex rest masses of the broken ledge two or three feet thick, which pass upward into a schistose conglomerate containing numerous well-rounded boulders and pebbles of the granite-gneiss and schist, in every respect like these rocks in the complex below. The matrix of the schist is sheared and crystalline, but the larger pebbles and boulders of granite have escaped any considerable crushing. There are several alternations of coarse conglomerate and fine siliceous schist before the conglomerate finally grades into the overlying quartzite. The geology has not been worked out in detail here, and that this formation is the lower quartzite of the Lower Huronian rests upon the authority of Brooks.[1, 11]

In the Penokee district no absolute contact is known between the Basement Complex and the Lower Huronian, but at Penokee Gap the cherty limestone of the Lower Huronian dips to the north at an angle of 65°, while about 50 feet to the southward is the typical hornblende-gneiss of the Basement Complex, its foliation having a dip of about 70° to the southward.[12]

Upper Huronian. —The Upper Huronian series consists lithologically of conglomerates, quartzites, graywackes, graywacke-slates, shales, mica-schists, ferruginous slates, cherts, jaspers, and schists, and igneous rocks, including both lava flows and volcanic fragmentals, as well as basic and acid intrusives. The series as a whole is very much less crystalline than the Lower Huronian, although locally the shales and graywackes have been transformed into mica-schists and even into gneisses.

The Upper Huronian immediately about Lake Superior is divisible into three formations, a lower slate, an iron-bearing formation, and an upper slate, the basis of separation being that of mechanical or non-mechanical detritus.

The inferior formation is mainly a quartzose slate or shale, but locally it passes into a quartzite, while the basal horizon is frequently a conglomerate. The nature of this conglomerate varies greatly, depending upon the character of the underlying formation, which may belong to the Basement Complex or to the Lower Huronian. In the first case the slate may rest upon the granite-gneiss, upon the schists, or upon the junction of the two.

The basal conglomerate corresponds in its character, being a recomposed granite or granite-conglomerate, a recomposed schist or schist-conglomerate, or finally a combination of the two. When the lowest member of the Upper Huronian rests upon the Lower Huronian series, the underlying formation may be any one of the three formations of the Lower Huronian. As a consequence the basal conglomerate may consist mainly of the fragments of any one of these formations, or of all of them together, and not infrequently mingled with this detritus is that also derived from the Basement Complex. However, as a con-

KEWEENAWAN

| Graywacke & Black Slate member | |

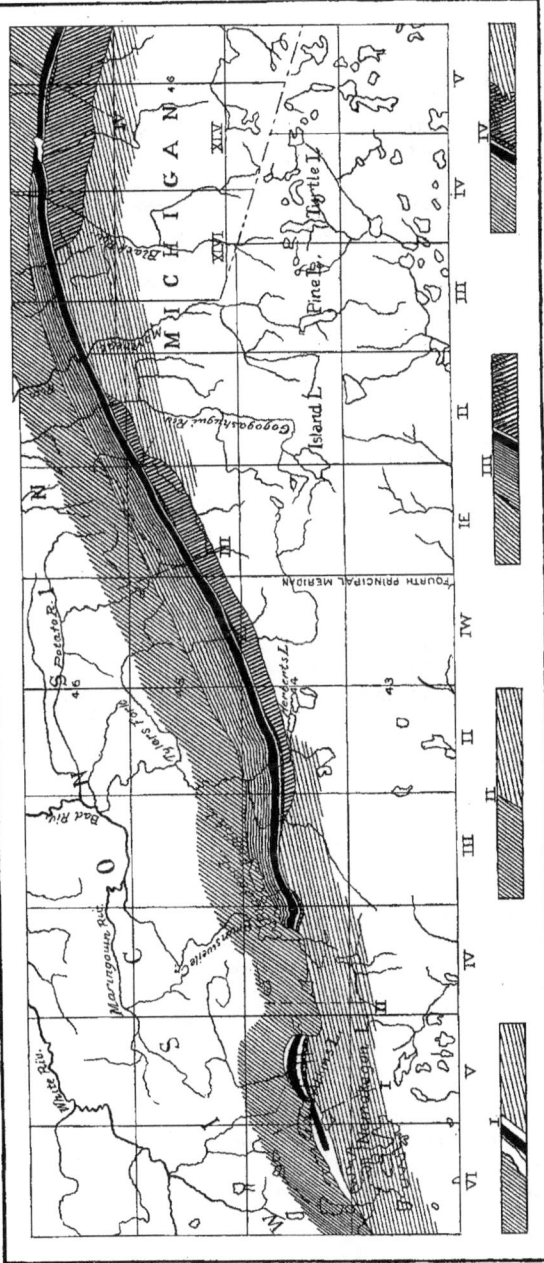

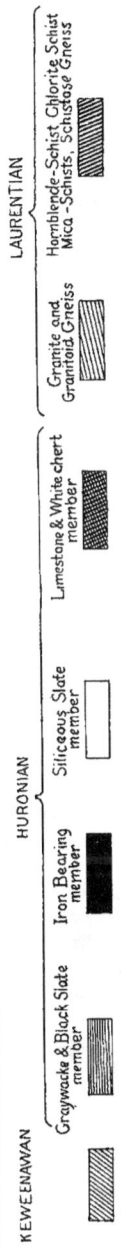

Scale of Map 1/507000

HURONIAN

| Iron Bearing member | | Siliceous Slate member | | Limestone & White chert member | |

Scale of Sections 1/326000

LAURENTIAN

| Granite and Granitoid Gneiss | | Hornblende-Schist, Chlorite Schist Mica-Schists, Schistose Gneiss | |

GEOLOGICAL MAP OF THE PENOKEE-GOGEBIC IRON REGION.

sequence of the resistant character of the jaspery iron-bearing formation of the Lower Huronian and of mining operations, the discovered contacts are most frequently between the Upper Huronian and this iron-bearing formation. In the basal conglomerate at these points the characteristic fragments are chert, jasper, and other ferruginous materials. This basal horizon is locally so rich in iron as to bear ore bodies. The uppermost horizon of the lower slate of the Upper Huronian is, in the Penokee district, a pure persistent layer of quartzite. The central mass of the formation is a graywacke or graywacke-slate passing in places into a shale or sandstone.

The iron-bearing formation consists of various ferruginous rocks, including cherts, jaspers, magnetite-actinolite schists, iron ores, and ferruginous carbonates. It has been shown that all these varieties have been derived directly or indirectly by transformations from an original lean iron-bearing carbonate, which was of chemical or organic origin or a combination of both. Mingled with these non-mechanical sediments is a greater or lesser quantity of mechanical detritus.

Above the iron-bearing formation is the upper slate formation. This is mainly composed of shales, frequently carbonaceous or graphitic, slates, greywackes, and mica-schists, often garnetiferous and staurolitic. The mica-schists are usually toward the upper part of the formation. All stages of the change have been made out between these crystalline rocks and plainly fragmental phases from which they are derived.

The lower slate formation is of variable thickness, but is generally less than a thousand feet. The iron-bearing formation is also of very variable thickness, its maximum being perhaps about the same as that of the lower slate, and from this it varies to disappearance, the horizon being usually represented, however, by carbonaceous and ferruginous shales and slates. The upper slate formation includes the great mass of the Upper Huronian series. Its maximum thickness is more than 10,000 feet.

In certain areas, during Upper Huronian time, there was great volcanic activity, as a result of which peculiar formations were piled up, wholly different from any of the ordinary members of the series. Also this volcanic activity greatly disturbed the regular succession, so that for each of the volcanic districts an independent succession exists, the sedimentary and volcanic formations being intimately interlaminated. The two areas which are best known are the Michigamme iron district north of Crystal Falls, and the east end of the Penokee district. In the Michigamme iron district is an extensive area of greenstones, greenstone-conglomerates, agglomerates, and surface lava flows, many of which are amygdaloidal. In the Penokee district the materials are almost identical. The typical succession for this district extends in unbroken order for 50 miles or more, but east of Sunday Lake this is suddenly disturbed by the appearance of the volcanics. The character of the rocks and their order soon become so different that if one were not able to trace the change there would be a great temptation to regard the part of the series bearing volcanics as earlier than or later than the Penokee series proper. But the continuity of the two can not be doubted. Thus this occurrence well illustrates that lithological character in pre-Cambrian, as in post-Cambrian, time is no certain guide to relative age. Finally, associated with the Lake Superior Upper Huronian rocks are many later intrusive dikes and interlaminated sills, chiefly diabases, gabbros, and diorites, but local granitic intrusives also occur, particularly in the Felch Mountain and Crystal Falls districts, and possibly also in the Menominee district.

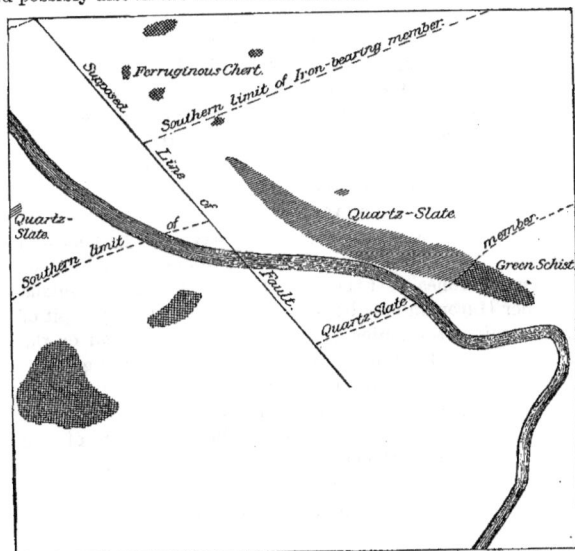

FIG. 34.—Map of exposures at Potato river.

The Upper Huronian can advantageously be studied in the Penokee district,[9, 12, 13] (Pl. XXI). Here the entire series extends for many miles in a nearly east and west direction. The lower slate is handsomely exposed south of the Aurora and Palms mines, and at the following rivers: West Branch of the Montreal, Potato, Tyler's Fork, and Bad.

Its upper horizon of quartzite is exposed at most of these localities and also at many of the mines. The various phases of the iron-bearing member—except the actinolite-magnetite schists—may be seen along the line of mines from the Sunday Lake on the east to the Pence on the west. The transformation of the original iron carbonate to other phases of rock is particularly well seen at the Palms mine railroad spur, just north of the mine, and along Sunday Lake outlet, just north of the railroad, about one-half mile east of the Black River bridge. The magnetite-actinolite-schist phase of this member is finely exposed at, as well as both east and west of, Penokee Gap. The upper slate member is largely exposed in the railroad cuts about half way between Hurley and Pence, and just north of Penokee Gap. In this vicinity may be seen the transition between plainly fragmental greywackes and mica-schists.

In the Marquette district (Pl. XX) good exposures of the Upper Huronian may be seen along the wagon roads between Humboldt and Clarksburg, and north of the latter place in Sec. 1, T. 47 N., R. 29 W., and in Secs. 6 and 7, T. 47 N., R. 28 W.; along the wagon roads north and northwest of Champion, in Secs. 29, 30, 31, 32, T. 47 N., R. 29 W.; south of Michigamme, and at other localities. These places are exceptional in that they afford opportunities to see, in continuous sections, both the Lower and Upper Huronian. Near Humboldt, at the Barron mine, is the basal conglomerate of the Upper Huronian, while to the east is the iron-bearing member of the Lower Huronian. To the east of the town are the slates and graywackes of the Upper Huronian. At the second locality these slates and graywackes may again be seen, and also the iron-bearing formation of the Upper Huronian at the Pascoe and North Champion mines. At the third locality the basal conglomerate of the Upper Huronian may be seen just south of the open pit of the Michigamme mine; the slates and graywackes are found on the promontory just south of the village, while the staurolitic and garnetiferous mica schists cut by granite veins are finely exposed along the south shore of Michigamme Lake opposite the village, and upon the islands to the eastward. Above the iron-bearing formation of the Lower Huronian at nearly all the mines of T. 47 N., R. 27 W., and also at Republic, may be seen the lowest formation of the Upper Huronian. At many of these places it is heavily ferruginous and contains important ore deposits.

In the Menominee district (Pl. XV) the lowest formation of the Upper Huronian may be seen above the Lower Huronian iron formation at Iron Mountain and Quinnesec. The series can, however, be best studied about the Commonwealth mine, and 2 or 3 miles to the westward in the vicinity of Lake Eliza.[1] The country to the north of the western half of the Menominee district presents a great area of Upper Huronian, extending to the mica schist of Michigamme Lake.

The unconformity at the base of the Upper Huronian.—It has been said that the Upper Huronian frequently rests directly upon the Basement Complex, the Lower Huronian not appearing, and at other places rests upon the Lower Huronian.

Contacts of the first kind are easily accessible in the Penokee district.

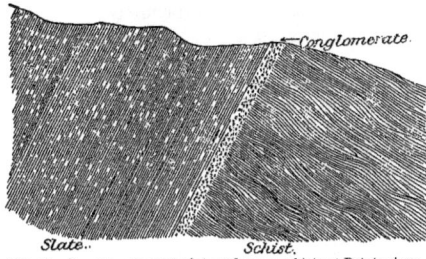

FIG. 35.—Junction of quartz-slate and green schists at Potato river.

That there is between the Penokee series and the Basement Complex a great un-conformity is indicated (1) by the completely crystalline condition of the Basement Complex as compared with the little altered charac ter of the Penokee series; (2) by the intricate relations of the schists and gneissoid granite of the Basement Complex and its complicated folding as contrasted

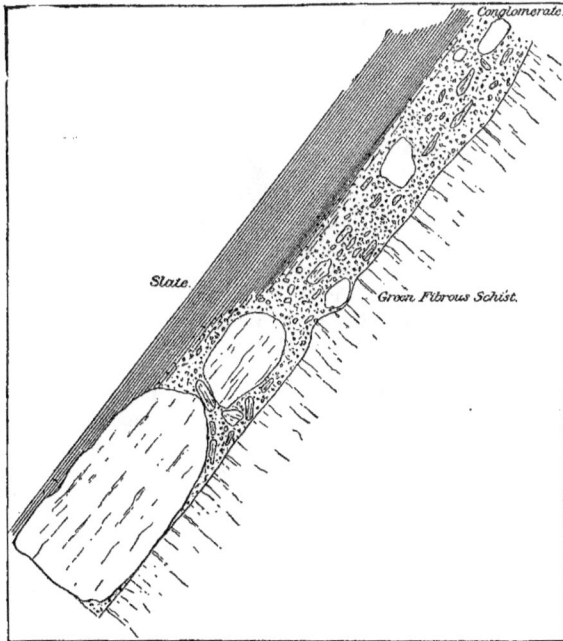

FIG. 36.—Junction of quartz-slate and green schists (large scale).

with the orderly succession and simple monoclinal tilting of the Penokee series; and (3) by unconformable contacts at various places.

An unconformable contact between the base of the Penokee series and the schists may be seen at Potato river [9] (figs. 34, 35, and 36). Here the schistose structure of the gneiss abuts perpendicularly against the bedding of the slates, and in the latter are numerous fragments of the former.

Contacts between the slates and the gneissoid granite may be seen just south of the Aurora and Palms mines. Here the slates contain abundant granitic débris, and detritus has filtered down into the cracks of the preexistent granite.

Near the center of Sec. 28, T. 47 N., R. 42 W. [9] is a contact (fig. 37), below which are both the schists and intrusive granites of the Basement Complex, and above which is the basal conglomerate of the Penokee series, its débris being derived from both the underlying rocks.

In the Marquette district the Upper Huronian rests unconformably upon the Basement Complex in the gorge of Plumbago brook, about 400 feet southwest of the southwest corner of Sec. 9, T. 49 N., R. 33 W. [2] In the Silver Lake area of this district a beautiful unconformable contact between these same series may be seen on the northwest shore of Silver Lake, [11] SE. ¼ NW. ¼ Sec. 8, T. 49 N., R. 28 W. Here in the Huronian slates are great blocks of the subjacent granite,

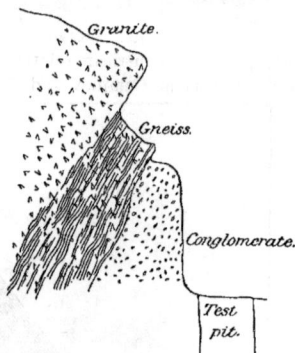

FIG. 37.—Basal conglomerate of Upper Huronian in contact with Basement Complex.

the conglomerate resting with visible contact upon the granite and filling its hollows and pre-Huronian clefts.

We now turn to the second class of contacts, those in which the Upper Huronian rests upon one of the formations of the Lower Huronian. In the Penokee district, at Penokee Gap, the inferior horizon of the lower slate is a magnetitic quartzite. This is in contact with the subjacent chert of the cherty limestone, from which the greater part of the débris is derived. East of Sunday Lake, near the east half of the south line of Sec. 10, T. 47 N., R. 45 W., Wisconsin, a chert conglomerate is well exposed. The underlying cherty limestone is not seen at the immediate vicinity but appears a short distance to the east. At neither of these places has a discordance in strike and dip of the two series been described.

In the Marquette district the basal conglomerate of the Upper Huronian, with its material mainly from the adjacent iron-bearing formation, may be seen at nearly all the large mines. This conglomerate is exposed on the grandest scale, however, at the Republic and Kloman mines, and along the Saginaw range, particularly at and near the old Goodrich mine. Frequently the lamination of the Lower Huronian iron formation and the bedding of the conglomerate nearly accord, as at the Republic and Kloman, where the discordance is not more than 5° or 10°, although the pre-Upper Huronian surface was plainly one of erosion. But in certain cases the bedding of the two series is nearly at right angles to each other,[23] as at one place just south of the Goodrich open pit (fig. 38). At the Republic the basal conglomerate is

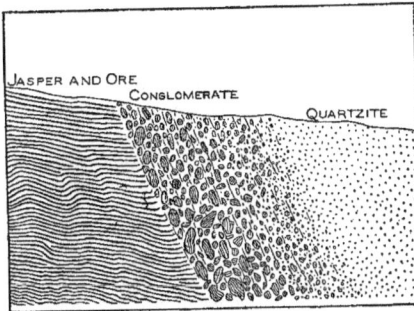

Fig. 38.—Contact of Upper and Lower Huronian at Goodrich mine.

best observed at the southwest corner of the horse-shoe, and along the edge and west slope of the east arm of the horse-shoe.

In the Menominee district the basal conglomerate of the Upper Huronian is not well exposed, but has been detected at the Chapin and Quinnesec mines by shafts and drill holes. In every respect the occurrences here are analogous with those at Marquette.

The Keweenaw series.—The Keweenaw series is composed lithologically of gabbros, diabases, porphyrites, amygdaloids, felsites, quartz-porphyries, etc., and of sandstones and conglomerates. The basic and acid rocks constituting the series are mainly surface flows. The gabbro flows are often of immense thickness. The diabase flows are usually much thinner, and frequently pass in their upper parts into porphyrites and amygdaloids. Many flows are porphyritic or amygdaloidal throughout. The beds of quartz-porphyry and felsite are abundant in certain districts but usually have no great lateral extent; but while a single flow may be traced but a little way, frequently a group of flows of the same general character may have a great extent and thickness. Even the groups of flows, however, can not be regarded as general formations for the whole of the Lake Superior basin.

In very large measure the sandstones and conglomerates of the Keweenawan derived their sediments from the volcanics of the series. Consequently they vary in character, as do the subjacent volcanics. Since fragments of the acid rocks are more resistant than those of the basic rocks, the former are relatively important as pebbles. The detrital formations, like the lava flows, are not universal for the Lake Superior basin, although many are much more persistent than the lava beds. A given detrital bed varies from a mere seam of narrow local extent to thick beds of sandstone and conglomerate. One conglomerate has been traced by Marvine as a continuous formation for more than 100 miles.

Since the number and thickness of the volcanic beds as well as the detritals vary greatly, the Keweenaw series as a whole is widely variable in different districts in its character and thickness. Structurally, Irving has divided the series into two parts, a lower division in which eruptives are present, and an upper division in which eruptives are absent. In any one section of the Keweenawan, at the lower part of the lower division are generally found numerous volcanic flows with few or no detrital beds. In passing toward the middle of the series the sandstones and conglomerates become more and more numerous and of greater thickness. Still higher the sandstones and conglomerates become predominant, and finally volcanic products disappear, the upper ten or fifteen thousand feet of the Keweenaw series being wholly composed of mechanical detritus.

The localities at which the Keweenaw rocks may be well studied are so numerous that only a few of the most conveniently situated will be pointed out. The lowest members of the Keweenawan are well and continuously exposed north of the Upper Huronian rocks of the Penokee series. They are particularly accessible north and east of Sunday lake; in the bed of Black river; north of Bessemer; and along the Wisconsin Central Railroad, north and east of Penokee Gap. At the first of these places the bedded character of the amygaloids is finely shown. At Black river, and in the bluff at the west side of the valley north of Bessemer, may be seen sandstones interstratified with the lavas. At Penokee Gap the lower members of the Keweenawan are gabbros. The Montreal river section presents the greatest known aggregate thickness for the series, estimated by Irving to be about 50,000 feet.

On Keweenaw Point, in the vicinity of Houghton and Calumet, the middle copper-bearing portion of the series can best be studied. The copper occurs both in the conglomerates and amygdaloids. The higher members of the lower division are finely exposed at the end of Keweenaw point, these detritals being particularly well shown at Copper and Eagle harbors.[7] The Eagle river section shows both the detritals and the interstratified basic eruptives.[18] The acid members of the Keweenawan are most accessible along the coast of Lake Superior in the vicinity of and east of Beaver Bay. At the Palisades between

Beaver bay and Two Harbors may also be seen both basic and acid volcanic rocks, including the somewhat rare bytownite rocks.[7]

Unconformity between Upper Huronian and Keweenawan.—The only district on the south shore in which the Keweenawan and Upper Huronian are in contact is the Penokee. In cross section the two are in apparent conformity, for their northward dips are about the same. However, the unconformity is clearly shown by the fact that the Keweenawan is here in contact with a high member of the Penokee series, there in contact with a lower member (Pl. XXI). This can not be due to overlapping because the inclination of both series is from 60° to 80°. The phenomena are satisfactorily explained by a gentle orographic movement and an erosion which in places removed 10,000 feet of the sediments of the Penokee series between the Upper Luronian and Keweenaw times.

The Cambrian.—The Cambrian rocks of the south shore of Lake Superior consist wholly of sandstones. These are in a horizontal position or nearly so, and are wholly unaffected by any subsequent igneous rocks. They are magnificently exposed along the east half of the south shore of Lake Superior, where for many miles they form bold cliffs. Their character can be seen at the sandstone quarries which are scattered from Fond du Lac above Duluth to Sault Ste. Marie. Some of the more important of these are those at or near Fond du Lac, Bayfield, Portage Entry, Huron Bay, and Marquette.

Unconformity between Keweenawan and Cambrian.—With one or two exceptions it is agreed by all that between the Keweenawan and Cambrian is a great unconformity. But since it is considered by some a mooted question, and as the evidence can not here be fully presented on one or both sides, no attempt will be made in this sketch to show the existence of this physical break. In order to comprehend the matter it will be necessary to refer to the original literature.[3, 9, 20, 22, 24, 26] The relations between the two series may be well studied along the contacts on Keweenaw point; in Douglas county, Wisconsin; in the St. Croix Falls district, Wisconsin; between the north and south trap ranges east of Gogebic Lake; and also at one point west of Gogebic lake.

The Cambrian may be seen resting unconformably upon the Huronian and on the Basement Complex at many points. Some of the more convenient of these are Black River falls, Wisconsin, Ableman's and other places about the Baraboo ranges; at Granite Point, Presque Isle, L'Anse, and Inertzite, a point south of Marquette, all on the south shore of Lake Superior; and at Norway and other places in the Menominee district.[a]

The Original Huronian region.—The rocks of this district are separable into a Basement Complex, a Lower Huronian, and an Upper Huronian. The Keweenawan does not here occur. The lithological character of the Basement Complex is in all respects like that of the south shore of Lake Superior. The Lower Huronian consists of semi-metamorphosed fragmental rocks. The exact succession of its formations is not so well known that it can be correlated with the Lower Huronian formations on the south shore of Lake Superior. The Upper Huronian consists of little metamorphosed, but interstitially cemented, clastic rocks. The formations here belonging are several and they can not be equated with those of the Upper Huronian of the south shore.

Between the three series there are physical breaks. That between the Basement Complex and the Lower Huronian is of the first order. This discordance may be seen at two localities. At the north end of a large bluff near Garden river, east of Sault St. Marie, the slate-conglomerate varies downward into a magnificent basal conglomerate, the fragments of which are in large part great blocks of granite, a solid ledge of which is adjacent. Four or five miles east of Thessalon, on the North Channel of Lake Huron, is an actual contact between the lowest member of the Lower Huronian and the Basement Complex (Fig. 39). Above the contact line may be seen a great development of basal conglomerate the water - worn fragments of which are derived from both the granitic and schistose members of the underlying rock complex.[16, 17]

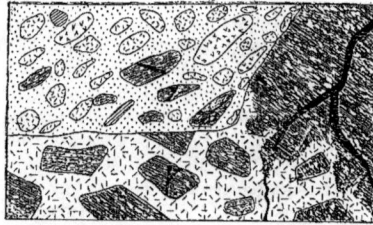

GRANITES   SCHISTS   PEGMATITE

FIG. 39.—Contact of Lower Huronian and Basement Complex.

The evidence of the physical break between the Lower Huronian and Upper Huronian while considerable is not demonstrative. The one actual contact described is near a marble quarry east of Garden river, where the basal conglomerate of the Upper Huronian rests upon the limestone of the Lower Huronian and carries abundant fragments of this rock.[17]

Northwest shore of Lake Superior.—Northwest of Lake Superior the full succession of the south shore appears, that is, a Basement Complex, Lower Huronian, Upper Huronian (Animikie), and Keweenawan. The essential lithological characters of these series are very like to those of the south shore, and in the case of the Upper Huronian (Animikie) the order of its formations and their character are the same as those of the Upper Huronian of the Penokee and Marquette districts.

The relations between the Basement Complex and the Lower Huronian have not been closely studied, although the Canadian geologists have held that the two are conformable. At Steep Rock lake, Ontario,[21] there is, however, the clearest evidence that the two are unconformable, the lowest horizon of the Steep Rock series lying upon and containing fragments of the fundamental complex of the district.

The unconformity between the Lower and Upper Huronian is of the most marked character. This is so evident that it has been fully understood since the earliest work of Logan and Murray. It may be well studied in the vicinity of Port Arthur, where the flat-lying and unmetamorphosed rocks of the Animikie are frequently found in close relations with the vertical schists and slates of the Lower Huronian; and at one place northeast of Port Arthur, an actual basal conglomerate of the Animikie was seen resting as a thin layer capping the vertical schists. At the Kaministiquia river, above Kakabikka falls, are the flat-lying beds of the Animikie, while at a short distance appears the vertical schists of the Lower Huronian.[15,16]

The relations which obtain between the Animikie and Keweenawan are identical with those of the Penokee district, that is, in section the two appear generally to be conformable, but in following along the contact line the Keweenawan is found, now adjacent to one member of the Animikie and now with another. However, near Grand Portage, at Sawyer's Bay on the southwest side of Thunder Cape, and at other places may be seen definite evidence of an erosion between the two.[7,16]

General conclusions.—It thus appears that north of Lake Huron, in the various districts south of Lake Superior, and northwest of Lake Superior the order of succession, lithological character, and the relations which obtain between the pre-Cambrian series are much the same. We therefore conclude that the Basement Complex, the Lower Huronian, and the Upper Huronian are synchronous series, and that the unconformities which separate them extended throughout this vast region and are general. How much farther to the westward these unconformities extend can not now be stated. The results announced by Lawson and Winchell in northeastern Minnesota and western Ontario do not appear at first sight to harmonize with the conclusions here given. Whether there is here a different order of series is too difficult a subject to discuss in this place. All unconformities must end somewhere, and it is not impossible that this occurs of one or more of those defined in this paper before the Lake of the Woods is reached. Abundant granitic intrusions later than the Basement Complex in this region farther complicate the geology. I think it is probable, that ultimately it will be found that this region and the Lake Superior region have a similar sequence of geological series.

BIBLIOGRAPHY.

1. Brooks, T. B.: "The Geology of the Menominee Iron Region." Geol. of Wisconsin, Vol. III, 1880, pp. 452, 465–468, 475–477, 481–492.

2. Brooks, T. B.: "Iron-Bearing Rocks (economic)." Geol. Survey of Michigan, 1873, Vol. I, pt. 1, pp. 126, 128, 156, 166.

3. Chamberlin, T. C.: Geol. of Wisconsin, Vol. I, 1883, pp. 122–124.

4. Foster, J. W., and Whitney, J. D.: Report on the Geology and Topography of the Lake Superior Land District. Part I, "Copper Lands," Ex. Docs., 1st Sess. 31st Cong., 1849–'50, Vol. IX, No. 69, pp. 244, with map. Part II, "The Iron Region," Senate Docs., Special Sess. 32d Cong., 1851, Vol. III, No. 4, 406 pp., with maps.

5. Irving, R. D.: "The Geological Structure of Northern Wisconsin." Geol. of Wisconsin, Vol. III, 1880, pp. 22–24.

6. Irving, R. D., and Chamberlin, T. C.: "Observations on the Junction between the Eastern Sandstone and the Keweenaw Series on Keweenaw Point, Lake Superior." U. S. Geol. Survey, Bull. No. 23, 1885.

7. Irving, R. D.: "The Copper-Bearing Rocks of Lake Superior." U. S. Geol. Survey, Mon. V, 1883, pp. 163–179, 298–323, 350–366, 376–377.

8. Irving, R. D.: "On the Classification of the Early Cambrian and pre-Cambrian Formations." U. S. Geol. Survey, Seventh Ann. Rept. 1888, pp. 399–411, 412–414.

9. Irving, R. D., and Van Hise, C. R.: "The Penokee Iron-Bearing Series of Michigan and Wisconsin." U. S. Geol. Survey, Tenth Ann. Rept. 1890, pp. 376–378, 450–452, 455–456.

10. Irving, R. D.: "Is there a Huronian Group?" Am. Jour. Sci., 3d ser., Vol. XXXIV, 1887, p. 204.

11. Irving, R. D.: Introduction to Bull. U. S. Geol. Survey, No. 62, 1890, pp. 20, 22, 29.

12. Irving, R. D.: "Geology of the Eastern Lake Superior District." Geol. of Wisconsin, Vol. III, 1880, pp. 94, 106, 107.

13. Irving, R. D., and Van Hise, C. R.: "The Penokee Iron-Bearing Series of Michigan and Wisconsin." U. S. Geol. Survey, Mon. XIX. 1892.

14. Lawson, A. C.: "Rainy Lake Region." Ann. Rep. Geol. Survey, Canada, 1887–8, pp. 130F.–139F.

15. Logan, W. E.: "Huronian Series." Geol. of Canada, 1863, p. 64.

16. McKellar, Peter: "Correlation of the Animikie and Huronian rocks of Lake Superior." Trans. Roy. Soc. Canada, Vol. V, sec. 4, 1887, pp. 68–72, 72–73.

17. Pumpelly, Raphael, and Van Hise, C. R.: "Observations upon the Structural Relations of the Upper Huronian, Lower Huronian, and Basal Complex on the north shore of Lake Huron." Am. Jour. Sci., 3d ser., Vol. XLV, 1892, pp. 224–231.

18. Pumpelly, Raphael: "Copper-Bearing Rocks." Geol. Survey of Michigan, Vol. I, part 2, 1873, pp. 117–140.

19. Rominger, C.: "Marquette and Menominee Iron Regions." Geol. of Michigan, Vol. IV, 1881, pp. 20–22, 63, 190–192.

20. Sweet, E. T.: "Geology of the Western Lake Superior District." Geol. Survey of Wisconsin, Vol. III, 1880, pp. 340–352.

21. Smyth, Henry Lloyd: "Structural Geology of Steep Rock Lake." Am. Jour. Sci., 3d ser., Vol. XLII, 1891, pp. 317-331.

22. Strong, Moses: "The Geology of the Upper St. Croix District." (Edited by T. C. Chamberlin.) Geol. of Wisconsin, Vol. III, 1880, pp. 415-423.

23. Van Hise, C. R.: "An attempt to Harmonize some apparently conflicting Views of Lake Superior Stratigraphy." Am. Jour. Sci., 3d ser., Vol. XLI, 1891, pp. 119-121.

24. Van Hise, C. R.: "Archean and Algonkian". U. S. Geol. Survey, Bull. No. 86, 1892.

25. Wadsworth, M. E.: Report of the State Board of Geological Survey for 1891-'92. Lansing, 1893, pp. 102-103.

26. Wadsworth, M. E.: "Notes on the Geology of the Iron and Copper Districts of Lake Superior." Bull. Museum Comp. Zool., Harvard College, whole series, Vol. VII, 1880, pp. 113-122.

27. Williams, G. H.: "The Greenstone Schist Area of the Menominee and Marquette Regions of Michigan." U. S. Geol. Survey, Bull. No. 62, 1890, pp. 96, 123-133, 146.

INDEX ALPHABÉTIQUE.

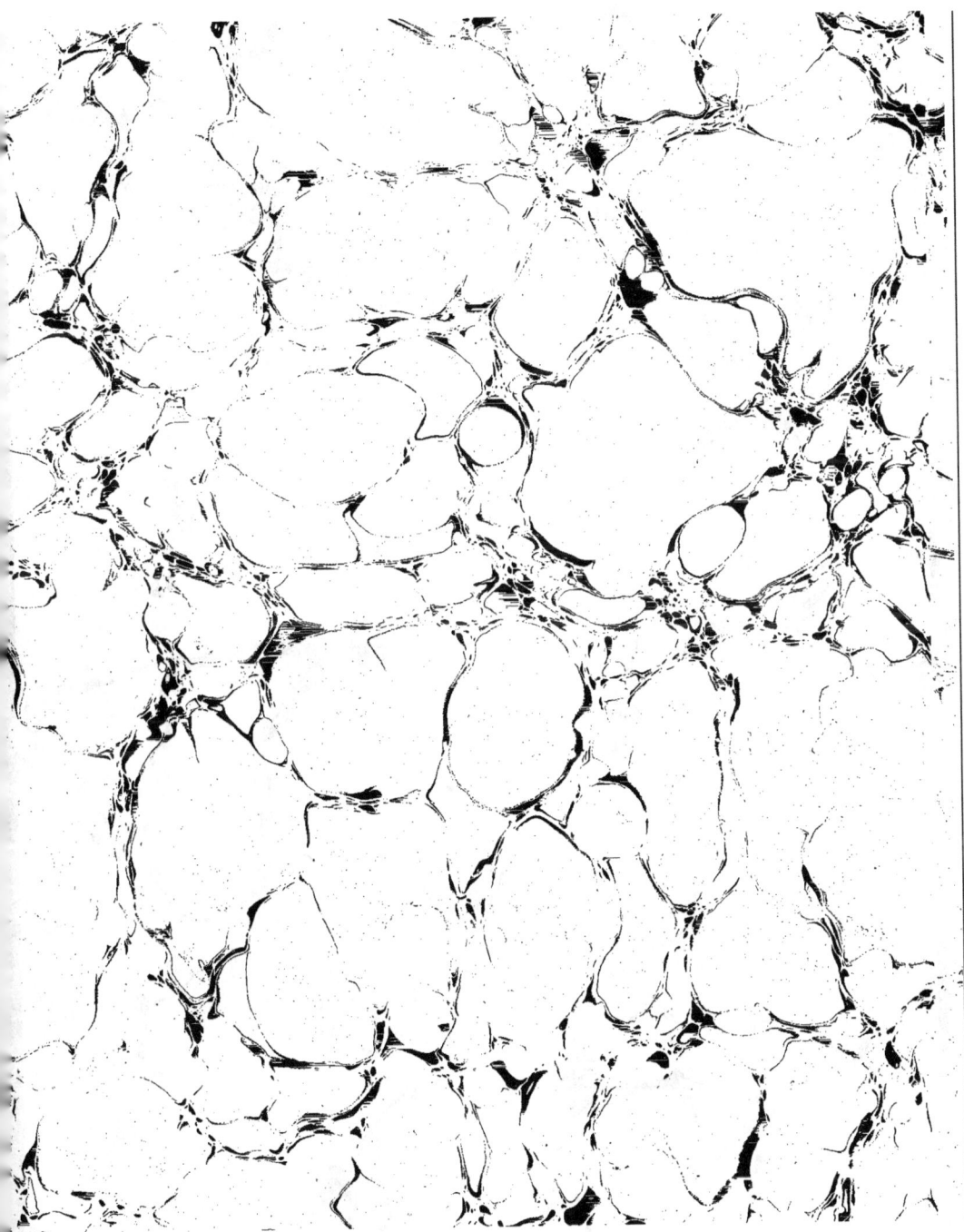

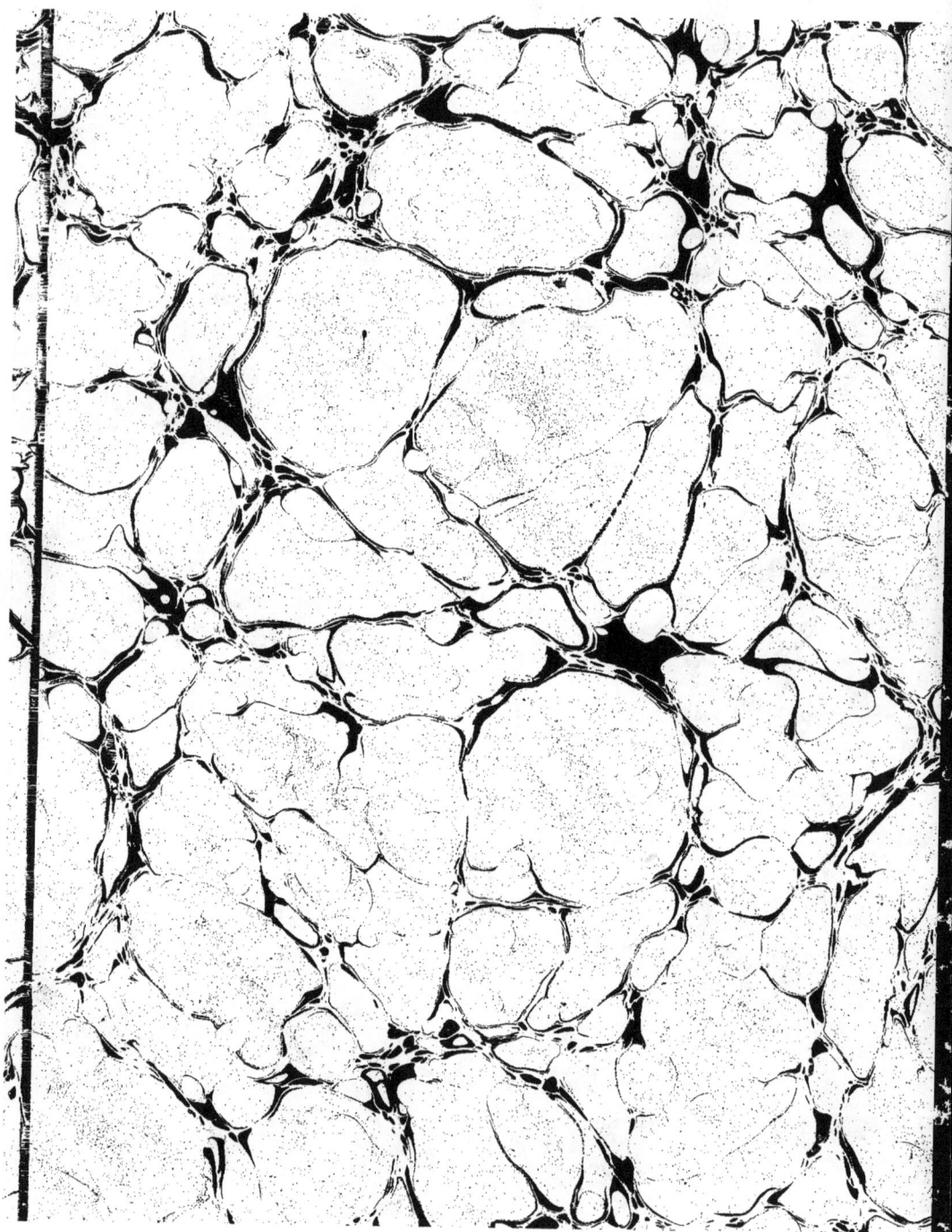

www.ingramcontent.com/pod-product-compliance
Lightning Source LLC
Chambersburg PA
CBHW031719210326
41599CB00018B/2444